Springer Series in

OPTICAL SCIENCES 129

founded by H.K.V. Lotsch

Springer Series in
OPTICAL SCIENCES

The Springer Series in Optical Sciences, under the leadership of Editor-in-Chief *William T. Rhodes,* Georgia Institute of Technology, USA, provides an expanding selection of research monographs in all major areas of optics: lasers and quantum optics, ultrafast phenomena, optical spectroscopy techniques, optoelectronics, quantum information, information optics, applied laser technology, industrial applications, and other topics of contemporary interest.
With this broad coverage of topics, the series is of use to all research scientists and engineers who need up-to-date reference books.

The editors encourage prospective authors to correspond with them in advance of submitting a manuscript. Submission of manuscripts should be made to the Editor-in-Chief or one of the Editors. See also www.springer.com/series/624

Laser Ablation and its Applications

edited by

Claude Phipps
Photonic Associates, LLC
Sante Fe, New Mexico

Springer

Claude Phipps
Photonics Associates, LLC
200A Ojo de la Vaca Road
Santa Fe, New Mexico 87508

Laser Ablation and its Applications

Library of Congress Control Number: 2006929445

ISBN 0-387-30452-5 e-ISBN 0-387-30453-3
ISBN 978-0387-30452-6

Printed on acid-free paper.

Printed in the United States of America.

9 8 7 6 5 4 3 2 1

springer.com

DEDICATION

This book is dedicated to Prof. Dr. Boris Luk'yanchuk. Boris is an eminent theorist in the field of laser interaction with materials, and his work has dealt with every aspect of laser desorption and ablation of materials. His work is well known in the fields of quantum electrodynamics, quantum electronics and laser physics. The breadth, depth and quality of his work are unique in the laser ablation field. He has been a visiting Professor at the Johannes Kepler University, Austria, the University of Lecce, Italy, the Royal Institute of Technology, Sweden and the Tokyo Institute of Technology, Japan. Currently, he works at the Data Storage Institute of the National University of Singapore.

Boris is a great scientist and gentleman as well as a longtime friend, whose encouragement was mainly responsible for my taking on the effort of organizing and editing this book.

Claude Phipps, Photonic Associates, LLC, April 30, 2006

Prof. Dr. B. Luk'yanchuk

PREFACE

This book arose from the SPIE series of High Power Laser Ablation Symposia which began in 1998. It is intended for a graduate course in laser interactions with plasmas and materials, but it should be accessible to anyone with a graduate degree in physics or engineering. It is also intended as a major reference work to familiarize scientists just entering the field with laser ablation and its applications.

Selecting topics which are representative of such a broad field is difficult. In this volume, we have tried to emphasize the wide range of these topics rather than - as is so often the case in advanced science – focusing on one specialty or discipline.

We have, somewhat arbitrarily, divided the book into four sections. The first emphasizes theory and modeling. Subsequent chapters cover ultrafast interactions, material processing and laser-matter interaction in novel regimes. The latter range from MALDI to ICF, SNOM's and femtosecond nanosurgery to laser space propulsion.

These sections are obviously not mutually exclusive, For example, ICF is a subset of laser ablation science at terrific intensity and energy, where laser ablation pressure can actually compress solid matter to a thousand times normal density. S. Nakai and K. Mima, representing the very successful Japanese ICF program, have graciously agreed to summarize ICF in chapter 14. Although this chapter appears in the "novel regimes" chapter, ICF is also based upon some of the most complex theory and modeling in the past 40 years. This theoretical effort began with Ray Kidder in the U.S., who predicted breakeven would require 1.6MJ in 1965, and Nakai nd Mima in Japan. Showing the importance of sound theory, Kidder's prediction was within a factor of two of the requirements for the National Ignition Facility (NIF) today. On the other extreme of laser intensity, but at the same microscopic scale, B. Chichkov summarizes the amazing results in microfabrication in chapter 6. Clearly, this work is part of "material processing" as well as "ultrafast interactions," but we have put it in the ultrafast section because fs-duration pulses are a unique aspect of the work. As you will see, this chapter also includes reviews of ultrafast measurements of the properties of shocked materials, time- and space-resolved spectroscopy and physical chemistry. Similarly, A. Vertes' chapter on MALDI, DIOS, and nanosctructures has aspects in common with material processing and with B. Chichkov's chapter. B. Luk'yanchuk's chapter on laser cleaning of nanoparticles has something in common with the chapter by Bounos, *et al.* regarding art restoration, etc.

We sincerely hope you enjoy this book, and that you adopt it as the valuable reference work we believe it is. We intend it to be your toolbox of references to this diverse and exciting field.

Finally, I would like to dedicate this book to Prof. Boris Luk'yanchuk, whose idea it was. He helped immeasurably during the process of its creation.

Claude Phipps
Santa Fe, NM
April 30, 2006

CONTRIBUTORS

C.N. Afonso
Laser Processing Group
Instituto de Optica, CSIC
28006 Madrid
Spain

Salvatore Amoruso
Coherentia CNR-INFM and Dipartimento di Scienze Fisiche
Università di Napoli Federico II
Via Cintia I-80126 Napoli
Italy

S.I. Anisimov
L.D. Landau Institute for Theoretical Physics
Russian Academy of Sciences
119334 Moscow
Russia

Craig B. Arnold
Dept. of Mechanical and Aerospace Engineering
Princeton University
Princeton, NJ 08544
USA

I.A. Artioukov
P. N. Lebedev Physcial Institute
117942 Moscow
Russia

M. Bittner
Institute of Physics
Academy of Sciences of the Czech Republic
182 21 Prague 8
Czech Republic

W.L. Bohn
DLR German Aerospace Center
Institute of Technical Physics
Pfaffenwaldring 38-40
70569 Stuttgart
Germany

Giannis Bounos
Institute of Electronic Structure and Laser
Foundation for Research and Technology – Hellas
71110 Heraklion Crete
Greece

Nadezhda M. Bulgakova
Institute of Thermophysics SB RAS
630090 Novosibirsk
Russia

Eleanor E.B. Campbell
Department of Physics
Göteborg University
SE-41296 Göteborg
Sweden

Boris N. Chichkov
Laser Zentrum Hannover e.V.
30419 Hannover
Germany

Anant Chimmalgi
Integrated Research Institute
Tokyo Institute of Technology
Ookayama Meguro-ku
Tokyo, Japan

T.C. Chong
Data Storage Institute
Singapore 117608
Republic of Singapore

H.-A. Eckel
DLR German Aerospace Center
Institute of Technical Physics
70569 Stuttgart
Germany

Costas Fotakis
Institute of Electronic Structure and Laser
Foundation for Research and Technology – Hellas
71110 Heraklion Crete
Greece

David J. Funk
Dynamic and Energetic Materials, MS Pg918
Los Alamos National Laboratory
Los Alamos, NM 87545
USA

Savas Georgiou
Institute of Electronic Structure and Laser
Foundation for Research and Technology – Hellas
71110 Heraklion Crete
Greece

J. Gonzalo
Laser Processing Group
Instituto de Optica, CSIC
28006 Madrid
Spain

J. Graf
Universität Konstanz
Fachbereich Physik 78457 Konstanz
Germany

Costas P. Grigoropoulos
Dept. of Mechanical Engineering
University of California, Berkeley
Berkeley, CA 94720-1740
USA

M.E. Grisham
NSF ERC for Extreme Ultraviolet Science & Technology
Dept. of Electrical and Computer Engineering
Colorado State University
Fort Collins, CO 80523
USA

Vitali E. Gruzdev
S.I. Vavilov State Optical Institute
St. Petersburg, 199034
Russia

Ingolf V. Hertel
Max-Born-Institut für Nichtlineare Optik und Kurzzeitspekroskopie
D-12489 Berlin
Germany

M.H. Hong
Data Storage Institute
Singapore 117608
Republic of Singapore

Ruth Houbertz
Fraunhoffer-Institut für Silicatforschung
97082 Würzburg
Germany

Gereon Hüttman
Institut für Biomedizinische Optik
Uiversität zu Lübeck
D-23562 Lübeck
Germany

David J. Hwang
Integrated Research Institute
Tokyo Institute of Technology
Ookayama Meguro-ku
Tokyo, Japan

N.A. Inogamov
L.D. Landau Institute for Theoretical Physics
Russian Academy of Sciences
119334 Moscow
Russia

L. Juha
Institute of Physics
Academy of Sciences of the Czech Republic
182 21 Prague 8
Czech Republic

Wolfgang Kautek
University of Vienna
Department of Physical Chemistry
Waehringer Strasse 42
A-1090 Vienna
Austria

Huengsoo Kim
Materials Science & Technology Division
Naval Research Laboratory
Washington, DC 20375-5342
USA

V.V. Kondratenko
Metal and Semiconductor Physics Department
NTU - Kharkov Polytechnical Institute
Kharkov 61002 Ukraine
Russia

P. Leiderer
Data Storage Institute
Agency for Science, Technology and Research
Singapore 117608
Republic of Singapore

Thomas Lippert
General Energy Research Department
Paul Scherrer Institut
5232 Villigen PSI
Switzerland

James R. Luke
Photonic Associates, LLC
200A Ojo de la Vaca Road
Santa Fe, NM 87508
USA

B.S. Luk`yanchuk
Data Storage Institute
Singapore 117608
Republic of Singapore

James G. Lunney
School of Physics
Trinity College
College Green
Dublin 2
Ireland

Shawn D. McGrane
Shock and Detonation Physics, MS P952
Los Alamos National Laboratory
Los Alamos, NM 87545
USA

C.S. Menoni
NSF ERC for Extreme Ultraviolet Science & Technology
Dept. of Electrical and Computer Engineering
Colorado State University
Fort Collins, CO 80523
USA

Kunioki Mima
Institute of Laser Engineering
Osaka University
Suita, Osaka 565-0871
Japan

David S. Moore
Shock and Detonation Physics, MS P952
Los Alamos National Laboratory
Los Alamos, NM 87545
USA

M. Mosbacher
Data Storage Institute
Singapore 117608
Republic of Singapore

Sadao Nakai
Institute of Laser Engineering
Osaka University
Suita, Osaka 565-0871
Japan

Austin Nevin
Institute of Electronic Structure and Laser
Foundation for Research and Technology – Hellas
71110 Heraklion Crete
Greece

K. Nishihara
Institute of Laser Engineering
Osaka University
Osaka 565
Japan

Joachim Noack
Institut für Biomedizinische Optik
Uiversität zu Lübeck
D-23562 Lübeck
Germany

Minoru Obara
Department of Electronics and Electrical Engineering
Keio University
Yokohama 223-8522
Japan

Takeshi Okato
Department of Electronics and Electrical Engineering
Keio University
Yokohama-shi 223-8522
Japan

Aleksandr Ovsianikov
Laser Zentrum Hannover e.V
Hollerithallee 8
30419 Hannover
Germany

Günther Paltauf
Institut für Physik
Karl-Franzens-Uiversität Graz
A-8010 Graz
Austria

Sven Passinger
Laser Zentrum Hannover e.V
30419 Hannover
Germany

Yu P. Pershyn
Metal and Semiconductor Physics Department
NTU - Kharkov Polytechnical Institute
Kharkov 61002 Ukraine
Russia

Yu V. Petrov
L.D. Landau Institute for Theoretical Physics
Russian Academy of Sciences
119334 Moscow
Russia

Alberto Piqué
Materials Science & Technology Division
Naval Research Laboratory
4555 Overlook Ave. SW
Washington, DC 20375-5342
USA

J.J. Rocca
NSF ERC for Extreme Ultraviolet Science & Technology
Dept. of Electrical and Computer Engineering
Colorado State University
Fort Collins, CO 80523
USA

Arkadi Rosenfield
Max-Born-Institut für Nichtlineare Optik und Kurzzeitspekroskopie
Max Born Str. 2a, D-12489 Berlin
Germany

W.O. Schall
DLR German Aerospace Center
Institute of Technical Physics
70569 Stuttgart
Germany

Jørgen Schou
Department of Optics and Plasma Research
Rise National Laboratory
DK-4000 Roskilde
Denmark

R. Sena
Laser Processing Group
Instituto de Optica, CSIC
28006 Madrid
Spain

J. Solís
Laser Processing Group
Instituto de Optica, CSIC
28006 Madrid
Spain

W.D. Song
Data Storage Institute
DSI Building, 5 Engineering Drive 1
Singapore 117608
Republic of Singapore

Razan Stoian
Laboratoire TSI (UMR 5516 CNRS)
Universite Jean Monnet
42000 Saint Etienne
France

Shigeaki Uchida
Integrated Research Institute
Tokyo Institute of Technology
Ookayama Meguro-ku
Tokyo, Japan

Lukas Urech
General Energy Research Department
Paul Scherrer Institut
5232 Villigen PSI
Switzerland

G. Vaschenko
NSF ERC for Extreme Ultraviolet Science & Technology
Dept. of Electrical and Computer Engineering
Colorado State University
Fort Collins, CO 80523
USA

Akos Vertes
Dept. of Chemistry
Institute for Proteomics Technology and Applications
The George Washington University
Washington DC 20052
USA

A.V. Vinogradov
P. N. Lebedev Physcial Institute
Moscow 117942
Russia

Alfred Vogel
Institute of Biomedical Optics
Uiversity of Lübeck
D-23562 Lübeck
Germany

Z.B. Wang
Data Storage Institute
DSI Building, 5 Engineering Drive 1
Singapore 117608
Republic of Singapore

Takashi Yabe
Integrated Research Institute
Tokyo Institute of Technology
Ookayama Meguro-ku
Tokyo, Japan

V.V. Zhakhovskii
Institute of Laser Engineering
Osaka University
Osaka 565
Japan

E.N. Zubarev
Metal and Semiconductor Physics Department
NTU - Kharkov Polytechnical Institute
Kharkov 61002 Ukraine
Russia

TABLE OF CONTENTS

Part 1: Basic Physics and Simulations

Part 2: Ultrafast Interactions

Part 3: Material Processing

Part 4: Laser-matter interaction in novel regimes

Chapter 1

NUMERICAL SIMULATION OF THE EXPANSION INTO VACUUM OF A CRYSTAL HEATED BY AN ULTRASHORT LASER PULSE

S.I. Anisimov, N.A. Inogamov, Yu.V. Petrov
L.D. Landau Institute for Theoretical Physics, Russian Academy of Sciences, 119334, Moscow, Russia

V.V. Zhakhovskii, K. Nishihara,
Institute of Laser Engineering, Osaka University, Osaka 565, Japan

1. INTRODUCTION

During the last few years, a considerable body of work has been published devoted to experimental study and numerical simulation of the interaction of ultrashort (picosecond and femtosecond) laser pulses with absorbing substances (see, e.g., von der Linde, 2000; Sokolowski-Tinten, 1998a, 1998b; Stuart, 1996; Anisimov, 1997; Inogamov, 1999; Anisimov, 2003; Zhakhovskii, 2000; Temnov, 2004; Hashida, 2002; Zeifman, 2002; Lorazo, 2003). It follows from the above papers that at short pulse length the process of laser-matter interaction is taking some new features. The most important of them lies in the fact that the processes of light absorption, electron heat transfer, and hydrodynamic ablation have different time scales and can be considered independently of one another.

The character of hydrodynamic motion and profiles of pressure differ dramatically in the cases of long, $\tau > d_T/c_S$, and short, $\tau < d_T/c_S$, pulses. Here, τ is the laser pulse length, d_T is the thickness of the layer heated by the electron thermal wave, and c_S is the sound speed. In the case of long pulse, the target is unloaded, the pressure is low and the ablation is determined by surface evaporation. In the case of short pulse, the situation is completely different. The ablation is governed by hydrodynamic motion caused by thermal expansion due to laser energy absorption. At the beginning of hydrodynamic stage, $t \leq \tau_H$, the pressure is relatively high because thermal stress is "confined" within a hot surface layer. The absorption of laser radiation occurs in a time τ, that in typical cases (von der Linde, 2000; Sokolowski-Tinten, 1998a, 1998b) comprises about 100 fsec. Laser energy is absorbed by electrons and transferred to cold material by electron heat conduction. Simultaneously energy exchange between electrons and the lattice takes place. Noticeable hydrodynamic motion occurs at the time of the order of τ_H, which exceeds in typical cases the electron-lattice relaxation time $\tau_{ei} \sim 1 - 10$ psec. Thus, during the hydrodynamic stage electron and lattice temperatures are approximately equal.

The simplifications that follow from the above described scheme have been used in theoretical works (Inogamov, 1999; Zhakhovskii, 2000; Anisimov, 2003), where the ablation was considered in the frame of a one-temperature model. In this connection one should note that several interesting attempts have been made (see, e.g., Zeifman, 2002; Lorazo, 2003) to consider laser absorption, electron-phonon relaxation, phase transitions, and hydrodynamic motion in the frame of unified "hybrid" model. It is apparent that the construction of such a model encounters some obstacles since the laser absorption and electron-phonon relaxation are quantum processes, while the ablation is a classical process. As far as is known to the authors of the present paper, the problem of unified description of all stages of laser-matter interaction was not successfully resolved in all "hybrid" models proposed. It is difficult to estimate real accuracy of "hybrid" models. One can expect that such models ensure, at best, a qualitative picture of interaction of ultrashort laser pulses with matter. For quantitative consideration of the problems of light absorption and electron-phonon relaxation, quantum kinetic equation should be solved (see, e.g., Lugovskoi, 1999; Rethfeld, 2004). It should be noted that the Umklapp processes play a determining role in lattice heating, which significantly reduces the electron-phonon relaxation time (Petrov, 2005).

It is natural to consider the expansion of hot surface layer in hydrodynamic approximation (Inogamov, 1999) since the thickness of this layer is much larger than the mean free path of atoms (which is of the order of interatomic distance). This brings up, however, the question relating to the role of metastable states in the two-phase system of condensed phase-vapor. Usually, this question is solved by addition of a semi-phenomenological kinetic equation to the set of hydrodynamic equations. The parameters of the kinetic model are taken from the comparison of calculations with experiment. More attractive is this different way based on the solution of equations of motion of individual atoms (molecular dynamics). This method does not require one to fit theoretical model to experimental data. This approach was employed in (Zhakhovskii, 2000) to check the validity of simplifications assumed in (Inogamov, 1999).

In the works (Inogamov, 1999; Zhakhovskii, 2000; Anisimov, 2003), the structure of expansion flow of instantly heated uniform layer is elucidated. It is shown that a thin shell of condensed matter is formed on the external edge of an expanding cloud. The shell is filled with two-phase gas-liquid mixture. The shell moves with constant velocity and its density remains constant. The two-phase mixture is subjected to uniform deformation. Its velocity is a linear function of the coordinate and average density decreases inversely proportional to the expansion time.

In the paper (Anisimov, 2003) the initial temperature close to the ablation threshold was considered. The papers (Inogamov, 1999; Zhakhovskii, 2000) are devoted to the situation where the initial temperature is significantly greater than the threshold temperature. In this case effects of metastability are of no importance because, at given entropy, the pressure at the spinodal, $p_s(S) - p_b(S)$, is much less than the pressure $p_o(S)$ at the isochore of initial density $\rho = \rho_o$. Here p_b and p_s are the pressures at the coexistence curve (binodal) and spinodal, respectively. The asymptotic (at $t \to \infty$) velocity distribution over the mass of expanding material is formed during the expansion along the segment of the adiabatic, from the initial point (ρ_o, p_o) at the isochore, up to the point (ρ_b, p_b) at the coexistence curve. In this case the set of hydrodynamic equations is complemented by the equation of state (Bushman, 1992), constructed in accordance with the Maxwell's rule (Hirschfelder, 1954). Note that at fluencies close to the ablation threshold, the metastability effects play an important role since the pressures $p_o - p_b(S)$ and $p_s(S) - p_b(S)$ are comparable to one another.

In the papers (Inogamov, 1999; Zhakhovskii, 2000; Anisimov, 2003) bilateral expansion of a uniform hot layer was considered. This model corresponds to thin film whose thickness is less than the thickness of layer d_T heated by thermal wave. In actual experiments (von der Linde, 2000; Sokolowski-Tinten, 1998a, 1998b), the distribution of temperature at the instant the hydrodynamic motion comes into play is non-uniform. The temperature $T_o(x)$ decreases with depth x (Fig. 1). This situation will be considered in the present paper.

2. ABSORPTION OF RADIATION, THERMAL WAVE, AND CHARACTERISTICS OF HEATED LAYER AT THE ONSET OF HYDRODYNAMIC MOTION

Expanding plume and crater are formed as the result of hydrodynamic motion. It is natural to consider the hydrodynamic problem in one-dimensional formulation because the size of focal spot D is much greater than the thickness of the layer heated by electron thermal conduction d_T. In a typical case, d_T ~100 nm, while D ~ 100 μm. In one-dimensional flow all variables depend on the coordinate x and time t. The velocity has single component $u(x, t)$ directed along the x – axis. Since laser fluence is non-uniformly distributed over the focal spot, the characteristics of the flow depend parametrically on the variable y normal to laser beam direction. In the experiments (von der Linde, 2000; Sokolowski-Tinten, 1998a, 1998b) laser fluence distribution is close to Gaussian:

$$F(y) = F\exp(-y^2 / D^2) \tag{1}$$

where F is the surface energy density of absorbed laser radiation (J/cm^2). Laser pulse is over before the hydrodynamic motion starts. Therefore, in studies of initial stages of laser-matter interaction one can neglect the mass motion.

Absorbed laser energy is transferred to the bulk of the target by electron heat conduction. This stage is usually described in the frame of the two-temperature model (Kaganov, 1957; Anisimov, 1974):

$$c_e \frac{\partial T_e}{\partial t} = \frac{\partial}{\partial x}\left(\kappa_e \frac{\partial T_e}{\partial x}\right) - \alpha(T_e - T_i) + Q \tag{2}$$

$$c_i \frac{\partial T_i}{\partial t} = \alpha(T_e - T_i) \tag{3}$$

where T_e and T_i are the temperatures of electrons and the lattice, respectively, c_e and c_i are corresponding specific heats, κ_e is the electron thermal conductivity, α is the energy exchange rate between electron and phonon subsystems, and Q is the laser power absorbed. Equation

$$Q(x, y, t\} = \frac{I(y)}{\delta}\exp\left(-\frac{t^2}{\tau^2}\right)\exp\left(\frac{x}{\delta}\right) \tag{4}$$

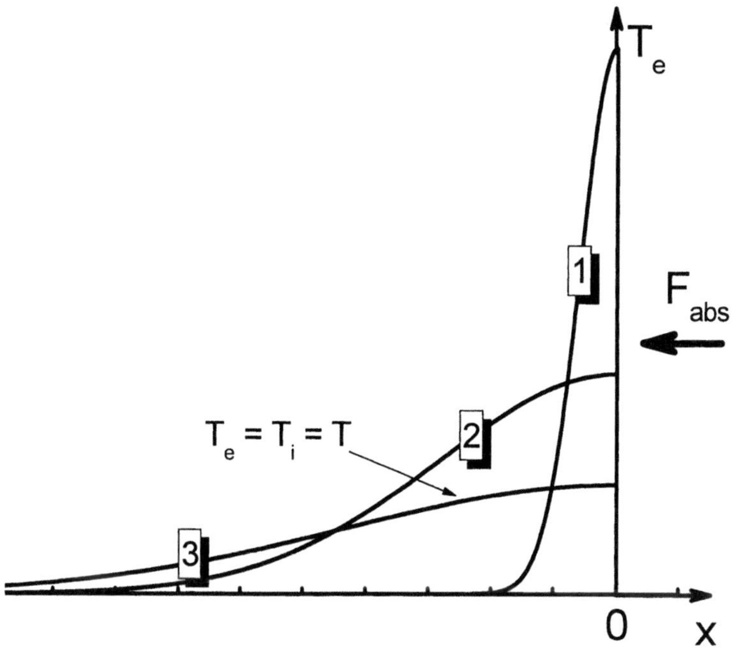

Figure 1. Qualitative profiles of electron temperature for three initial stages. (1) –absorption of radiation, (2) – thermal wave, (3) – electron-lattice relaxation.

is commonly used for Q. In (4) the skin layer thickness is denoted by 2δ.

The ablation of material takes place at the stage of hydrodynamic motion. This stage is preceded by three "electronic" stages: (i) the light absorption, (ii) the electron thermal wave, and (iii) the electron-lattice relaxation. During all these stages the electron temperature is significantly higher than the lattice temperature, and the motion of material is negligible. The duration of the first stage is of the order of laser pulse length τ. Laser radiation is absorbed in the surface layer δ. Second stage is the electron thermal wave. Its initial velocity is supersonic. At this stage, losses of energy by electron subsystem due to electron-lattice relaxation increase rapidly with the thickness of layer heated by electron heat conduction.

The electron thermal wave moves according to the equation $d_T \sim \sqrt{\chi_e t} \sim v_F \sqrt{\tau_e t}$, where v_F is the Fermi velocity (we assume that $k_B T_e << \varepsilon_F$), χ_e is the heat diffusivity, and τ_e is the electron mean free time. As the electron thermal wave propagates, its velocity decreases and becomes of the order of sound speed. The rarefaction wave moves as $d_H \sim c_s t$, where c_s is the sound velocity. The rarefaction wave overtakes the thermal wave $(d_T \approx d_H)$ at the time instant $t^* \sim \tau_e v_F^2/c_s^2$. The thickness of heated layer at this instant equals $d_T^* \sim \chi_e/c_s$. If the laser fluence is not very high, this thickness depends only weakly on the fluence and comprises about 100 nm.

It follows from Eq. (3) that the characteristic time of lattice heating, due to the energy exchange between electrons and the lattice, is of the order of $\tau_i = c_i/\alpha$. According to (Kaganov, 1957), the energy exchange rate between electrons and the lattice equals $\alpha = \pi^2 m n_e c_s^2/6T_i\tau_e$. Using this equation we obtain the estimate

$$\frac{t^*}{\tau_i} \approx \frac{\varepsilon_F}{k_B T_i} >> 1 \,. \tag{5}$$

Thus, at the beginning of the hydrodynamic stage the temperature of lattice becomes approximately equal to that of electrons, and the expansion of hot layer can be considered, to a first approximation, in the frame of a one-temperature model.

It should be noted that simple estimate (5) holds if the electron temperature is not very high, and electron thermal conduction is determined by electron-phonon collisions. At higher electron temperatures the electron-electron collision frequency increases and reaches its maximum at temperatures of the order of ε_F/k_B. On further increase of the electron temperature, the collision frequency decreases as $T_e^{-3/2}$. As a result, the thickness d_T^* of the layer heated at the beginning of the hydrodynamic stage appears to be a non-monotonic function of laser fluence.

At hydrodynamic stage the energy transfer is due to the motion of material. At this stage in materials with fast electron-lattice relaxation, the electron thermal energy becomes small as compared to the phonon energy. However, if the energy exchange between electrons and lattice proceeds slowly and total energy absorbed is sufficiently large, a significant part of absorbed laser energy remains in the electron subsystem at the beginning of the hydrodynamic stage. In this paper, we shall consider moderate fluences for which the hydrodynamic stage begins at equal electron and ion temperatures. In this case, the distribution of temperature $T_e(x) = T_i(x) = T_o(x)$ at the instant t^* is the initial condition for the simulation of material expansion.

3. FORMATION OF THE SHELL AND CRATER

In actual experimental conditions the heated surface layer is non-uniform. Its temperature depends on the coordinate x normal to the surface of the target (the scale is d_T^*), and on the coordinate y lying in the plane of the target (the scale is $D >> d_T^*$). The results of molecular dynamics simulation of the expansion of a surface layer whose temperature depends on x will be presented below. This simulation shows that the qualitative picture of the flow remains the same as in the case of uniformly heated layer. Specifically, the liquid (or solid) shell moving at a constant velocity also exists in the non-uniform case. Experimental study and molecular-dynamics simulation show that the ablation takes place at fluences exceeding a certain threshold value F_{th}.

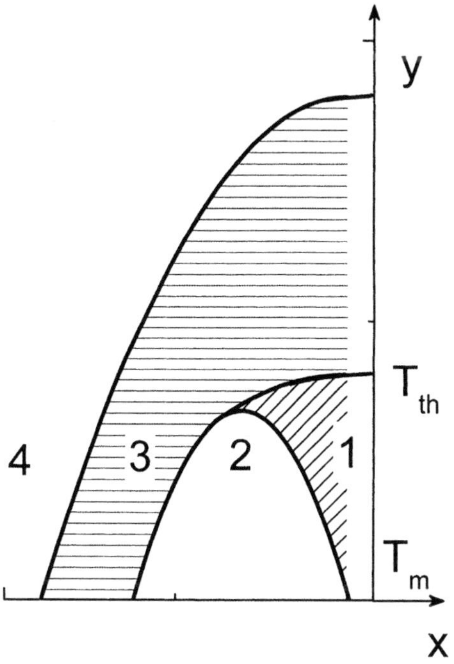

Figure 2. Scheme of crater formation. Initial surface of the target is denoted by the vertical straight line. 1 – shell, 2 – two-phase material, 3 – liquid phase, 4 – solid phase.

If the fluence is higher than F_{th}, a crater is formed with a sharp edge. This edge is located between adjacent areas 1 and 3 (Fig. 2). The interface separating the areas 1 and 3 form the side wall of the crater, whereas the interface between the areas 2 and 3 form its bottom. The formation of high-density shell is responsible for finite depth of the crater at the fluence slightly higher than F_{th}. The thickness of the shell 1 depends on the initial temperature of the surface. As is seen from the results of simulation, the thickness decreases with increasing temperature. Due to non-uniform distribution of laser intensity, and hence initial temperature along the y-axis, the thickness of the shell appears to be dependent on y (Fig. 2).

The regions 1 and 2 in Fig. 2 are filled with material that expands into vacuum in the course of ablation. The regions 3 and 4 represent the part of the target not degraded. Because of the growth of laser fluence from periphery to center of focal spot, local temperature increases and shell thickness decreases. At a certain local temperature, the thickness of shell reduces to zero. Thus, the shell exists only in certain range of laser fluencies, between upper and lower boundaries, F_m and F_{th}.

Molecular dynamic simulations of the expansion of non-uniformly heated layers have been performed at various initial temperatures of the surfaces $x = 0$: $T_o(0) = 1, 1.25, 2, 3,$ and 5. At $T_o(0) = 1$ there is no ablation. The value $T_o(0) = 1.25$ corresponds to ablation threshold F_{th}. Upper boundary F_m, when the shell disappears, lies between the temperatures 3 and 5. Note that all variables that occur in MD-simulation are measured in dimensionless MD-units defined in terms of the Lennard-Jones (6 - 12) potential used for the simulation of the hydrodynamic stage:

$$u(r) = 4\varepsilon\left[\left(\frac{\sigma}{r}\right)^{12} - \left(\frac{\sigma}{r}\right)^{6}\right] \tag{6}$$

In particular, the unit of temperature is ε and the unit of length is σ.

Formation of a crater takes several characteristic acoustic times τ_H. During this time interval the shell (area 1 in Fig. 2) moves at a constant rate. This rate increases monotonically with surface temperature T_o, which, in turn, is a monotonic function of laser fluence $F(y)$. At the ablation threshold the shell velocity equals zero. As a result, the shell takes the shape depicted in Fig. 2.

4. INTERFERENCE

To study the dynamics of ablation a probe laser beam is used. This beam is retarded for some time relative to the pump beam (the so-called pump-probe technique). The probe beam is reflected in part by the shell and is partially passed through the shell and reflected by the bottom of the crater (Fig. 3). The interference of the two reflected beams produces a well-known picture – Newton rings. This picture was observed in papers (von der Linde, 2000; Sokolowski-Tinten, 1998a, 1998b). Consider the interference more in detail.

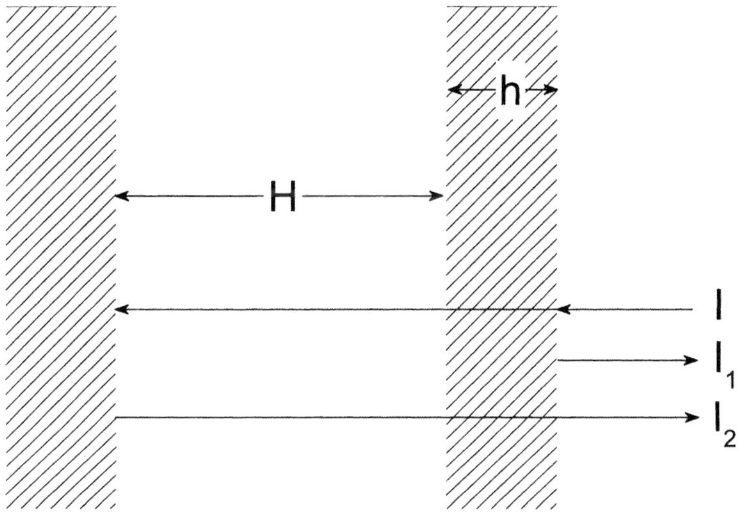

Figure 3. The formation of Newton rings as the result of interference of beams I_1 and I_2 reflected by the shell and by the bottom of the crater. The shell and bottom are separated by the gap H filled with two-phase material, its refraction index being close to unity.

The scheme of formation of the interference picture is shown in Fig. 3. A liquid shell is separated from the target by a gap filled with two-phase mixture. The thickness of the shell equals h, and the thickness of the gap is H. A probe beam I is incident on the shell, and reflected from the shell (I_1) and from the bottom of crater (I_2). Total reflectivity depends on the parameters h and H, the refraction indices of shell $\tilde{n}_1$, material of the target $\tilde{n}_2$, and the wavelength of probe beam λ. The two-phase medium bridging the gap H between the shell and the bottom of crater has a low density; hence its refraction index may be considered equal to unity.

The results of calculations using the Fresnel formulae and taking into account an infinite number of reflections are presented in Figs 4 to 6. The calculation has been performed for aluminum and probe wavelength $\lambda =$ 700 nm. It is assumed that aluminum in the crater and shell is in liquid state. Optical properties of liquid aluminum are taken from (Drits, 1985). An important parameter in these calculations is the thickness of skin layer. For liquid aluminum it comprises 17 nm for a wavelength of 700 nm. As is seen from Figs. 4 and 5, the oscillations of reflectivity vary in the amplitude $\Delta R = (R_{max} - R_{min})/(R_{max} + R_{min})$ and shape with a change of shell thickness h. Here R_{max} and R_{min} are, respectively, the maximum and minimum amplitudes of relative intensity of reflected light (see Fig. 4).

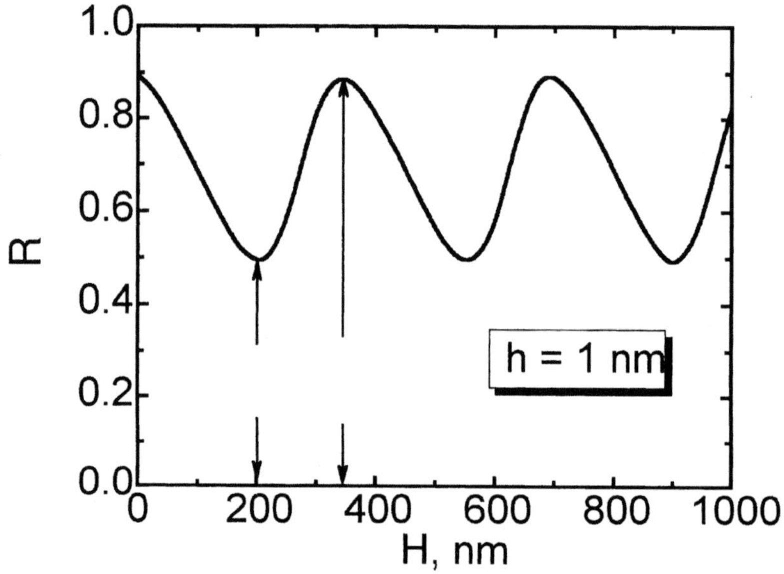

Figure 4. Interference oscillations of the intensity of reflected light as a function of distance between the shell and bottom of the crater.

The variation of relative amplitude and shape of oscillations as a result of a change of shell thickness is illustrated in Fig. 5 (h = 10 nm). The dependence of normalized amplitude on shell thickness is presented in Fig. 6. Note that the normalized amplitude $\Delta R(h)$ approaches zero for both thick and thin shells. In the case of the thin shell, $h/\delta \rightarrow 0$, and the contribution R_1 in the sum $R = R_1 + R_2$ is lowered. In the case of the thick shell,

$h/\delta \to \infty$, and the contribution R_2 decreases, because in this case light is absorbed by the shell.

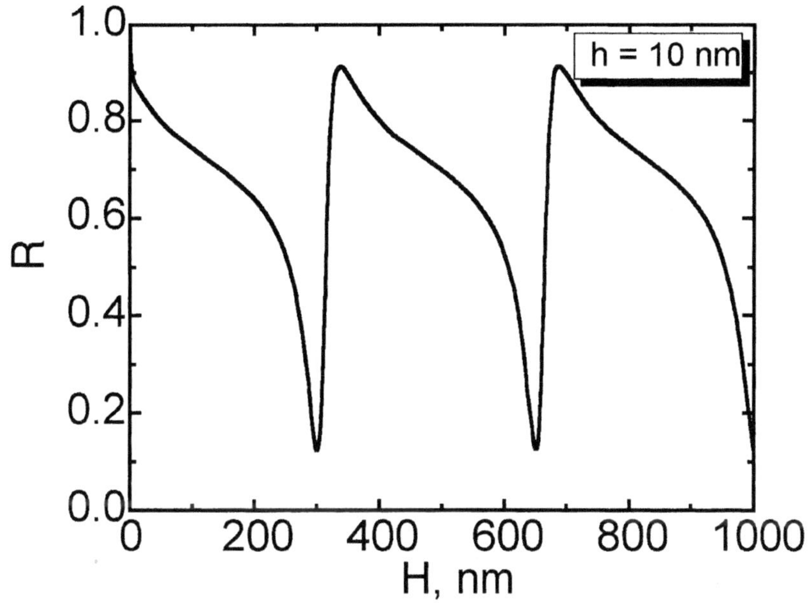

Figure 5. Variation of the shape and relative amplitude of oscillations resulting from the increase of shell thickness.

5. SIMULATION OF HYDRODYNAMIC STAGE

Hydrodynamic stage of expansion of non-uniformly heated layer is studied using the method of molecular dynamics (MD) simulation. Since earlier MD-calculations (Anisimov, 2003; Zhakhovskii, 2000) were performed using the Lennard-Jones (6 - 12) potential and the intention of our simulation is to describe the qualitative picture of expansion, we employ the same interatomic potential. The problem is formulated as follows. The expansion of surface layer of a crystal with exponential or Gaussian initial temperature profile is simulated. The expansion proceeds in the negative direction of the $x-$ axis. Periodic boundary conditions are imposed in the $y-$ and $z-$ directions. At the left "cold" boundary of the simulation box a Langevin thermostat that simulates undisturbed crystal is placed. At this boundary constant temperature $T_\infty = 0.25$ MDU is maintained.

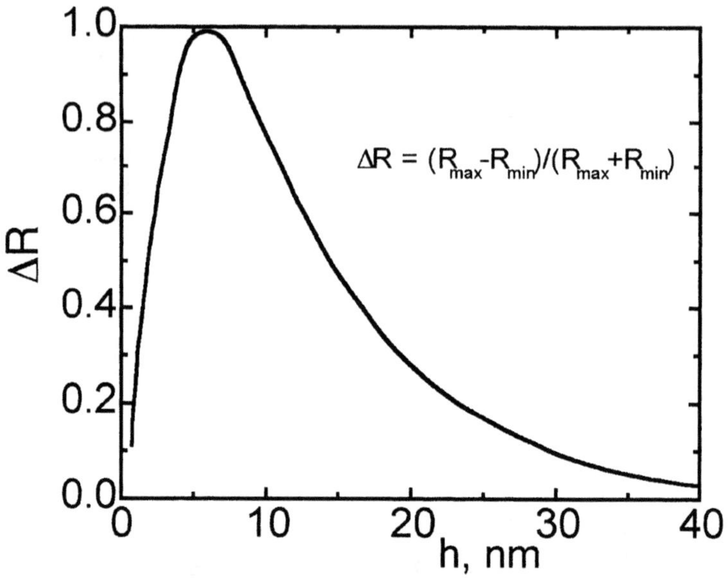

Figure 6. Normalized oscillation amplitude as a function of shell thickness.

At the initial time instant $t = t^*$, atoms are packed in the fcc lattice with average density n_o = 1.09 MDU. The plane $x = 0$ corresponds to initial free boundary of a crystal. The expansion takes place along the direction <110>. The simulation includes the solution of equations of motion and calculation of macroscopic parameters of the system. The method of solution is similar to that used in (Zhakhovskii, 1997; Zhakhovskii, 1999). The simulation was performed using dynamical domain decomposition (MPD[3]) described in (Zhakhovskii, 2004). Total number of atoms used in a typical simulation was $8 \cdot 10^6$. The dimensions of the simulation box were L_X = 1420.373, L_y = 35.50933, L_Z = 150.6533 MDU. Initial surface temperature varied between 1 and 5 MDU. Calculations show that the value T_0 = 1.25 corresponds to ablation threshold. Density distribution in the expanding matter is presented in Figs. 7 and 8.

In Fig. 7, which corresponds to the threshold, the shell is clearly seen. After detachment of the shell, a crater of finite depth is formed. The thickness of the shell decreases with increasing surface temperature. At T_0= 3 the shell still persists. However, its thickness is less than 30 MDU. At T_0= 5 the shell is missing (see Fig. 8) and the flow resembles usual rarefaction wave. When the shell disappears, the interference structure disappears as well. This occurs at a fixed initial temperature $T_0 = T_m$, which lies in the range between 3 and 5 MDU.

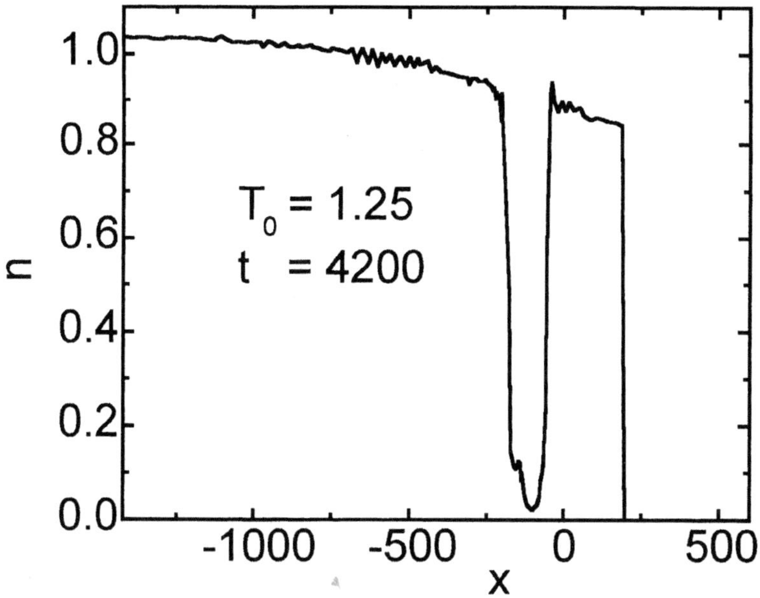

Figure 7. Asymptotic density profile. Initial surface temperature corresponds to the ablation threshold.

Using the results of MD-simulation one can determine the profile of the shell, make an estimate of the size of the crater, and calculate a fluence at which Newton rings disappear. Let us assume that the reflectivity R does not depend on laser fluence F_{in}. Then the function $F(y)$ within a constant factor repeats the profile of laser beam $F_{\text{in}}(y)$. Take the value of threshold fluence F_{th} as a scale for measurements of absorbed laser fluence F. The one-parametric set of laser beams $\hat{F} = F(y)/F_{th} = \hat{F}_c \exp(-y^2/D^2)$ is specified by non-dimensional ratio $\hat{F}_c = F_{\max}/F_{th}$. Hence, it follows that the radius of a crater equals $y_{th} = D\sqrt{\ln \hat{F}_c}$.

Each specific version of molecular-dynamic simulation is determined by two parameters: (i) initial surface temperature T_0 and (ii) initial thickness of heated layer d_T. The simulation of hydrodynamic stage was performed in one-temperature approximation. As already noted, this approximation is valid if the electron temperature is not too high, that is at moderate fluence. In this case the thickness of the layer heated by electron thermal conduction by the beginning of hydrodynamic stage, d_T^*, depends only slightly on the laser fluence.

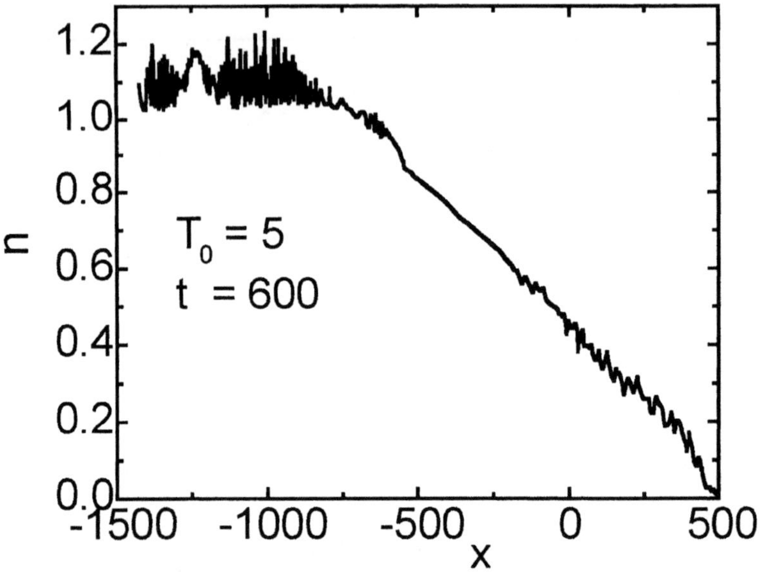

Figure 8. Evolution of density profile with increase of initial temperature. At sufficiently high initial temperature the shell disappears.

On the assumption that d_T^* is constant in the range of fluence under consideration, the transverse profile of initial temperature is coincident with the transverse profile of laser fluence, $F(y)$. Therefore, if the initial temperature T_0 is prescribed, one can determine the distance from the center y/y_{th} corresponding to this temperature. Further, using the results of MD-simulation one can determine the profile of the shell $h(y)$. The fluence at which the Newton rings disappear can be estimated from the relation: $F_m \approx F_{th}T_m / T_{th}$.

6. CONCLUSION

The above described qualitative picture of interaction of an ultra-short laser pulse with matter gives an insight into main features of the process and agrees well with experiment. It is possible to explain that the ablation is a threshold phenomenon and a crater of finite depth is formed when the fluence reaches its threshold value. The mechanism of formation of interference Newton rings is proposed, connected with special features of expansion into vacuum of instantly heated condensed matter. The existence of upper and lower boundaries of fluence range, where the Newton rings are observed, is explained.

The model proposed in this paper is applicable at fluences that are not too much higher than the ablation threshold. At very high fluences, two-temperature effects at hydrodynamic stage and energy transfer by fast electrons on the stage of thermal wave should be taken into account. These phenomena have been studied in connection with the problems of inertial confinement fusion (Atzeni, 2004).

The authors are indebted to Prof. B. Luk'yanchuk for collaboration. Support of this work by the Russian Foundation for Basic Research, under grant 04-02-16972, is gratefully acknowledged.

REFERENCES

Anisimov S.I., Kapeliovich B.L., Perelman T.L., 1974, Electron emission from metal surfaces exposed to ultrashort laser pulses, Sov. Phys.- JETP, **39**, 375

Anisimov S.I., Rethfeld B., 1997, On the theory of ultrashort laser pulse interaction with a metal, Proc. SPIE, **3093**, 192

Anisimov S.I., Zhakhovskii V.V., Inogamov N.A., Nishihara K., Oparin A.M., Petrov Yu.V., 2003, Destruction of a solid film under the action of ultrashort laser pulse, JETP Lett., **77**, 606

Atzeni S., Meyer-ter-Vehn J., 2004, *The Physics of Inertial Fusion*, Clarendon Press, Oxford

Bushman A.V., Lomonosov I.V., Fortov V.E., 1992, *Equations of State of Metals at High Energy Densities*, Chernogolovka (Russian)

Drits M.E., Ed., 1985, *Properties of Elements*, Metallurgizdat, Moscow (Russian)

Hashida M, Semerok A.F., Gobert O., Petite G., Izawa Y., Wagner J.F., 2002, Ablation threshold dependence on pulse duration for copper, Appl. Surf. Sci., **197-198**, 862

Hirschfelder J.O.,Curtiss Ch.F., Bird R.B., 1954, *Molecular Theory of Gases and Liquids*, John Wiley, NY, Ch. 5

Inogamov N.A., Petrov Yu.V., Anisimov S.I., Oparin A.M.,.Shaposhnikov N.V., von der Linde D., Meyer-ter-Vehn J., 1999, Expansion of matter heated by an ultrashort laser pulse, JETP Lett., **69**, 310

Kaganov M.I., Lifshitz I.M., Tanatarov L.V., 1957, Relaxation between electrons and the crystalline lattice, Sov. Phys.- JETP, **4**, 173

Linde von der D., Sokolovski-Tinten K., 2000, The physical mechanisms of short-pulse laser ablation, Appl. Surf. Sci., **154-155**, 1

Lorazo P., Lewis L.J., Meunier M., 2003, Short-pulse laser ablation of solids: from phase explosion to fragmentation, Phys. Rev. Lett., **91**, 225502

Lugovskoy A.V., Bray I., 1999, Ultrafast electron dynamics in metals under laser irradiation, Phys. Rev. B, **60**, 3279

Petrov Yu.V., 2005, Energy exchange between the lattice and electrons in a metal under femtosecond laser irradiation, Laser and Particle Beams, **23**, N 3

Rethfeld B., 2004, Unified model for the free-electron avalanche in laser-irradiated dielectrics, Phys. Rev. Lett., **92**, 187401

Sokolowski-Tinten K., Bialkowski J., Cavalleri A., von der Linde D., Oparin A.M., Meyer-ter-Vehn J., Anisimov S.I., 1998a, Transient states of matter during short pulse laser ablation, Phys. Rev. Lett., **81**, 224

Sokolowski-Tinten K., Bialkowski J., Cavalleri A., von der Linde D., 1998b, Observation of a transient insulating phase of metals and semiconductors during short-pulse laser ablation, Appl. Surf. Sci., **127-129**, 755

Stuart B.C., Feit M.D., Herman S., Rubenchik A.M., Shore B.W., Perry M.D., 1996, Optical ablation by high-power short-pulse lasers, J. Opt. Soc. Am. **B13**, 459

Temnov V.V., Sokolowski-Tinten K., Zhou P., von der Linde D., 2004, Femtosecond time-resolved interferometric microscopy, Appl. Phys. A, **78**, 483

Zeifman M.I., Garrison B.J., Zhigilei L.V., 2002, Combined molecular dynamics-direct simulatiom Monte-Carlo computational study of laser ablation plume evolution, Journ. Appl. Phys., **92**, 2181

Zhakhovskii V.V., Anisimov S.I., 1997, Molecular-dynamics simulation of evaporation of a liquid, JETP, **84**, 734

Zhakhovskii V.V., Nishihara K., Anisimov S.I., Inogamov N.A., 2000, Molecular-dynamics simulation of rarefaction waves in media that can undergo phase transitions, JETP Lett., **71**, 162

Zhakhovskii Vasilii, Nishihara Katsunobu, Fukuda Yuko, and Shimojo Shinji, 2004, *A New Dynamical Domain Decomposition Method for Parallel Molecular Dynamics Simulation on Grid*, Annual Progress Report 2003, Institute of Laser Engineering, Osaka University, 2004, p. 147 (aRxiv: cs.DC/0405086)

Zhakhovskii V.V., Zybin S.V., Nishihara K., Anisimov S.I., 1999, Shock wave structure in Lennard-Jones crystal via molecular dynamics, Phys. Rev. Lett., **83**, 1175

Chapter 2

FAST ELECTRONIC TRANSPORT AND COULOMB EXPLOSION IN MATERIALS IRRADIATED WITH ULTRASHORT LASER PULSES

Nadezhda M. Bulgakova[1], Razvan Stoian[2], Arkadi Rosenfeld[3], Ingolf V. Hertel[3,4], and Eleanor E.B. Campbell[5]

[1]*Institute of Thermophysics SB RAS, 1 Lavrentyev Ave., 630090 Novosibirsk, Russia* [2]*Laboratoire TSI (UMR 5516 CNRS), Universite Jean Monnet, 10 rue Barrouin, 42000 Saint Etienne, France,* [3]*Max-Born-Institut für Nichtlineare Optik und Kurzzeitspektroskopie, Max-Born Str. 2a, D-12489 Berlin, Germany,* [4]*Department of Physics, Free University of Berlin, Amimallee 14, 14195 Berlin, Germany,* [5]*Department of Physics, Göteborg University, SE-41296 Göteborg, Sweden*

1. INTRODUCTION

Ultrafast laser ablation of dielectrics and semiconductors proceeds through a sequence of physical processes such as nonlinear absorption, non-equilibrium effects related to electronic and vibrational excitations, and avalanche breakdown, generating dense, overcritical electron-hole plasmas during the action of the laser pulse. One of the most intriguing and complicated features of laser heating and ablation of dielectric materials is the strong temporal and spatial variation of the optical properties, ranging from a fully transparent, non-reflecting substance to a metal-like state. In dielectrics, metal-like behavior develops in a narrow surface layer. This results in strong localization of the energy coupling since the electron thermal conductivity is restricted to the high-density region. This aspect accounts for one of the main distinctions between metals and dielectrics under pulsed laser irradiation, the possibility of Coulomb explosion (CE) as an ablation mechanism alternative or additional to phase explosion.

Coulomb explosion is one of the electronic mechanisms of material removal from interfaces or surfaces sputtered by photons or charged particles (Fleischer, et al., 1965; Bitenskii et al., 1979; Schneider and Briere, 1996; Cheng and Gillaspy, 1997; Varel, et al., 1998; Stoian, et al., 2000a,

2000b, 2000c, 2002a; Henyk, et al., 2000a, 2000b; Vanagas, et al., 2003; Costache, et al. 2003; Costache and Reif, 2004). The occurrence of this particular phenomenon was intensely discussed during the last decades for various materials with radically different electronic properties. It has been postulated that CE can be exploited in different applications such as surface nanostructuring (Stoian, et al., 2002a; Vanagas, et al., 2003) and nanoparticle formation (Dong and Molian, 2004). Note that when referring to Coulomb explosion, we imply removal of at least several monolayers (Stoian, et al., 2000c; Bulgakova, et al., 2004a). We are not referring to electrostatic desorption, which is of a stochastic nature and results from localization of electronic excitation energy at specific atomic sites (Kanasaki and Tanimura, 2002). The main features of CE are the following. Energetic ions of different species, e.g., aluminum and oxygen in the case of ultrafast laser ablation of sapphire (Stoian, et al., 2000c), detected by time-of-flight mass spectrometry, have the same momenta but not the same energy. Also, doubly-charged ions have velocities twice as high as singly-charged ones. This indicates that the ions could be extracted and accelerated in the electric field generated in the target following a laser-induced neutrality breakdown. The electric field can be generated due to intensive electron photoemission, leading to accumulation of positive charge in a superficial target layer. Thus, *the concept of Coulomb explosion* is based on the fact that, due to photoemission, the irradiated surface gains a high positive charge, so that the repulsive force between ions exceeds the lattice binding strength, resulting in surface layer disintegration. An interesting CE aspect, in view of practical applications, is that the exploded material leaves behind a smooth, perfect surface that makes Coulomb explosion attractive for nanoscale applications (Stoian, et al., 2000c).

The present article gives an account of the underlying physical phenomena of the CE mechanism of laser ablation and presents a unified model to describe the processes that can influence the development and decay of charge accumulation on the surfaces of laser-irradiated materials of different classes (dielectrics, semiconductors, and metals). We will confine our considerations to gold, silicon, and sapphire as typical representatives of their classes.

2. EXPERIMENTAL EVIDENCE IN FAVOR OF LASER-INDUCED COULOMB EXPLOSION

The ultrashort pulse laser ablation of Al_2O_3 has been the subject of a large number of studies due to its useful mechanical, optical, and electrical properties. This material exhibits two completely different etch phases under laser irradiation (Tam, et al., 1989; Brand and Tam, 1990; Ashkenasi, et al.,

1997; Varel, et al., 1998; Stoian, et al., 2000a, 2000b, 2000c): a "gentle" ablation with material removal of a few nanometers per pulse and a "strong" ablation phase characterized by an order of magnitude higher ablation rate. "Gentle" ablation is characteristic for the irradiation regimes slightly above the ablation threshold and for a low number of laser shots irradiating the same surface area. Under the "gentle" ablation phase, the ablated material leaves behind an extremely smooth surface (sometimes even smoother than the initial state) (Fig. 1a). With increasing laser fluence and/or increasing number of shots, the "strong" ablation phase takes over and the irradiated surface begins to show the characteristics of "phase explosion" (Fig. 1b).

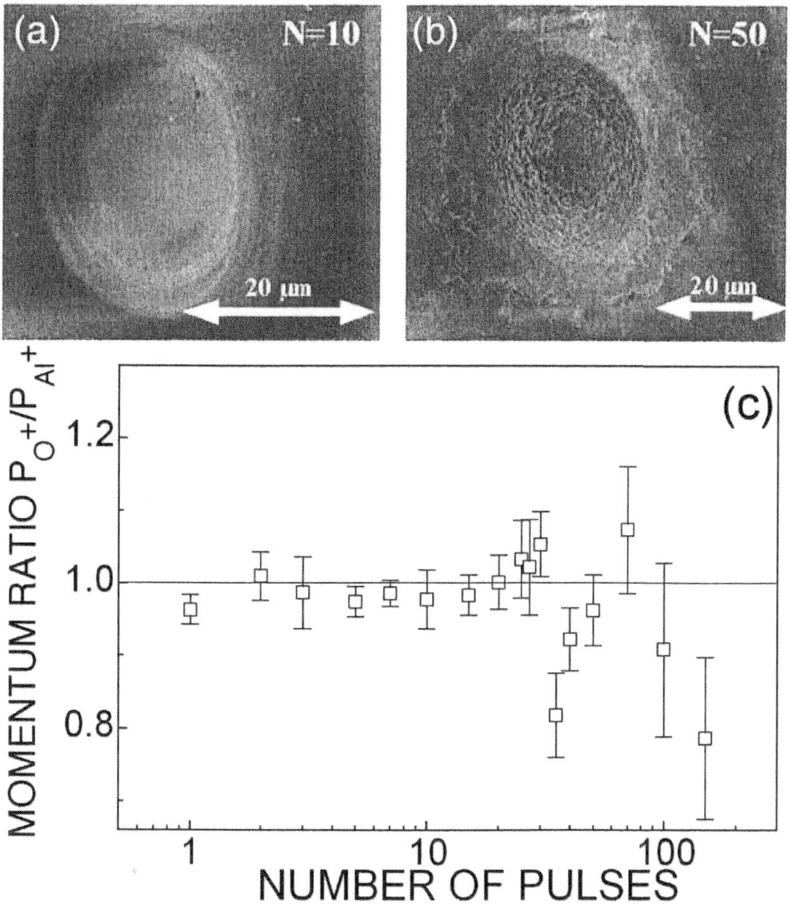

Figure 1. Transition from "gentle" (a) to "strong" (b) etch phase under ultrashort pulse laser ablation of Al_2O_3 (200 fs, 4 J/cm^2) with increasing number of shots. (c) Ratio of ion momenta, calculated from maxima of the velociy distributions. Reproduced from Stoian, et. al., 2000b.

Experimental evidence shows that under ultrafast irradiation of dielectric materials a significant component of the ablation products at intensities close to the material removal threshold is generated via a mechanism of electrostatic nature. We present below a short review of the experimental results.

The observed ionization degree and ion velocity distributions under the two etch phases differ considerably. Under gentle ablation, the ions, constituting the main component of the plume, have a clearly non-thermal nature exhibiting a most probable velocity of $\sim 2 \cdot 10^4$ m/s and a narrow angular velocity distribution directed along the normal to the irradiated targets. With the transition to the strong ablation stage, the ion velocity distributions become broader and slower, and thermal plasma ions ($\sim 1.2 \cdot 10^4$ m/s) are seen predominantly in the time-of-flight signals (Stoian, et al., 2000c). Moreover, ions of different species (Al^+ and O^+) observed under the gentle etch phase have mass-independent momenta (Fig. 1c), while on the transition to the strong phase they tend to have mass-independent quasi-equal energies (Stoian et al., 2002a, 2000c). These features indicate that, under gentle ablation, the ions are subjected to the action of a mass-independent acceleration mechanism, most probably due to a transient electric field generated in the target. The origin of the laser-induced neutrality breakdown is based on two main consequences of the interaction: a relatively high photoemission yield and a reduction of the transport properties for the laser-generated carriers. Even more striking is the difference in the behavior of F^+ and F^- ions, observed under barium fluoride ablation (Henyk, et al., 2000b), where the negative ions are retarded by the generated electric field.

3. CRITERION FOR COULOMB EXPLOSION

The occurrence of surface Coulomb explosion generating macroscopic material removal and high ion kinetic energies has been observed for dielectric materials, as mentioned above, while for semiconductors and metals the subject remains controversial (Gamaly, et al., 2002; Bulgakova, et al., 2005a). This raises the question: What processes are responsible for the CE occurrence or its inhibition? The answer to this question is based on a detailed study of the processes taking place in different types of laser-irradiated materials and their intricate interplay. Consequently, we introduce here a criterion for the occurrence of CE.

It is assumed that the electric field generated in the irradiated target can reach extremely high values so that the atomic bonds are broken and a surface layer of the material is disintegrated, resulting in the occurrence of Coulomb explosion. The estimation of the threshold electric field with respect to CE can be made by assuming that the electrostatic energy density per individual atom exceeds a threshold value related to the required sublimation energy per atom. The threshold electric field can be approximated as $E_{th}\big|_{x=0} = \sqrt{2\Lambda_{at} n_0 / \varepsilon\varepsilon_0}$, where Λ_{at} is the sublimation energy per atom, n_0 is the lattice atomic density, and ε is the dielectric permittivity. For sapphire we obtain $E_{th} \sim 5{\cdot}10^{10}$ V/m, while for gold and silicon the estimated threshold electric fields are smaller, $2.76{\cdot}10^{10}$ and $2.65{\cdot}10^{10}$ V/m, respectively. This assumption is valid for the case of a cold lattice, roughly during femtosecond laser exposure. For longer pulses or longer times, one should take into account that the heated lattice atoms, having a higher vibrational energy, can escape from the surface with higher probability than the cold ones. Thus, thermal reduction of the threshold field can be expressed as $E_{th}\big|_{x=0} = \sqrt{2(\Lambda_{at} - 3kT_l) n_0 / \varepsilon\varepsilon_0}$.

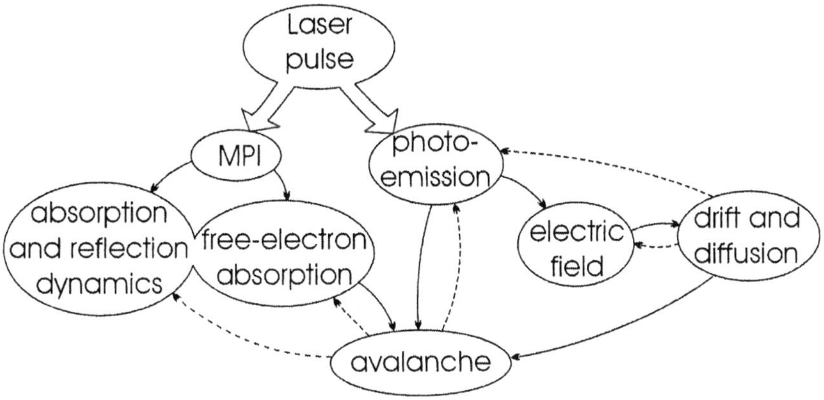

Figure 2. Modeling scheme. Major interconnections between different processes are shown by the arrows. Direct effects are indicated by solid lines and feedback effects are shown with dashed lines.

4. CARRIER DYNAMICS AND MODELING APPROACHES TO CARRIER TRANSPORT

The importance of carrier dynamics under pulsed laser ablation (PLA) conditions, especially for ultrashort laser pulses, has been demonstrated in numerous studies (Petite, et al., 1996; Li, et al., 1999; Stoian, et al., 2002b). Electronic transport differs strongly in various classes of materials irradiated under similar conditions at moderate energy densities (Stoian, et al., 2002a; Bulgakova, et al., 2004a). We discuss a situation where, apart from the photoemission yield, the mobility of the charge carriers, electrons and holes, is the main parameter influencing the rate of charge redistribution in self-consistent electric fields generated under the action of pulsed laser radiation. Mobility determines both drift and diffusion of charge carriers, including thermal diffusion in the presence of temperature gradients.

The theoretical description of carrier dynamics in dielectric and semiconductor targets under short pulsed laser irradiation is a complicated task which implies consideration of a broad spectrum of interrelated processes. A simplified scheme of the processes triggered by ultrafast laser irradiation in dielectrics and semiconductors is shown in Fig. 2. A laser pulse impinging on a target causes photoionization (with the order of the process depending on the relation between the photon energy and the material energy band-gap) and photoemission (with the order of nonlinearity determined by the ratio of the work function to the photon energy). The electrons excited to the conduction band absorb further laser radiation and can produce secondary electrons by collisional ionization. In addition, they change the dielectric response of the material, thus explicitly influencing the optical properties. As soon as electronic avalanche is developed, absorption and reflection dynamics vary dramatically. Photoemission depletes a superficial target layer arresting further development of collisional multiplication and, in turn, avalanche regulates photoemission through its influence on the available free-electron density. Another drastic consequence of electron photoemission is the violation of the target quasi-neutrality that results in positive charging of the target as a whole and thus in generation of an ambipolar electric field. The electric field forces the charge carriers to relocate in order to neutralize the excess positive charge. The electric current is a superposition of two components: charge-carrier drift in the electric field and diffusion under the action of the gradients, both in density and temperature. Redistribution of free electrons due to the electric current strongly influences both electron photoemission and the avalanche multiplication process.

Existing models treat most, or at least some, of the processes described above and may, by convention, be divided into three groups. One of the dominating approaches to describe carrier dynamics in silicon targets is based on ambipolar diffusion with an implicit assumption of an equal number of electrons and holes in the solid and the preservation of local quasi-neutrality of the sample (Yoffa, 1980; van Driel, 1987; Mao, et al., 1998). Another group of models, developed for semiconductors irradiated by laser pulses (Held, et al., 1991) and dielectrics under the action of electron beams (Melchinger and Hofmann, 1995; Miotello and Dapor, 1997), takes into account the generation of local electric fields inside the target with the assumption that the target remains neutral as a whole. This implies the absence of electron photoemission (Held, et al., 1991) or relies on secondary electron emission equal to the absorbed electron flux (Melchinger and

Hofmann, 1995; Miotello and Dapor, 1997). In this approach, the concept of a double layer can be applied to describe the spatial charge arrangement in the bulk of the irradiated materials (Melchinger and Hofmann, 1995).

A third approach (Ribeiro, et al., 1997, 1998), proposed for the case of a dielectric target (MgO) irradiated by a laser pulse of nanosecond duration, may be labeled as the drift-diffusion approach. The authors studied the self-consistent generation of an electric field as a result of laser heating and thermionic emission of the electrons excited to the conduction band and their diffusion and drift in the locally established fields. It was found that the self-consistent electric field could reach values exceeding 10^8 V/m under normal ablation conditions. In several recent works (Stoian, et al., 2002a, Bulgakova, et al., 2004a, 2004b, 2005a, 2005b) the authors used the same simplified approach to model charging of different types of materials under ultrashort pulsed laser irradiation, taking into account specific properties of the materials. This implies calculations of energy deposition into the electronic subsystem and subsequent heating of the lattice through electron-lattice energy exchange. For the latter task we used the two-temperature model which is mostly utilized to describe metals heated by ultrashort laser pulses (Kaganov, et al., 1957; Anisimov, et al., 1974; Wellershoff, et al., 1999).

5. A UNIFIED APPROACH

Under high power ultrafast laser irradiation, dense plasmas are generated in dielectrics and semiconductors (Stuart, et al., 1996; Tien, et al., 1999; Nolte, et al., 1999; Sokolowski-Tinten and von der Linde, 2000), thus justifying a unified approach based on an intrinsic or a laser-induced metallic behavior for the three classes of materials investigated. The following equations are used as the building blocks for a common simplified frame applicable to different kinds of materials:

(1) The continuity equations for the evolution of the laser-generated charge carriers:

$$\frac{\partial n_x}{\partial t} + \frac{1}{e}\frac{\partial J_x}{\partial x} = S_x + L_x, \qquad (1)$$

where S_x and L_x are the source and loss terms describing the free carrier populations, n_x denotes the carrier densities, with subscript $x = e, i$ representing electrons and ions, respectively;

(2) The expression for the electric current density J_x includes drift and diffusion terms (Ribeiro, et al., 1998) and can be considered as the equation of motion:

$$J_x = |e| n_x \mu_x E - e D_x \nabla n_x . \qquad (2)$$

Here the time and space dependent diffusion coefficient D_x is calculated according to the Einstein relation as $D_x = kT_x\mu_x/e$, where T_x represents the carrier temperature, k is the Boltzmann constant, and μ_x is the carrier mobility. We assume that the charge-carrier flows are caused by quasi-neutrality violation on and below the target surface due to electron photoemission and strong density and temperature gradients.
(3) Poisson equation to calculate the electric field generated as a result of breaking quasi-neutrality in the irradiated target:

$$\frac{\partial E}{\partial x} = \frac{e}{\varepsilon \varepsilon_0}(n_i - n_e) . \qquad (3)$$

(4) Energy conservation equations to account for the heating of electronic and lattice subsystems. We assume that laser-excited metals and strongly ionized insulators and semiconductors can be considered as dense plasmas so that the two-temperature model (Kaganov, et al., 1957; Anisimov, et al., 1974; Wellershoff, et al., 1999) may be applied for the description of the energy balance:

$$C_e\left(\frac{\partial T_e}{\partial t} + \frac{J}{en_e}\frac{\partial T_e}{\partial x}\right) = \frac{\partial}{\partial x} K_e \frac{\partial T_e}{\partial x} - g(T_e - T_l) + \Sigma(x,t) \qquad (4)$$

$$(C_l + L_m \delta(T_l - T_m))\frac{\partial T_l}{\partial t} = \frac{\partial}{\partial x} K_l \frac{\partial T_l}{\partial x} + g(T_e - T_l) . \qquad (5)$$

Even if complete equilibration does not take place in the subsystems, the values T_e and T_i can be considered as measures for the average energies of electrons and lattice. In the Equations (4) and (5), indexes e, l refer to the electron and lattice parameters; C_e, C_l, K_e, K_l are the heat capacities and the thermal conductivities of electrons and lattice respectively; g is the electron-lattice coupling constant; and $\Sigma(x,t)$ is the energy source term. All these parameters will be specified below for every type of material. The term $L_m\delta(T - T_m)$ allows calculations of the liquid–solid interface (Zvavyi and Ivlev, 1996) having the temperature T_m, where L_m is the latent heat of fusion. The electron energy equation accounts for both heat conductivity and direct bulk or across the vacuum interface electronic transport (e.g. due to photoemission).

The system of equations (1)–(5) was solved numerically for targets irradiated with a laser pulse having a Gaussian temporal profile. The calculations were performed for a laser radiation wavelength $\lambda = 800$ nm and a pulse duration of $\tau_L = 100$ fs (FWHM) (Stoian, et al., 2000c, 2002a). A one-dimensional model, justified by the much larger transverse lateral dimensions for the laser spot compared to the absorption depth, has been employed. Particular situations for each class of material will be discussed below.

5.1 Metals

The continuity equation for free electrons (Eq. (1)) in a metal target (gold in our case) has no source terms. Photoemission was considered in the form of a surface boundary condition for the electron current density (Logothetis and Hartman, 1969; Bechtel, et al., 1977), describing electron flow into the vacuum. The three-photon photoemission cross-section c was determined empirically (Logothetis and Hartman, 1969) and was corrected here for the 800 nm irradiation wavelength. Ultrafast, sub-ps irradiation of metals can induce high electronic temperatures, ~1 eV, while the lattice remains cold during the irradiation time. Assuming temperature-dependent effects, we corrected the photoemission cross-section, based on the generalized Fowler-DuBridge theory for multiphoton photoemission at high temperatures (Bulgakova, et al., 2004a). The thermal emission of high-temperature electrons can also play a significant role. Thus, the photoemission term is written in the form:

$$J_s = c\frac{2(kT_e)^2}{(3\hbar\omega-\varphi)^2}F\left(\frac{3\hbar\omega-\varphi}{kT_e}\right)(1-R)^3 I^3 + A_0 T_e^2 \exp\left(-\frac{e\varphi}{kT_e}\right). \qquad (6)$$

Here A_0 is the Richardson coefficient, φ is the work function [4.25 eV (Samsonov, 1965)], $\hbar\omega$ is the incident photon energy (1.55 eV), and F is the Fowler function (Bechtel, et al., 1977).

The parameters for the energy equations (4)-(5) for gold are similar to the parameters reported by Wellershoff, et al. (1999), as well as the energy source accounting for ballistic effects of electronic energy transport. Optical properties for gold are taken for 800 nm wavelength. Electron mobility was estimated from data on gold conductivity (Grigoryev, et al., 1996) and found to be $5.17 \cdot 10^{-3}$ $m^2/(V \cdot s)$.

5.2 Dielectrics

In dielectric materials, free electrons are generated in the processes of multiphoton (MPI) and collisional ionization, and their recombination usually proceeds through a trapping-like phenomenon. For sapphire as an example, the source and loss terms in Eq. (1) for electrons can be written as:

$$S_e = (W_{\text{mph}} + Q_{\text{av}})\frac{n_a}{n_a + n_i}, \quad L_e = -R_e - PE. \tag{7}$$

Here $W_{\text{mph}}=\sigma_6 I^6$ is the rate of a 6-photon ionization process corresponding to an energy forbidden band E_g of ~9 eV, n_a is the density of neutral atoms, $Q_{\text{av}}=\alpha I n_e$ is the avalanche term (Stuart, et al., 1996), R_e represents the decay term taken in the form n_e/τ with τ = 1 ps, and PE denotes photoelectron emission. To compensate for the mobility decay with temperature, we use a simplified, time-independent diffusion coefficient, assuming that the average electron energy in the conduction band is ~5 eV (Arnold and Cartier, 1992; Stuart et al., 1996). The values of $\sigma_6 = 8\cdot 10^9$ $\text{cm}^{-3}\text{ps}^{-1}(\text{cm}^2/\text{TW})^6$ and $\alpha = 6$ cm^2/J were estimated by fitting the experimental determination of the optical damage threshold at different pulse durations (Ashkenasi, et al., 2000).

Attenuation of the laser power inside the dielectric target is determined by loss mechanisms involving free electron generation and by the optical response of a free-electron plasma through the Fresnel formulas. The complex dielectric function can be seen as a mutual contribution of the unexcited solid and the response of the laser–induced free electron gas (Driscoll and Vaughan, 1978; Sokolowski-Tinten and von der Linde, 2000). The local intensity $I(x,t)$ is given by the superposition of the direct irradiation and back-scattered radiation:

$$\frac{\partial}{\partial x} I(x,t) = -W_{\text{mph}} \frac{n_a}{(n_a + n_i)} \hbar \omega n_{\text{ph}} - a_{\text{ab}}(x,t) I(x,t) \; . \tag{8}$$

Here n_{ph} is the number of photons required for a MPI event (n_{ph} = 6). The free electron absorption coefficient $a_{\text{ab}}(x,t)$ is calculated from the complex dielectric response (Stuart, et al., 1996).

For a quantitative estimation of photoemission, we assume a statistical distribution of free electronic momenta in a wide bandgap dielectric where the vacuum level lies close to the conduction band minimum, and only electrons with a momentum component normal to and in the direction of the surface can escape into the vacuum. Thus, we assume that, on average, half of the electrons generated in the MPI and avalanche processes are emitted from the target. Maximum photoemission occurs from the surface with an exponentially decreasing probability with bulk depth (Bulgakova, et al., 2004a):

$$PE(x,t) = \frac{1}{2}\left(W_{\text{mph}} + Q_{\text{av}}\right)\frac{n_a}{n_a + n_i}\exp(-x/l_{PE}) \tag{9}$$

where the electronic escape depth l_{PE} is 1 nm (Jones, et al., 1988). The integral photoemitted charge calculated as above is in good agreement with reported experimental values for dielectric materials (Siekhaus, et al., 1986).

The ion density is calculated based on Eq. (1), disregarding photoemission and neglecting hole transport in the bulk. The electron mobility is taken as $\mu_e = 3 \cdot 10^{-5}$ m^2/(V·s) (ten times lower than reported by Hughes, 1979) for a better match of the observed diffusivities (Miotello and Dapor, 1997) in laser-induced dense plasmas. The source term in the energy equation (4) was constructed to account for the energy of the electrons generated via photoionization, energy expenses for avalanche multiplication, free electron absorption, and the energy localized in the strained lattice via trapping processes (Bulgakova, et al., 2005a). The thermodynamic parameters for the electronic subsystem were taken according to van Driel (1987). The value g can be defined via the characteristic electron-lattice relaxation time τ_r as $g = C_e/\tau_r$ ($\tau_r \sim 1$ ps).

5.3 Semiconductors

The model for silicon is based on a similar approach as for sapphire. The reflection of radiation at the vacuum interface and light absorption inside the bulk of a strongly excited semiconductor are described based on the mutual contribution of the unexcited material and the Drude response of the laser-generated free electron gas (Sokolowski-Tinten and von der Linde, 2000). In Eq. (1), one and two-photon ionization terms and avalanche were considered (Mao, et al., 1998; Sokolowski-Tinten and von der Linde, 2000):

$$S_e = \left[\left(\sigma_1 + \frac{1}{2}\sigma_2 I\right)\frac{I}{\hbar\omega} + \delta n_e\right]\frac{n_a}{n_a + n_i}, \qquad L_e = -R_e - PE \tag{10}$$

where σ_1 and σ_2 are one- and two-photon ionization cross-sections [σ_1 = 1021 cm^{-1} (Choi and Grigoropoulos, 2002), σ_2 = 10 cm/GW (Sjodin, et al. 1998)], the total atomic density is $n_0 = (n_a + n_i) = 5 \cdot 10^{22}$ cm^{-3}, and δ is the avalanche coefficient (van Driel, 1987). The loss term at low electronic densities is mainly determined by Auger recombination which saturates at electronic densities approaching 10^{21} cm^{-3} (Bok and Combescot, 1981). Consequently, the decay rate can be written as $R_e = n_e/(\tau_0 + 1/Cn_e n_i)$, with

$\tau_0 = 6\cdot10^{-12}$ s and $C = 3.8\cdot10^{-31}$ cm^6/s (van Driel, 1987). The equation for hole generation takes into account the hole transport process. The mobilities for electrons and holes are taken to be (Kuhn and Mahler, 1989) μ_e = 0.015 m^2/(V·s) and μ_h = 0.0045 m^2/(V·s), ten times reduced compared to the low electron-hole density values (Hummel, 1993). Diffusion coefficients for electrons and holes are calculated according to the Einstein relation.

Electron photoemission (Kane, 1962) was considered in analogy with gold [Eq. (6)] taking into account a three-photon photoemission from the conduction band with a coefficient corrected for the wavelength and the accessible density of states (Bulgakova, et al., 2004a). Instead of the Si work function (φ = 4.6 eV), an effective potential barrier $\varphi_{\rm eff}$ = 4.05 eV measured from the bottom of the conduction band (Mihaychuk, 1999) was introduced as the relevant parameter for the Fowler function.

The spatial and temporal distribution of the laser intensity in the sample was calculated taking into account one- and two-photon ionization, and free-electron absorption of the light. The source term for the heat transport equation (4) was calculated considering the balance of the laser energy deposited in the free electronic system (Bulgakova, et al., 2005a). The electron-lattice relaxation time was taken as $\tau_T = \tau_R(1 + (n_e/n_{cr})^2)$ with τ_R = 240 fs (Sjodin, et al. 1998).

6. RESULTS AND DISCUSSION

The model was first tested by calculating the damage thresholds for Au, Si, and Al_2O_3 and comparing them with previously published data on laser melting and damage of thin gold films (Wellershoff, et al., 1999) and optical breakdown in sapphire and silicon (Li, et al., 1999; Ashkenasi, et al., 2000; Cavalleri, et al., 2001). The results of the modeling are in good agreement with experimental observations (Bulgakova, et al., 2005a).

In Figure 3 the temporal profiles of the net positive charge resident on the surface for samples representative of different classes of materials (Al_2O_3, Si, and Au) are plotted for laser pulses of 100 fs duration at 800 nm wavelength. The laser fluences are chosen to be slightly above the experimental ion emission thresholds (Wellershoff, et al., 1999; Li et al., 1999; Quèrè, et al., 1999; Ashkenasi, et al., 2000; Sokolowski-Tinten and von der Linde, 2000), namely 4, 0.8, and 1.2 J/cm^2 for sapphire, silicon, and gold respectively. Under these specific irradiation conditions the electronic temperature reaches values ranging from around 1 eV in gold, to approximately 6 eV in silicon and more than 10 eV in sapphire. It is obvious that the net charge is significantly higher for the dielectrics than for the metal or for the semiconductor target.

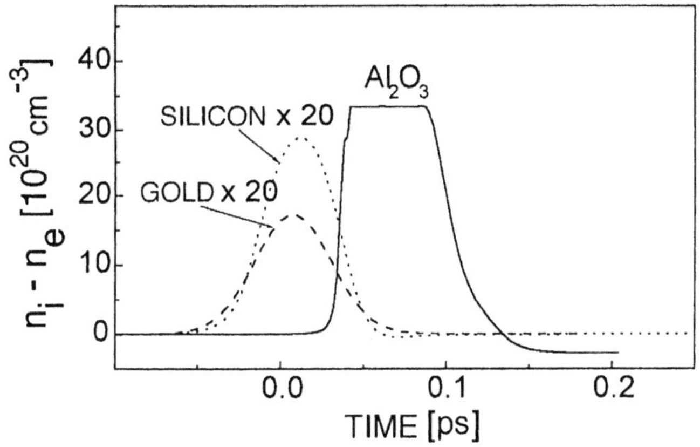

Figure 3. Temporal behavior of the net surface charge density (the difference between the hole and electron populations) for different classes of materials. Laser fluences are chosen to be above the ion emission threshold for each material (4 J/cm^2, 0.8 J/cm^2, and 1.2 J/cm^2 for Al_2O_3, Si, and Au, respectively). The laser pulse of 100 fs duration is centered at $t = 0$.

It should be emphasized that strong charging of sapphire (Fig. 3) is not a result of higher photoemission as compared to silicon and gold. According to modeling, the photoemission yield is approximately $6.8 \cdot 10^8$ electrons for the sapphire target over the whole laser pulse duration from an irradiated spot of 470 μm^2, whereas $6.2 \cdot 10^{11}$ and $3.5 \cdot 10^{11}$ electrons are removed for Si and Au targets, respectively. At the same time, the largest electric field is generated in the sapphire target. Figure 4 shows the temporal behavior of the electric field developed at the sapphire surface in comparison with the fields induced in other types of materials. The electric field exceeds the critical value and reaches a value of $8.4 \cdot 10^{10}$ V/m beneath the surface. The above threshold electric field exists for a few tens of fs. The spatial distribution of the electric field in the near-surface layers is given in Fig. 5 for times when the generated electric field is at its maximum. For sapphire, the layer with overcritical electric field where electrostatic disintegration of the lattice should occur is approximately 40 Å wide, in excellent agreement with the experimental estimation of the Coulomb exploded region (Stoian, et al., 2000c).

As discussed above, the electric field in the first cell below the surface reaches the value of ~8.4×10^{10} V/m for sapphire. An external field $E_{ex}=\varepsilon E_{in}$ is established in front of the surface, with ε being the relative permittivity. The accumulated electrostatic stress determines the surface disruption and the emitted Al^+ ions will be driven by the field for about a few tens of fs (characteristic time of the electric field "pulse" in Fig. 4), and subsequently accelerated.

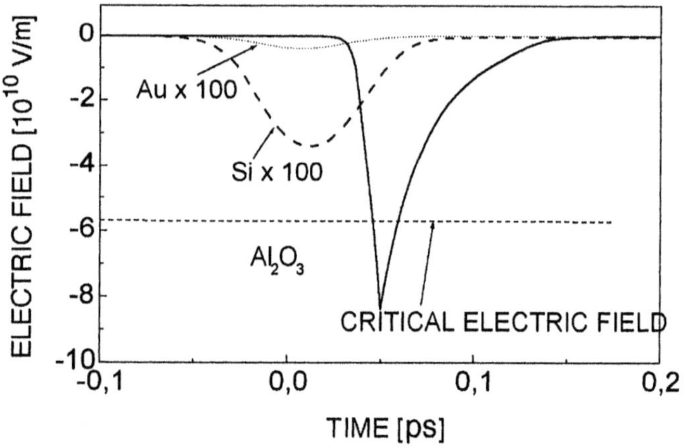

Figure 4. Temporal profiles of the laser-induced electric field at the surface region of the targets. Irradiation conditions as in Fig. 3.

The final momentum obtained by the ions subject to the action of the electric field E_{ex}, during time τ, is written as $m_{Al} \cdot v = eE_{ex}\tau$, where m_{Al} is the ion mass. This gives an estimate of the maximum velocity acquired by the ion of $v \geq 10^4$ m/s that closely agrees with the value detected in the time-of-flight experiments (Stoian, et al., 2000c). During the time of considerable surface charging, the ions travel a distance of the order of a few tens of Å. Thus, the charged surface layer is destroyed within an interval of several tens of femtoseconds.

With semiconductors and metals, the higher electron mobility and higher density of free electrons ensure effective screening and a much smaller net positive charge accumulated during the laser pulse, in spite of the fact that for the Si sample, supercritical carrier densities are reached. This is not sufficient to induce a macroscopic electrostatic breakup of the outer layers of the substrate. The maximum values of the electric field are only 4.1×10^7 V/m and 3.4×10^8 V/m for gold and silicon, respectively (Figs. 4 and 5).

We note here that recent studies (Roeterdink, et al., 2003) have indicated the emission of energetic ions for high-fluence irradiated silicon. The possible electrostatic mechanisms of ion ejection from laser-exposed regions where ultrafast phase-transitions (to transient states with low carrier transport properties) occur are not discussed in this work.

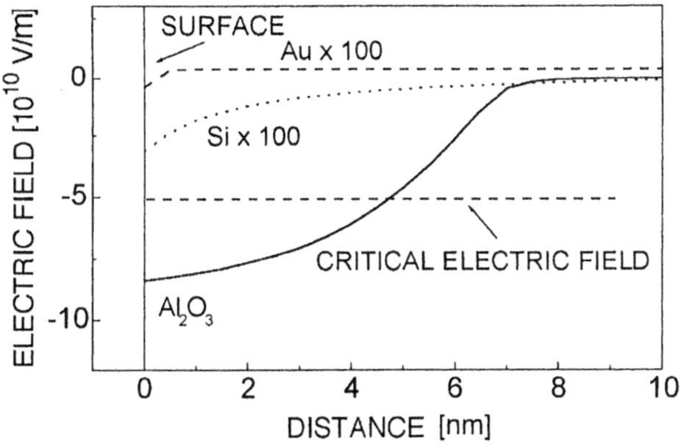

Figure 5. Spatial bulk profiles of the electric field induced by ultrafast laser radiation in metals, semiconductors, and dielectrics at time moments corresponding to the maximum values of the electric field for each material.

The charge dynamics are strongly correlated with the absorption characteristics for each material (see Fig. 3). In the case of sapphire, the charge maximum is strongly retarded, instead of roughly following the laser pulse envelope, as for metals and semiconductors. The effect is explained by the fact that electron heating and collisional multiplication take place during the tail of the laser pulse, thus being directly linked to the number of free electrons in the surface layers. The calculations show that a sapphire sample irradiated by a laser pulse with a fluence of 4 J/cm^2, accumulates multiphoton-generated seed electrons during the first half of the pulse. Only when the electron density reaches a value of order of 10^{17}–10^{18} cm^{-3} does avalanche start to dominate over MPI that leads to the subsequent dielectric breakdown. At this point, very efficient electron heating and photoemission occurs, resulting in supercritical surface charging.

A mention should be made about spatial distributions of the accumulated charge in the near-surface regions for the sapphire target (Bulgakova et al., 2004a, 2005a). During the laser pulse and approximately 100 fs after the laser pulse termination, only a 50–60 Å thick surface layer is positively charged. Later the charging process reaches a quasi-equilibrium between the opponent drift and diffusion terms, with a slight variation in time due to possible electronic decay channels (recombination at traps, self-trapping, Auger recombination) and electron supply from the bulk. The surface layer acquires a negative net charge (Fig. 3) while a certain quantity of net positive charge still exists in the subsurface region. Due to the attraction generated by the narrow positive subsurface layer, a region with electron excess arises in the less-ionized interface region, so the picture of a classic double layer develops similar to that generated at the boundary of expending plasmas (Bulgakova, et al., 2000).

7. CONCLUSIONS

We have reviewed a unified continuum approach to describe the electronic transport in materials of different classes under the action of pulsed laser radiation. The approach is applied to model the charging of metallic, dielectric, and semiconductor targets irradiated by femtosecond, *near-infrared* laser pulses at intensities slightly above the material removal threshold. For dielectrics (sapphire as an example) the laser irradiation induces conditions for the electrostatic explosion of the target surface layer, in good agreement with the experimental observations (Stoian, et al., 2000c). This effect appears to be strongly inhibited for metals and semiconductors in the studied conditions as a consequence of their superior carrier transport properties.

The model can be applied to any material by taking into consideration the particular material properties. The model can also be applied for other irradiation conditions. Thus, for ultraviolet laser pulses, it has been recently shown that the Coulomb explosion mechanism can also be induced in silicon for nanosecond irradiation regimes while, for fs, UV pulses, phase explosion is the dominant process from the onset of ablation (Marine, et al., 2004; Bulgakova, et al., 2005a). However, a quantification of the Coulomb explosion behavior of silicon calls for further studies in view of uncertainties in the photoemission efficiency (Bulgakova, et al., 2005a).

Another intriguing aspect is the possibility of Coulomb explosion in conductors or semiconductors exposed to high laser fluences. It has been argued that an electric field sufficient for electrostatic disintegration of the crystalline lattice can be generated even for metals under extremely high laser fluences (Borghesi, et al., 2003). At moderate laser fluences, Coulomb explosion of dielectrics can be realized on a fs-time scale preceding the volume ablation via thermal mechanisms involving phase transitions. In this case, the observed ion energies are found to change from highly energetic non-thermal to thermal distributions (Stoian, et al., 2000c). However, under high laser fluences, strong gradients of the electron density considerably reduce the possibility to apply continuum approaches. Other formal schemes, e.g. combinations of the continuum approach for generation of free-carriers and electric field with the Boltzmann equation for the electronic transport, and energy relaxation in the electron subsystem, are desirable.

REFERENCES

Anisimov, S. I., Kapeliovich, B. L., and Perel'man, T. L., 1974, Electron emission from metal surfaces exposed to ultrashort laser pulses, *Sov. Phys. JETP*, **39**: 375-377.

Arnold, D., and Cartier, E., 1992, Theory of laser-induced free-electron heating and impact ionization in wide-ban-gap solids, *Phys. Rev. B* **46**:15102-15115.

Ashkenasi, D., Rosenfeld, A., Varel, H., Wärmer, M., and Campbell, E. E. B., 1997, Laser processing of sapphire with picosecond and subpicosecond pulses, *Appl. Surf. Sci.* **120**:65–80.

Ashkenasi, D., Stoian, R., and Rosenfeld, A., 2000, Single and multiple ultrashort laser pulse threshold of Al_2O_3 (corundum) at different etch phases, *Appl. Surf. Sci.* **154-155**:40-46.

Bechtel, J. H., Lee Smith, W., and Bloembergen, N., 1977, Two-photon photoemission from metals induced by picosecond laser pulses, *Phys. Rev. B* **15**:4557-4563.

Bitenskii, L. S., Murakhmetov, M. N., and Parilis, É. S., 1979, Sputtering of nonmetals by intermediate-energy multiply charged ions through a Coulomb "explosion", *Sov. Phys. Tech. Phys.* **24**:618-620.

Bok, J., and Combescot, M., 1981, Comment on the "Evidence for a self-confined plasma" in laser annealing, *Phys. Rev. Lett.* **47**:1564.

Borghesi, M., Romagnani, L., Schiavi, A., Campbell, D. H., Haines, M. G., Willi, O., Mackinnon, A. J., Galimberti, M., Gizzi, L., Clarke, R. J., and Hawkes, S., 2003, Measurement of highly transient electrical charging following high-intensity laser-solid interaction, *Appl. Phys. Lett.* **82**:1529-1531.

Brand, J.L., and Tam, A.C., 1990, Mechanism of picosecond ultraviolet-laser sputtering of sapphire at 266 nm, *Appl. Phys. Lett.* **56**:883-885.

Bulgakova, N. M., Bulgakov, A. V., and Bobrenok, O.F., 2000, Double layer effects in laser-ablation plasma plumes, *Phys. Rev. E* **62**:5624-5635.

Bulgakova, N. M., Stoian, R., Rosenfeld, A., Hertel, I. V., and Campbell, E. E. B., 2004a, Electronic transport and consequences for material removal in ultrafast pulsed laser ablation of materials, *Phys. Rev. B*, **69**:054102(1-12).

Bulgakova, N. M., Stoian, R., Rosenfeld, A., Campbell, E. E. B., and Hertel, I. V., 2004b, Model description of surface charging during ultra-fast pulsed laser ablation of materials, *Appl. Phys. A* **79**:1153-1155.

Bulgakova, N. M., Stoian, R., Rosenfeld, A., Hertel, I. V., Marine, W., and Campbell, E. E. B., 2005a, A general continuum approach to describe fast electronic transport in pulsed laser irradiated materials: The problem of Coulomb explosion, , *Appl. Phys. A* **81**:345-356.

Bulgakova, N. M., Stoian, R., Rosenfeld, A., Hertel, I. V., and Campbell, E. E. B., 2005b, Surface charging under pulsed laser ablation of solids and its consequences: studies with a continuum approach, *Proc. SPIE* **5714**:9-23.

Cavalleri, A., Siders, C. W., Rose-Petruck, C., Jimenez, R., Tóth, Cs., Squier, J. A., Barty, C. P. J., and Wilson, K. R., 2001, Ultrafast x-ray measurement of laser heating in semiconductors: Parameters determining the melting threshold, *Phys. Rev. B* 63:193306 (1-4).

Cheng, H.-P., and Gillaspy, J. D., 1997, Nanoscale modification of silicon surfaces via Coulomb explosion, *Phys. Rev. B* **55**:2628-2636.

Choi, T. Y., and Grigoropoulos, C. P., 2002 Plasma and ablation dynamics in ultrafast laser processing of crystalline silicon, *J. Appl. Phys.* **92**: 4918-4925.

Costache, F., Henyk, M., and Reif, J., 2003, Surface patterning on insulators upon femtosecond laser ablation, *Appl. Surf. Sci.* **208–209**:486–491.

Costache, F., and Reif, J., 2004, Femtosecond laser induced Coulomb explosion from calcium fluoride, *Thin Solid Films* **453-454**:334-339.

Dong, Y., and Molian, P., 2004, Coulomb explosion-induced formation of highly oriented nanoparticles on thin films of 3C-SiC by the femtosecond pulsed laser, *Appl. Phys. Lett.* **84**:10-12.

van Driel, H.M., 1987, Kinetics of high-density plasmas generated in Si by 1.06-and 0.53 μm picosecond laser pulses, *Phys. Rev. B* **35**:8166-8176.

Driscoll, W. G., and Vaughan, W., (Eds), 1978, *Handbook of Optics*, McGraw-Hill Book Company, New York.

Drits, M.E., (Ed.), *Properties of Elements: Handbook* (Moscow, Metallurgiya, 1985) (in Russian).

Fleischer, R. L., Price, P. B., and Walker, R. M., 1965, Ion explosion spike mechanism for formation of charged-particle tracks in solids, *J. Appl. Phys.* **36**:3645-3652.

Gamaly, E. G., Rode, A. V., Luther-Davies, B., and Tikhonchuk, V. T., 2002, Ablation of solids by femtosecond lasers: Ablation mechanism and ablation thresholds for metals and dielectrics, *Phys. Plasmas* **9**:949-957.

Grigoryev, I. S., Meilikhov, and E. Z., Radzig, A.A., (Eds), 1996, *Handbook of Physical Quantities*, CRC Press, Boca Raton, FL.

Held, T., Kuhn, T., and Mahler, G., 1991, Influence of internal electric fields and surface charges on the transport of an optically generated electron-hole plasma, *Phys. Rev. B* **44**: 12873-12879.

Henyk, M., Mitzner, R., Wolfframm, D., and Reif, J., 2000a, Laser-induced ion emission from dielectrics, *Appl. Surf. Sci.* **154-155**:249-255.

Henyk, M., Wolfframm, D., and Reif, J., 2000b, Ultra short laser pulse induced charged particle emission from wide bandgap crystals, *Appl. Surf. Sci.* **168**:263-266.

Herrmann, R., Gerlach, J., and Campbell, E. E. B., 1998, Ultrashort pulse laser ablation of silicon: an MD simulation study, *Appl. Phys. A* **66**(1):35-42.

Hughes, R.C., 1979, Generation, transport, and trapping of excess charge carriers in Czochralski-grown sapphire, *Phys. Rev. B* **19**:5318-5328.

Hummel, R. E., 1993, *Electronic Properties of Materials*, Springer-Verlag, Berlin, Heidelberg.

Jones, S. C., Fischer, A. H., Braunlich, P., and Kelly, P., 1988, Prebreakdown energy absorption from intense laser pulses at 532 nm in NaCl, *Phys. Rev. B* **37**:755-770.

Kaganov, M. I., Lifshitz., I. M., and Tanatarov, M. V., 1957, Relaxation between electrons and crystalline lattices, *Sov.Phys. JETP* **4**:173-178.

Kanasaki, J., and Tanimura, K., 2002, Laser-induced electronic desorption of Si atoms from Si(111)-(7×7), *Phys. Rev. B* **66**:125320(1-5).

E.O. Kane, 1962, Theory of photoelectric photoemission from semiconductors", *Phys. Rev.* **127**:131-141.

Kuhn, T., and Mahler, G., 1989, Carrier capture in quantum wells and its importance for ambipolar transport, *Solid-State Electron.* **32**:1851-1855.

Li, M., Menon, S., Nibarger, J. P., and Gibson, G.N., 1999, Ultrafast electron dynamics in femtosecond optical breakdown of dielectrics, *Phys. Rev. Lett.* **82**:2394-2397.

Logothetis, E. M., and Hartman, P. L., 1969, Laser-induced electron emission from solids: many-photon photoelectric effects and thermionic emission, *Phys. Rev.* **187**:460-474.

Mao, S. S., Mao, X.-L., Greif, R., and Russo, R.E., 1998, Simulation of infrared picosecond laser-induced electron emission from semiconductors, *Appl. Surf. Sci.* **127-129**:206-211.

Marine, W., Bulgakova, N. M., Patrone, L., Ozerov, I., 2004, Electronic mechanism of ion expulsion under UF nanosecond laser excitation of silicon: experiment and modeling, *Appl. Phys. A* **79**:771-774.

Mihaychuk, J. G., Shamir, N., and van Driel, H. M., 1999, Multiphoton photoemission and electric-field-induced optical second-harmonic generation as probes of charge transfer across the Si/SiO_2 interface, *Phys. Rev. B* **59**: 2164-2173.

Miotello, A., and Dapor, M., 1997, Slow electron impinging on dielectric solids. 2. Implantation profiles, electron mobility, and recombination processes, *Phys. Rev. B* **56**:2241-2247.

Melchinger, A., and Hofmann, S., 1995, Dynamic double-layer model description of time-dependent charging phenomena in insulators under electron-beam irradiation, *J. Appl. Phys.* **78**:6224-6232.

Nolte, S., Chichkov, B. N., Welling, H., Shani, Y., Lieberman, K., and Terkel, H., 1999, Nanostructuring with spatially localized femtosecond laser pulses, *Opt. Lett.* **24**,:914-916.

Petite, G., Daguzan, P., Guizard, S., and Martin, P., 1996, Conduction electrons in wide-bandgap oxides: a subpicosecond time-resolved optical study, *Nucl. Instr. Meth. B* **107**:97-101.

Quèrè, F., Guizard, S., Martin, P., Petite, G., Gobert, O., Meynadier, P., and Perdrix, M., 1999, Ultrafast carrier dynamics in laser-excited materials: subpicosecond optical studies, *Appl. Phys. B* **68**:459-463.

Ribeiro, R. M., Ramos, M. M. D., Stoneham, A. M., and Correia Pires, J. M., 1997, Modelling of surface evaporation by laser ablation, *Appl. Surf. Sci.* **109-110**:158-161.

Ribeiro, R. M., Ramos, M. M. D., and Stoneham, A. M., 1998, Mesoscopic modeling of laser ablation, *Thermophysics and Aeromechanics* **5**:223-234.

Roeterdink, W. G., Juurlink, L. B. F., Vaughan, O. P. H., Dura Diez, J., Bonn, M., and Kleyn, A.W., 2003 Coulomb explosion in femtosecond laser ablation of Si(111), *Appl. Phys. Lett.* **82**: 4190-4192.

Samsonov, G.V., (Ed.), 1965, *Physicochemical Properties of the Elements*, Naukova Dumka, Kiev (in Russian).

Schneider, D. H. G., and Briere, M. A., 1996, Investigations of the interactions of highest charge state ions with surfaces, *Phys. Scr.* **53**:228-242.

Siekhaus, W. J., Kinney, J. H., Milam, D., Chase, L. L., 1986, Electron emission from insulator and semiconducor surfaces by multiphoton excitation below the optical damage threshold, *Appl. Phys. A* **39**:163-166.

Sjodin, T., Petek, H., and Dai, H-L., 1998, Ultrafast carrier dynamics in silicon: a two-color transient reflection grating study on a (111) surface, *Phys. Rev. Lett.* **81**: 5664-5667.

Sokolowski-Tinten, K., and von der Linde, D., 2000, Generation of dense electron-hole plasmas in silicon, *Phys. Rev. B* **61**:2643-2650.

Stoian, R., Varel, H., Rosenfeld, A., Ashkenasi, D., Kelly, R., and Campbell, E. E. B., 2000a, Ion time-of-flight analysis of ultrashort pulsed laser-induced processing of Al_2O_3, *Appl. Surf. Sci.* **165**:44–55.

Stoian, R., Ashkenasi, D., Rosenfeld, A., Wittmann, M., Kelly, R., and Campbell, E. E. B., 2000b, The dynamics of ion expulsion in ultrashort pulse laser sputtering of Al_2O_3, *Nucl. Instrum. Methods Phys. Res. B* **166-167**:682-690.

Stoian, R., Ashkenasi, D., Rosenfeld, A., and Campbell, E. E. B., 2000c, Coulomb explosion in ultrashort pulsed laser ablation of Al_2O_3, *Phys. Rev. B* **62**(19):13167-13173.

Stoian, R., Rosenfeld, A., Ashkenasi, D., Hertel, I.V., Bulgakova, N.M., and Campbell, E. E. B., 2002a, Surface charging and impulsive ion ejection during ultrashort pulsed laser ablation, *Phys. Rev. Lett.* **88**:097603(1-4).

Stoian, R., Boyle, M., Thoss, A., Rosenfeld, A., Korn, G., Hertel, I. V., and Campbell, E. E. B., 2002b, Laser ablation of dielectrics with temporally shaped femtosecond pulses, *Appl. Phys. Lett.* **80**:353-355.

Stuart, B.C., Feit, M.D., Herman, S., Rubenchik, A.M., Shore, B.W., and Perry, M.D., 1996, Nanosecond-to-femtosecond laser-induced breakdown in dielectrics, *Phys. Rev. B* **53**:1749-1761.

Tam, A.C., Brand, J.L., Cheng, D.C., and Zapka, W., 1989, Picosecond laser sputtering of sapphire at 266 nm, *Appl. Phys. Lett.* **55**:2045-2047.

Tien, A.-C., Backus, S., Kapteyn, H., Murnane, M., and Mourou, G., 1999, Short-pulse laser damage in transparent materials as a function of pulse duration, *Phys. Rev. Lett.* **82**:3883-3886.

Vanagas, E., Kudryashov I, Tuzhilin D, Juodkazis, S., Matsuo, S., and Misawa, H., Surface nanostructuring of borosilicate glass by femtosecond nJ energy pulses, *Appl. Phys. Lett.* **82**(17):2901-2903.

Varel, H, Wähmer, M., Rosenfeld, A., Ashkenasi, D., and Campbell, E. E. B, 1998, Femtosecond laser ablation of sapphire: time-of-flight analysis of ablation plume, *Appl. Surf. Sci.* **127-129**:128-133.

Wellershoff, S.-S., Hohlfeld, J., Güdde, J., and Matthias, E., 1999, The role of electron-phonon coupling in femtosecond laser damage of metals, *Appl.Phys. A* **69**: S99-S107.

Yoffa, E.J., Dynamics of dense laser-induced plasmas, 1980, *Phys. Rev. B* **21**:2415-2425.

Zvavyi, S. P., and Ivlev, G. D., 1996, Influence of the initial silicon temperature on crystallization of a layer melted by nanosecond laser heating, *Inzh.-Fiz. Zh.* **69**:790-793 (in Russian).

Chapter 3

NEW METHODS FOR LASER CLEANING OF NANOPARTICLES

B. S. Luk`yanchuk, W. D. Song, Z. B. Wang, M. H. Hong and T. C. Chong
Data Storage Institute, Agency for Science, Technology and Research, Singapore 117608

J. Graf, M. Mosbacher and P. Leiderer
University of Konstanz, Department of Physics, 78457 Konstanz, Germany

1. INTRODUCTION

There are several types of laser cleaning, based on different underlying physical processes. The first one is called "ablative cleaning", where the undesirable particle or thin film is deleted from the surface due to direct ablation. Arthur Schawlow carried out the first demonstration of this type of cleaning. He proposed the "laser eraser" in 1965, using different absorptivities of paper and ink to remove the ink without damaging the underlying paper (Schawlow, 1965). Ablative cleaning is the basis of an important technological process related to laser cleaning of organic contamination on microelectronic devices (Hong, 2002). Another important application of the ablative mechanism is in laser cleaning of artwork, e.g. painting restoration and cleaning of antique marble (Zafiropulos, 2003). However, the application of this method for a selective ablation of nanoparticles from the surface is questionable because of the strong decrease in absorption of nanoparticles with decreasing size.

The second type of laser cleaning is the so-called "Steam Laser Cleaning (SLC)" (the term was suggested by A. C. Tam (Zapka, 2002)). A high particle removal efficiency can be achieved with this method when a thin liquid film is present on the substrate surface during laser irradiation (Assendel'ft, 1988; Petrov, 1989; Zapka, 1991). SLC is achieved with the momentum transfer during the explosive vaporization of the liquid layer. This is quite a complicated process and many papers were devoted to the analysis of this process (Lang, 2004; Lang, 2003; Leiderer, 2002). It was found that the cleaning threshold for SLC is practically "universal" (Lang, 2003; Mosbacher, 2000). It does not at all (or at most weakly) depend on the particle size and material. The process of bubble nucleation in liquid is assumed to play the basic role here. This method can provide the removal of at least 100 nm sized particles from a solid surface (Lang, 2003; Mosbacher, 2000; Zapka, 1991).

The IBM group found another important regime of laser cleaning – the so called "Dry Laser Cleaning (DLC)." Initially it was assumed that the acceleration of the particle arises due to thermal expansion of material during nanosecond pulse laser heating (Tam, 1992) and the small particles do not perturbate the local field around the particle. However, recent publications (Münzer, 2002; Mosbacher, 2002b; Graf, 2005) have proven that in DLC, scattering of radiation by contaminant particles plays an important role in the cleaning process. For example, a small transparent particle can work as a near-field lens, which leads to a strong field enhancement. Depending on particle size, material and geometry the field enhancement may be up to a factor of 100 (Luk'yanchuk, 2000b). This produces a nonstationary 3D distribution of temperature which can give rise to local ablation. It has great potential in technological applications of nanostructuring of surfaces (Denk, 2003; Huang, 2002; Lu, 2000; Mosbacher, 2002a; Mosbacher, 2001; Münzer, 2001) and imaging of near-field distributions (Leiderer, 2004; Münzer, 2002; Wang, 2005).

The fourth successful idea in laser cleaning utilizes the excitation of nonlinear surface acoustic waves and phonon focusing (Kolomenskii, 1991, 1998). However, this method also leads to surface destruction and cannot be applied for cleaning of microelectronic devices.

Lately the method of small particle removal, based on the action of pulsed laser-induced plasma and shock waves, attracted a lot of attention (Cetinkaya, 2002; Hooper, 2003; Lee, 2001; Lim, 2004, 2005; Wanderwood, 2003). In this method the laser initiated plasma and shock waves during optical breakdown above the surface in the surrounding media lead to particle removal. Particles over 1 μm can be efficiently removed by this method.

The efficiency of different laser cleaning methods was summarized in (Kane, 2002). From the tables in that paper it can be clearly seen that the smallest size of removed particles was in the order of 100 nm in diameter (papers published before 2002). This is comparable to the capabilities of megasonic cleaning systems (Vereecke, 2005), which are used at present in semiconductor device fabrication lines, but do not meet the future requirements in this field. The International Technology Roadmap for Semiconductors (ITRS, 2004), is considering a 1/2 pitch of 50 nm which should be in mass production in 2009. It means that during the next three years a satisfactory solution for cleaning of 35-50 nm particles should be developed. Megasonic cleaning has limitations caused by the fundamental physics of the cleaning process, which do not allow improvement of the cleaning efficiency in the desired size range. Thus, several alternative techniques have been developed at present: snow-jet cleaning, plasma cleaning, laser cleaning, etc.

Recently several breakthroughs have been achieved with the application of laser cleaning methods for removal of nanoparticles with sizes smaller than 100 nm. The contact-free removal of 60-nm latex particles from silicon (Si) wafers with laser-induced plasma was reported in (Cetinkaya, 2004; Varghese, 2004). Works (Song, 2004a; Song, 2004b) reported on removal of 50 nm Polystyrene (PS) particles from the Si wafers by "Wet Laser Cleaning (WLC)" (cleaning in a liquid near the meniscus). A paper (Luk'yanchuk, 2005b) reported about cleaning of metallic particles using a plasmon resonance technique. Gold particles with sizes of 40 nm were successfully removed during laser cleaning of Si wafers by this method. Finally, the novel technique of "Matrix Laser Cleaning (MLC)" was developed and reported in (Graf, 2004). Experimentally, PS particles of 50 nm were removed by this technique without destroying the Si wafer surface. This paper is devoted to the basic physics of laser cleaning of nanoparticles with these new techniques. It is the result of collaboration between the Singapore and Konstanz groups.

2. LASER CLEANING IN LIQUID NEAR MENISCUS: "MINI-TSUNAMI" EFFECT

Conventional laser cleaning in liquid uses the explosive vaporization of water to produce the momentum necessary for particle removal. The highest cleaning efficiency with this method was found with a transparent liquid and strong substrate absorption (Tam, 1992). The basic idea of the removal force formation is illustrated in Fig. 3(a) in the paper of (Tam, 1992) (this picture is reproduced in Fig. 1). The degree of overheating and related pressure of the expanding vapor, however, are limited by the laser fluence which produces material damage.

The force of particle adhesion is proportional to particle radius, $F_a \propto a$ (Tam, 1992), while the removal force, created by vapor pressure, is proportional to particle surface, i.e. $F_v \propto a^2$. This implies a critical particle size r_c for SLC. Particles with $a < r_c$ cannot be removed by this method. The practical limit, which was reached so far with SLC, is above the 100 nm particle size (Kane, 2002).

One can think about removal of a particle from its initial position by rolling it due to a tangential flow of the liquid (this is a typical situation for brush scrubber cleaning (Hu, 2004)). This tangential flow can be created, e.g. by cavitation of bubbles and other methods. However, the basic limitation here arises from the thickness of the interface boundary layer originated by the liquid viscosity (Landau, 2004). Within the megasonic cleaning technique the limitation of boundary layer is overcome by increasing the sound frequency. With 1.2 MHz megasonic technique it is possible to clean particles down to at least 150 nm (Witter-Publishing-Corporation, 2004).

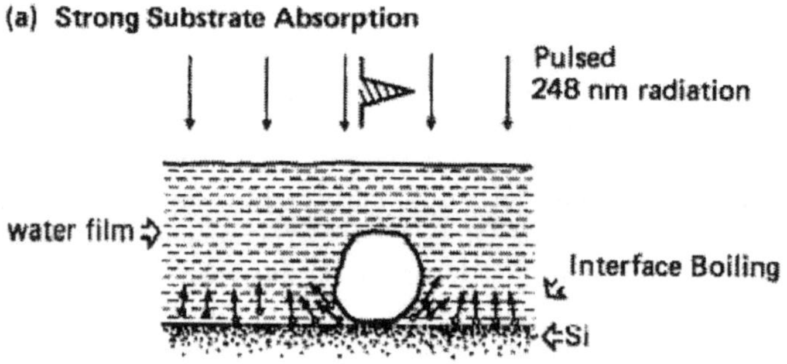

Figure 1. Illustration of Steam Laser Cleaning for a strong absorbing substrate (Tam, 1992).

In order to overcome the mentioned limitations, we suggested a new technique for laser cleaning, for which we use the term "Wet Laser Cleaning (WLC)" (Song, 2004a). The basic idea of this technique is illustrated in Fig. 2. When a high-power laser beam is focused into liquid, it results in a shock wave emission and cavitation bubble generation. In fact, cleaning action induced by cavitation is the basic mechanism of ultrasonic and megasonic techniques (Cheeke, 2002), as these techniques involve the processes in the liquid volume (on the bottom under the liquid). The key issue of our technique is, however, that we discuss the process near the liquid meniscus. It involves some additional cleaning factors.

Upon inserting a rigid substrate into the liquid, the bubbles migrate towards the substrate due to the Bjerknes attractive force. Due to bubble–substrate and/or bubble–free-surface interaction, a high-speed liquid jet is formed during bubble collapse, and a collapse shock wave is generated from the moment of bubble collapse near the substrate. These shock waves and the liquid jet induce large forces acting on the substrate to remove particles from it.

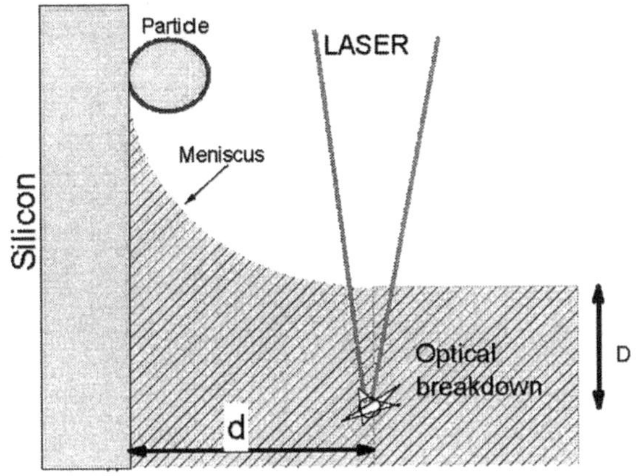

Figure 2. The schematic diagram of Wet Laser Cleaning system. An excimer laser with a wavelength of 248 nm and pulse duration of 23 ns was used as a light source. A laser beam was focused into liquid. The focus area of the laser beam is about 2 mm^2 and laser fluence at the focus point is between 0 and 13.5 J/cm^2. One or few bubbles were observed for each laser pulse.

There are different effects which arise due to bubble interaction with the rigid surface and the free surface of liquid. A collapse of the bubble near the rigid surface leads to deformation of the initially spherical bubble to toroidal form and formation of a high-speed jet directed to the surface. The theory of this type of collapse was discussed in many papers (Best, 1993; Blake, 1986; Pearson, 2004; Zhang, 2001). Cavitation damage and erosion of the solid surface is caused by action of this high-speed jet, see, e.g. (Shaw, 2000). This jet is also responsible for the cleaning effect by ultrasonic technique (Lamminen, 2004).

Near the free surface, a collapse of the bubble yields a high-speed liquid jet directed away from the free surface. The theory of this jet was developed in (Blake, 1987; Robinson, 2001; Wang, 2004a). The velocity of this jet is much higher than the tangential velocity of the liquid which can be achieved within the boundary layer. This jet can produce the necessary momentum to clean the particle located near the meniscus. It was found experimentally in (Song, 2004a) that cleaning efficiently arises near the meniscus region, which means that the asymmetrical oblique liquid jet works like a “mini-tsunami”, which washes off coastal nanoparticles (Luk'yanchuk, 2003a).

There are four important parameters for this mechanism: the distances d and D, shown in Fig. 2, the laser fluence, and the surface tension. The cleaning efficiency increases with increasing laser fluence and decreases with an increase of distance between substrate surface and laser beam focus point or depth below the liquid surface.

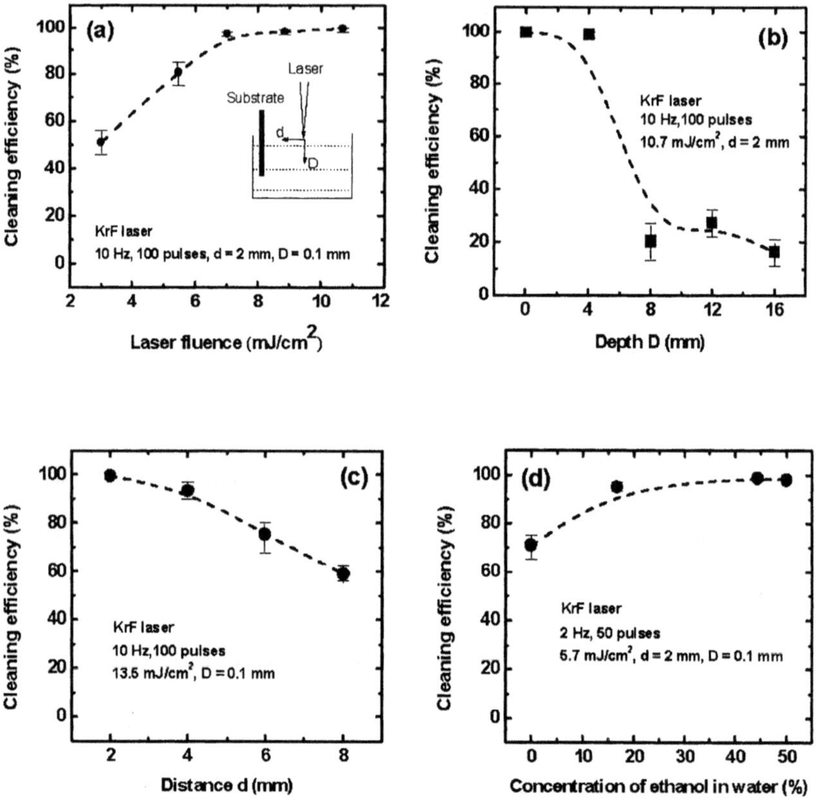

Figure 3. The cleaning efficiency dependence on laser fluence (a); on the depth of laser beam focus point below liquid surface (b); on the distance between the laser beam and substrate (c); on the concentration of ethanol in water (d).

A liquid, such as alcohol or commercial washing solution, enhances the cleaning efficiency. These dependences are illustrated in Fig. 3.

It was found that not only micrometer sized particles could be removed from the solid surfaces by WLC, but also nanoparticles with sizes down to 50 nm. Figure 4 shows SEM images of Si substrates with 51 and 110 nm PS particles before and after WLC in the mixture of water and ethanol. The laser fluence is 13.3 J/cm^2 and pulse number is 200. The laser beam is focused into the liquid at a depth of 0.1 mm from the liquid surface and a distance of 2 mm from the Si substrate. At the original area, 110 and 51 nm particles can be observed on Si substrates as shown in Fig. 4(a) and 4(c), respectively. After laser cleaning, the 110 and 51 nm particles on Si substrate have been removed from the substrates, as shown in Fig. 4(b) and 4(d), respectively.

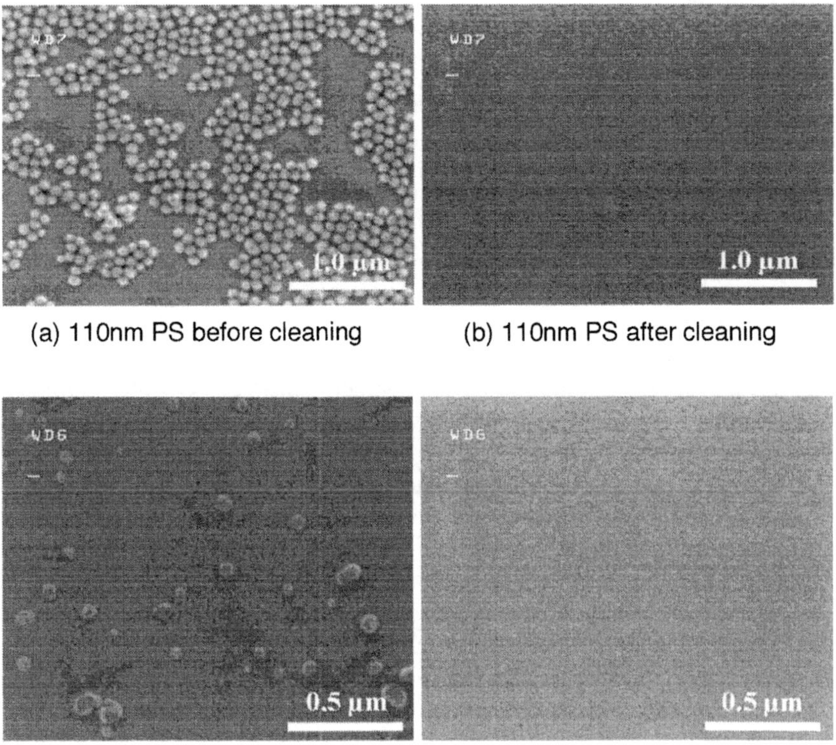

(a) 110nm PS before cleaning (b) 110nm PS after cleaning

(c) 51nm PS before cleaning (d) 51nm PS after cleaning

Figure 4. SEM images of Si substrates with 110 and 51 nm PS particles before ((a) and (c)) and after ((b) and (d)) wet laser cleaning in mixture of water and ethanol at a laser fluence of 13.3 J/cm^2, repetition rate of 10 Hz, pulse number of 200, depth of 0.1 mm from the liquid surface, and distance of 2 mm from the Si substrate.

Another important parameter in WLC is the angle between the substrate and the surface of liquid. We did not analyze this influence experimentally, but results of theoretical simulations demonstrated the importance of surface inclination (Wang, 2004a) and surface curvature (Tomita, 2002) for the direction of the cleaning jet and collapse dynamics.

3. PLASMONIC EFFECT IN LASER CLEANING OF SMALL METAL PARTICLES

With DLC the thermal expansion of the particle (and/or substrate expansion/ablation) produces particle acceleration, sufficient for particle removal (Bäuerle, 2000). However, with smaller particles one needs to produce a higher surface temperature for particle removal. This is illustrated in Fig. 5 which shows calculated surface temperature versus particle size (Luk'yanchuk, 2003b, 2004).

These calculations imply that it is not possible to clean small transparent particles by DLC. The necessary temperature exceeds the melting and boiling temperatures, which leads to a change in the removal mechanism; it becomes ablative cleaning which is in agreement with the experimental findings (Arnold, 2004; Luk'yanchuk, 2003b; Münzer, 2002; Wang, 2005). One of the reasons for difficulties in cleaning small particles is related to their small optical enhancement; for 100 nm sized particles the enhancement is close to one. Sub-50 nm transparent particle generally produces "shadowing effect" instead of enhancement effect under most kinds of available laser irradiations.

However, one can think about significant field enhancement for metallic nanoparticles, using radiation which excites localized surface plasmon (Zayats, 2003). Typical values for the enhancement of the field in the vicinity of the particle vary from several to several ten times depending on the particle properties. There are three effects which one can expect in the discussed problem: 1) Absorption of the energy by the particle itself may play an important role; 2) The intensity on the substrate surface can enhance the substrate heating due to the coupling of surface plasmon; and 3) A pronounced angular effect of the radiation incidence may play an important role.

In Fig. 6(a) we show the extinction, scattering, and absorption cross-sections for a spherical Au particle of 20 nm in radius as a function of the wavelength λ. One can see the plasmon resonance at $\lambda = 498$ nm. This resonance corresponds to the dipole excitation that can be seen clearly from the field distribution in Fig. 6(b). The left peak at $\lambda = 207$ nm is also a dipole Mie resonance, but without the formation of a localized plasmon due to the condition $\mathrm{Re}\varepsilon > 0$. For the case of particles with weak dissipation one can see quadrupole and octupole plasmon resonances (Luk'yanchuk, 2005a).

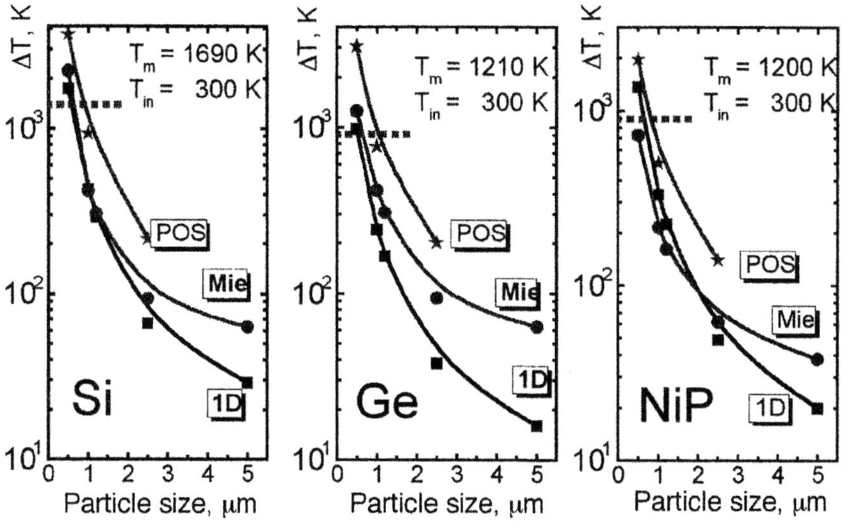

Figure 5. Maximum surface temperature at threshold fluences, calculated for an excimer laser with a wavelength of 248 nm, and a pulse duration 23 ns. The removal of SiO_2 particles on Si, Ge and NiP substrates were investigated. Three curves in the pictures are calculated with different approximations. The 1D curves present the results of one-dimensional theory (Arnold, 2002), which neglects variation of the intensity under the particle, i.e, no enhancement effect; The Mie-curves show the result of calculations for the case when a near-field focusing effect is taken under the approximation of the Mie theory (Luk'yanchuk, 2003b, 2002); The POS-curves are calculated on the basis of "particle on surface" theory, which takes into account the secondary scattering of radiation reflected from the surface of substrate (Bobbert, 1986; Luk'yanchuk, 2004, 2000b).

From Fig. 6(b) one can see that the highest field enhancement arises on the particle "equator", while for laser cleaning enhancement underneath the particle is needed. The natural way to increase enhancement under the particle is to apply the angular incident radiation which is non-perpendicular to the sample surface. To illustrate this effect we performed calculations with a particle on surface problem (here we only considered a single particle case. A more accurate model taking into account the coupling effect between aggregated particles will be used in future modeling). The preliminary results of these calculations are shown in Fig. 7. One can see from the figure that the field enhancement under the particle with $\alpha = 45^\circ$ is about two times higher than that for $\alpha = 0^\circ$ due to a more efficient coupling.

It was the basic idea to enhance the efficiency of laser cleaning with surface plasmon excitation and inclined laser radiation. We performed an experimental study of laser cleaning of sub-50 nm Au particles from Si substrate by a 532 nm laser radiation and found that these nanoparticles can be removed efficiently.

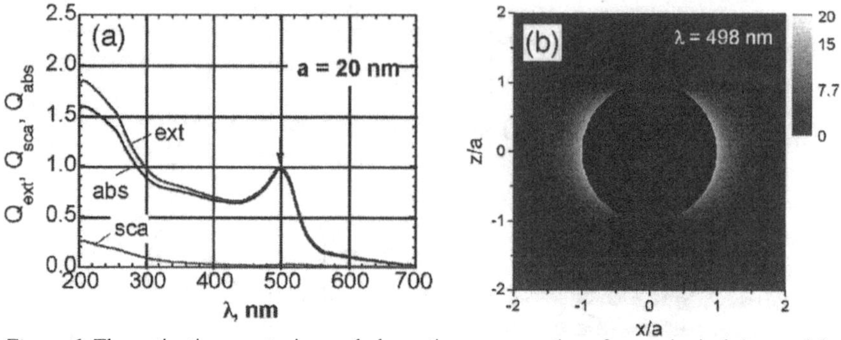

Figure 6. The extinction, scattering and absorption cross-sections for a spherical Au particle of 20 nm radius sphere as a function of the laser wavelength (a). The distributions of field $\mathbf{E}^2$ around the Au particle at an exact dipole resonance at 498 nm (b).

In the experiment, an n-type polished Si wafer was used as the substrate. The sample was cleaned with acetone in an ultrasonic bath for 5 minutes. After that, the sample was rinsed with DI water and dried with N_2 gas. The used suspension of 40 nm Au spherical particle has a size deviation of 5%. The particles were applied on Si surface by a small dispenser. The solvent was dried due to evaporation and consequently the self-assembled particles were left on the surface. A Q-switched 2nd harmonic Nd:YAG laser (BMI industry Series 5000) was used as the laser source. The wavelength was 532 nm with pulse duration of 7 ns. The laser spot size was about 6 mm. The repetition rate varied from 1 to 10 Hz. The output laser beam was linearly polarized.

Fig. 8 shows the SEM images of 40 nm Au nanoparticles on the Si substrate surface before (a) and after (b) 300 laser pulses irradiation at a laser fluence of 50 mJ/cm^2 and at an incident angle of 45°. It can be seen from Fig. 8(a) that an as-deposited mask of Au nanoparticles on Si surface have arranged in different forms: region 1 is free from the particles; in region 2 particles form a monolayer; in region 3 particles form a multilayer; and in region 4 one can see isolated particles.

After laser cleaning, one can see from Fig. 8(b) that the majority of Au nanoparticles were removed from the surface; the total cleaning efficiency was estimated to be 80% in this case. A part of the removed particles from regions 2 and 3 was found to be redeposited in regions 1 and 4. These re-deposited particles mainly stayed in region 4 instead of region 1.

One can see from Fig. 6 that Au particles with a diameter of 40 nm are highly absorbing for radiation of 532 nm ($Q_{abs} = 0.43$). This means that the free particle can be heated efficiently. The corresponding temperature rise can be estimated from the energy balance

$$\mathrm{T} = \frac{3}{4} \frac{Q_{abs} F}{c \rho a} \tag{1}$$

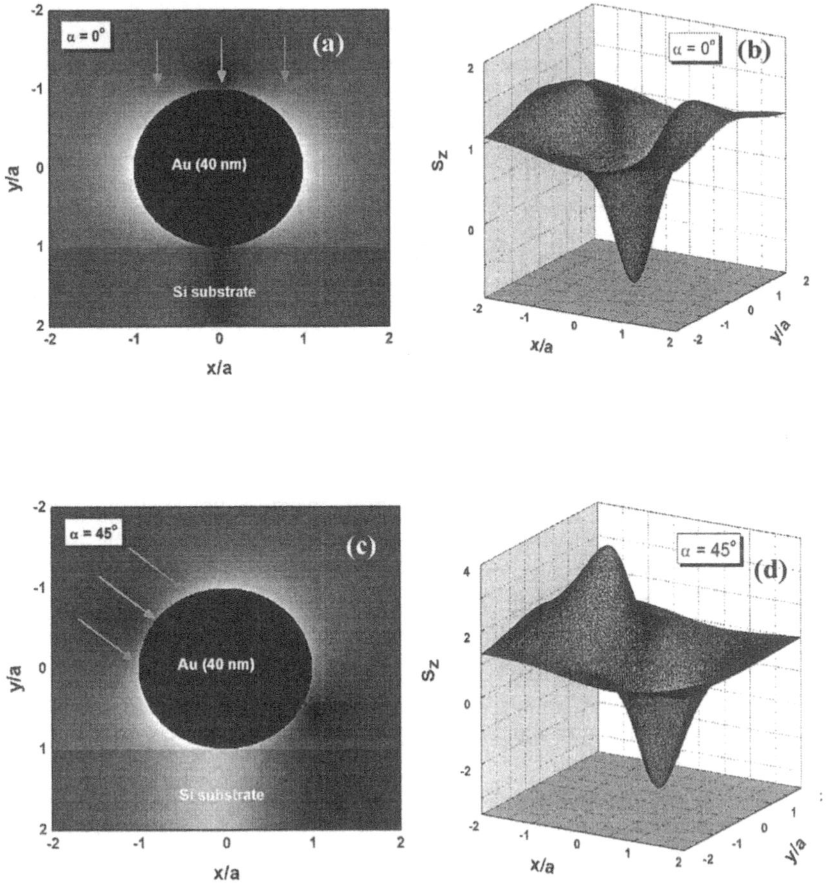

Figure 7. Contour plots for the intensity distribution in the xz-plane (a, c) and the normalized intensity (*z*-component of the Poynting vector) under the 40 nm gold particle on a n-Si surface (b, d) at different incidence angles: $\alpha = 0^{\circ}$ (a, b) and $\alpha = 45^{\circ}$ (c, d).

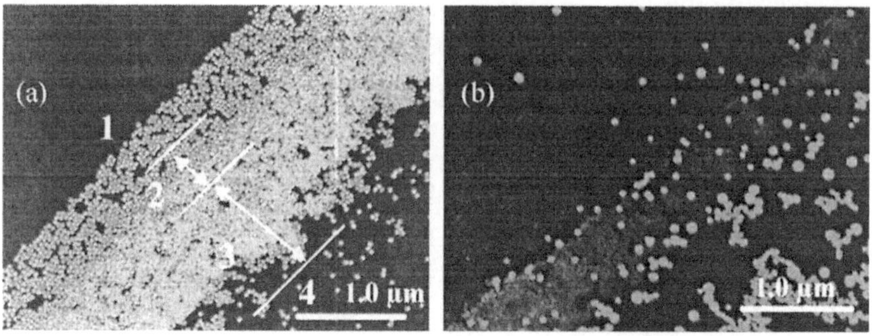

Figure 8. SEM images of 40 nm gold nanoparticles on a n-Si substrate surface before (a) and after (b) 300 pulses (532 nm, 7 ns) at a laser fluence of 50 mJ/cm^2 and an incidence angle of 45°.

where F is the (homogeneous) laser fluence. For example for $F = 50$ mJ/cm^2 formula (1) yields a very high temperature above 6000 K for a 40 nm ($a = 20$ nm) Au particle. Although this temperature cannot be reached for the particle on the surface due to heat conductivity of the substrate, it is clear that effects related to the direct heating of the particle may play an important role. Heating of the substrate arises due to radiation enhancement under the particle and due to thermal contact of the particle and the substrate.

As it was shown in Fig. 7, under angular incidence of light the peak position of intensity field under the particle is not at the contacting point but at a point some distance away from it. In other words, the substrate thermal deformation (and/or ablative) force repels the particle away from the surface at some angle. This could explain why most of the removed particles were redeposited in region 4.

One can see another phenomenon from Fig. 8(b) - the appearance of big Au nanoparticles after laser irradiation. The biggest nanoparticle size that has been observed in our experiments was 200 nm (5 nanoparticles together). As we mentioned above this can be due to the considerable heating of particles up to the melting temperature. One should also remember that nanoparticles have a lower melting temperature as compared to the bulk material. The low melting temperature of nanoparticles is due to the large ratio of surface atoms to inner atoms, in which the surface energy of the surface atoms is reduced. The melting point of bulk Au is about 1064°C, while it is 600 – 800°C for several tens nanometer sized Au nanoparticles (Buffat, 1975). In experiments, it was found that multilayered Au nanoparticles (particles in region 3) can be melted more easily to form a bigger sized nanoparticle than monolayer particles (particles in region 2).

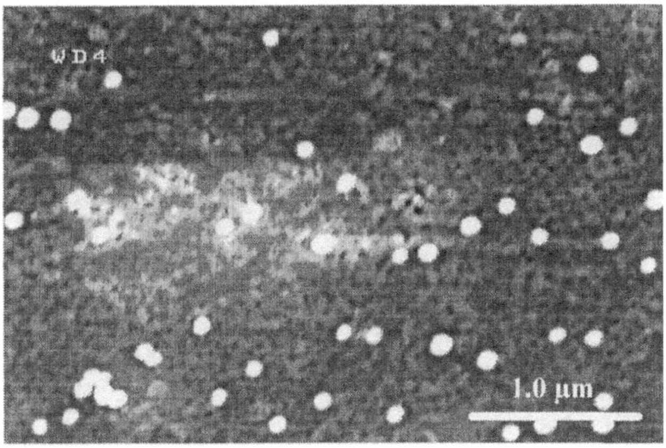

Figure 9. SEM image of damage patterns under 40 nm gold nanoparticles after 532 nm laser processing. A total of 300 laser pulses was used and the laser beam is incident on sample at 45°. The used laser fluence is 50 mJ/cm^2.

Fig. 9 shows the SEM image of damage patterns under multilayer particles (region 3) after 300 laser pulses irradiation at a laser fluence of 50 mJ/cm^2 and at an incidence angle of 45°. The damage sites are in irregular shapes with a lateral size in a range from 20 to 40 nm. In contrast, no damage was observed under monolayer particles in region 2. The reason could be attributed to the electromagnetic (EM) energy coupling between different layers of Au nanoparticles which lead to a more efficient energy coupling with the substrate. From the viewpoint of nanofabrication, this small Au nanoparticle aggregation permits to produce an array of nanostructures with a feature size smaller than 40 nm by single/multiple laser shots. Compared to other techniques such as E-beam lithography and near-field scanning optical microscopy (NSOM), nanopatterning with small particles is rather cheap and easier to implement in a parallel manner (Wang, 2004b). It should be emphasized that a metal plasmon particle should be used in order to obtain sub-50 nm pattern features by this method. Meanwhile, an appropriate laser source should be employed whose frequency must be close to the plasmon resonance frequency of the metal particles.

Fig. 10 presents the cleaning efficiency as a function of the incidence angles for Au particles. As it can be seen, the cleaning efficiency increases smoothly with the incidence angles.

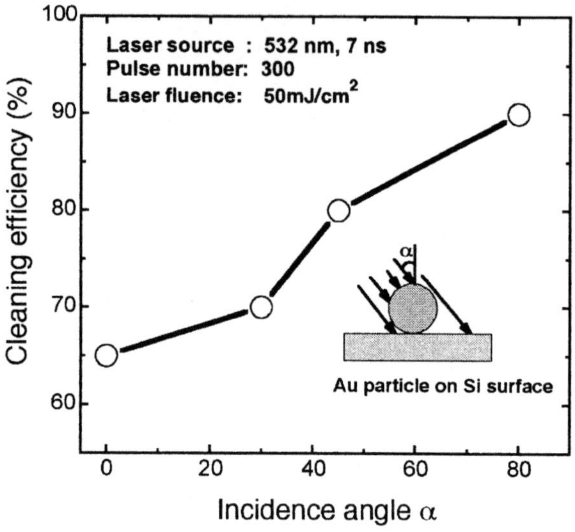

Figure 10. Cleaning efficiency as a function of the incidence angle.

This tendency is quite different from that in laser cleaning of transparent micrometer sized particles on the surface, in which a steep decline of cleaning efficiency appears with increasing incidence angles (Zheng, 2001). Effective removal of metallic particles (Cu) by angular laser cleaning was previously discussed in (Lee, 2000). However, the authors demonstrated this effect for sufficiently big particles (diameter about 10 μm) and they suggested a different mechanism in efficiency increasing.

4. MATRIX LASER CLEANING

The third idea for a new cleaning method for the removal of nanoparticles involves the deposition of an additional solid layer of material at a temperature and pressure below the triple-point (in our case a layer of quench condensed CO_2) and its subsequent sublimation by the laser pulse. This has an analogy with the Matrix-Assisted Laser Desorption Ionization (MALDI) technique (Berkenkamp, 1998; Karas, 1988). MALDI utilizes laser evaporation of a host organic crystal matrix, which was rapidly vaporized, launching the big embedded biomolecules into vacuum. The plume dynamics, mechanism of ejection and the initial velocities of the ejecta, the extent of collisions between the matrix vapor and biomolecules during expansion, and the times to establish final velocity distributions were discussed in many papers, see e.g. (Karas, 2003; Zhigilei, 2003). In particular it was shown that the sharpening effect arises in plume dynamics, namely the "heavy plume" of embedded biomolecules has a very narrow angular distribution when compared to the light plume of the host matrix material (Luk'yanchuk, 2000a; Puretzky, 1999). As the MALDI technique shows an efficient launching of very heavy biomolecules (up to several hundred thousand Daltons), one can expect that the expanding vapor of condensed CO_2 can also remove contaminant particles from the surface. However our analysis reveals that MLC works in a different way compared to MALDI. Due to the application of a solid film condensed onto the substrate from the gaseous phase, MLC has some advantages over cleaning techniques involving a liquid: MLC does not face wetting problems and thus it is beneficial for cleaning of different and structured substrates, including structured ones. Furthermore, the watermark problem can be avoided.

The schematic of the MLC process is presented in Fig. 11. The layer of dry ice, with thickness h, is deposited on a Si surface with contaminant nanoparticles. The CO_2 matrix is considered to be nonabsorbing for the used laser wavelength. Thus, the laser energy is only absorbed by the Si substrate, and the energy transfer to the matrix is achieved solely by the heat transfer due to thermal conductivity. Due to the contact heating, the CO_2 layer in the vicinity of the substrate surface sublimates, producing a compressed CO_2 vapor. For a fixed value of laser fluence this thermal effect does not depend on the thickness h as long as this thickness h is bigger than the heat penetration depth, $\ell_T \approx 2\sqrt{D_T t_\ell}$ (D_T is heat diffusivity, and t_ℓ is the pulse duration). Thus, the thermal energy E_T of the heated area does not depend (or weakly depend) on the film thickness. Due to the expansion of compressed gas, the solid CO_2 layer obtains a velocity v_0, and a corresponding kinetic energy E_k. Due to energy conservation, the absorbed laser energy E should be equal to $E_T + E_k$. When E and E_T are constant, this means that $E_k \propto h v_0^2$ is also constant and the characteristic initial velocity of the film should be inversely proportional to the square root of the film thickness

$$v_0 \propto h^{-1/2}. \tag{2}$$

Clearly, this is a simple model, but it shows the main tendency of the initial velocity of the film versus thickness in a correct way.

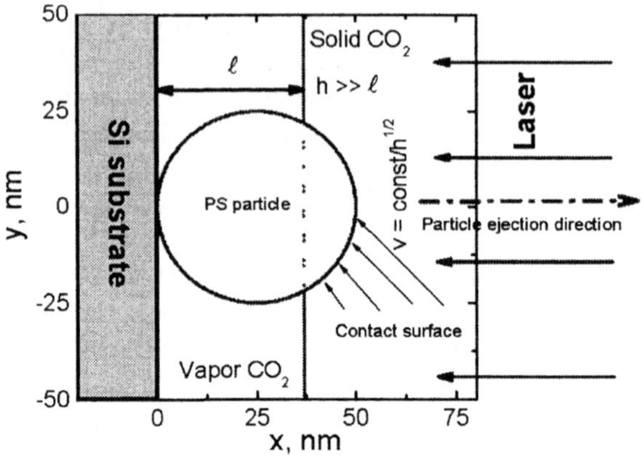

Figure 11. Schematic illustration of the Matrix Laser Cleaning process

Now we can consider two cases. When the thickness of the gaseous region is bigger than the particle diameter, we return to a typical situation for MALDI; the particle is embedded into the moving gas. It is important to say, that for typical conditions for the matrix vaporization by a nanosecond laser pulse, the characteristic time scale to establish the pressure within the vapor $t_p \approx \ell_T / c_v$ is much shorter than the pulse duration. Thus, the gas expansion arises without pressure gradient just due to the boundary conditions, namely the normal velocity of the gas is equal to zero on the substrate surface and it is equal to the velocity v_0 on the surface of the solid matrix at position $x = L(t) \approx \ell_{Tt} + v_0 t$. The gas velocity distribution for this case can be estimated from the Euler equation:

$$\frac{\partial v}{\partial t} + v\frac{\partial v}{\partial x} = 0, \quad v\Big|_{x=0} = 0, \quad v\Big|_{x=L(t)} = v_0 . \tag{3}$$

This equation yields a linear profile of velocity

$$v(x,t) = v_0 \frac{x}{L(t)} . \tag{4}$$

The driving force arises due to collision of vapor molecules with the particle. This force is proportional to the velocity of the gas at distance $x = a$ (a is the particle radius):

$$F_{\max} \propto \nabla v\, a = v_0 \frac{a}{\ell_T}. \qquad (5)$$

The situation when the big biomolecules are carried by the matrix vapor is well-known in MALDI (Luk'yanchuk, 2000a; Puretzky, 1999). Because of Eq. (2) one can enhance the driving force using thin films (however, the condition $h >> \ell_T$ should be fulfilled).

Another situation is different from MALDI. When the gas layer thickness is smaller than the particle diameter, the upper part of the particle remains "immured" into the part of the matrix that remains solid. For example, for solid CO_2 and a laser pulse length of $t_\ell = 9$ ns, the value of $\ell_T \approx$ 38 nm. Such a situation is shown in Fig. 11 for a 50 nm PS particle. As the contact area between the PS particles and the solid part of the matrix is significantly bigger than the contact zone of the particle and the Si substrate, the adhesion force between particle and matrix exceeds the adhesion between particle and substrate. Consequently the particles start to move together with the ablating matrix. The momentum of the matrix layer $p \propto hv_0 \propto h^{1/2}$ increases with increasing film thickness. However, according to the energy criterion of the particle removal (Arnold, 2002; Luk'yanchuk, 2002) the kinetic energy of the particle should exceed the work necessary for the particle removal from the adhesion potential. For the case of a particle embedded into the matrix this kinetic energy does not depend on the film thickness. This clearly indicates the difference between the two regimes discussed above. It is useful to note that using a shorter laser pulse (when $\ell_T \approx 2\sqrt{D_T t_\ell}$ would be smaller), one should be able to remove smaller particles. Considering that for 9 ns laser pulse it was able to delete 50 nm particles, one can expect that for 100 ps laser pulse it should be able to delete 5 nm particles.

The simplified consideration given above illustrates the main tendencies in MLC. We also performed a detailed experimental analysis of this technique. So far all experiments on MLC were carried out using a matrix of quench condensed CO_2. Therefore, the samples were cooled down below the condensation temperature, whereon a defined amount of gaseous CO_2 was blown over the surface. In order to reduce condensation of unwanted species while cooling down the sample, the experiments were carried out in a custom built vacuum chamber. The scheme of this vacuum chamber is shown in Fig. 12. It basically consists of a sample holder, which can be cooled down by liquid nitrogen, a nozzle to supply the CO_2 gas, and windows on three sides of the chamber to enable optical access to the sample.

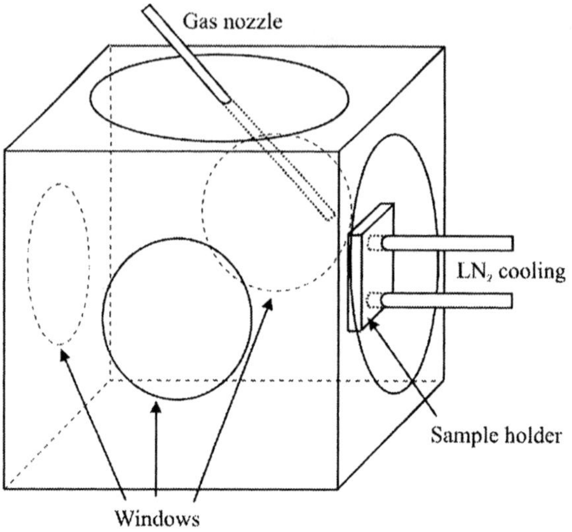

Figure 12. Scheme of the vacuum chamber built for the Matrix Laser Cleaning experiments

The temperature of the sample was measured in parallel inside the sample holder close to the sample position and on the surface of the sample itself.

The optical setup is shown in Fig. 13. A frequency doubled Nd:YAG laser (λ = 532 nm, pulse duration 9 ns FWHM) was used as the cleaning laser. A HeNe laser was used for an online monitoring of the film thickness by reflectometry with an accuracy of 5 nm. Furthermore, the scattered signal of the HeNe laser was used to control the morphology of the quench condensed films. Obviously their microscopic structure varies for different substrate and gas temperatures, as well as the gas flow applied to the sample (Müller, 1987). However, these parameters were kept constant to enable reproducible results. In particular the substrate temperature was kept at 175 K. As no scattered light was detected during the condensation process, we assume that for these parameters the condensed films are homogenous well below the wavelength of the probe laser. The laser fluences necessary for particle removal have been determined from the diameters of the cleaned areas in *x*- and *y*- direction for different pulse energies under the assumption of a Gaussian laser beam profile. In all measurements PS colloidal particles (IDC, Portland, Oregon) with diameters between 50 nm and 810 nm have been used as model contaminants. They have been applied on the Si samples by a spin coating technique (Mosbacher, 2000). The spherical shape and narrow size distribution of the particles enable investigations of size dependent effects and comparisons with theoretical models.

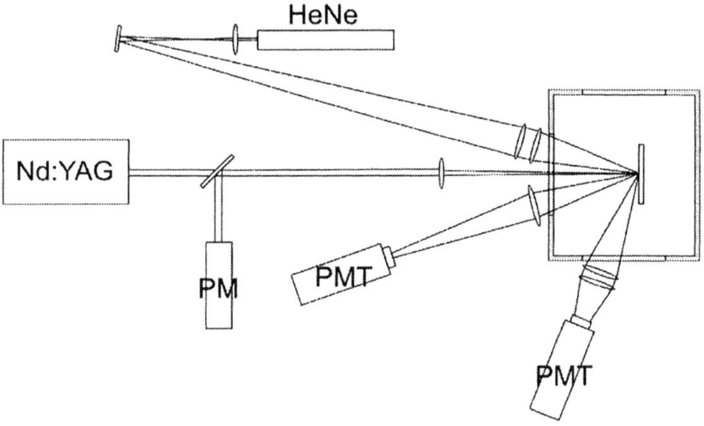

Figure 13. Optical setup with a frequency doubled Nd:YAG laser (9 ns FWHM), a HeNe laser (8 mW) as probe laser, photo multiplier tubes (PMT) to detect the reflected and scattered light, and a power meter (PM) for the energy calibration.

In the first experiments on MLC the laser fluences necessary for particle removal have been measured for different CO_2 film thickness and particle sizes, respectively. As shown in Fig. 14 for a particle diameter of 810 nm, the cleaning fluences are not constant at all, but show an oscillatory behavior in dependence of the applied film thickness. This can be explained by the fact that the CO_2 matrix is almost transparent for the used laser wavelength (according to (Warren, 1986) optical penetration depth for 532 nm is 5.3 cm, which is big compared to film thicknesses in the nanometer range) and the applied energy gets solely coupled into the matrix via heat transfer from the substrate. So the energy coupled into the Si sample is the important parameter, which is of course a function of the film thickness d dependent transmission $T(h)$ into the substrate due to multiple interferences at the interfaces of the film. And indeed the amplitude, as well as the period of the oscillation in the cleaning fluences versus the CO_2 film thickness h, agrees with the expected values. So the applied laser fluences have to be compensated by the particular transmission $T(h)$, what will in the following be determined as the "effective fluence" $F_{Eff} = T(h) \cdot F = [1\text{-}R(h)]F$ (here R is the sample reflectivity).

Fig. 15 shows this effective cleaning fluence for the same particle size of 810 nm. Indeed, the oscillatory behavior cancelled out, which confirms the role of the effective fluence on the substrate as the relevant parameter. On the other hand, it can be seen that there is no decrease in the effective cleaning fluence when applying a CO_2 matrix compared to DLC (film thickness $h = 0$ nm). So, there is no visible cleaning effect of the CO_2 matrix.

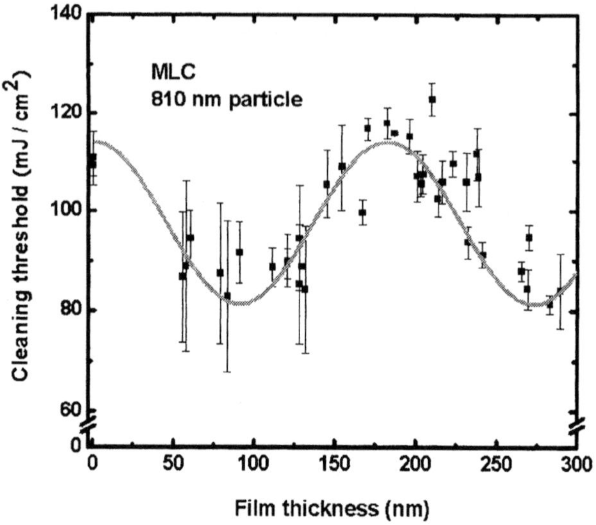

Figure 14. Laser fluence necessary for particle removal dependent on the CO_2 film thickness. The particle size was 810 nm.

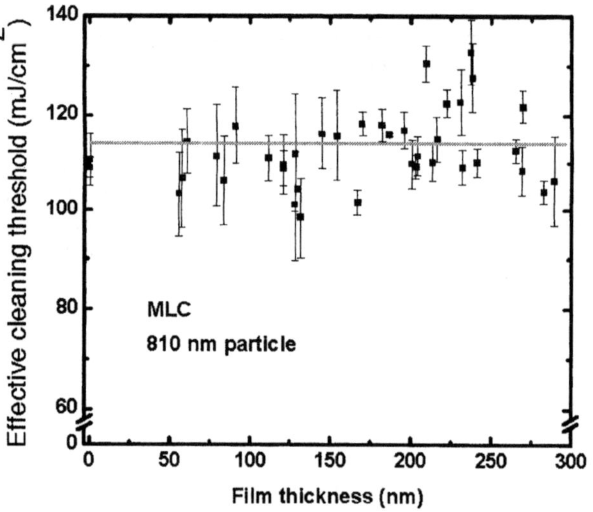

Figure 15. Effective laser fluence on the surface of the sample necessary for particle removal, neglecting film thickness dependent coupling of the applied laser fluence. The particle size was 810 nm.

A different picture arises for smaller particles, the more relevant particles to be cleaned e.g. in semiconductor industry. It has been shown previously that particles below 170 nm cannot be removed by DLC anymore (Mosbacher, 2002b). If these particles could be removed by MLC, this would clearly indicate an additional cleaning effect of the applied CO_2 matrix. And indeed, MLC is able to remove particles with diameters below 170 nm, even with 50 nm in diameter, as can be seen in Fig. 16 and Fig. 17. The latter shows a Scanning Electron Microscope image of a Si wafer contaminated with 50 nm sized particles after a single shot MLC experiment. In the left part of the image, the applied laser fluence was below the cleaning threshold so that the particles remained on the sample. However, on the right side of the image the particles have been clearly removed. The cleaning efficiency is close to 100 % within a single shot experiment. This leads us to the assumption that we have not yet reached the limit of this technique and even smaller particles should be removable. Furthermore the possibility to gain a high cleaning efficiency with just a single laser shot is an important point regarding possible applications, e.g., in the semiconductor industry, where a high throughput is a crucial prerequisite. Additionally, from a scientific point of view, single shot experiments allow us to directly link the outcome of the cleaning process (efficiency, redeposition, damage formation) with the parameters of the specific laser shot (exact fluence, pulse profile etc.), whereas multishot experiments inevitably average over these parameters.

As already shown for larger particles, Fig.16 shows the effective cleaning threshold F_{eff} for the 50 nm sized particles dependent on the thickness of the applied CO_2 matrix. As written before, these particles can't be removed by DLC, which is the reason that there are no data points for small film thickness values. However, when exceeding a minimum film thickness of about 60 nm, cleaning sets in. More detailed investigations reveal that this minimum film thickness shows a slight increase with the diameter of the contaminants. It can also be seen that, in agreement with the energy criterion discussed above, the minimum effective fluence is independent of the applied film thickness. We also analyzed the initial velocity of the starting film by interferometry (typical velocities were found on the order of some ten meters per second). For this case a time resolved measurement of the reflected signal of a probe beam was used. When a homogenous sheet of CO_2 leaves the surface, one can see oscillations in the reflectivity signal due to interference between the reflected beams at the film and the substrate (Lang, 2004) . From the periodicity of these oscillations the velocity can be calculated. It was found that this velocity follows an inverse square root law as Eq. (2).

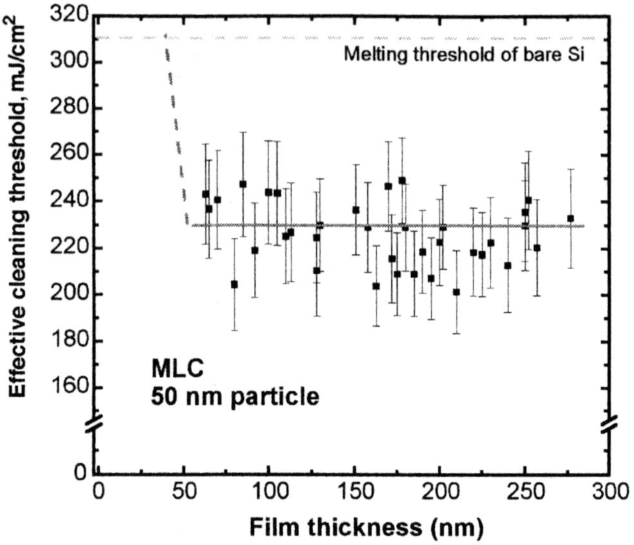

Figure 16. Effective laser fluence on the surface of the sample necessary for removal of 50 nm sized particles, neglecting film thickness dependent coupling of the applied laser fluence. This size particle cannot be removed by Dry Laser Cleaning.

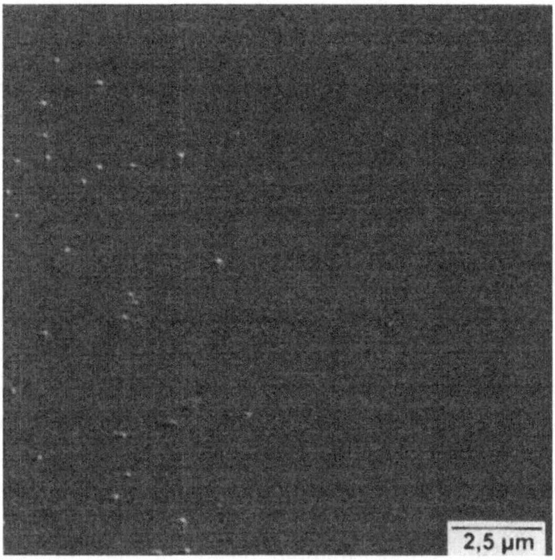

Figure 17. SEM picture of a Si sample contaminated with 50 nm sized polystyrene particles, from (Graf, 2004). The right half of the image has been cleaned by Matrix Laser Cleaning. As can been seen, the cleaning efficiency is close to 100 % within a single shot experiment. Particles with 50 nm diameter are, to our knowledge, the smallest transparent particles ever cleaned by laser based techniques (Song, 2004a).

An important question regarding possible applications is whether the particles can be removed damage free, or if near-field induced damages are created, as is the case in DLC (Arnold, 2004). Therefore, we analyzed the cleaned areas of the sample by Scanning Electron Microscopy (SEM) and Atomic Force Microscopy (AFM). It turns out that for particles larger than 300 nm such defects can be found, as is shown in Fig. 18.

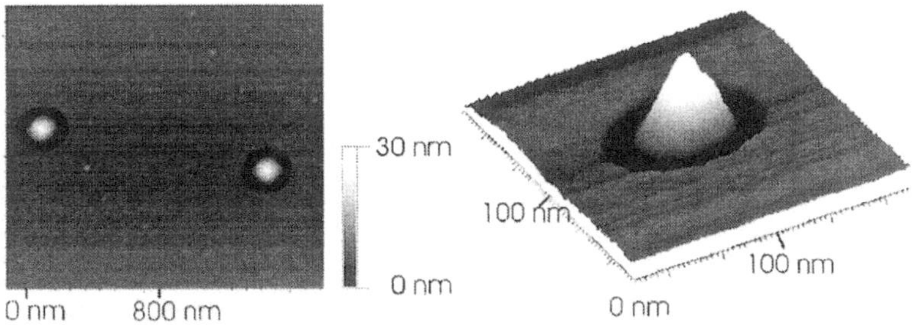

Figure 18. AFM picture of damages created by near-field focusing under the particles during a Matrix Laser Cleaning process. For particle sizes above 300 nm, these damages resemble those found in Dry Laser Cleaning. For particle sizes below 300 nm, no damage has been found.

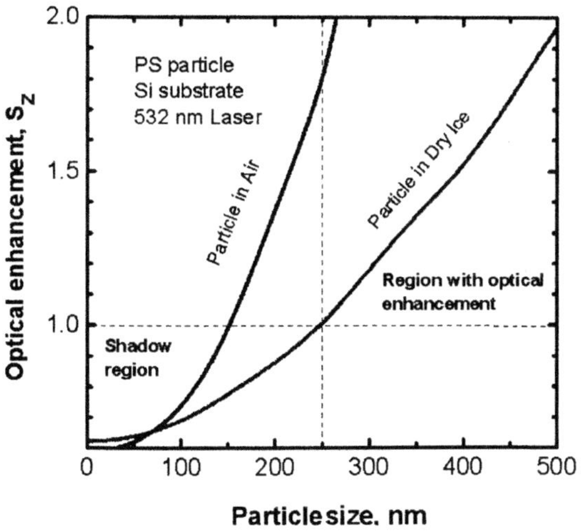

Figure 19. Optical enhancement (*z*-component of the Poynting vector) under Polysterene particle immersed in dry ice on Si substrate for radiation of 532 nm. The following optical constants were used in calculations: solid CO_2 - $n = 1.414$ and $\kappa = 8 \cdot 10^{-7}$; PS particle - $n = 1.6$ and $\kappa = 0$; Si substrate $n = 2.115$ and $\kappa = 0.208$. Enhancement for the particle in air is also shown.

Their shape resembles those found in DLC experiments. For particles smaller than 300 nm, where a MLC effect has been observed, no created defects have been found. This of course cannot be an accurate proof for the absence of defects, since the surface was only checked on selective areas.

However, a comparison between the examined area and the former particle densities strongly suggests that in MLC damage free removal of small particles is possible. This result is also supported by the theory of near-field focusing which shows that for 300 nm PS particles in dry ice, the optical enhancement is close to one (see Fig. 19). Sub-50 nm transparent particle generally produces a "shadowing effect" instead of an enhancement effect under most kinds of laser irradiation. In addition, a better matching between the particle's refractive index and the medium's refractive index decreases the near-field enhancement in MLC, as compared to the DLC case in which the particle is immersed in air (see Fig. 19).

From these first experimental results, we conclude that there seems to be no MLC effect for particles larger than 300 nm, while for the smaller particles a clear effect of the applied CO_2 matrix is visible. This result is quite surprising since, in general, particles are harder to remove the smaller they get. The reason for this is that for smaller particles a higher fraction of the particle volume is located close to the surface and contributes to the short ranged adhesion forces. Here however, cleaning only occurs for the smaller particles.

To explain this behavior, we have to take a closer look at the underlying mechanism, the ablation of the CO_2 matrix.

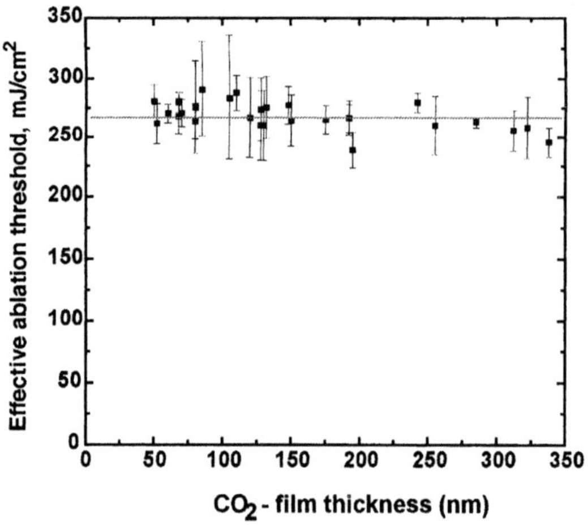

Figure 20. Effective ablation threshold of a quench condensed CO_2 film on a bare Si substrate, dependent on the particular film thickness.

Therefore, we conducted experiments on the ablation of quench condensed CO_2 films on bare Si substrates. Fig. 20 shows the ablation threshold of a quench condensed CO_2 film, depending on the particular film thickness. As in the case of cleaning of small particles, the effective ablation threshold is independent of the film thickness. This leads to the conclusion that only a small and constant fraction of the film at the interface of the substrate gets transferred into the gaseous phase. If the whole matrix material underwent a phase transition, the necessary latent heat would result in a linear dependence of the ablation threshold on the film thickness. The pressure induced by this phase transition and the subsequent expansion of the gaseous CO_2 also leads to the ablation of the material lying beyond.

In Fig. 21, the ablation threshold of the CO_2 matrix is compared to the measured effective cleaning fluences of the MLC experiments and the DLC threshold. As mentioned above, for particles larger than 300 nm there was no visible effect of the applied CO_2 matrix. The effective cleaning threshold stayed at the same level as in the DLC case, where no CO_2 matrix is applied. In Fig. 21, it can be seen that for these particle sizes the DLC threshold is below the ablation threshold of the CO_2 matrix. This leads to the situation where the particles already get removed by DLC at even lower fluences, while the surrounding CO_2 film stays on the surface. Since there are no capillary forces involved, the surrounding matrix does not seem to affect the DLC process remarkably.

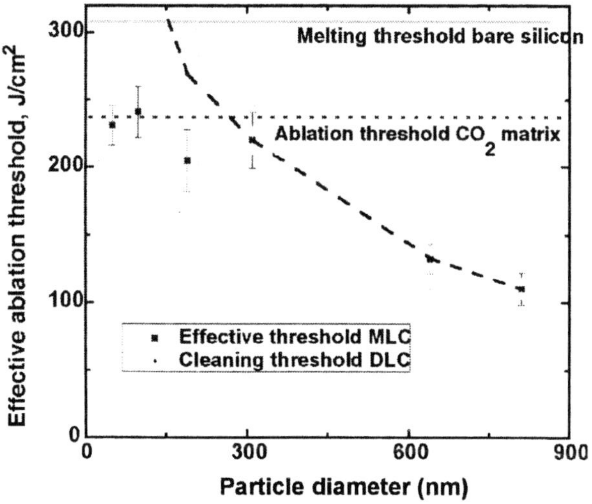

Figure 21. Comparison of the measured ablation threshold (red dotted) of a quench condensed CO_2 matrix, the measured effective cleaning fluencies in Matrix Laser Cleaning and the Dry Laser Cleaning threshold (blue) dependent on the particle size.

For smaller particle sizes, the DLC threshold increases until it intersects with the ablation threshold of the CO_2 matrix at particle sizes of about 300 nm.

For this particle size, the CO_2 matrix ablates at fluences where no DLC occurs, and leads to particle removal. So the comparison between the ablation threshold of the CO_2 film and the DLC threshold is the key as to why a MLC only occurs for small particles.

For the particles below 300 nm the MLC threshold gets independent of the particle size, as it has already been shown in the case of SLC (Mosbacher, 2000). The threshold arises only from the ablation threshold of the matrix material in use.

5. CONCLUSIONS

Modern microelectronics technology needs methods to clean particles from Si surfaces in the diameter range from 50 to 100 nm in the near future (ITRS, 2004). This range cannot be achieved by existing ultrasonic and megasonic techniques, which are restricted at present by 150 nm particles (Witter-Publishing-Corporation, 2004). However, the cleaning target can be achieved by new laser based cleaning methods, which are different from the already known dry, steam, and ablative laser cleaning. The present work discussed three different novel techniques which solve the cleaning problem for sub- 100 nm particles.

The first method utilizes the effect of a directed jet produced by an optical breakdown near meniscus (mini-tsunami effect). It was demonstrated experimentally that this method permits removal of 50 nm PS particles from the Si surfaces without any damage to the surface.

The second method uses the enhancement in optical coupling of metallic particles with the surface due to the surface plasmon excitation. The feature of this technique is related to specific wavelengths which excite the surface plasmon. Another peculiarity of this technique is the use of inclined laser radiation (the laser beam is close to parallel to the surface, as seen in Fig. 10). This plasmonic effect permits the cleaning of metallic nanoparticles from the substrate surface. We demonstrated experimentally the ability to clean 40 nm Au particles from the Si substrate. However, this technique produces nanoscopic damage patterns under aggregated Au particles.

The third technique uses an additional layer of solid CO_2 deposited on the Si surface prior to the laser pulse. The ablation of this layer permits the removal of PS particles down to at least 50 nm in diameter. It is important that the removal of small particles can be achieved without any damage of the substrate. Furthermore, with respect to a high process productivity, the capability to reach high cleaning efficiencies within single shot experiments is of great interest for possible industrial applications.

ACKNOWLEDGEMENT

We wish to thank Prof. S. I. Anisimov, Dr. N. Arnold and F. Lang for discussions. B. L. is thankful to Russian Basic Research Foundation (grants 04-02-17225, 04-02-16972) and discussions with Prof. V. G. Mikhalevich. The work in Konstanz was supported by the Center for Applied Photonics CAP and the Deutsche Forschungsgemeinschaft (SFB 513). M. Mosbacher would like to thank the Regierungspräsidium Freiburg for support of his research.

REFERENCES

Arnold, N., 2002. In: B.S. Luk'yanchuk (Editor), Laser cleaning. World Scientific, New Jersey, London, Singapore, Hong Kong, pp. 51-102.

Arnold, N., Schrems, G. and Bäuerle, D., 2004. Ablative thresholds in laser cleaning of substrates from particulates. Appl. Phys. A, 79(4-6): 729.

Assendel'ft, E.Y., Beklemyshev, V.I., Makhonin, I.I., Petrov, Y.N., Prokhorov, A.M., Pustovoy, V.I. and , v., 444 (1988); vol. 14(8), 650 (1988). 1988. Sov. Tech. Phys. Lett., 14(6): 444.

Bäuerle, D., 2000. Laser Processing and Chemistry. Springer, Berlin.

Berkenkamp, S., Kirpekar, F. and Hillenkamp, F., 1998. Science, 281: 260.

Best, J.P., 1993. J. Fluid. Mech., 251: 79.

Blake, J.R., Taib, B.B. and Doherty, G., 1986. J. Fluid. Mech., 170: 479.

Blake, J.R., Taib, B.B. and Doherty, G., 1987. J. Fluid. Mech., 181: 197.

Bobbert, P.A. and Vlieger, J., 1986. Light scattering by a sphere on a substrate. Physica, 137A: 209-242.

Buffat, P. and Borel, J.P., 1975. Size effect on the melting temperature of gold particles. Phys. Rev. A, 13(6): 2287.

Cetinkaya, C. and Peri, M.D.M., 2004. Non-contact nanoparticle removal with laser induced plasma pulses. Nanotechnology, 15: 435.

Cetinkaya, C., Wanderwood, R. and Rowell, M., 2002. Nanoparticle removal from substrates with pulsed-laser induced plasma and shock waves. J. Adhesion Sci. Technol., 16(9): 1201.

Cheeke, J.D.N., 2002. Fundamentals and Applications of Acoustic Waves. CIP Press, Boca Raton.

Denk, R., Piglmayer, K. and Bäuerle, D., 2003. Laser-induced nano-patterning by means of interference subpatterns generated by microspheres. Appl. Phys. A, 76(1): 1.

Graf, J., Luk`yanchuk, B.S., Mosbacher, M., Hong, M.H., Chong, T.C. and Leiderer, P., 2004. Optimization of the energy transfer medium in laser cleaning: a new strategy. Abstracts of 4th International Workshop on Laser Cleaning, Sydney, Australia, 14-17 December 2004, http://www.physics.mq.edu.au/IWLC4/program.htm.

Graf, J., Lang, F., Mosbacher, M., Leiderer, P., 2005. Laser Cleaning of Particles from Silicon Wafers: Capabilities and Mechanisms. Solid State Phenomena, 103-104: 185-188.

Hong, M.H., Song, W.D., Lu, Y.F., Luk'yanchuk, B.S. and Chong, T.C., 2002. In: B.S. Luk'yanchuk (Editor), Laser cleaning. World Scientific, New Jersey, London, Singapore, Hong Kong, pp. 433-464.

Hooper, T. and Cetinkaya, C., 2003. Efficiency studies of particle removal with pulsed-laser induced plasma. J. Adhesion Sci. Technol., 17(6): 763-776.

Hu, K., Vos, R., Vereecke, G., Doumen, G., Fyen, W., Mertens, P.W., Heyns, M.M., Vinckler, C. and Fransaer, J., 2004. Particle adhesion and removal mechanisms during brush scrubber cleaning. J. Vac. Sci. Technol. B, 22(6): 2844-2852.

Huang, S.M., Hong, M.H., Luk'yanchuk, B.S., Zheng, Y.W., Song, W.D., Lu, Y.F. and Chong, T.C., 2002. Pulsed laser-assisted surface structuring with optical near-field enhanced effects. J. Appl. Phys., 92(5): 2495-2500.

ITRS, 2004. The International Technology Roadmap for Semiconductors (http://public.itrs.net).

Kane, D.M., Fernandes, A.J. and Halfpenny, D.R., 2002. In: B.S. Luk'yanchuk (Editor), Laser cleaning. World Scientific, New Jersey, London, Singapore, Hong Kong, pp. 181-228.

Karas, M. and Hillenkamp, F., 1988. Laser desorption ionization of proteins with molecular masses exceeding 10,000 daltons. Analytical Chemistry, 60(20): 2299.

Karas, M. and Kruger, R., 2003. Ion Formation in MALDI: The Cluster Ionization Mechanism. Chem. Rev., 103: 427.

Kolomenskii, A.A. and Maznev, A.A., 1991. Sov. Tech. Phys. Lett., 17: 62.

Kolomenskii, A.A., Schuessler, H.A., Mikhalevich, V.G. and Maznev, A.A., 1998. Interaction of laser-generated surface acoustic pulses with fine particles: Surface cleaning and adhesion studies. J. Appl. Phys., 84(5): 2404-2410.

Lamminen, M.O., Walker, H.W. and Waevers, L.K., 2004. Mechanisms and factors influencing the ultrasonic cleaning of particle-fouled ceramic membranes. Journal of Membrane Science, 237(1-2): 213-223.

Landau, L.D. and Lifshitz, E.M., 2004. Fluid Mechanics, Butterworth-Heinemann, (Chapter IV) pp.

Lang, F., Georgiou, S. and Leiderer, P., 2004. Phase transition dynamics measurements in superheated liquids by monitoring the ejection of nanometer-thick films. Appl. Phys. Lett., 85(14): 2759-2761.

Lang, F., Mosbacher, M. and Leiderer, P., 2003. Near field induced defects and influence of the liquid layer thickness in Steam Laser Cleaning of silicon wafers. Appl. Phys. A, 77: 117-123.

Lee, J.M. and Watkins, K.G., 2001. Removal of small particles on silicon wafer by laser-induced airborne plasma shock waves. J. Appl. Phys., 89(11): 6496-6500.

Lee, J.M., Watkins, K.G. and Steen, W.M., 2000. Angular laser cleaning for effective removal of particles. from a solid surface. Appl. Phys. A, 71(6): 671.

Leiderer, P., Bartels, C., Koenig-Birk, J., Mosbacher, M. and Boneberg, J., 2004. Imaging optical near-fields of nanostructures. Appl. Phys. Lett., 85(22): 5370-5372.

Leiderer, P., Mosbacher, M., Dobler, V., Schilling, A., Yavas, O., Luk'yanchuk, B.S. and Boneberg, J., 2002. In: B.S. Luk'yanchuk (Editor), Laser cleaning. World Scientific, New Jersey, London, Singapore, Hong Kong, pp. 255-310.

Lim, H., Jang, D., Kim, D., Lee, J.W. and Lee, J.-M., 2005. Correlation between particle removal and shock-wave dynamics in the laser shock cleaning process. J. Appl. Phys., 97: 054903.

Lim, H. and Kim, D., 2004. Optical diagnostics for particle-cleaning process utilizing laser-induced shockwave. Appl. Phys. A, 79(4-6): 965.

Lu, Y.F., Zhang, L., Song, W.D., Zheng, Y.W. and Luk'yanchuk, B.S., 2000. Laser Writing of a Subwavelength Structure on Silicon (100) Surfaces with Particle-Enhanced Optical Irradiation. JETP Lett., 72(9): 457-459.

Luk'yanchuk, B.S., 2003a. Cleaning Unlimited? Storage Unlimited (DSI E-Newsletter), 262: 262/10/2002, http://www.dsi.a-star.edu.sg/Library/Files/PDF/Newsletter/January2003.pdf.

Luk'yanchuk, B.S., Arnold, N., Huang, S.M., Wang, Z.B. and Hong, M.H., 2003b. Three-dimensional effects in dry laser cleaning. Appl. Phys. A, 77(2): 209.

Luk'yanchuk, B.S., Kirichenko, N.A., Puretzky, A.A. and Geohegan, D.B., 2000a. Proc. SPIE, 4070: 166-181.

Luk'yanchuk, B.S., Mosbacher, M., Zheng, Y.W., Münzer, H.-J., Huang, S.M., Bertsch, M., Song, W.D., Wang, Z.B., Lu, Y.F., Dubbers, O., Boneberg, J., Leiderer, P., Hong, M.H. and Chong, T.C., 2002. In: B.S. Luk`yanchuk (Editor), Laser Cleaning. World Scientific, New Jersey, London, Singapore, Hong Kong, pp. 103-178.

Luk'yanchuk, B.S. and Tribelsky, M.I., 2005a. Anomalous Light Scattering by Small Particles and inverse hierarchy of optical resonances. Collection of papers devoted to memory of Prof. M. N. Libenson. The St.-Petersburg Union of the Scientists, Russia: 101-117.

Luk'yanchuk, B.S., Wang, Z.B., Song, W.D. and Hong, M.H., 2004. Particle on surface: 3D-effects in dry laser cleaning. Appl. Phys. A, 79(4-6): 747.

Luk'yanchuk, B.S., Wang, Z.B., Zhou, Y., Hong, M.H., Song, W.D. and Chong, T.C., 2005b. In: D.M. Kane (Editor), Laser cleaning II. World Scientific, New Jersey, London, Singapore, Hong Kong.

Luk'yanchuk, B.S., Zheng, Y.W. and Lu, Y.F., 2000b. Proc. SPIE, 4065: 576.

Mosbacher, M., Bertsch, M., Munze, H.-J., Dobler, V., Runge, B.-U., Bäuerle, D., Boneberg, J. and Leiderer, P., 2002a. Proc. SPIE, 4426: 308.

Mosbacher, M., Dobler, V., Boneberg, J. and Leiderer, P., 2000. Universal threshold for the steam laser cleaning of submicron spherical particles from silicon. Appl. Phys. A, 70(6): 669.

Mosbacher, M., Münzer, H.-J., Bertsch, M., Dobler, V., Chaoui, N., Siegel, J., Oltra, R., Bäuerle, D., Boneberg, J. and Leiderer, P., 2002b. In: K.L. Mittal (Editor), Particles on surfaces 7: Detection, Adhesion and Removal. VSP, Netherland.

Mosbacher, M., Münzer, H.-J., Zimmermann, J., Solis, J., Boneberg, J. and Leiderer, P., 2001. Optical field enhancement effects in laser-assisted particle removal. Appl. Phys. A, 72(1): 41-44.

Müller, K.H., 1987. Surf. Sci., 184: 375.

Münzer, H.-J., Mosbacher, M., Bertsch, M., Dubbers, O., Burmeister, F., Pack, A., Wannemacher, R., Runge, B.-U., Bäuerle, D., Boneberg, J. and Leiderer, P., 2002. Proc. SPIE, 4426: 180.

Münzer, H.J., Mosbacher, M., Bertsch, M., Zimmermann, J., Leiderer, P. and Boneberg, J., 2001. Local field enhancement effects for nanostructuring of surfaces. Journal of Microscopy, 202(1): 129.

Pearson, A., Blake, J.R. and Otto, S.R., 2004. Jets in bubbles. Journal of Engineering Mathematics, 48(3-4): 391.

Petrov, Y.N., 1989. SPIE, 1352: 266.

Puretzky, A.A., Geohegan, D.B., Hurst, G.B., Buchanan, M.V. and Luk'yanchuk, B.S., 1999. Imaging of Vapor Plumes Produced by Matrix Assisted Laser Desorption: A Plume Sharpening Effect. Phys. Rev. Lett., 83(2): 444.

Robinson, P.B., Blake, J.R., Kodama, T., Shima, A. and Tomita, Y., 2001. Interaction of cavitation bubbles with a free surface. J. Appl. Phys., 89(12): 8225-8237.

Schawlow, A.L., 1965. Lasers. Science, 149: 13.

Shaw, S.J., Schiffers, W.P., Gentry, T.P. and Emmony, D.C., 2000. The interaction of a laser-generated cavity with a solid boundary. J. Acoustic. Soc. Am., 107(6): 3065-3072.

Song, W.D., Hong, M.H., Luk'yanchuk, B. and Chong, T.C., 2004a. Laser-induced cavitation bubbles for cleaning of solid surfaces. J. Appl. Phys., 95(6): 2952.

Song, W.D., Hong, M.H. and Luk'yanchuk, B.S., 2004b. US Patent: No. 6,777, 642 B2.

Tam, A.C., Leung, W.P., Zapka, W. and Ziemlich, W., 1992. Laser-cleaning techniques for removal of surface particulates. J. Appl. Phys., 71(7): 3515-3523.

Tomita, Y., Robinson, P.B., Tong, R.P. and Blake, J.R., 2002. Growth and collapse of cavitation bubbles near a curved rigid boundary. J. Fluid. Mech., 466: 259.

Varghese, I. and Cetinkaya, C., 2004. Non-contact removal of 60-nm latex particles from silicon wafers with laser-induced plasma. J. Adhesion Sci. Technol., 18(7): 795.

Vereecke, G., Holsteyns, F., Arnauts, S., Beckx, S. Jaenen, P., Kenis, K. Lismont, M. Lux, M., Vos, R., Snow, J. and Mertens, P.W., 2005. Solid State Phenomena, 103-104: 141-146.

Wanderwood, R. and Cetinkaya, C., 2003. Nanoparticle removal from trenches and pinholes with pulsed-laser induced plasma and shock waves. J. Adhesion Sci. Technol., 17(1): 129-147.

Wang, Q.X., 2004a. Numerical simulation of violent bubble motion. Physics of Fluids, 16(5): 1610.

Wang, Z.B., 2005. Optical Resonance and Near-Field Effects: Small Particles under Laser Irradiation, PhD. thesis, National University of Singapore, Singapore.

Wang, Z.B., Hong, M.H., Luk'yanchuk, B.S., Huang, S.M., Wang, O.F., Shi, L.P. and Chong, T.C., 2004b. Parallel nanostructuring of GeSbTe film with particle mask. Appl. Phys. A, 79(4-6): 1603.

Warren, S.G., 1986. Optical constants of carbon dioxide ice. Appl. Opt., 25(16): 2650.

Witter-Publishing-Corporation, 2004. Megasonic Cleaning Systems (http://www.cleantechcentral.com/knowledgebase/technologyspotlight/megasonic.aspx).

Zafiropulos, V., Balas, C., Manousaki, A., Marakis, Y., Maravelaki-Kalaizaki, P., Melesanaki, K., Pouli, P., Stratoudaki, T., Klein, S., Hildenhagen, J., Dickmann, K., Luk'yanchuk, B.S., Jujat, C. and Dogariu, A., 2003. Yellowing effect and discoloration of pigments: experimental and theoretical studies. Journal of Cultural Heritage, 4(3): 249-256.

Zapka, W., 2002. In: B.S. Luk'yanchuk (Editor), Laser cleaning. World Scientific, New Jersey, London, Singapore, Hong Kong, pp. 23-48.

Zapka, W., Ziemlich, W. and Tam, A.C., 1991. Efficient pulsed laser removal of 0.2 μm sized particles from a solid surface. Appl. Phys. Lett., 58(20): 2217-2219.

Zayats, A.V. and Smolyaninov, I.I., 2003. Near-field photonics: surface plasmon polaritons and localized surface plasmons. J. Opt. A: Pure Appl. Opt., 5(4): S16.

Zhang, Y.L., Yeo, K.S., Khoo, B.C. and Wang, C., 2001. 3D Jet Impact and Toroidal Bubbles. Journal of Computational Physics, 166(2): 336-360.

Zheng, Y.W., Luk'yanchuk, B.S., Lu, Y.F., Song, W.D. and Mai, Z.H., 2001. Dry laser cleaning of particles from solid substrates: Experiments and theory. J. Appl. Phys., 90(5): 2135-2142.

Zhigilei, L.V., Leveugle, E., Garrison, B.J., Yingling, Y.G. and Zeifman, M.I., 2003. Computer Simulations of Laser Ablation of Molecular Substrates. Chem. Rev., 103(2): 321-348.

Chapter 4

PLUME DYNAMICS

Jørgen Schou[1], Salvatore Amoruso[2] and James G. Lunney[3]

[1]Department of Optics and Plasma Research, Risø National Laboratory, DK-4000 Roskilde, Denmark; [2]Coherentia CNR-INFM and Dipartimento di Scienze Fisiche, Università di Napoli Federico II, Complesso Universitario di Monte S. Angelo, Via Cintia, I-80126 Napoli, Italy; [3]School of Physics, Trinity College, Dublin 2, Ireland.

1. INTRODUCTION

The ejection of particles from a surface induced by irradiation with a high intensity laser beam leads to the formation of a cloud of ablated material moving rapidly away from the surface. Usually, such a laser ablation cloud consists of excited or ground-state neutrals, electrons, and ions. In the early history of laser ablation, photographs showed a distinct visible plume produced by fluorescence from excited atoms in the cloud (Harris, 1963; Geohegan, 1992; Kelly,, et al., 1998). An example of a plume as a strongly light-emitting cloud growing in all directions, but preferentially along the target normal, is shown in Fig. 1. In the following discussion the word "plume" will denote the expanding cloud of material derived from an ablation process, independent of whether or not the cloud is visible.

The physical parameters in the plume, e.g. the mass distribution, ion and atom velocity, and the angular distribution of the plume species, play an important role for applications of laser ablation in mass analysis of laser-induced plasma (Vertes,, et al., 1993) and in the production of thin films by pulsed laser deposition (PLD) (Hubler, 1992; Chrisey and Hubler, 1994; Gorbunoff, 2002). In particular, the thickness distribution in film deposition on a substrate is determined by the plume shape that has evolved during the expansion from the target surface to the substrate (Saenger, 1994; Amoruso, et al., 1999; Schou, 2006).

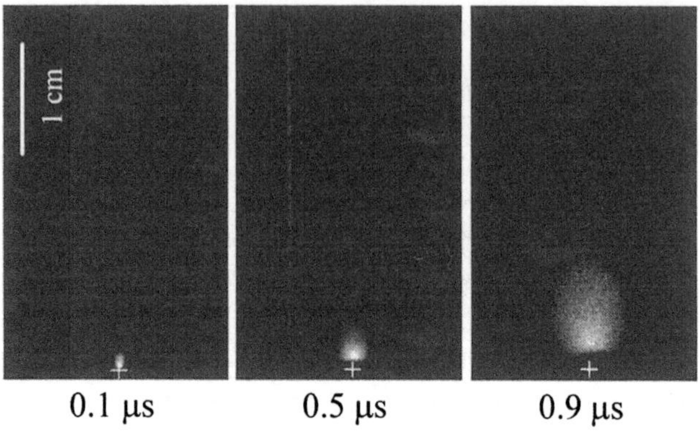

Figure 1. Images of the ablation plume produced during laser ablation of a $LaMnO_3$ target by 20 ns excimer laser pulse at 351 nm with a fluence of 1.5 J/cm^2 in vacuum, at three different time delays after the laser pulse. The white cross marks the point where the laser beam hits the target surface (Amoruso, et al., 2006a).

The complicated interactions between the species in the plume and the interaction between the species and the incident laser light mean that a detailed theoretical or computational treatment has not yet appeared for the comparatively simple case of a plume expanding in vacuum (Sibold and Urbassek, 1991; Lowndes, 1998; Willmott and Huber, 2000; Gorbunoff, 2002). An even more complex case occurs for a laser ablation process of a solid in a background gas, which is often used in PLD. Therefore, in principle, any optimisation of the film deposition process requires a comprehensive knowledge of the plume processes during the expansion. Surprisingly, the optimisation of the film deposition is still mostly done on an empirical basis - almost 20 years after the breakthrough of PLD for films of high-temperature superconductors in 1987 (Dijkkamp, et al.).

The ablation process using lasers with pulses of *ns* or longer duration can be thought of as occurring in four different stages, though these stages do overlap to some extent in time. A schematic representation of these stages is shown in Fig. 2.

1) In the first stage, the laser light strikes the solid and is absorbed by the electrons in the solid. After a period of tens of *ps* the electrons and atoms in the solid equilibrate, which leads to a strong heating of the irradiated volume. In this stage, laser-solid interactions are dominant.
2) In the second stage the material from the heated volume is ejected from the solid but continues to absorb energy from the laser, resulting in the formation of a thin layer ionized vapour on the

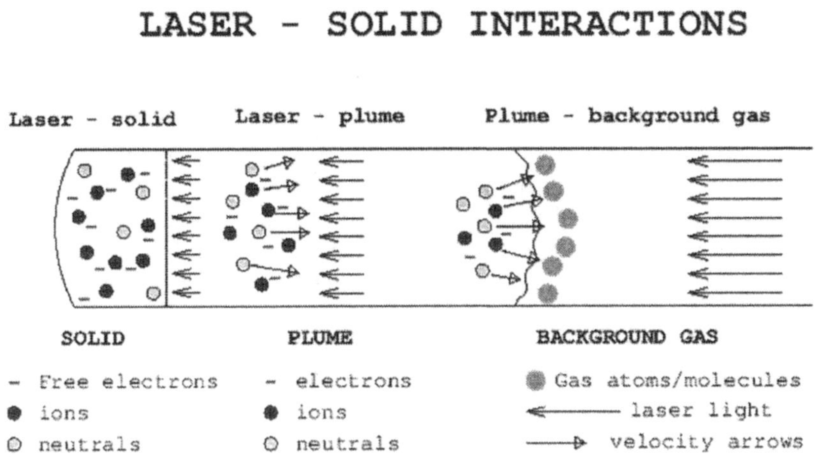

Figure 2. A schematic view of the processes that take place during ablation by a ns laser pulse. From left to right: light absorption in the solid (stage 1), ejection of the ablated material in a plasma plume (stage 3) and the interaction of the light with the plume (stage 2), and the expansion in a background gas (stage 4).

surface of the target. In this stage, laser-gas or laser-plasma interactions prevail. For *fs* lasers the duration of the pulse is so short that any significant movement of the atoms from the lattice happens after the pulse has terminated.

3) The third stage begins after the termination of the laser pulse. Here the plume expands adiabatically in three dimensions. If the expansion takes place in vacuum, the shape and velocity distribution in the plume will reach asymptotically constant values.
4) If the ablation takes place in a background gas, the high plume pressure initially drives the expansion as if it were occurring in vacuum. After several μs, the plume expansion is entirely determined by the interaction of the plume atoms with the atoms and molecules of the background gas (stage 4).

Already, during the light absorption in the solid (in stage 1) the initial heating of the solid may lead to a strong evaporative ejection of material. Since the heating is extremely fast, surface temperatures close to the thermodynamically critical temperature can be reached. At these high temperatures the material ejection processes may change from evaporation and boiling to explosive boiling, by which nano and microparticles can also be ejected. In stage 2 the expanding plume is heated by various absorption processes which attenuate the laser light that is transmitted through the plume to the solid/liquid surface. A simplified picture is that the initial plume expands in a dynamical equilibrium in such a way that the absorption of laser light provides the kinetic energy for the expansion (Phipps, et al. 1988; Lowndes, 1998).

In the plume the laser is primarily absorbed by inverse bremsstrahlung (IB) processes and direct single-photon processes (Zel'dovich and Raizer, 2001). IB involves absorption of photons by free electrons which are accelerated during collision with neutral or ionized atoms. The cross-section for IB via electron-neutral collisions is much smaller than that via electron-ion collisions, but can be important for the initial plume of a weakly ionized gas. Initially, there may be very few "seed" electrons present, produced by thermal emission from the solid or multiphoton ionization processes. The contribution from multiphoton processes increases with decreasing wavelength, but is particularly important for ultrafast lasers. For plumes induced by IR lasers, e.g laser light at 1064 nm, absorption by IB is dominant and ionization breakdown may occur even at intensities near the ablation threshold (Amoruso, et al., 1999). For visible and UV laser light photon absorption also takes place by photoionization processes. In this case the photon energy is comparable to the ionization energy of excited atoms which are produced in the plume by electron-atom collision processes. A number of studies of UV ablation of metal targets have shown that the plasma plume can achieve temperatures of a few eV and an ionization degree that exceeds 0.1 for moderate irradiation intensities ($I \approx 0.1 - 1$ GW/cm^2). The comparatively high temperature which cannot be reached with IB processes alone, indicates that direct photoionization processes play an important role (Amoruso, et al.,1999; Jordan and Lunney, 1998).

After termination of a *ns* laser pulse, i.e. after first and second stage, the vapourised material exists as a 10-100 μm thick layer of a dense, partly ionized gas which is expanding away from the target surface. For typical PLD conditions, with a fluence of 1-10 J/cm^2 on a laser spot of 0.01-0.1 cm^2, the number of ablated particles from a metal is in the range of 10^{15} to 10^{16} atoms/pulse. This leads to a plume density of 10^{19}-10^{20} atoms/cm^3 and a temperature of 2-3 eV, depending on the fluence and the volatility of the metal (Lunney and Jordan, 1998; Amoruso, 1999). These values agree with those obtained by others (Amoruso, et al., 1999; Schou, 2006). The plume is typically moving with a velocity of 10-20 km/s (Amoruso, 1999; Thestrup, et al., 2000; Ashfold, et al., 2004).

In the present chapter the plume expansion originating from such a dense, initial plume will be described and discussed. Our starting point is at the beginning of stage 3. However, as discussed below, the subsequent plume dynamics do depend critically on the conditions which pertained in stages 1 and 2.

2. PLUME DYNAMICS IN VACUUM

After the termination of the laser pulse the initial gas layer expands in all directions. The expansion is driven by the energy which is accumulated as thermal energy and energy which is stored as excitation and ionization in the initial layer. This energy is converted to kinetic energy of the atoms in the plume, and eventually all atoms will move with an asymptotic, constant velocity distribution. As soon as the laser pulse ends, there is little further transfer of energy and mass to the ablation plume, and the plume propagation can essentially be considered as an adiabatic expansion.

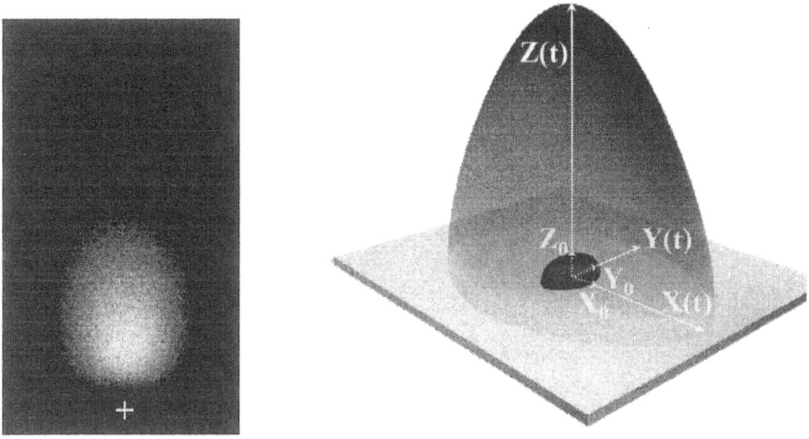

Figure 3. A photo of a plume (see text in Fig. 1) and the geometry for Anisimov's model. The initial plume is seen as a small "bubble" in the center (left panel).

The theoretical description of the adiabatic expansion has been considered by Anisimov, et al. (1993, 1996) and Singh and Narayan (1990). Singh and Narayan's treatment assumes that the plume is isothermal, i.e. $\nabla T = 0$, and that the spatial variation of the plasma density is Gaussian throughout the laser irradiation and the subsequent expansion. The assumption of a temperature gradient equal to zero in the plume is questionable, as pointed out by Anisimov, et al. (1993). Instead, Anisimov, et al. consider the expansion to be isentropic, which essentially means that there is no heat conduction between different parts of the plume. Less attractive features of the model of Singh and Narayan (1990) are that the results cannot easily be generalized to other parameters and that they are obtained for a specific multicomponent system, YBaCuO.

The light emission from an expanding plume usually indicates an ellipsoidal shape (Fig. 3) (Li, et al., 2002, Harilal, et al., 2003; Amoruso, et al., 2006a). The model of Anisimov, et al. (1993) is based on Lie group transformation theory which simplifies the solutions of the gas dynamic equations. In the model the plume is approximated by a semi-ellipsoid with a front determined by the axes $X(t)$, $Y(t)$, and $Z(t)$ at a time t, where the z-axis is normal to the surface (see Fig. 3). The characteristic properties of the expansion, such as the density and the pressure (see below), are constant on ellipsoidal surfaces, i. e. $x^2/X(t)^2 + y^2/Y(t)^2 + z^2/Z(t)^2$ = constant. The complete hydrodynamic motion of all particles in the plume is determined by self-similarity of the elliptical surfaces, so that the instantaneous velocity is determined by the relative position of the front (see Anisimov, et al., 1993, 1996; Anisimov and Luk'yanchuk, 2002).

In Anisimov's theory the initial conditions for this semi-ellipsoidal gas bubble with the axes X_0, Y_0, and Z_0, which starts the expansion after the termination of the laser pulse at $t = \tau_L \approx 0$, are

$$X(0) = X_0,\ Y(0) = Y_0,\ Z(0) = Z_0, \tag{1}$$

and

$$\frac{dX}{dt}(0) = 0, \frac{dY}{dt}(0) = 0, \frac{dZ}{dt}(0) = 0. \tag{2}$$

The underlying assumption is that the thermal energy of the initial plume is much larger than the kinetic energy. For a given value of the adiabatic constant $\gamma = c_p/c_v$, the mass density and pressure profiles for the adiabatic expansion have the form:

$$\rho(x,y,z) = \frac{M_p}{I_1(\gamma)XYZ}\left[1-\frac{x^2}{X^2}-\frac{y^2}{Y^2}-\frac{z^2}{Z^2}\right]^{1/(1-\gamma)} \tag{3}$$

$$p(x,y,z) = \frac{E_p}{I_2(\gamma)XYZ}\left[\frac{X_0Y_0Z_0}{XYZ}\right]^{\gamma-1}\left[1-\frac{x^2}{X^2}-\frac{y^2}{Y^2}-\frac{z^2}{Z^2}\right]^{\gamma/(1-\gamma)} \tag{4}$$

where E_p is the initial energy of the plume and M_p is the plume mass. The expressions for $I_1(\gamma)$ and $I_2(\gamma)$ are given in Anisimov, et al. (1993, 1996). This leads to the following temperature profile during the adiabatic expansion:

$$T(x,y,z) = (5\gamma-3)(\gamma-1)/(2\gamma)\frac{E_p m}{k_B M_p}\left(\frac{X_0Y_0Z_0}{XYZ}\right)^{\gamma-1}\left[1-\frac{x^2}{X^2}-\frac{y^2}{Y^2}-\frac{z^2}{Z^2}\right], \tag{5}$$

where k_B is Boltzmann's constant and m is the mass of the plume atom (Hansen, et al., 1999). Fig. 4 shows the radial variation of the temperature, density, and flow velocity according to the isentropic model of plume expansion.

The sharp edges of the pressure and density profiles at the front, at which the density is equal to zero, mean that the temperature at the edge approaches zero as well. This is in contrast to Singh and Narayan's model, in which the temperature T of the plume is considered constant.

With the density profiles from Eqs. (3-4) and the velocity determined from the self-similarity principle, Anisimov, et al. (1993) as well as Singh and Narayan (1990), arrive at the differential equations which control the expansion of the plume:

$$X(t)\frac{d^2X}{dt^2} = Y(t)\frac{d^2Y}{dt^2} = Z(t)\frac{d^2Z}{dt^2} = A\left[\frac{X_0Y_0Z_0}{XYZ}\right]^{\gamma-1} = \left[\frac{V_0}{V}\right]^{\gamma-1} \quad (6)$$

where Anisimov's model gives

$$A = (5\gamma\text{-}3)E_p/M_p \quad (7)$$

and Singh and Narayan's model,

$$A = k_BT/m. \quad (8)$$

Since γ in many cases is just slightly larger than unity, the parenthesis on the right-hand side is practically constant. It means that the acceleration of the front in any point, $(d^2X/dt^2, d^2Y/dt^2, d^2Z/dt^2)$, is inversely proportional to the instantaneous position of the point $(X(t), Y(t), Z(t))$. The underlying dynamics of Eqs. (6-8) is that the expansion is driven by the pressure gradients of the plume (Zel'dovich and Raizer, 2001). For large plume volumes the front acceleration goes asymptotically to zero, such that the asymptotic velocity of the front is constant.

Three well-known effects from plume dynamics can all be explained on the basis of the equations: i) the plume sharpening in forward direction (Saenger, 1994), ii) the flip-over effect (see below) (Anisimov, et al., 1996; Toftmann, et al., 2003), and iii) the forward peak from a broad beam spot on target (Weaver and Lewis, 1996).

After the termination of the laser pulse, the plume thickness is usually much smaller than the lateral dimensions of the plume. The acceleration is thus much stronger in forward direction. This is also clearly seen from the left panel of Fig. 3. The flip-over effect occurs for noncircular beam spots, so that the faster expansion is in that lateral direction where the initial plume was narrower. For a rectangular or elliptical beam spot, it means that the plume turns in such a way that the major axis lies 90° to that of the beam spot.

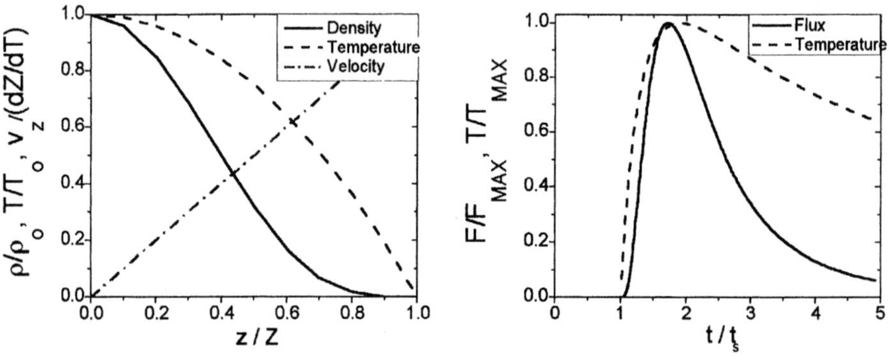

Figure 4. (Left panel) Spatial variation of flow velocity, density, and temperature, from Eqs. (3) and (5), in the z-direction in the plume. (Right panel) Normalized variation of the particle flux, Eq. (10), and temperature in the plume at a position from the target $z_s = (dz/dt)t_s$.

This is again seen from equation (6), since the smallest value of the initial front position results in the largest acceleration of the front. Finally, a small beam spot, i.e. small initial, lateral values of the front, $X = X_0$ and $Y = Y_0$, for the same volume of the plume (same ablation yield), means a larger value of the acceleration in the lateral directions and thus a broadening of the plume.

An expression for the signal onto a flux detector can be found directly from the treatment by Anisimov, et al. (1993). In the asymptotic limit of constant plume front velocity, the time t_s, at which the front reaches a planar substrate parallel to the flat probe surface at a distance z_s and an angle θ in the z-x plane with respect to the normal, is determined by

$$t_s = z_s/(dZ/dt)\,[1 + k_z^2 \tan^2\theta]^{1/2} \qquad (9)$$

where dZ/dt is the velocity of the front along the z-axis, and $k_z = Z_{inf}/X_{inf}$ is the ratio of the limiting value of the position along the z-axis as well as the value of the front in horizontal direction along the front. For a hemispherical substrate with a radial distance R_s from the target t_s can be found from

Eq. (9) by replacing z_s with $R_s/(1+\tan^2\theta)^{1/2}$. The time-of-flight distribution for a plume can be expressed in units of t/t_s:

$$F(\theta,t)\,dt = \frac{M_p k^2}{m I_1(\gamma) R_s^2}\left(\frac{1+\tan^2\theta}{1+k^2\tan^2\theta}\right)^{3/2} \times \frac{t_s^4}{t^4}\left[1-\frac{t_s^2}{t^2}\right]^{\frac{1}{\gamma-1}} \frac{dt}{t_s}, \qquad (10)$$

and is plotted in Fig. 4 (right panel). This figure also shows the expected variation of plume temperature at R_s for an isentropic expansion as the plume expands beyond that point.

3. EXPERIMENTAL METHODS

There is now a very extensive literature describing a wide range of experimental techniques which have been used to diagnose laser ablation plumes, and these have been reviewed by various authors (Chrisey and Huber, 1994; Lowndes, 1998; Amoruso, et al., 1999). It is of interest to measure the following parameters in a laser ablation plume:

1. the shape and velocity of the plume at various times after the ablating laser pulse,
2. the spatial distribution of density and temperature at various times after the ablating laser pulse, and
3. the spatial variation of the plume composition, in terms of the atoms, molecules, excited states, and clusters present.

Broadly speaking, the primary diagnostic techniques can be divided in optical and electrical probe methods.

Firstly, we will look briefly at some of the more commonly used optical diagnostic techniques. Gated intensified charge-coupled devices (ICCDs) can be used to record images of the self-emission of the plume with time resolution of a few ns (see Fig. 1). Time-lapse images of the plume can be recorded by changing the time delay between the laser and the gating pulse to the ICCD, and these give a very good impression of the plume dynamics. The optical emission spectrum indicates which electronically excited species are present in the plume. Space and time resolved spectra can be recorded using spectrometers with photomultiplier and ICCD detector. Fig. 5 (left panel) shows the emission spectrum emitted during the laser ablation of MgB_2 from a region 5 mm in front of the target. Line emission from both neutral and ionized species of both Mg and B are observed. Fig. 5 (right panel) shows a TOF record of emission on two lines from this region. By recording TOF spectra at various distances, the velocity of specific excited species can be obtained.

Rather than relying on the self-emission of the plume, a second pulsed tuneable laser may be used to excite a specific transition at well-defined time and distance from the target, such that the fluorescence can be recorded. This technique of laser induced fluorescence (LIF) has been used by various investigators to study the ground state atoms and molecules in laser ablation plumes (Geohegan, 1992).

When LIF is done with a narrow linewidth laser it is possible to exploit the motional Doppler effect to map out species that are moving parallel, or antiparallel, to the probe laser direction (AlWazzan, et al., 1996).

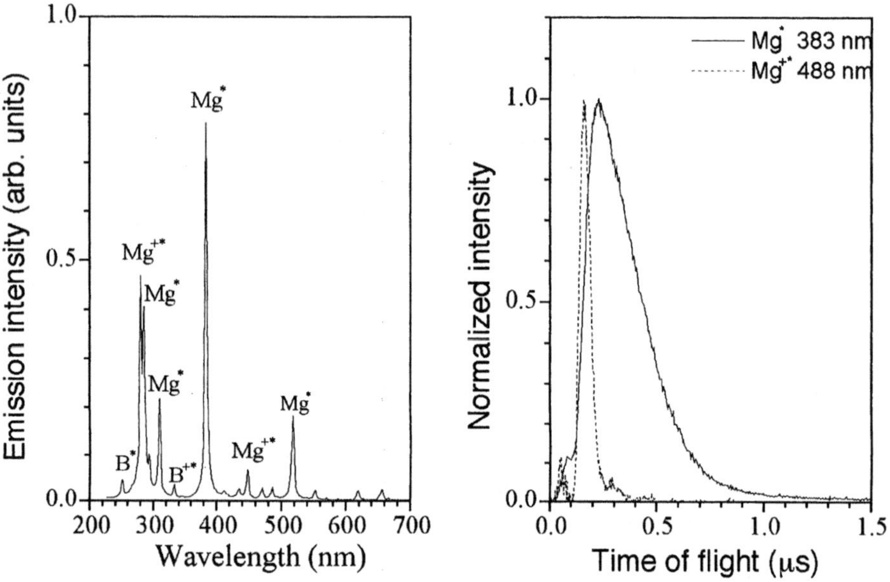

Figure 5. (Left panel) Emission spectrum of the plasma plume produced during excimer laser ablation of MgB_2 at 351 nm in vacuum. (Right panel) Optical time of flight profiles of the Mg* 383 nm and Mg^{+*} 488 nm emission line recorded at a distance of 5 mm from the target surface in a vacuum.

The emission spectrum may be analysed to find the electron temperature and density in the plume. If two excited states in the same atom or ion are in partial local thermodynamic equilibrium, then the intensity ratio of two spectral lines, I_1/I_2, derived from these states is related to the electron temperature T_e as:

$$I_1 / I_2 = \frac{f_1 g_1 \lambda_2^3}{f_2 g_2 \lambda_1^3} \exp\left(-\frac{E_1 - E_2}{k_B T_e}\right), \tag{11}$$

where E and g denote the energy and multiplicity of the upper levels and f and λ denote the oscillator strength and wavelength of the transitions. The technique can be extended to several lines by making what is known as a Boltzmann plot of the intensities of the lines as is shown in Fig. 6; the electron temperature is given by the slope of the line. This analysis assumes that the plasma is optically thin for the lines being measured. Thus it is necessary to check that self-absorption is not significantly reducing the intensity of any of the measured lines.

There is some potential to use Stark broadening of optical spectral lines to measure the electron density in the plume. A summary of this technique and some results are presented in the review by Amoruso, et al. (1999).

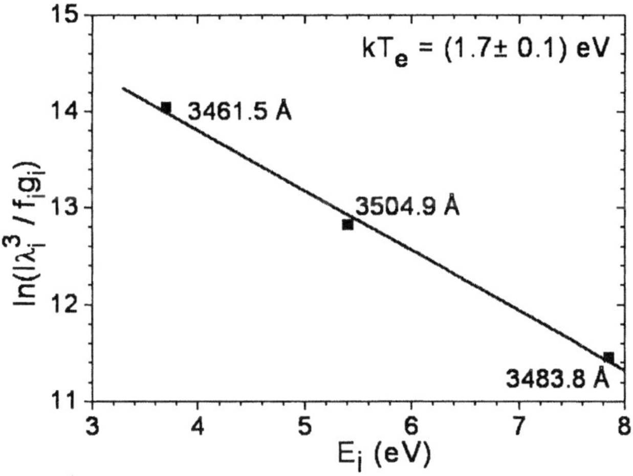

Figure 6. A Boltzmann plot of Ti II lines emitted at 150 ns after the laser pulse, from a region 2 mm above the target surface for 308 nm irradiation of Ti at 500 MW/cm^2. An electron temperature of 1.7 ± 0.1 eV was deduced. Reused with permission from J. Hermann, A. L. Thomann, C. Boulmer-Leborgne, B. Dubreuil, M. L. De Giorgi, A. Perrone, A. Luches, and I. N. Mihailescu [Hermann, et al. (1995)]. Copyright 1995, American Institute of Physics.

The particle density can also be measured using interferometry (Lindley, et al., 1994). Far from any absorption lines in the plume only the free electrons contribute to the refractive index, while near an absorption line there is contribution from the resonant transition. Thus it is, in principle, possible to use interferometry to measure both the electron density and the density of ground-state atoms or ions. By using pulsed laser illumination, the density distribution can be measured at various stages of plume expansion (Lindley, et al., 1994).

Since the early days of laser ablation, Langmuir probes have been used to investigate the expansion dynamics of ablation plumes (Koopman, 1971).

Essentially, a Langmuir probe consists of a small, bare electrical conductor which is immersed in a plasma and can be biased with respect to the plasma potential. If the probe is biased negatively, electrons moving towards the probe are rejected and the current to the probe is given by the ion thermal flux. On the other hand, if the probe is biased positively a net electron current will be recorded. Starting at the plasma potential and changing the probe bias to more negative values leads to a reduction of the electron current as electrons of increasingly higher energy are unable to reach the probe. This region of the *I-V* characteristic can be used to find the electron temperature (Chen, 1965).

In a laser plume expanding into vacuum, the behaviour of a Langmuir probe is somewhat different from a probe in a stationary plasma. Since the flow is supersonic, the ion velocity is larger than the thermal velocity, and the ion current is mainly determined by the flow velocity. The ion current I to a negatively biased flat probe of area A lying normal to the flow is given by:

$$I = en_i v_i A \quad , \tag{12}$$

where n_i is the ion density and v_i is the flow velocity. This equation assumes that the ions are mainly singly-charged, which is usually the case in the ablation plumes of interest here.

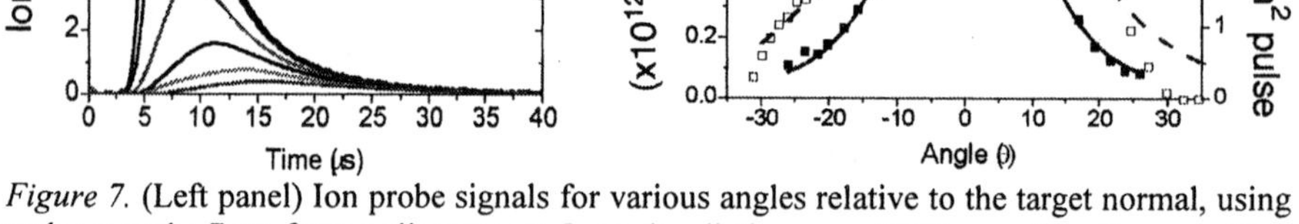

Figure 7. (Left panel) Ion probe signals for various angles relative to the target normal, using a planar probe 7 cm from a silver target. Laser irradiation was at 248 nm with a fluence of 1.5 J/cm^2. The probe bias was –15 V and the laser angle of incidence was 45°. (Right panel) Angular distribution of ions and deposition (ions + neutrals) on a hemispherical surface 6.5 cm from a silver target, for a fluence of 1.2 J/cm^2. Fitting the data to Eq. (13) yields a value of k_z= 4.5 for the ion distribution and k_z= 2.7 for the deposition (Doggett and Lunney, unpublished).

The left panel of Fig. 7 shows the angular variation of ion probe signals for laser ablation of silver recorded with an ion probe, which can be moved on a radius of 7 cm about the ablation spot on the target (Doggett and Lunney, unpublished). The shape of this TOF signal is very similar to the adiabatic prediction shown in the left panel of Fig. 4. The signal amplitude is maximised and the TOF minimised in the direction normal to the target according to Eq. (10). The full angular distribution in the z-x plane on a hemispherical surface can be found by integration of Eq. (10) with respect to t from t_s to infinity, yielding:

$$F(\theta) = F(0)(1+\tan^2\theta)^{3/2}/(1+k_z^2\tan^2\theta)^{3/2}, \quad (13)$$

where $F(0)$ is the areal number density in normal direction $\theta = 0$. Similarly, the TOF signals in the left panel of Fig. 7 can be integrated to yield the angular distribution of ion production, and this is plotted in the right panel of Fig. 7. Fitting this data set to Eq. (13) yields a k_z of 4.5, which is similar to, or somewhat larger than, the values reported earlier (Toftmann, et al., 2003; Hansen, et al., 1999; Thestrup, et al., 2002). Recording the angular variation of the net deposition gives some indication of the total ablation flow, i.e. ions and neutrals. This is also plotted in the left panel of Fig. 7; the k_z value for the deposition is 2.7, indicating that the ion component is more peaked in forward direction than the total ablation flow, as had been reported earlier (Hansen, et al., 1997).

For a planar substrate, the angular distribution is obtained by omitting the geometrical factor $(1+\tan^2\theta)^{3/2}$ from Eq. (13). This formula is also given explicitly by Anisimov, et al. (1993):

$$F(\theta) = F(0)/(1+k_z^2\tan^2\theta)^{3/2}. \quad (14)$$

The full area profile on a planar substrate given by the variables $\theta_x = \arctan(x/z_s)$ and $\theta_y = \arctan(y/z_s)$ (Anisimov, et al., 1996):

$$F(\theta_x, \theta_y) = F(0,0)\,[1 + k_z^2\tan^2\theta_x + (k_z^2/k_y^2)\tan^2\theta_y]^{-3/2}, \quad (15)$$

where $k_y = Y_{inf}/X_{inf}$ is the ratio of the limiting value of the front along the y-axis and the x-axis in analogy to the quantity k_z. Similar expressions were used for analysis by Tyunina and Leppävuori (2000) and for calculating the angular profile for fs-laser irradiation by Komashko, et al. (2000).

Langmuir probes have been used to measure electron temperature in laser ablation plumes (Koopman, 1971). Since the ion energies are much greater than the electron temperature and the small bias voltages in the retarding region of the *I-V* characteristic, the ion flow to the probe is scarcely changed and will reduce the apparent electron current.

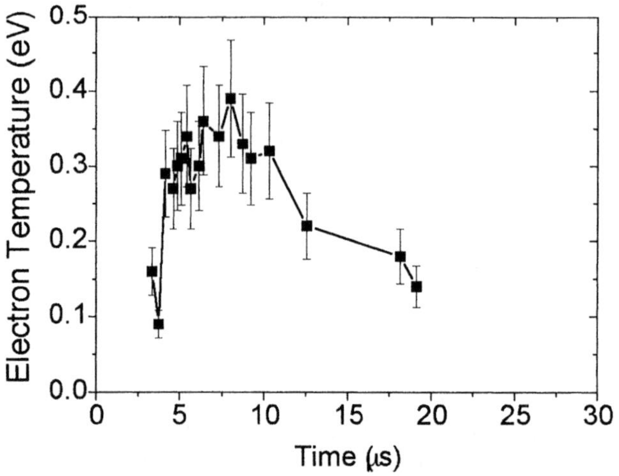

Figure 8. Laser ablation at 248 nm with a fluence of 1.5 J/cm^2. The electron temperature is determined with a planar probe parallel to the flow for a range of times showing a maximum of 0.35 eV, occurring at ~ 7 μs. (Doggett and Lunney, unpublished).

Both Koopman (1971) and Toftmann, et al. (2000) have adopted the procedure of correcting the electron region of the *I-V* characteristic by subtracting the ion current due to the flow and then extracting T_e. More recently, Doggett, et al. (2005) have explored the possibility of measuring T_e with a flat probe lying parallel to the plasma flow in order to avoid the current due to ion flow. The electron temperature shown in Fig. 8 has been determined from a plot of log *I* vs. *V* of the *I-V* characteristic similar to the procedure used by Toftmann, et al. (2000). Fig. 8 shows how T_e varies as the plume flow passes the probe (Doggett and Lunney, unpublished). It can be seen that the temperature first rises to a maximum value and then falls, as is expected in an isentropic expansion (see left panel of Fig. 4). In contrast, the isothermal model predicts a monotonic decrease of temperature.

TOF mass spectrometry has also been widely used to measure the velocity distributions of the ion, atomic, and molecular species produced in laser ablation. This technique will not be discussed here, but several variations of the technique are described in the review by Amoruso, et al. (1999).

4. PLUME DYNAMICS IN BACKGROUND GASES

The initial expansion of an ablation plume in a background gas does not differ much from the expansion in vacuum, since the driving pressure (~ 1 kbar) usually is much higher than that of a low-pressure background gas (< 1 mbar).

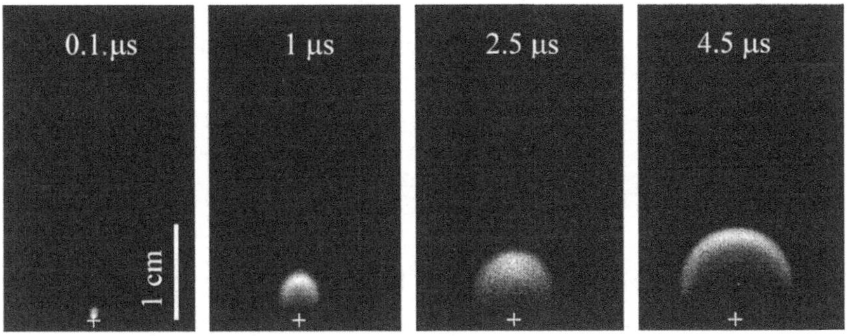

Figure 9. Images of an ablation plume expansion in an oxygen background pressure of ≈ 10^{-1} mbar at different time delays, after the laser irradiation of a $LaMnO_3$ target at 351 nm with a fluence of 1.5 J/cm^2. From Amoruso, et al. (2006a).

However, the expansion (stage 3 in Fig. 2) is soon largely controlled by the interaction processes between the ablated species and the background gas atoms/molecules (stage 4 in Fig. 2). Essentially, the plume expansion continues until the internal pressure is equal to the background pressure. We will mainly treat the plume propagation in a background gas on the basis of the general models by Anisimov, et al. (1996) and Arnold, et al. (1999). Most of the other existing treatments (e.g. Gonzalo, et al., 1995; Bulgakov and Bulgakova, 1998; Wood, et al.,1998; Itina, et al., 2002) have been performed for a specific choice of target and background gas and cannot readily be extended to other target-gas combinations.

The plume-gas interaction is well demonstrated by the sequence of time-lapse pictures shown in Fig. 9. In a 0.1 mbar atmosphere of oxygen the plume is seen to expand, rather like the vacuum case in the first μs, and then to progressively slow down thereafter. The plume continues to slow down, until it stops after approximately 5 μs (not seen in the figure). The photographs also show the strongly emitting crescent-shaped periphery with a contact surface to the background gas. During the expansion, a fast component of the ablation plume moves through the background gas as an expansion in free space, while a slower component is acting like a piston on the surrounding background gas. The collisions between the atoms in the expanding (slower) component with the background gas atoms/molecules lead to collisional excitation of the ablated atoms at the boundary between the plume and the background gas, which clearly is seen as a strongly emitting layer in Fig. 9, and also has been reported in the literature.

The slow component pushes a compressed layer of background gas further away from the target. This releases an internal as well as an external shock wave. According to Zel'dovich and Raizer (2001), shock wave formation becomes important when the mass of the displaced gas is comparable to the mass M_p of the plume. For a hemispherical expansion it occurs for a front radius R_{sw} :

$$(2/3)\pi R_{SW}^{3}\rho_{g} = M_{p}, \quad)\pi R^{3}_{sw}\rho_{g} \approx M_{p}, \tag{16}$$

where ρ_g is the mass density of the background gas. This approximation demonstrates that for a given target (i. e. the same plume mass M_p), the shock wave production starts much closer to the target for a background gas with a high pressure than for one with low pressure. Similarly, the shock production also starts closer to the target for a heavier gas than for a light one (Toftmann, et al., 2006).

The knowledge of plume slowing down has advanced tremendously from the many recent spectroscopic studies with fast imaging techniques (Geohegan, 1992; Harilal, et al., 2002, 2003; Amoruso, et al., 2004a, 2005a), but Langmuir probe studies have also contributed to the understanding of plume slowing down (Geohegan, 1992; Wood, et al., 1997; Hansen, et al., 1999; Amoruso, et al., 2004b, 2005a).

The shock wave formation does not appear abruptly, but is the result of a growing hydrodynamic interaction between the plume and the background gas. At a certain distance from the target, the fraction of atoms from the plume that penetrate the surrounding gas as a freely expanding plume decreases strongly with the increasing gas pressure.

This decrease is accompanied by a large enhancement of the slow component which leads to the so-called plume splitting. This plume behavior is seen in Fig. 10, and was first observed by Geohegan (1992) and Geohegan and Puretzky (1996), and later explained by Wood, et al. (1997 and 1998) by a hydrodynamic model that includes multiple scattering.

Results for a silver ion plume expanding in a background atmosphere of argon and xenon are shown in Fig. 10. There is a distinct plume splitting in a narrow pressure regime (the row in the middle) in the argon as well as the xenon gas. The second peak appears for a much lower pressure in the heavier xenon than in argon gas in agreement with Eq. (16). The two ion components in the time-of-flight distribution can be separated by a fitting procedure for the time of arrival of the peaks in background gases with increasing pressure (Amoruso, et al., 2004a). For the fast peak, the velocity of the ions is constant and equal to the "vacuum" expansion velocity v_f, up to a relatively high pressure:

$$z = v_f t \,. \tag{17}$$

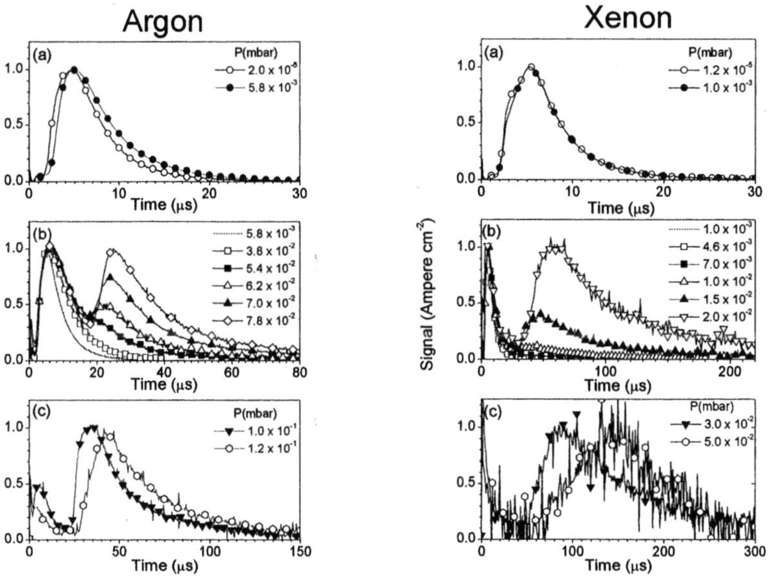

Figure 10. Time-of-flight signals from a silver plume in an Ar and Xe background gas of different pressure measured by a Langmuir probe 75 mm from target. Laser irradiation was at 355 nm and with a fluence of 2.5 J/cm^2. The signals in each panel have been normalized to the peak, so that the double-peak structure is clearly seen. From Toftmann, et al. (2006).

The slow component which is responsible for the shock wave production can be approximated by a point blast wave (Zel'dovich and Raizer, 2001; Arnold, et al., 1999; Amoruso, et al., 2003), which appears to fit the data well up to 0.2 mbar for a silver plume in an argon gas (Amoruso, et al., 2004a):

$$z = \alpha t^{2/5} \,, \tag{18}$$

where α is a constant that depends on the plume geometry.

A plume with an initial volume V_0 and pressure p_0 can expand adiabatically until the plume pressure has decreased to the pressure p_g of the background gas. Since pV^γ is constant, the final plume volume V_p in this simple picture is:

$$V_p = V_0\,(p_0/p_g)^{1/\gamma}, \tag{19}$$

leading to a final plume length L_p in the z-direction :

$$L_p \sim V_0^{1/3}(p_0/p_g)^{1/3\gamma}. \tag{20}$$

Anisimov's model for an adiabatic expansion, which works well for many aspects of plume dynamics in vacuum, cannot be extended to expansion in a background gas in the original framework. However, the theory can be modified to describe an expansion into a background gas with a low pressure p_g, where $p_g < (E_p/V_0)$ (Anisimov, et al., 1996). The plume expands until a contact surface described by

$$p(x,y,z) = p_g \tag{21}$$

is established. The final plume length L_p is determined by

$$L_p = \xi X_o \left[\frac{E_p}{p_g X_o^3}\right]^\beta, \tag{22}$$

where ξ and β (Anisimov, et al., 1996) are both functions of (Y_0/X_0), (Z_0/X_0), and γ. For the simple case of a spherical expansion $(X_0 = Y_0 = Z_0 = R_0)$, Anisimov, et al. obtained $\beta = 1/(3\gamma)$, similar to the simple adiabatic expansion discussed above. A similar result was also obtained with another geometry by Dyer, et al. (1990) and Proyer and Stangl (1995). However, since γ can vary between 1 and 5/3, the exponent β varies between 1/3 and 1/5.

In a realistic expansion in a background gas, a part of the initial plume energy is spent on shockwave formation, and the expansion is no longer adiabatic. Arnold, et al. (1999) have derived an analytical model for the plume dynamics during the expansion into a background gas. The results of the model are based on a spherical expansion in forward and backward direction from a fictive target surface without any assumptions of a background gas of low pressure. Even though the plume in Fig. 9 is not exactly semispherical, the model is very convenient for analysis of plume data (Amoruso, et al., 2003, 2005a).

The natural parameter for the stopping distance, i.e. the final plume length L_p, is $(E_p/p_g)^{1/3}$. Arnold, et al. (1999) found that the stopping distance for the spherical expansion is

$$L_p = \xi (2E_p/p_g)^{1/3}, \tag{23}$$

where ξ was determined from experimental results of the expansion of plumes from steel and YBCO to be $\xi = 0.4$, and from a numerical solution of the expansion model, with $\gamma = 5/3$ for the plume gas as well as the ambient gas to $\xi = 0.39$. These authors also found an expression for the stopping time t_{st}, expressed by the stopping distance and the sound velocity c_g of the ambient gas:

$$t_{st} = \xi_{st} (2E_p/p_g)^{1/3}/c_g. \tag{24}$$

The equations in the model of Arnold, et al. (1999) also imply that the plume dynamics can be described by dimensionless variables length as $\rho_A = R(2E_p/p_g)^{-1/3}$, and time as $\tau_A = t\, c_g(2E_p/p_g)^{-1/3}$. Signals measured at different distances from target can thus be compared in the model. This is shown in Fig. 11 (Amoruso, et al., 2005a), where data from a multicomponent ($LaMnO_3$) target obtained from optical, as well as from ion probe measurements, are shown. The striking point is that the neutrals and ions behave similarly, and that points from different elements all lie on the same curve. Initially, the front moves as an expansion in vacuum, $\rho_A \sim \tau_A$, and later it follows the behavior for the point-blast model $\rho_A \sim \tau_A^{2/5}$. For ions, a clear transition zone with two discernable peaks which reflect the plume splitting, is also seen in Fig. 10.

Thus, plume splitting occurs not only for ions and elemental targets (Amoruso, et al., 2004a; Harilal, et al., 2002), but also for neutrals and species from multicomponent targets.

Figs. 10 and 11 also show the three regimes which occur for the propagation of a plume in a background gas with increasing pressure (Amoruso et. al, 2004b; Schou, 2006). For the lowest pressures (up to 0.01 mbar), the behavior of the plume is essentially *vacuum-like*, then there is a transition regime with plume splitting into a fast and slow component. At the highest pressure the plume is slowed down and eventually has been stopped in the *diffusion-like* regime as a result of shock wave formation.

Eventually, the plume stops and the particles diffuse out of the plume volume and become thermalized (Amoruso, et al., 2003, 2004b; Arnold, et al.,1999; Dyer, et al.,1990; Bulgakov and Bulgakova, 1998; Itina, et al., 1997). The deviations from the point-blast wave behavior seen in Fig. 11 for large values of τ_A, indicate the start of thermalization and diffusion of ablated particles away from the "confined" plume.

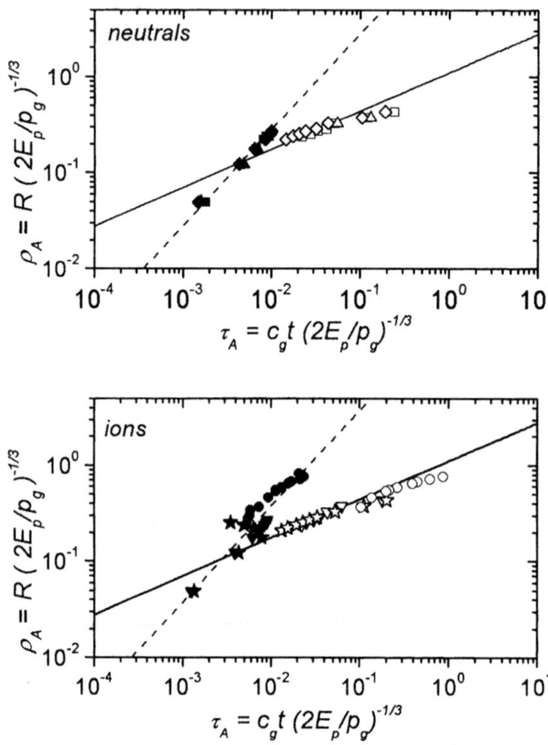

Figure 11. Distance-time plots of excited neutrals (upper panel) and ions (lower panel) dynamics in dimensionless variables from Arnold, et al. (1999) (see text for details). Upper panel: first (▲) and second peak (△) of the O I 646 nm emission line; first (◆) and second peak (◇) of the Mn I 403.3 nm emission line; first (■) and second peak (□) of the La I 624.9 nm emission line. Lower panel: first (▼) and second peak (▽) of the Mn II 294.1 nm emission line; first (★) and second peak (☆) of the La II 394.7 nm emission line; first (●) and second peak (○) of the ion TOF profiles. The dashed and solid lines represent a free plume (r~t) and a point-blast wave(r~$t^{2/5}$) expansion behavior, respectively. (From Amoruso, et al., 2005a).

5. PLUMES FROM ULTRASHORT LASER ABLATION

Laser ablation with ultrashort pulses (<1 ps) has attracted a lot of attention, not only because of the access to ultrafast physical and chemical processes, but also because of the new prospects for applications in materials processing and thin film deposition (Banks, et al., 1999; Nolte, et al., 2000; Ausanio, et al., 2005; Amoruso, et al., 2005b). Also, for fs laser ablation the initial plume dynamics is important for the physical parameters in the late expansion phase. The main difference between fs laser and ns ablation is that stage 2 (in Fig. 2) does not exist, i. e. the laser pulse is so short that the atoms in the irradiated volume hardly move from their positions during the pulse, and the laser intensity typically is much larger than in the ns case. Thus the heated volume may reach temperatures well in excess of the thermodynamic limit for stability of the liquid/solid, which can be the origin of nanoparticle production (Eliezer, et al., 2004; Amoruso, et al., 2005c).

The experimental analysis of fs laser ablation plumes has built on experience acquired in the studies on plasma plumes produced by ns laser pulses, and most of the techniques employed are similar to those applied in that case.

In the case of ultrashort laser ablation, optical emission spectroscopy and fast photography have identified the presence of various populations in the laser-induced ablation plume, which are characterized by different expansion dynamics and temporal evolution of the emission intensity (Albert, et al., 2003; Amoruso, et al., 2004a; Grojo, et al., 2005). As an example, Fig 12(a) shows an image of the laser-induced plasma plume emission produced during ablation of a silicon target by ≈ 900 fs Nd:Glass laser pulses at 1055 nm, in vacuum, at a time delay of few tens of ns after the laser pulse hit the target (Amoruso, et al., 2006b).

Three different components can be distinguished from the emission intensity profile on the left-hand side of the image, and the velocity distributions are reported in Fig. 12(b): i) a fast population characterized by a faint emission, which expands along the normal to the target surface with an average velocity, v, of the order of ≈ 100-200 km/s; ii) a more intense, second population whose maximum emission is located at a distance of ≈1 mm from the target surface in Fig. 12(a), moving away from the target at a typical velocity $v \approx$ 20-30 km/s,; and iii) a third, low-velocity component of very slow particles (v < 1 km/s), characterized by a quite large velocity distribution which is still located very close to the target surface, and whose expansion takes place only after a longer time delay (see Fig. 13(a)). The faster components are both characterized by a strongly forward peaked expansion, while the slow population shows a much larger angular divergence.

Spectrally resolved emission analysis of the laser-produced plume revealed that the luminescence from the fast components is dominated by emission lines of the atoms of the target material, while the emission from the slow, delayed plume population is characterized by structureless, broadband continuum spectra.

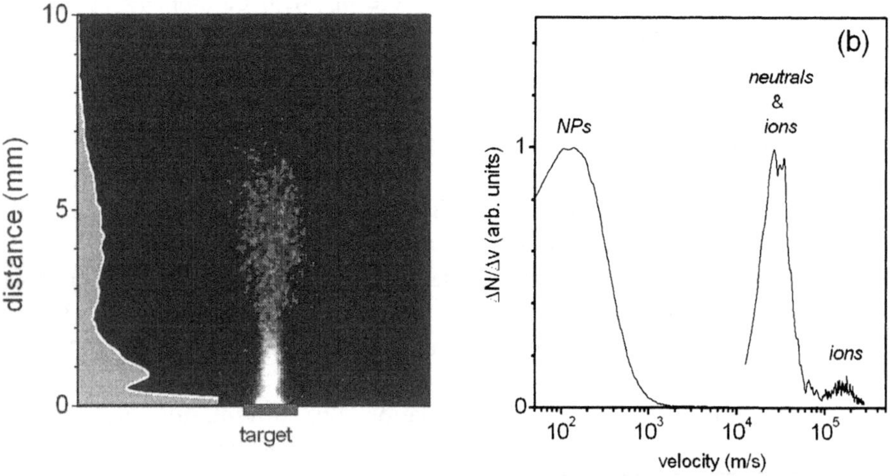

Figure 12. (a) Image of the plume expansion of a silicon plume produced by a 0.9 ps, 1055 nm laser pulse at a fluence of 0.75 J/cm^2 ($I = 8.3x10^{11}$ W/cm^2) at a time delay of 25 ns. Distance 0 mm defines the front face of the target, which is shown as a grey box. The profile on the left-hand part of the image shows the intensity vs distance plot obtained by integrating the emission along a direction parallel to the target surface; (b) Velocity distribution of the ejected species obtained from the time-resolved images of the silicon plasma plume. The velocity distribution profiles of the fast and slow part of the plume have been normalized to facilitate the comparison. Reprinted from Amoruso, et al. (2006b) with the permission of Elsevier.

This demonstrates that the fast components are mainly formed by atoms and ions of the target material, while the slow components essentially contain nanoparticles with radii ranging from few to few tens of nm (see below). These features have been observed for different laser pulse durations and wavelengths (Albert, et al., 2003; Amoruso, et al., 2004c; Grojo, et al., 2005; Amoruso and Vitiello, 2005). Moreover, the presence of a very fast ionic component in the ablation plume produced by ultrashort laser ablation of different materials has been observed by different authors using different experimental techniques, e.g. optical emission spectroscopy (Albert, et al., 2003; Amoruso, et al., 2004d), time-of-flight mass spectrometry (Ye and

Grigoropoulos, 2001), ion probe (Qian, et al, 1999; Amoruso, et al., 2000), and electrostatic energy analyzer mass analysis (VanRompay, et al., 1998).

The observation of structureless continuum emission with the presence of nano-sized particles in the plume has been explained by considering that hot nanoparticles behave as a black body radiator, and their emission is mainly determined by the temperature T of the system (Heszler, et al., 2001; Amoruso, et al., 2004c). For the spectral region of concern here ($hc/\lambda >> k_B T$), by taking into account the emissivity of the nanoparticles and that the CCD detector counts photons, the emission intensity $I(\lambda)$ is given by:

$$I(\lambda) \propto \lambda^{-5} \exp(-\frac{hc}{\lambda k_B T}) \; , \qquad (25)$$

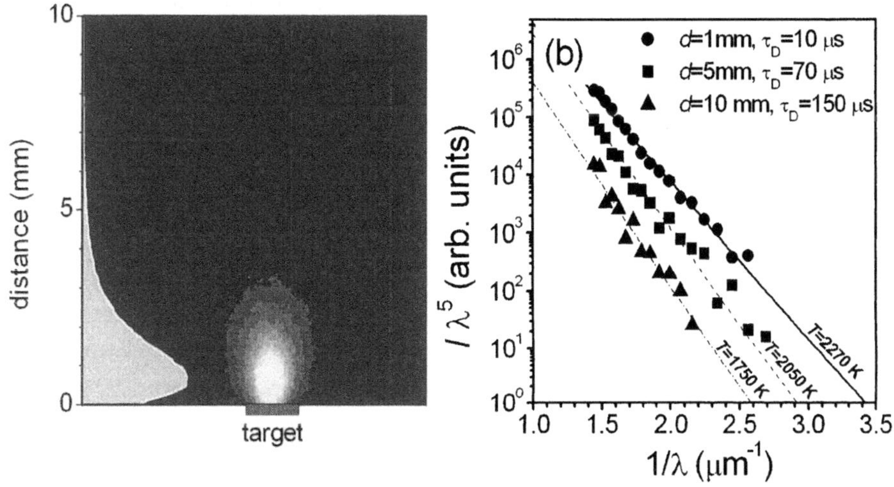

Figure 13. (a) Typical image of the plume expansion of nanoparticles. The profile on the left-hand part of the image shows the intensity vs. distance plot obtained by integrating the emission along a direction parallel to the target surface; (b) Emission spectra of silicon nanoparticles in $I\,\lambda^5$ - λ^{-1} coordinates on a semilogarithmic plot. Each spectrum refers to the emission recorded at the distance d reached by the nanoparticles after a time delay τ_D. The straight lines are fitting curves according to Eq. (25), the slope of which gives the temperature T of the nanoparticles. Reprinted from Amoruso, et al., (2006b), with the permission of Elsevier.

where λ is the emission wavelength, h is the Planck constant, and c is the speed of light. Fig. 13 (b) shows typical power spectra of silicon nanoparticles presented in the form of log-linear plots of $I(\lambda)\lambda^5$ vs λ^{-1}, at different distances and time delays. The experimental points are well approximated by straight lines whose slopes, according to Eq. (25), provide an estimate of the temperature T of the emitting particles. Fig. 13(b) shows the emission spectra for a selected group of particles travelling with the same velocity, measured at increasing time delays and distances. The lowering of the temperatures with time demonstrates the radiation cooling of the emitting nanoparticles during expansion into vacuum (Amoruso, et al., 2004e).

The presence of nanoparticles in the ablation plume produced by the ultrashort laser ablation process has also been studied by deposition experiments, and their size distribution was measured by atomic force microscopy or scanning electron microscopy analysis of less than one layer deposits (Dinh, et al., 2002; Eliezer, et al., 2004; Amoruso and Vitiello, 2005; Scuderi, et al., 2005). As an example, a size distribution of silicon nanoparticles produced by 100 fs, Ti:sapphire laser pulses is shown in Fig. 14, demonstrating that particles with a radius ranging from ≈ 5 to ≈ 25 nm can be produced in this case. The experimental analysis has shown that ablation with fs laser pulses results in the generation of nanoparticles for different materials, both in elemental and multicomponent form (Millon, et al., 2003; Dinh, et al., 2002; Amoruso, et al., 2005b). The formation of nanoparticles is likely to happen during the relaxation of the material state induced by the intense, ultrashort laser irradiation of the target.

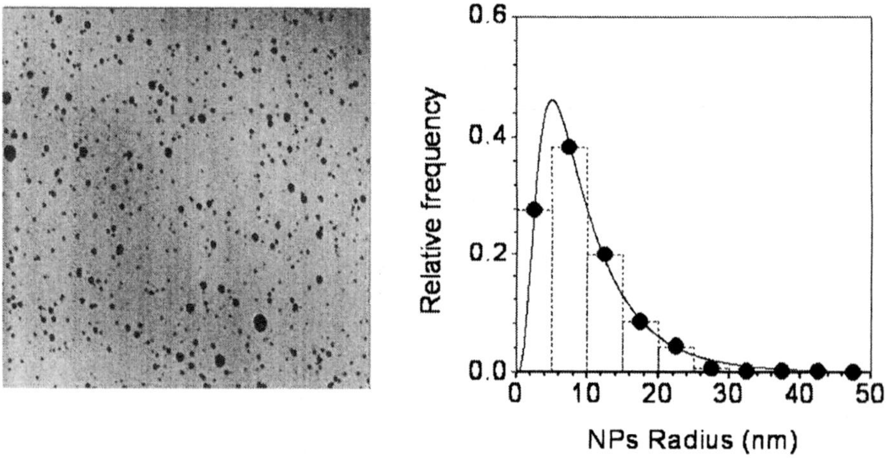

Figure 14. (Left panel), AFM image of Si nanoparticles (NP) deposited onto mica substrates in a vacuum by 100 fs Ti:sapphire laser pulses; (Right panel), size histogram of the Si nanoparticles. From Amoruso, et al., (2006b).

In particular, different theoretical studies suggested that the rapid expansion and cooling of solid-density matter irradiated by a fs laser pulse may result in the generation of nanoparticles via different mechanisms, such as liquid phase ejection and fragmentation, homogeneous nucleation and decomposition, or spinodal decomposition, depending on the different heating regimes (Sokolowski-Tinten, et al., 1998; Vidal, et al., 2001; Glover, 2003; Perez and Lewis, 2003). The appearance of a visible, slow peak in the emission spectrum originating from nanoparticles is a feature which so far has only been observed for fs-laser ablation.

The results reported above show a complex structure of the plume produced by ablation with ultrashort laser pulses, both in terms of composition and velocity distribution of the ejected species. The variety of species present in the ablation plume originates from the different mechanisms involved into the relaxation of the state of matter created during irradiation of a solid target with ultrashort laser pulses, which is still under intensive investigation.

REFERENCES

Albert, O., Roger, S., Glinec, Y., Loulergue, J.C., Etchepare, J., Boulmer-Leborgne, C., Perrière, J., and Millon, E., 2003, Time-resolved spectroscopy measurements of a titanium plasma induced by nanosecond and femtosecond lasers, *Appl. Phys. A* **76**: 319-323.

AlWazzan, R.A., Lewis, C.L.S. and Morrow, T., 1996, A technique for mapping three-dimensional number densities of species in laser produced plasmas, *Rev. Sci Instrum.* **67**: 85-88.

Amoruso, S., 1999, Modeling of UV pulsed-laser ablation of metallic targets, *Appl. Phys. A* **69**: 323–332.

Amoruso, S., Bruzzese, R., Spinelli, N. and Velotta,R., 1999, Characterization of laser-ablation plasmas, *J. Phys. B* **32:** R131-R172.

Amoruso, S., Wang., X., Altucci, C., de Lisio, C., Armenante, M., Bruzzese and R., Velotta, R., 2000, Thermal and non-thermal ion emission during high-fluence femtosecond laser ablation of metallic targets, *Appl. Phys. Lett.* **77**: 3728-3730.

Amoruso, S., Bruzzese, R., Spinelli, N., Velotta, R., Vitiello, M. and Wang, X., 2003, Dynamics of laser-ablated MgB_2 plasma expanding in argon probed by optical emission spectroscopy, *Phys. Rev. B* **67**: 224503-1-224503-11.

Amoruso, S., Toftmann, B., Schou, J., Velotta, R. and Wang, X., 2004a, Diagnostics of laser ablated plasma plumes, *Thin Solid Films* **453-454**: 562-572.

Amoruso, S., Toftmann, B. and Schou, J., 2004b, Thermalization of a UV laser ablation plume in a background gas: from directed to diffusionlike flow, Phys. Rev. E **59:** 056403-1-056403-6.

Amoruso, S., Bruzzese, R., Spinelli, N., Velotta, R., Vitiello, M., Wang, X., Ausanio, G., Iannotti, V. and Lanotte, L., 2004c, Generation of silicon nanoparticles via femtosecond laser ablation in vacuum, *Appl. Phys. Lett.* 84: 4502-4504.

Amoruso, S., Altucci, C., Bruzzese, R., de Lisio, C., Spinelli, N., Velotta, R., Vitiello and M., Wang, X., 2004d, Study of the plasma plume generated during near IR femtosecond laser irradiation of silicon targets, *Appl. Phys. A* **79**: 1377-1380.

Amoruso S., Bruzzese R., Spinelli N., Velotta R., Vitiello M. and Wang X., 2004e, Emission of nanoparticles during ultrashort laser irradiation of silicon targets, *Europhys. Lett.* **67**, 404-410.

Amoruso, S. and Vitiello, M., 2005, Characterization of plumes produced during ultrashort laser ablation of metals and semicoductors, in *Laser Physics and Applications*, Atanasov, P. A., Gateva, S.V., Avramov, L. A. and Serafetinides, A. A. eds., Proceedings SPIE vol. 5830,, Bellingham, USA, pp. 11-20.

Amoruso, S., Bruzzese, R., Velotta, R., Spinelli, N., Vitiello, M. and Wang, X., 2005a, Characterization of $LaMnO_3$ laser ablation in oxygen by ion probe and optical emission spectroscopy, Appl. Surf. Sci. **248**: 45-49.

Amoruso, S., Vitiello, M. and Wang, X., 2005b, Femtosecond laser ablation and deposition, in *Pulsed Laser Deposition of Optoelectronic Films*, Popescu, M. ed., Series Optoelectronic Materials and Devices, INOE, Bucharest, pp. 41-80.

Amoruso, S., Bruzzese, R., Vitiello, M., Nedialkov, N. N. and Atanasov, P. A, 2005c, Experimental and theoretical investigation of femtosecond laser ablation of aluminium in vacuum, *J. Appl. Phys.* **98**: 044907-1–044907-7.

Amoruso, S., Sambri, A., Vitiello, M. and Wang, X., 2006a, Propagation of $LaMnO_3$ laser ablation plume in oxygen gas, *Appl. Surf. Sci.* (in press).

Amoruso, S. , Ausanio, G., Bruzzese, R., Gragnaniello, L., Lanotte, L., Vitiello, M. and Wang, X., 2006b, Characterization of laser ablatioin of solid targets with near-infrared pulses of 100 fs and 1 ps duration, *Appl. Surf. Sci.* (in press).

Anisimov, S. I., Bäuerle, D. and Luk'yanchuk, B. S., 1993, Gas dynamics and film profiles in pulsed-laser deposition of materials, *Phys. Rev. B* **48**: 12076-12081.

Anisimov, S. I., Luk'yanchuk, B. S., and Luches, A.., 1996, An analytical model for three-dimensional laser plume expansion into vacuum in hydrodynamic regime, *Appl. Surf. Sci.* **96-98**: 24-32.

Anisimov, S. I., and Luk'yanchuk, B. S., 2002, Selected problems of laser ablation theory, *Physics-Uspekhi*, **45**:293-324.

Arnold, N., Gruber, J. and Heitz, J.,1999, Spherical expansion of the vapor plume into ambient gas: an analytical model, Appl. Phys. A **69**: S87- S93.

Ashfold, M. N. R., Claeyssens, F., Fuge, G. M. and Henley, S. J., 2004, Pulsed Laser Ablation and Deposition of Thin Films, *Chem. Soc. Rev.* **33**: *23-31.*

Ausanio, G., Barone, A. C., Iannotti, V., Lanotte, L., Amoruso, S., Bruzzese, R. and Vitiello, M., 2004, Magnetic and morphological characteristics of nickel nanoparticles films produced by femtosecond laser ablation, *Appl. Phys. Lett.* **85**: 4103-4105.

Banks P.S., Dinh L., Stuart B.C., Feit M.D., Komashko A.M., Rubenchik A.M., Perry M.D. and McLean W., 1999, Short-pulse laser deposition of diamond-like carbon thin films, *Appl. Phys. A* **69**: S347-S353.

Bulgakov, A.V. and Bulgakova, N. M., 1998, Gas-dynamic effects of the interaction between a pulsed laser-ablation plume and the ambient gas: analogy with an underexpanded jet, *J. Phys. D. Appl. Phys.* **31**: 693-703.

Chen, F. F. 1965, Electric probes, *Plasma Diagnostic techniques*, Huddlestone, R. H. and Leonard, S. L., eds., Academic Press, New York, pp. 113-200.

Chrisey, D. B. and Hubler, G. K., 1994, *Pulsed Laser Deposition of Thin Films,* eds, Wiley, New York.

Dijkkamp, D. , Venkatesan, T., Wu, X.D, Shafeen, S.A., Jishraw, N., Minley, Y. H., McLean, W. L. and Croft, M., 1987, Prepation of Y-Ba-Cu oxide superconductor thin-films using pulsed laser evaporation from high-T_c bulk material, *Appl. Phys. Lett.* **51**: 619-621.

Dinh, L. N., Hayes, S. E., Wynne, A. E., Wall, M. A., Saw, C. K., Stuart, B. C., Balooch, M., Paravastu, A. K. and Reimer, J. A., 2002, Properties of GaAs nanoclusters deposited by femtosecond laser, *J. Mat. Sci.* **37**: 3953-3958.

Doggett, B. and Lunney, J.G., unpublished.

Doggett, B., Budtz-Joergensen, C., Lunney, J.G., Sheerin, P. and Turner, T.T., 2005, Behaviour of a planar Langmuir probe in a laser ablation plasma, *Appl. Surf. Sci.***247**: 134-138.

Dyer, P. E., Issa, A. and Key, P. H., 1990, Dynamics of excimer laser ablation of superconductors in an oxygen environment, *Appl. Phys. Lett.* **57**: 186-188.

Eliezer, S., Eliaz, N., Grossman, E., Fisher, D., Gouzman, I., Henis, Z., Pecker, S., Horovitz, Y., Fraenkel, M., Maman, S. and Lereah Y., 2004, Synthesis of nanoparticles with femtosecond laser pulses, *Phys. Rev. B* **69**: 144119-1 – 144119-6.

Geohegan, D. B., 1992, Physics and diagnostics of laser ablation plume propagation for high-T_c superconductor film growth, *Thin Solid Films* **220**:138-145.

Geohegan, D. B. and Puretzky, A: A.,1996, Laser ablation plume thermalization dynamics in background gases: combined imaging, optical absorption and emission spectroscopy, and ion probe measurements, *Appl. Surf. Sci.* **96-98**: 131-138.

Glover, T. E., 2003, Hydrodynamics of particle formation following femtosecond laser ablation, *J. Opt. Soc. Am. B* **20**: 125-131.

Gonzalo, J., Vega, F. and Afonso, C. N., 1996, Plasma expansion dynamics in reactive and inert atmospheres during laser ablation of Bi(2)Sr(2)Ca(2)O(7-y), *J. Appl. Phys.* **77**:6588-6593.

Gorbunoff, A., 2002, *Laser-Assisted Synthesis of Nanostructured Materials*, Fortschritt-Berichte VDI 357, VDI verlag, Düsseldorf.

Grojo, D., Hermann, J. and Perrone, A., 2005, Plasma analysis during femtosecond laser ablation of Ti, Zr, and Hf, *J. Appl. Phys.* **97**: 063306- 1 – 063306-9.

Hansen, T. N., Schou, J. and Lunney, J. G., 1997, Angular distributions of silver ions and neutrals emitted in vacuum by laser ablation, *Europhys. Lett.* **40**: 441-446.

Hansen, T. N., Schou, J. and Lunney, J. G., 1999, Langmuir probe study of plasma expansion in pulsed laser ablation, *Appl. Phys. A* **69**: S601- S604.

Harilal, S. S., Bindhu, C. V., Tillack, M. S., Najmabadi, F. and Gaeris, A. C., 2002, Plume splitting and sharpening in laser-produced aluminium plasma, *J. Phys. D. Appl. Phys* **35**: 2935-2938.

Harilal, S. S., Bindhu, C. V., Tillack, M. S., Najmabadi, F. and Gaeris, A. C., 2003, Internal structure and expansion dynamics of laser ablation plumes into ambient gases, *J. Appl. Phys.* **93**: 2380-2388.

Harris, T. J., 1963, High-speed photographs of laser-induced heating, *IBM J. Res. Develop.* **7**: 342-345.

Hermann, J., Thomas, A. L., Boulmer-Leborgne, C., Dubreil, B., Giorgi, M. L. De, Perrone, A. , Luches, A. and Mihailescu, I. N. 1995, Plasma diagnostics in pulsed laser TiN layer deposition, *J. Appl. Phys.* **77**: 2928-2936.

Heszler P., Landström, L., Lindstam M. and Carlsson J.-O., 2001, Light-emission from tungsten nanoparticles during laser-assisted chemical vapour deposition of tungsten, *J. Appl. Phys.* **89**: 396-3970.

Hubler, G. K., 1992, Pulsed laser deposition, *MRS Bulletin*, Vol. XVII, No. 2, Feb. 92.

Itina, T. E., Marine, W. and Autric, M., 1997, Monte Carlo simulation of pulsed laser ablation from two-component target into diluted ambient gas, *J. Appl. Phys.* **82**: 3536-3542.

Itina, T. E., Hermann, J., Delaporte, P.and Sentis, M., 2002, Laser-generated plasma plume expansion: Combined continuous-microscopic modelling, *Phys. Rev. E* **66**: 066406-1 – 066406-12.

Kelly, R., Miotello, A., Mele, A. and Giardini Giudoni, A., 1998, Plume formation and characterization in laser-surface interactions, *Laser Ablation and Desorption*, Miller, J. C. and Haglund, R. F. eds., Experimental Methods in the Physical Sciences vol. *30*. Academic Press, New York, pp. 225 – 289.

Komashko, A. M., Feit, M. D. and Rubenchik, A. M., 2000, Modeling of long term behavior of ablation plumes produced with ultrashort laser pulses, SPIE Vol. 3935, *Laser plasma generation and diagnostics*, pp. 97-103.

Koopman, D. W., 1971, Langmuir probe and microwave measurements of the properties of streaming plasmas generated by focused laser pulses, Phys. Fluids **14**: 1707- 1716.

Li, P., Lim, D. and Mazumder, J., 2002, Diagnostics of nanosecond dynamics of the plasma produced during KrF excimer laser ablation of zirconia in vacuum, *J. Appl. Phys.* **92**:666-671.

Lindley, R. A., Gilgenbach, R. M., Ching, C. H., Lash, J. S., Doll, G. L., 1994, Resonant holographic interferometry measurements of laser ablation plumes in vacuum, gas, and plasma environments, *J. Appl. Phys.* **76**: 5457-5472.

Lowndes, D. H., 1998, Growth and doping of compound semiconductor films by pulsed laser ablation, *Laser Ablation and Desorption*, Miller, J. C. and Haglund, R. F. eds., Experimental Methods in the Physical Sciences vol. *30*. Academic Press, New York, pp. 475 - 571.

Lunney, J. G. and Jordan, R., 1998, Pulsed laser ablation of metals, *Appl. Surf. Sci.*, **127-129**: 941-946.

Millon, E., Perrière, J., Déforneau, R. M., Déforneau, D., Albert, O. and Etchepare, J., 2003, Femtosecond pulsed-laser deposition of $BaTiO_3$, *Appl. Phys. A* **77**:73-80.

Nolte, S., Kamlage, G., Korte, F., Bauer, T., Wagner, T., Ostendorf, A., Fallnich, C. and Welling, H., 2000, Microstructuring with femtosecond lasers, *Adv. Eng. Mat.* **2**: 23-27.

Perez, D. and Lewis, L. J., 2003, Molecular-dynamics study of ablation of solids under femtosecond laser pulses, *Phys. Rev. B* **67**: 184102-1 –184102-15.

Phipps, Jr.,C. R., Turner, T. P., Harrison, R. F., York, G. W., Osborne, W. Z., Anderson, G. K., Corlis, X. F., Haynes, L. C., Steele, H. S., Spicochi, K. C. and King, T. R., 1988, Impulse coupling to targets in vacuum by KrF, HF, and CO_2 single-pulse lasers, *J. Appl. Phys*: **64**, 1083-1098.

Proyer, S. and Stangl, E., 1995, Time-integrated photography of laser-induced plasma plumes, *Appl. Phys. A* **60**: 573-580.

Qian, F., Craciun, V., Singh, R. K., Dutta, S. D. and Pronko, P. P., 1999, High intensity femtosecond laser deposition of diamond-like carbon thin films, *J. Appl. Phys.* **86**: 2281-2290.

Saenger, K. L. 1994, Angular distribution of ablated material, *Pulsed Laser Deposition of Thin Films,* Chrisey, D. B. and Hubler, G. K. eds., Wiley, New York, pp. 199-227.

Schou, J. 2006, Laser beam-solid interactions: Fundamental aspects, in: *Materials Surface Processing by Directed Energy Techniques,* Pauleau, I. ed., Elsevier, pp. 33 – 62.

Scuderi, D., Albert, O., Moreau, D., Pronko, P. P. and Etchepare, J., 2005, Interaction of a laser plume with a second time delayed femtosecond pulse, *Appl. Phys. Lett.* **86:** 071502-1 - 071502-3.

Sibold, D. and Urbassek, H. M., 1991, Kinetic study of pulsed desorption flows into vacuum, *Phys. Rev. A* **43**: 6722- 6734.

Singh, R. K. and Narayan, J., 1990 Pulsed-laser evaporation technique for deposition of thin films: Physics and theoretical model, *Phys. Rev. B* **41**: 8843-8859.

Sokolowski-Tinten, K., Bialkowski, J., Cavalleri, A., von der Linde, D., Oparin, A., Meyer-ter-Vehn, J. and Anisimov, S. I., 1998, Transient states of matter during short pulse laser ablation, *Phys.Rev. Lett.* **81**:224-227.

Thestrup, B., Toftmann, B. ,Schou, J., Doggett, B. and Lunney, J. G., 2002, Ion dynamics in laser ablation plumes from selected metals at 355 nm, *Appl. Surf. Sci.* **197-198**: 175-180.

Toftmann, B., Schou, J., T. N. Hansen and Lunney, J. G., 2000, Angular distribution of electron temperature and density in a laser-ablation plume, *Phys. Rev. Lett.* **84**: 3998-4001.

Toftmann, B., Schou, J. and Lunney, J. G., 2003, Dynamics of the plume produced by nanosecond ultraviolet laser ablation of metals, *Phys. Rev. B* **67:** 104101-1–104101-5.

Toftmann, B., Amoruso, S., Schou, J. and Lunney, J. G., 2006, The propagation of laser ablation plume ions in a background gas *(unpublished).*

Tyuina, M., and Leppävuori, S., 2000, Effects of laser fluence, size, and shape of the laser focal spot in pulsed laser deposition using a multielement target, *J. Appl. Phys.* **87**: 8132-8142.

VanRompay, P. A., Nantel, M. and Pronko, P. P., 1998, Pulse-contrast effects on energy distributions of C^{1+} to C^{4+} ions for high-intensity 100-fs laser-ablation plasmas, *Appl. Surf. Sci.* **129**: 1023-1028.

Vertes, A., Gijbels, R. and Adams, F., 1993, *Laser Ionization Mass Analysis,* Chemical Analysis Series, vol 124, John Wiley, New York.

Vidal, F., Johnston, T. W., Laville, S., Barthelemy, O., Chaker, M., Le Drogoff, B., Margot, J. and Sabsabi, M., 2001, Critical-point phase separation in laser ablation of conductors, *Phys. Rev. Lett.* **86**: 2573-2576.

Weaver, I. and Lewis, C. L. S.,1996, Polar distribution of ablated atomic material during pulsed laser deposition of Cu in vacuum: Dependence on foused laser spot size and power density, *J. Appl. Phys.* **79**: 7216-7222.

Willmott, P. R. and Huber, J. R., 2000, Pulsed laser vaporization and deposition, *Rev. Mod. Phys.* **72**: 315 – 328.

Wood, R. F.,Chen, K. R., Leboeuf, J. N., Puretzky, A. A. and Geohegan, D. B. 1997, Dynamics of plume propagation and splitting during pulsed-laser ablation, *Phys. Rev. Lett.* **79**: 1571-1574.

Wood, R. F., Leboeuf, J. N., Geohegan, D. B., Puretzky, A. A. and Chen, K.R. 1998, Dynamics of plume propagation and splitting during pulsed-laser ablation of Si in He and Ar, *Phys. Rev. B* **58:** 1533-1543.

Ye, M. and Grigoropoulos, C. P., 2001, Time-of-flight and emission spectroscopy study of femtosecond laser ablation of titanium, *J. Appl. Phys.* **89** : 5183-5190.

Zel'dovich, Ya. R. and Raizer, Yu. P., 2001, *Physics of Shock Waves and High-Temperature Hydrodynamic Phenomena*, Dover, Cambridge, Massachusetts.

Chapter 5

NEW ASPECTS OF LASER-INDUCED IONIZATION OF WIDE BAND-GAP SOLIDS

Vitali E. Gruzdev
S.I.Vavilov State Optical Institute, Birzhevaya Liniya 12, St.Petersburg, 199034, Russia

1. INTRODUCTION

Laser ionization plays an important role in ablation of transparent solids connected with at least three reasons. First, the ionization is widely assumed to be almost inertia-free, being the fastest process taking place under action of laser radiation in the solids. Second, laser-induced dynamics of electrons influences the initial stages of the ablation since the electrons are much less inertial than crystal-lattice ions, giving the fastest response to laser action. Third, the ionization determines the value of ablation threshold since if it is successful, then density of free electrons in irradiated areas becomes high enough to provide effective absorption of radiation energy and its transfer to the lattice ions, resulting in destruction of the crystal lattice and the ablation.

Thus, understanding of the laser ionization is of great importance for study and many applications of laser ablation. Due to this, we devote this chapter to the problem of the ionization of transparent solids. We consider only one mechanism of the ionization connected with electron transitions from valence to conduction band induced by direct action of electric field of laser radiation and referred to as photoionization. The problem is very traditional and has been studied from the early days of laser epoch (see, for example, a review by Nathan, et al., 1985), but there are several reasons to return to it.

First, recent fast and intensive development of femtosecond laser systems has induced a new wave of interest to laser-solid interactions (Stuart, et al.,

1996; von der Linde and Schüller, 1996; Ashkenasi, et al., 1996; Kautek et.al., 1996; Lenzner et.al., 1998; Tien, et al., 1999; Stoian, et al., 2000; von der Linde and Sokolovski-Tinten, 2000; Schaffer, et al., 2001; Quere, et al., 2001), in particular, ionization (Audebert, et al., 1994; Guizard, et al., 1996; Du, et al., 1996; Li, et al., 1999). Specific features of femtosecond interactions are 1) short pulse duration comparable or even less than the characteristic time of energy transfer from electrons to phonons, and 2) possibility to reach high laser intensity (about 10 TW/cm^2) without damaging the transparent solids. Due to these features, experimental data on the femtosecond action on the solids differ much from those obtained in the case of longer pulses. Together with that, traditional theoretical models of the photoionization (Keldysh, 1965; Bychkov and Dykhne, 1970; Kovarskii and Perlin, 1971; Jones and Reiss, 1977) and the avalanche ionization (Stuart, et al., 1996; Apostolova and Hahn, 2000) with small or even no modifications are utilized for interpreting and understanding the experimental results (Stuart, et al., 1996; von der Linde and Schüller, 1996; Ashkenasi, et al., 1996; Kautek et al., 1996; Lenzner et al., 1998; Tien, et al., 1999; Stoian, et al., 2000; von der Linde and Sokolovski-Tinten, 2000; Schaffer, et al., 2001; Quere, et al., 2001; Audebert, et al., 1994; Guizard, et al., 1996; Du, et al., 1996; Li, et al., 1999;). The models were developed for longer pulses and lower intensities, and their validity is questionable in the case of the femtosecond interactions due to the mentioned features. That situation forces us to return to the problem of description of fundamental mechanisms of laser ionization in nonmetallic crystals in order to spread them correctly to the domain of femtosecond laser pulses. We concentrate on the influence of one of the features of the interactions – high laser intensity.

Second, current understanding of the photoionization is based on the Keldysh formula for the ionization rate (Keldysh, 1965) and its modifications (Bychkov and Dykhne, 1970; Kovarskii and Perlin, 1971; Jones and Reiss, 1977). Certain assumptions (Gruzdev, 2004, 2006a) underlie the corresponding calculations making them valid only in the case of laser intensity below approximately 1 TW/cm^2 for visible radiation and wide band-gap materials (Gruzdev, 2006a). It means that attempts of utilizing the formulae in cases of higher intensities can lead to unpredictably wrong results, and the Keldysh approach should be properly modified. Some modifications of the traditional avalanche-ionization model have already been done (for example, in Kaiser, et al., 2000; Rethfeld, 2004, 2005), but almost nothing (Nathan, et al., 1985; Gruzdev, 2004) has been done to improve the photoionization model utilized in the same formulation as it was presented some 40 years ago by Keldysh (1965). In this chapter we present an attempt to modify the Keldysh approach and extend it to the range of high intensities. The Keldysh model is not chosen occasionally; it is based on a very powerful physical idea that can be fruitful for general understanding of the photo-ionization in a wide range of laser parameters. Due to this, we start our considerations with analysis of the physical processes underlying the Keldysh calculations.

Some notes must be made before further discussions. First, theoretical investigations of the photoionization are connected with complicated and extensive mathematical calculations. We try to avoid utilizing huge math and concentrate on discussion of physical background of the theoretical model. Expressions for the ionization rates, together with calculation details, can be found in several papers (Keldysh, 1965; Bychkov and Dykhne, 1970; Gruzdev, 2005, 2006a, 2006b), but they are quite complicated and can hardly help in analysis of the considered problems. Second, to make our consideration transparent and illustrative, we utilize some simplifications: the crystals are assumed to be direct-gap (i.e., minimum of forbidden band lies in the center of the first Brillouin zone) with cubic symmetry. Influence of these assumptions is discussed in Section 4. Third, we consider motion of electrons under action of laser radiation while, strictly speaking, one should analyze motion of an electron-hole pair with the mass equal to reduced effective electron-hole mass. It is done so in calculations underlying our discussion (Gruzdev, 2006b), but for simplicity and brevity we mention electrons only.

Before continuing analysis of the problem mentioned above we should specify the term "band structure" that is often referred to in this chapter and is one of the key points for our considerations. The band structure $\varepsilon(\vec{p})$ is to be understood as energy–momentum relation for an electron-hole pair in absence of any perturbation to the crystal (Keldysh, 1965; Bonch-Bruevich and Kalaschnikov, 1982), i.e., it is the sum of energies of an electron $\varepsilon_e(\vec{p})$ in the conduction band and the corresponding hole $\varepsilon_h(\vec{p})$ in the valence band, considered as a function of quasi-momentum $\vec{p}$:

$$\varepsilon(\vec{p}) = \varepsilon_e(\vec{p}) - \varepsilon_h(\vec{p}). \qquad (1)$$

2. PHYSICAL MODEL

The effective result of any ionization model for solids is a formula for the ionization rate describing its dependence on laser and material parameters. The formula can be derived in the framework of various approaches among which two are the most popular and developed. The first one is based on the classical high-order perturbation method (for example, see Nathan, 1985;

Braunstein and Ockman, 1964; Yee, 1971, 1972). Its main drawback is connected with fast increasing of calculation problems with increasing of the number of photons simultaneously absorbed by a crystal electron. Therefore, the classical perturbation approach has never been applied to the case of multiphoton ionization of the order of 5 and higher (Yee, 1971, 1972).

The second approach was first proposed by L. V. Keldysh (1965). It provides the multiphoton-ionization rate of any order in the first approximation of a very specific formulation of the perturbation theory. The most important point of all the Keldysh-type approaches (Keldysh, 1965; Bychkov and Dykhne, 1970; Kovarskii and Perlin, 1971; Jones and Reiss, 1977) is the special form of functions describing initial and final electron states for the interband transitions. The functions are essentially Bloch ones, but they do not describe steady states and depend on time through field-modified quasi-momentum. They have a deep physical meaning to be analyzed in this section. To do this, we start with the simplest case of interaction of crystal's electrons with constant homogeneous electric field. It allows understanding of a more complicated case of time-dependent field as well as getting some important qualitative results and conclusions.

2.1 Constant electric field

The constant electric field modifies the quasi-momentum of electrons $\vec{p}$ at constant rate:

$$\vec{p}(t) = \vec{p}(0) - e\vec{F}t, \tag{2}$$

where $\vec{F}$ is electric field strength. Equation (2), first derived by Bloch (1928), holds for the whole Brillouin zone (BZ) except its boundaries with no respect to a particular form of electron functions (Bloch, 1928; Houston, 1940; Wannier, 1960, 1962; Adams, 1952; Adams and Argyres, 1956; Avron, 1976), and not only for the wave packets but for single Bloch functions due to the space periodicity of the crystal potential (Bloch, 1928; Houston, 1940). It is similar to the classical equation of motion with constant acceleration, but in the case of an electron in a crystal it corresponds to oscillatory variations of electron state and quasi-momentum as it is illustrated by Fig. 1. Really, all values of quasi-momentum corresponding to different physical states of the electron in an energy band (e.g., valence) of the crystal are inside the first BZ. The number of the states in the band is huge, but it is limited. The electron can move from one of the states to another under action of the field according to Eq. (2) as long as its quasi-momentum varies inside the BZ. On the other hand, field-induced variation of electron quasi-momentum is not limited (Fig. 1, line 1-1') and, increasing at a constant rate, its value can go out of the first BZ near one of its boundaries at certain instant.

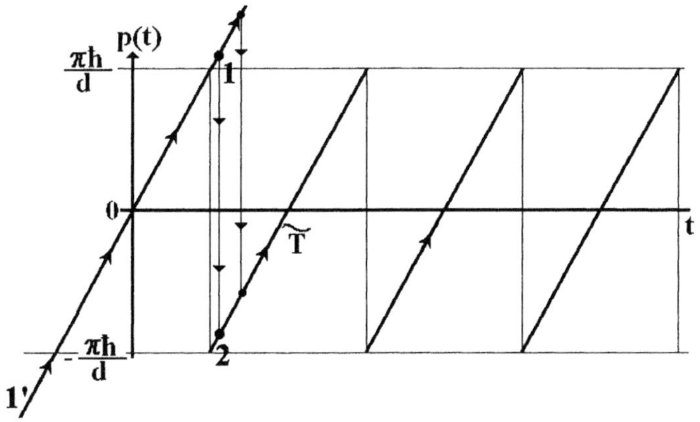

Figure 1. Illustration of the unlimited variation of electron quasi-momentum at constant rate (line 1-1') according to Eq. (2), and the equivalent saw-type oscillations in the reduced BZ. Electric field is directed along the principle axis of the crystal with lattice constant *d*.

It means that the electron has passed all the states inside the first BZ and has come to a state outside the first BZ (state 1 in Fig. 1) which is absolutely equivalent to the state inside the zone near its opposite boundary (state 2 in Fig. 1), if electric field is directed along one of the principle crystal axes. In this case the transition to the state beyond one of the BZ boundaries can be considered as a jump to the physically equivalent state inside the zone near its opposite boundary. This is usually associated with Bragg reflection of electrons at the boundaries of the first BZ (Bloch, 1928; Houston, 1940) occurring each time when electron reaches the edges of the first BZ if the component of its quasi-momentum normal to the field direction is very small or zero.

As a result, the electron states and energies oscillate in time, as if the quasi-momentum varied periodically within the first BZ in a saw-like manner with the slope given by Eq. (2) (Fig. 1). In other words, the continuous increasing of electron quasi-momentum in time at constant rate is equivalent to the saw-like variations with the same slope since both of them give similar time variations of the electron states. In this meaning the field-induced variations of electron quasi-momentum can be represented by the saw-type oscillations. Their period can easily be found if *d* is the lattice constant corresponding to the principal axis along which the field is directed. Then the size of the first BZ is $2\Delta_B = 2\pi\hbar/d$, and the period reads as

$$\tilde{T} = \frac{2\Delta_B}{\Delta_{FS}} = \frac{2\pi\hbar}{eFd}, \tag{3}$$

where we have introduced amplitude of field-induced variations of quasi-momentum $\Delta_{FS} = eF$. Thus, acceleration of crystal electrons by constant electric field are coupled to the saw-type oscillations of their quasi-momentum in the first BZ. Both the effects can be taken into account by introducing the time-dependent quasi-momentum (2) into Bloch functions (Bloch, 1928; Houston, 1940; Wannier, 1960, 1962; Adams, 1952):

$$\Psi_n(\vec{r},t) = u_n(\vec{p}_s[t],\vec{r})\exp\left[\frac{i}{\hbar}\left\{\vec{p}_s[t]\vec{r} - \int_0^t \varepsilon_n(\vec{p}_s[\tau])d\tau\right\}\right]. \tag{4}$$

The functions (4) show that the oscillations can be realized as periodical time variations of both the wavelength and the amplitude of the Bloch waves i.e., the electron states are nonsteady, but the electron oscillations in the BZ are not always accompanied by oscillatory motion of electrons in the coordinate space (Lyssenko, et al., 1997; Sudzius, et al., 1998).

Coming back to the ionization, we note that among all types of ionization only tunneling can take place in the framework of the considered model, since electron collisions and avalanche ionization are excluded from our considerations. Thus, the tunneling ionization must be associated with the saw-type electron oscillations in the first BZ. Their shape does not change with increasing of electric-field strength, while their period (3) decreases.

We should note that the considerations of this section were performed for the case of electric field directed along one of the principal crystal axes. We imply this field orientation in all considerations related to the ionization in the following sections to make our explanations more transparent. On the other hand, Keldysh (1958) has shown that expressions for the ionization rate are not influenced by orientation of electric field in the framework of his approach, while the motion of the electrons can be complicated and even aperiodic if the field is not directed along one of the crystal axes.

2.2 Time-dependent electric field

Now it is easier to understand how the electrons oscillate under the action of time-dependent electric field and how that effect is related to ionization. In this case, the quasi-momentum oscillations are described as follows:

$$\vec{p}(t) = \vec{p}_0 - \frac{e\vec{F}}{\omega}\sin(\omega t), \tag{5}$$

where we assumed the electric field to vary in time as cosine:

$$\vec{F}(t) = \vec{F}\cos(\omega t). \tag{6}$$

The corresponding electron functions

$$\psi(\vec{r},t) = u_n(\vec{p}(t),\vec{r})\exp\left\{\frac{i}{\hbar}\left[\vec{p}(t)\vec{r} - \int_0^t \varepsilon_n[\vec{p}(\tau)]d\tau\right]\right\}, \tag{7}$$

are similar to (4) except that eq. (5) is utilized in them. Both functions (4) and (7) are approximate solutions of the Schrödinger equation (Houston, 1940; Gruzdev, 2006a) corresponding to negligibly small probability of interband electron transitions.

Considering the ratio of the amplitude $\Delta_F = eF/\omega$ of field-induced variations of quasi-momentum to the half-width of the BZ $\Delta_B = \pi\hbar/d$, one can distinguish the following types of the oscillations. First, if $\Delta_F/\Delta_B \leq 1$, electron quasi-momentum varies harmonically with frequency equal to the input laser frequency ω (Fig. 2, curves 1, 2). It is obvious from Fig. 2 that the maximum amplitude of external electric field for which the quasi-momentum oscillations are harmonic reads as follows:

$$F_{TR} = \frac{\pi\hbar\omega}{e\,d}. \tag{8}$$

To get the feeling of the magnitude of laser intensity corresponding to (8) one can, for example, take lattice constant $d = 0.5628$ nm and refractive index $n_0 = 1.535$, characteristic of NaCl (Poole, et al., 1975; Vaidyanathan, et al., 1980) at laser wavelength of 800 nm and get the transitional intensity

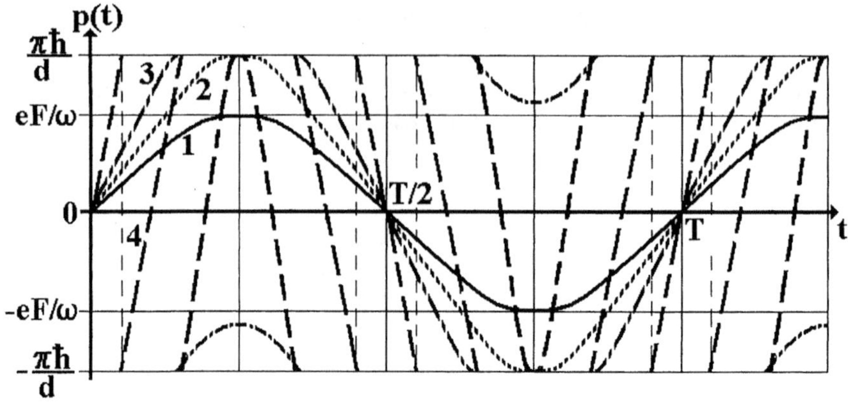

Figure 2. Oscillations of electron quasi-momentum induced by harmonic electric field (6). Solid curve 1 – electric-field amplitude is below the level (8); dotted curve 2 – it is equal to the value (8); dash-dotted curve 3 – it slightly exceeds the level (8); dashed curve 4 – electric field exceeds the level (8) much.

I_{TR} = $5.087 \cdot 10^{12}$ W/cm^2. The value of the adiabatic parameter γ (Keldysh, 1965), corresponding to (8), reads as follows:

$$\gamma_{TR} = \frac{d\sqrt{m\Delta}}{\pi\hbar}. \qquad (9)$$

For the considered material parameters ($m = 0.6m_{oe}$, $\Delta = 8.97$eV, see, for example, Vaidyanathan, et al., 1980), it gives $\gamma_{TR} = 1.505$.

The character of the electron oscillations changes if electric field exceeds the value F_{TR} given by Eq. (8). Electron quasi-momentum varies harmonically, but its amplitude exceeds the size of the first BZ. Electron states change, as if the quasi-momentum varied in a non-harmonic way (depicted by curve 3 in Fig. 2) due to the effect of Bragg-type reflections at the edges of the BZ. Further increasing of electric-field strength results in transition to the oscillations very similar to those of the saw-type discussed in case of constant electric field, since quasi-momentum varies in saw-type manner, for the most part, of the period of field oscillation $T = 2\pi/\omega$ (compare Fig. 1 and curve 4 in Fig. 2). Qualitative conditions for this transition between the different types of the oscillations can be derived (Gruzdev, 2004, 2006a) by introducing eq. (3) into Eq. (5). For example, the following condition determines the saw-type regime:

$$\omega\widetilde{T} << 1 . \tag{10}$$

From Fig. 2 and the discussion above it is obvious that (10) can be expressed through the ratio of electric field F to the transitional electric field:

$$\frac{F_{TR}}{F} << 1 . \tag{11}$$

One should also note that the same transition between the oscillation regimes can be induced by variations of laser frequency at fixed electric field. In analogy to (8) one can introduce (Gruzdev, 2004, 2006a) the transitional frequency ω_{TR} at fixed laser intensity.

Thus, in contrast to the case of constant electric field, the type of the considered oscillations depends on amplitude of electric field and frequency, and can vary qualitatively (Fig. 2). If the amplitude is below the level (8), electron quasi-momentum oscillates harmonically, and the only dominating type of ionization is the multiphoton one. It implies the harmonic oscillations of quasi-momentum to be associated with the multiphoton ionization. On the other hand, in section 2.1 we have shown the saw-type oscillations satisfying the eqs. (10) and (11) to be associated with the tunneling ionization. Thus, the presented analysis allows attributing each of the considered regimes of ionization to the corresponding type of oscillations. Then, the transition from the harmonic to the saw-type oscillations should be associated with the transition from the multiphoton to the tunneling ionization.

Since the functions (7) are utilized for description of the initial and the final states of electrons for interband transitions in Keldysh-type approaches (Keldysh, 1965; Bychkov and Dykhne, 1970; Kovarskii and Perlin, 1971; Jones and Reiss, 1977), we come to a natural conclusion that the physical model underlying the approaches is based on taking into account the influence of laser-induced electron oscillations in the first BZ on probability of interband transitions. Correspondingly, the type of the oscillations determines the regime of ionization. This deep physical idea leads to several important qualitative predictions to be discussed in the next section.

2.3 Some qualitative results

Several qualitative conclusions result from the discussed associating of the ionization regimes with the corresponding types of electron oscillations. For example, it implies increasing of ionization rate with coming from the multiphoton to the tunneling ionization, due to increasing of laser intensity. Really, the rate of the interband transitions is proportional to the product of probability of a single multiphoton transition (or tunneling through slowly varying triangle barrier) and the number of oscillations per second (Keldysh, 1958). The higher the laser intensity, the larger the probability of the multiphoton transition (or tunneling), and the smaller is the thickness of the potential barrier $s = \Delta/(eF)$ through which the electrons tunnel. The number of oscillations per second is the inverse period of the oscillations which stays constant in the multiphoton regime, or increases in the tunneling regime according to (3).

Our qualitative consideration allows estimating the upper limit of laser intensity for which Keldysh-type models of ionization (Keldysh, 1965; Bychkov and Dykhne, 1970; Kovarskii and Perlin, 1971; Jones and Reiss, 1977) are valid. It follows from Eq. (8) for the transitional electric field. Indeed, if energy-momentum relation (1) chosen for calculations of the ionization rate does not describe the band structure throughout the whole BZ, one cannot properly account for contribution of electron oscillations with amplitude Δ_F approaching and exceeding the half-width of the BZ. It means that electric field of laser radiation in those calculations must not exceed the value given by (8) what leads to the following upper estimation of the allowed intensity:

$$I < I_{TR} = \left(\frac{\pi \hbar \omega}{e d} \right)^2 \frac{n_0 c_0 \varepsilon_0}{2}, \tag{12}$$

where $\varepsilon_0 = 8.8542 \cdot 10^{-12}$, F/m is the electric constant, n_o is the refractive index at laser wavelength, and $c_o = 2.9979 \cdot 10^8$ m/s is the speed of light in vacuum. Utilizing data from Poole, et al. (1975) and Vaidyanathan, et al., (1980) it is easy to show that I_{TR} is about a few TW/cm^2 for wide-band gap materials and visible radiation (see estimation after Eq. (8)). Thus, Eq. (12) means that the Keldysh-type formulae obtained by Keldysh (1965), Bychkov and Dykhne (1970), Kovarskii and Perlin (1971), and Jones and Reiss (1977) for the ionization rate are valid only for intensities well below the I_{TR}, since the authors utilized Kane-type (Keldysh, 1965; Bychkov and Dykhne, 1970) or parabolic (Jones and Reiss, 1977) energy-momentum relations in deriving their formulae. The relations describe only the central part of the first BZ (Kane, 1957) and are not valid near its edges. It implies that the tunneling ionization by time-dependent field of laser radiation is not described properly by any of the existing theoretical models of ionization in solids.

Obviously, the probability of multiphoton ionization is determined by the band structure in the center of the first BZ since it corresponds to the case

$F << F_{TR}$ and electron oscillations near the center of the zone. On the other hand, the transitional and tunneling regimes of ionization must depend on the band structure of the whole BZ since they correspond to the condition $F \geq F_{TR}$, and oscillations of quasi-momentum over the whole BZ. The latter means that approximation of constant effective mass utilized in the Keldysh-type models (Keldysh, 1965; Bychkov and Dykhne, 1970; Kovarskii and Perlin, 1971; Jones and Reiss, 1977) is not valid for reliable description of the tunneling regime which is influenced by details of energy band near the edges of the BZ. Moreover, it is reasonable to expect that contribution of the Bragg-type reflections at the edges of the BZ must become more significant with increasing of laser intensity and passing to the tunneling ionization.

The contribution can be understood from the following analysis. Let us consider crystal electrons accelerated by a relatively weak electric field (6). The electrons oscillating near the center of the first BZ (between points A and B in Fig. 3) have an initial (i.e., in absence of the field) value of quasi-momentum close to zero. They are mainly accelerated by the field, obtaining values of quasi-momentum larger than the initial one, for most of the oscillation period. The electric field increases their initial energy by average energy of oscillations, i.e., pondermotive potential (Fig. 3, rich dotted area near the line AB). This situation absolutely differs from that for electrons having large initial values of quasi-momentum, i.e., occupying initial states close to the edges of the BZ. They oscillate near the edges (lines CD and C'D' in Fig. 3) and are mainly slowed down by the field since they obtain values of quasi-momentum smaller than the initial one for most of the oscillation period. The pondermotive potential is negative for them, and their average energy is decreased by the field (rarely dotted area near the lines CD and C'D'). Thus, the field-induced oscillations of the electrons result in specific distribution of the pondermotive potential – it is positive near the center of the zone and negative near its edges, i.e., the initial large energy of fast electrons decreases while the initial small energy of the slow electrons increases under action of the electric field. Obviously, there can be a value of the field for which the pondermotive potential compensates the initial energy difference between the fast and the slow electrons and make the distribution of the total energy (i.e., the sum of the initial energy in absence of the field and the pondermotive potential) homogeneous over the BZ. This situation can be referred to as the flattening effect (Gruzdev, 2005, 2006b). Obviously, it is impossible without the Bragg-type reflections at the edges of the BZ since in that case the fast electrons are not reflected back to the zone and can move to higher states with larger energy. Correspondingly, their total energy can either decrease slower than it is required for the flattening, or increase.

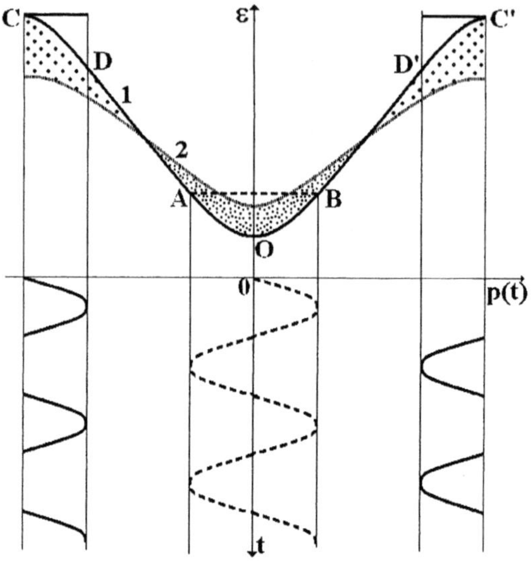

Figure 3. A sketch of distribution of pondermotive potential among the fast and the slow electrons in the BZ. Solid curve in the lower part below the lines CD and C'D' depicts oscillations of a fast electron with initial value of quasi-momentum lying exactly at the left edge of the BZ. Dashed curve in the lower part below the line AB depicts oscillations of a slow electron with zero initial quasi-momentum. Solid curve 1 depicts the initial band structure. Dotted curve 2 depicts the distribution of total energy of oscillating electrons.

Thus, the Bragg reflections can lead to the flattening of the energy-momentum distribution of the oscillating electrons. It must depend on the initial potential, i.e., band structure, in which the electrons oscillate and can result in specific behavior of ionization rate at intensity approaching the level corresponding to the flattening effect. The key questions are a) how large is the intensity? and b) what is the specific feature of the ionization rate resulting from the flattening effect? To answer the first question one has to do calculations since the value of the "flattening effect" intensity can hardly be predicted from the qualitative analysis. On the other hand, one should expect increasing of the ionization rate at that intensity. If the total energy of the oscillating electrons is the same over the whole BZ, probability of the interband transition must be the same for all the electrons with no respect to their initial quasi-momentum.

3. CALCULATIONS OF THE IONIZATION RATE

To answer the questions we analyze the ionization rate derived in the framework of the Keldysh procedure (Keldysh, 1965; Gruzdev, 2005, 2006b) for three band structures. The Keldysh procedure is utilized due to its strong advantage over other approaches; it is based on the powerful physical idea discussed above, and it allows obtaining analytical expressions for the ionization rate. Utilizing of the Keldysh approach implies the following assumptions (Gruzdev, 2004, 2006a) to be accepted:

- All collisions of electrons with other particles resulting in variations of electron quasi-momentum are neglected,
- All perturbations of band structure and material parameters induced by ionization and electric field of laser radiation are negligibly small,
- Ionization rate is small enough (Gruzdev, 2004, 2006a),
- Contributions of excitons are neglected,
- Radiation is considered to be monochromatic.

The ionization rates are calculated for the following bands (Fig. 4):

1) non-parabolic Kane's band (Keldysh, 1965; Kane, 1957)

$$\varepsilon_{NP}(\vec{p}) = \Delta \sqrt{1 + \frac{p^2}{m\,\Delta}}, \tag{13}$$

utilized by Keldysh in derivation of his formula and corresponding to narrow-gap semiconductors (Keldysh, 1965);

2) the parabolic band

$$\varepsilon_P(\vec{p}) = \Delta\left(1 + \frac{p^2}{2\,m\,\Delta}\right), \tag{14}$$

widely utilized for description of the central part of the first Brillouin zone and having become a standard approximation in many problems of solid-state physics (Bonch-Bruevich and Kalaschnikov, 1982); and

3) the cosine model

$$\varepsilon_{COS}(\vec{p}) = \Delta\cdot\left(1 + A - A\cos\left(\frac{d}{\hbar}p_x\right)\cos\left(\frac{d}{\hbar}p_y\right)\cos\left(\frac{d}{\hbar}p_z\right)\right). \tag{15}$$

Here m is reduced effective electron-hole mass, Δ is band gap, and d is lattice constants along the principle axes of the cubic crystal lattice (all are in SI units). Isotropic bands are utilized for simplicity since this approximation does not qualitatively change the results presented below. The Kane and the parabolic models are utilized to compare the ionization rate derived for the

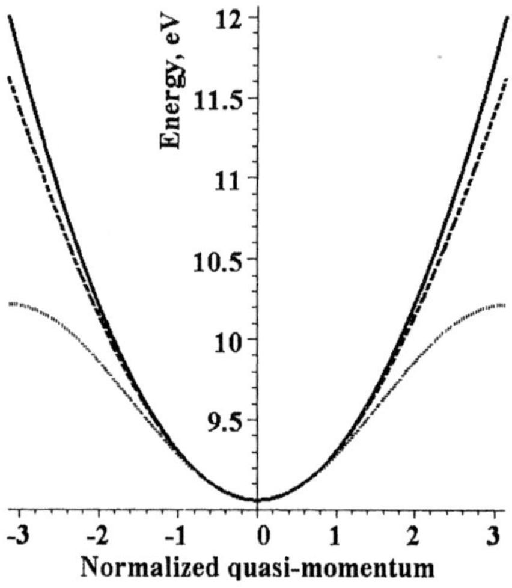

Figure 4. Solid curve – the parabolic band structure (14), dashed curve – non-parabolic structure (13), dotted curve – the cosine band structure (15) over the first Brillouin zone. The normalized quasi-momentum varies from -π to π. Band gap is 8.97 eV.

cosine band with the well-known results. The cosine model is constructed so as to meet the following requirements (Bonch-Bruevich and Kalaschnikov 1982):

- It turns into the parabolic model (14) for small values of p.
- It is periodic with space period of the crystal lattice.
- It is smooth at each point of the first Brillouin zone.

The key feature of the cosine relation is that it describes the energy band over the whole BZ, and allows correct treating of the Bragg-type reflections at the edges of the zone since it satisfies the necessary condition at the edges of the BZ along the principle crystal axes (Bloch, 1928; Bonch-Bruevich and Kalaschnikov 1982):

$$\vec{\nabla}_p \varepsilon(\vec{p}) = 0 \,. \tag{16}$$

The cosine model has other important characteristics: a) It is close to the band structure of alkali halides having almost flat top of the valence band and almost cosine-like bottom of the conduction band (Poole, et al., 1975). b) The cosine model corresponds to the first two terms of general Fourier series of the band structure (Bonch-Bruevich and Kalaschnikov 1982). The cosine band structure corresponds to the approximation of strongly bonded electrons and is more suitable for description of dielectrics and wide band-gap semiconductors with a dominating ionic type of chemical binding.

We should stress that all the other band models utilized by authors of earlier papers (Keldysh, 1965; Bychkov and Dykhne, 1970; Kovarskii and Perlin, 1971; Jones and Reiss, 1977) do not meet the condition (16) and, thus, do not allow correct describing of the Bragg-reflection effect.

Expressions (13) and (15) turn into the parabolic relation (14) for small values of quasi-momentum. The difference between them is connected with higher order corrections to the parabolic model, playing a significant role far from the center of the first BZ (Fig. 4), and influencing the ionization rate at different values of laser intensity depending on particular band structure. Performing simple calculations (Gruzdev, 2005, 2006b), one gets the upper limit of laser intensity and the lower limit of the Keldysh parameter γ for which the models (13) and (14) do not differ much from one another:

$$I << I^{*}_{LIM} \propto F^{2}_{LIM} = \frac{4 \cdot \omega^{2} \cdot m \cdot \Delta}{e^{2}}, \text{ and } \gamma >> \gamma^{*}_{LIM} = 0.5 . \quad (17)$$

The corresponding limits for the models (14) and (15) are as follows:

$$I << I^{**}_{LIM} \propto \frac{12\, \hbar^{2}\, \omega^{2}}{d^{2}\, e^{2}}, \text{ and } \gamma >> \gamma^{**}_{LIM} = \frac{d\sqrt{m\,\Delta}}{12\,\hbar} . \quad (18)$$

Here the adiabatic parameter is given by the following expression (Keldysh, 1965):

$$\gamma = \frac{\omega\sqrt{m\,\Delta}}{e\,F} . \quad (19)$$

Assuming, for example (Poole, et al., 1975; Vaidyanathan, et al., 1980), d =5.628 A, m=0.6 of free-electron mass m_{0e} and band gap $\Delta = 8.97$ eV (parameters of NaCl), one gets γ^{**}_{LIM}=0.39.

Coming to the final result, we omit details of calculations, referring readers interested in them to our papers (Gruzdev, 2005, 2006b). Ionization rates for the band models (13) – (15) are depicted in Fig. 5 as functions of laser intensity. They coincide with excellent accuracy for low intensity (i.e., in the multiphoton regime) since in that case all the models are reduced to the parabolic one. The most interesting feature of the cosine dispersion relation is occurring of a singularity – the corresponding ionization rate increases up to infinity as laser intensity approaches a limited value I_{TH}, given as follows:

$$I_{TH} = \left(\frac{\hbar \cdot \omega}{e \cdot d} \cdot \xi_1 \right)^2 \cdot \frac{c_0 \cdot \varepsilon_0 \cdot n_0}{2}, \tag{20}$$

where ξ_1=2.4048256 is the first zero of Bessel function J_0. The intensity (20) is referred to as singularity threshold. Substituting parameters corresponding to NaCl ((Poole, et al., 1975; Vaidyanathan, et al., 1980)), one gets $I_{TH} = 8.404 \cdot 10^{12}$ W/cm^2. We should note that the value given by (20) is well below the characteristic intensity I_{MAX} corresponding to unperturbed periodic electric fields induced by crystal ions (Gruzdev, 2005, 2006b). For example, for NaCl one gets $I_{MAX} = 5.176 \cdot 10^{13}$ W/cm^2.

Results of the calculations (Gruzdev, 2005, 2006b) also suggest that in case of the cosine model, the ionization rate depends on laser intensity through modified adiabatic parameter χ instead of the Keldysh adiabatic parameter (19). The modified parameter is determined by electric field strength F, laser frequency ω, and crystal constant d in the following way:

$$\chi = \frac{e F d}{\hbar \omega}, \tag{21}$$

which is very natural from the view-point of the discussions in Section 2.2, since the parameter (21) can be expressed through Δ_F and Δ_B:

$$\chi = \pi \frac{\Delta_F}{\Delta_B}, \tag{22}$$

and is proportional to the ratio of amplitude of field-induced variations of quasi-momentum to half-width of the BZ. It plays the role absolutely similar to the Keldysh parameter (19).

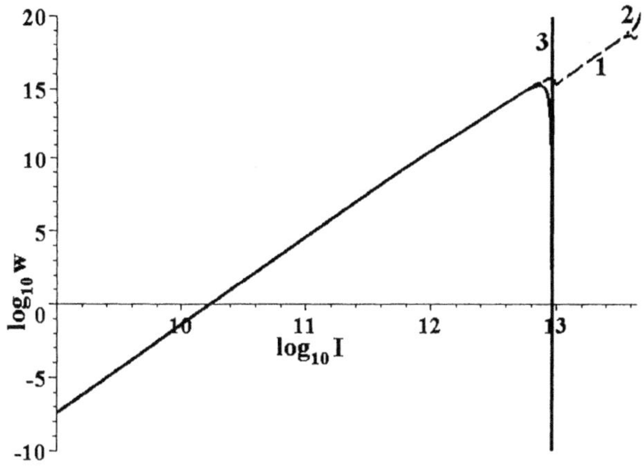

Figure 5. A log-log plot of the ionization rates for the three band structure models (13) – (15) vs. laser intensity for parameters corresponding to NaCl. Dashed curve 1 – non-parabolic band (13), dash-dotted curve 2 – parabolic band (14), solid curve 3 – cosine band (15). Laser wavelength is 825 nm, refractive index at input laser wavelength is 1.5356. Ionization rate w is given in 1/(fs cm^3), and intensity I is given in W/cm^2.

The difference between them is that χ increases with increasing of laser intensity or decreasing of laser frequency while the Keldysh parameter γ decreases in both cases, i.e., the parameters vary in opposite ways and, roughly speaking, are connected by inverse proportionality. Then small values of the modified parameter $\chi \ll 1$ correspond to the multiphoton regime, while its large values $\chi \gg 1$ correspond to the tunneling regime. In particular, one can see that the singularity of the ionization rate takes place at $\chi = \xi_1 = 2.4048256$.

4. DISCUSSION AND CONCLUSIONS

To understand the obtained results one should calculate the total energy of the oscillating electrons and its variation with increasing of laser intensity. The energy includes the pondermotive potential of the oscillating electrons and the initial energy determined by the initial value of quasi-momentum. It is referred to as "effective band structure" below and is calculated according to the following general relation (Keldysh, 1965):

$$\varepsilon_{eff}\left(\vec{p}_0\right) = \frac{1}{T}\int_0^T \varepsilon\left[\vec{p}_0 - \frac{e\vec{F}}{\omega}\sin\left(\omega t\right)\right] dt . \qquad (23)$$

One can calculate effective structures of all three considered band models and compare them (Figs. 6 – 8). In case of the parabolic model, the band stays parabolic for any intensity while it moves to the area of higher energies with increasing of laser intensity (Fig. 6).

In case of the Kane band, the effective structure also moves to higher energies, i.e., the effective band gap increases (Fig. 7), but also the band becomes flatter with increasing of laser intensity since the fast electrons get negative pondermotive potential as was discussed in Section 2.3. The speed of growth of the effective band gap in this case is close to that of the parabolic band only for low intensity when both models coincide. Exact flattening of the effective band is never reached at any limited laser intensity.

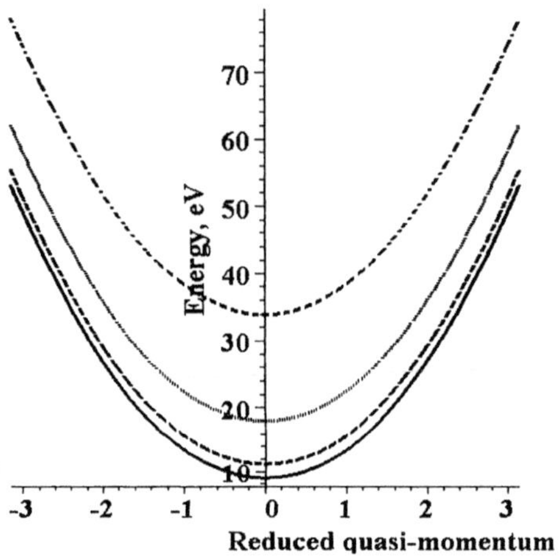

Figure 6. Variations of the effective band structure for the parabolic model (14). Solid curve is for $\gamma=\infty$ (initial band structure), dashed curve is for $\gamma=1.0$, dotted curve is for $\gamma=0.5$, dash-dotted curve is for $\gamma=0.3$. Quasi-momentum is normalized to vary from $-\pi$ to π.

The case of the cosine model looks like a natural developing of the Kane band case – the central part of the effective band structure moves to higher energies, and the whole band structure becomes flatter with increasing of laser intensity (Fig. 8). The characteristic feature of the cosine relation is that the effective band gap increases, but stays limited at any intensity, and the band becomes rigorously *flat over the whole BZ* at laser intensity given by (20). It is exactly the flattening effect we expected to discover!

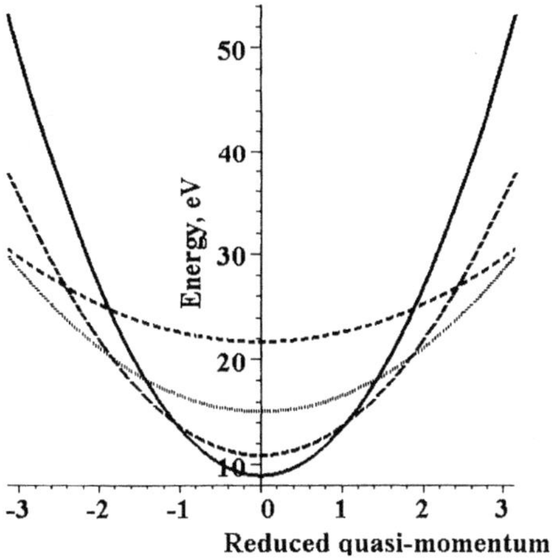

Figure 7. Variations of the effective band structure for the Kane model (13). All notations and values of the parameter γ are the same as in Fig. 7.

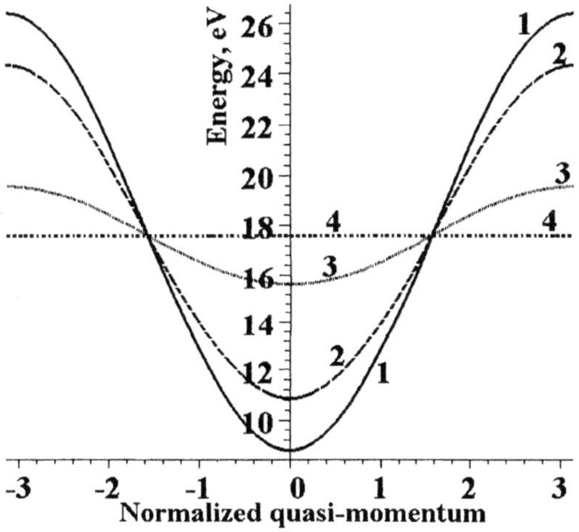

Figure 8. Variations of the effective band structure for the cosine band (15) with increasing of laser intensity: solid curve 1 is for $\chi = 0$ (initial band structure); dashed curve 2 is for $\chi = 1.0$; dotted curve 3 is for $\chi = 2.0$; dash-dotted curve 4 is for $\chi = \xi_1 = 2.4048256$.

The meaning of the flattening effect is obvious: since the effective band structure does not depend on quasi-momentum, it means that the two-band quantum system degenerates into a two-level system under action of the electric field of laser radiation! This is a result of synchronization of the electron oscillations by the radiation leading to the considered specific distribution of the pondermotive potential between the electrons. Due to this, all valence electrons are concentrated on a single effective level in the valence band, and they can jump to the only effective level in the conduction band. They do it within a few periods of their oscillations. Thus, the amount of electrons appearing in the conduction band is close to the total number of valence electrons.

There are a few important points to be clarified in connection with the presented results. First, we considered ideal coherent electron oscillations not perturbed by any collisions. A natural question arises in this connection: can the described ionization regime be killed by electron-particle collisions varying electron quasi-momentum? Generally speaking, yes, if characteristic time of the collisions is close or smaller than the period of electron oscillations. The latter can be found from Eq. (3) at intensity close to the threshold (20) since the singularity takes place in the tunneling regime. Then, substituting (3) into (22) and putting $\chi = \xi_1 = 2.4048256$, one can find the number of electron oscillations per one period of field oscillation at threshold intensity:

$$N = \frac{2\pi}{\omega\widetilde{T}} = 2\chi = 2\xi_1 \approx 4.8\,. \qquad (24)$$

Thus, the period of the electron oscillations is about 0.5 fs for laser wavelength 800 nm and the value of electric field given by (20). In case of a shorter wavelength it is even smaller, i. e., one period of electron oscillations is much less than 1 femtosecond for visible radiation which is smaller than the characteristic time τ_c of electron-particle collisions resulting in variations of electron quasi-momentum. This gives a serious reason to expect the singularity regime to be observable in real situations in spite of the perturbing influence of the collisions. The collisions can only decrease the number of the electrons making the interband jumps by factor $\left(1-\widetilde{T}/\tau_c\right)$, due to breaking the field-induced synchronization of electron oscillations.

Second, possibility of the considered effects was demonstrated for a particular band model, but a deeper analysis (Gruzdev, 2006c) points out that it is a quite general effect characteristic of a wide class of band structures corresponding to the approximation of strongly bonded electrons including the considered one. Moreover, careful consideration (Gruzdev, 2006b, 2006c) shows that the described effect does not depend on the mathematical approximations made in calculating the ionization rate. On the other hand, it is much influenced by a particular band model and the type of electron functions that are just approximate solutions to the Schrödinger equation. Exactly speaking, the singularity turns into a limited resonance-type increasing of the ionization rate if higher order corrections to the functions

(7) are taken into account, and it can disappear completely for some particular band structures. This point requires further analysis.

Thus, we have answered both questions asked at the end of Section 2. The flattening effect exists and the corresponding flattening intensity is not extremely large – it is close to 10 TW/cm^2 for most wide band-gap materials. The effect is coupled with singularity on intensity dependence of the ionization rate occurring exactly at the same value of intensity. Actually, we have discovered a new regime of ionization taking place under action of high-intensity laser radiation in wide band-gap dielectrics. Its specific features have been studied for the particular cosine band model, but we have shown very general and critical roles of the Bragg-type reflections at boundaries of the first BZ in the photoionization. This regime results in a huge value of the ionization rate (in the considered ideal case – singularity of the ionization rate) for which the number of electrons jumping over the forbidden band within one period of field oscillations is comparable to the total amount of valence electrons. Due to this, it can be referred to as collective ionization. Thus, instead of the traditional individual interband electron jumps through the minimum point of the forbidden band, one should expect collective jumps of the electrons at laser intensity given by (20) (Gruzdev, 2006b, 2006c).

Returning to the role of ionization in laser-solid interactions, we should note that many traditional opinions regarding the interactions should be reviewed and modified for the case of femtosecond laser pulses since the opinions are based on the traditional models of ionization developed for low-intensity approximation of constant effective electron-hole mass. In particular, the presented results show that the photoionization can provide huge electron density in the conduction band of wide band-gap dielectrics without any (or with minimum) additional assistance of the avalanche ionization. Moreover, instead of traditionally utilized criteria for threshold of intrinsic laser-induced damage and ablation of transparent materials (Stuart, et al., 1996), a very natural criterion for the threshold results from our considerations: it is exactly the singularity threshold determined by laser frequency and material parameters according to (20).

REFERENCES

Adams, E.N., 1952, Motion of an electron in a perturbed periodic potential, *Phys. Rev.* **85**(1): 41-50.

Adams, E.N., and Argyres, P.N., 1956, Acceleration of electrons by an external force field, *Phys. Rev.* **102**(3): 605-606.

Apostolova, T., and Hahn, Y., 2000, Modeling of laser-induced breakdown in dielectrics with subpicosecond pulses, *J. Appl. Phys.* **88**(2): 1024-1034.

Ashkenasi, D., Varel, H., Rosenfeld, A., Noack, F., Campbell, E.E.B., 1996, Pulse-width influence on the laser-induced structuring of CaF_2 (111), *Appl. Phys. A* **63**: 103-107.

Audebert, P., Daguzan, Ph., Dos Santos, A., et al., 1994, Space-time observation of an Electron Gas in SiO_2, *Phys. Rev. Lett.* **73**(14): 1990-1993.

Avron, J.E., 1976, Model calculation of Stark ladder resonances, *Phys. Rev. Lett.* **37**(23): 1568-1571.

Bloch, F., 1928, Quantum mechanics of electrons in crystals, *Zeitschrift fuer Physik* **52**: 555-599.

Bonch-Bruevich, V.L., and Kalaschnikov, S.G., 1982, *Halbleiterphysik*, VEB Deutscher Verlag Wissenschaften, Berlin, Chapters III, IV.

Braunstein, R., and Ockman, N., 1964, Optical Double-Photon Absorption in CdS, *Phys.Rev.* **134**: A499-A507.

Bychkov, Yu.A., and Dykhne, A.M., 1970, Breakdown in semiconductors in an alternating electric field, *Sov. Phys. – JETP* **31**: 928-9337 [transl. from *Zh. Eksp. Teor. Phys.* (USSR) **58**(5), 1734-1743 (1970)].

Du, D., Liu, X., Mourou, G., 1996, Reduction of multi-photon ionization in dielectrics due to collisions, *Appl. Phys. B* **63**: 617-621.

Gruzdev, V.E., 2004, Analysis of the transparent-crystal ionization model developed by L.V.Keldysh, *J. of Opt. Technology* **71**(8): 504-508 [transl. from *Opticheskii Zhurnal* **71**(8), 14-20 (2004)].

Gruzdev, V.E., 2005, Laser-induced collective ionization in wide band-gap crystalline dielectrics, in "ICONO 2005: Ultrafast Phenomena and Physics of Superintense Laser Fields", Proc. SPIE (to appear).

Gruzdev, V.E. , 2006a, "Analysis of the Keldysh model for laser-induced ionization of transparent solids", *J of Phys .C: Cond. Matter*, (submitted).

Gruzdev, V.E., 2006b, Influence of band structure on photo-ionization of non-metallic crystals by high-intensity laser radiation, *Phys. Rev. B*, (submitted).

Gruzdev, V.E., 2006c, Effect of collective ionization in wide band-gap crystalline dielectrics, *Phys. Rev. B* (in preparation).

Guizard, S., Martin, P., Petite, G., D'Oliveira, P., Meynadier, P., 1996, Time-resolved study of laser-induced colour centres in SiO_2, *J. Phys.: Cond. Matter*. **8**: 1281-1290.

Houston, W.V., 1940, Acceleration of Electrons in a Crystal Lattice, *Phys. Rev.* **57**: 184-186.

Jones, H.D., and Reiss, H.R., 1977, Intense-field effects in solids, *Phys. Rev. B* **16**(6): 2466-2473.

Kaiser, A., Rethfeld, B., Vicanek, M., Simon, G., 2000, Microscopic processes in dielectrics under irradiation by subpicosecond laser pulses, *Phys. Rev. B.* **61**(17): 11437-11450.

Kane, E.O., 1957, Band structure of indium antimonide", *J. Phys. Chem. Solids* **1**: 249-261.

Kautek, W., Krueger, J., Lenzner, M., Sartania, S., Spielmann, Ch., Krausz, F., 1996, Laser ablation of dielectrics with pulse durations between 20 fs and 3 ps, *Appl. Phys. Lett.* **69**(21): 3146-3148.

Keldysh, L.V., 1958, Behavior of non-metallic crystals in strong electric fields, *Sov. Phys. – JETP* **6**(4): 763-770 [transl. from *Zh. Eksp. Theor. Phys.* (USSR) **33**, 994-1003 (1957)].

Keldysh, L.V., 1965, Ionization in the field of a strong electromagnetic wave, *Sov. Phys. - JETP*, **20**(5): 1307-1314 [transl. from *Zh. Eksp. Theor. Phys.* (USSR) **47**, 1945-1957 (1964)].

Kovarskii, V.A., and Perlin, E.Y., 1971, Multi-photon interband optical transitions in crystals, *Phys. Status Solidi (b)* **45**: 47-56.

Lenzner, M., Krueger, J., Sartania, S., Cheng, Z., Spielmann, Ch., Mourou, G., Kautek, W., Krausz, F., 1998, Femtosecond optical breakdown in dielectrics, *Phys. Rev. Lett.* **80**(18): 4076-4079.

Li, M., Menon, S., Nibarger, J.P., Gibson, G.N., 1999, Ultrafast electron dynamics in femtosecond optical breakdown of dielectrics, *Phys. Rev. Lett.* **82**(11): 2394-2397.

Lyssenko, V.G., Valusis, G., Löser, F., et al., 1997, Direct measurement of the spatial displacement of Bloch-oscillating electrons in semiconductor superlattices, *Phys. Rev. Lett.* **79**(2): 301-304.

Nathan, V., Guenther, A.H., Mitra, S.S., 1985, Review of multiphoton absorption in crystalline solids, *J. Opt. Soc. Am. B* **2**(2): 294-316.

Poole, R.T., Jenkin, J.G., Liesegang, J., Leckey, R.C.G., 1975, Electronic band structure of the alkali halides. I. Experimental parameters, *Phys. Rev. B.* **11**(12): 5179-5189.

Quere, F., Guizard, S., Martin, Ph., 2001, Time-resolved study of laser-induced breakdown in dielectrics, *Europhys. Lett.* **56**(1): 138-144.

Rethfeld, B., 2004, Unified Model for the Free-Electron Avalanche in Laser-Irradiated Dielectrics, *Phys. Rev. Lett.* **92**(18): 187401.

Rethfeld, B., 2005, Free electron generation in laser-irradiated dielectrics, in "ICONO 2005: Nonlinear Optical Phenomena VII", Proc. SPIE (to appear).

Schaffer, C., Brodeur, A., Mazur, E., 2001, Laser-induced breakdown and damage in bulk transparent materials induced by tightly focused femtosecond laser pulses, *Meas. Sci. Technol.* **12**: 1784-1794.

Stoian, R., Ashkenasi, D., Rosenfeld, A., Campbell, E.E.B., 2000, Coulomb explosion in ultrashort laser ablation of Al_2O_3, *Phys. Rev. B* **62**(19): 13167 – 13173.

Stuart, B.C., Feit, M.D., Herman, S., Rubenchik, A.M., Shore, B.W., Perry, M.D., 1996, Nanosecond-to-femtosecond laser-induced breakdown in dielectrics, *Phys. Rev. B* **53**(4): 1749-1761.

Sudzius, M., Lyssenko, V.G., Löser, F., et al., 1998, Optical control of Bloch-oscillation amplitudes: From harmonic spatial motion to breathing modes, *Phys. Rev. B* **57**(20): R12693-R12696.

Tien, A.-C., Backus, S., Kapteyn, H., Murnane, M., Mourou, G., 1999, Short-pulse laser damage in transparent materials as a function of pulse duration, *Phys. Rev. Lett.* **82**(19): 3883-3886.

Vaidyanathan, A., Walker, T., Guenther, A.H., Mitra, S.S., Narducci, L.M., 1980, Two-photon absorption in several direct-gap crystals, *Phys. Rev. B* **21**(2): 743-748.

Von der Linde, D., and Schüller, H., 1996, Breakdown threshold and plasma formation in femtosecond laser–solid interaction, *J. Opt. Soc. Am. B* **13**(1): 216-222.

Von der Linde, D., and Sokolovski-Tinten, K., 2000, The physical mechanism of short-pulse laser ablation, *Appl. Surface Science* **154-155**: 1-10.

Wannier, G.H., 1960, Wave functions and effective Hamiltonian for Bloch electrons in an electric field, *Phys. Rev.* **117**(2): 432-439.

Wannier, G.H., 1962, Dynamics of band electrons in electric and magnetic fields, *Rev. Mod. Phys.* **34**(4): 645-655.

Yee, J.H., 1971, Four-Photon Transition in Semiconductors, *Phys. Rev. B* **3**(2): 355-360.

Yee, J.H., 1972, Three-Photon Absorption in Semiconductors, *Phys. Rev. B* **5**(2): 449-458.

Chapter 6

THREE DIMENSIONAL MATERIAL PROCESSING WITH FEMTOSECOND LASERS

Aleksandr Ovsianikov, Sven Passinger, Ruth Houbertz[1], and Boris N. Chichkov

Laser Zentrum Hannover e.V., Hollerithallee 8, 30419 Hannover, Germany

[1]Fraunhoffer-Institut für Silicatforschung, Neunerplatz 2, 97082 Würzburg, Germany

1. INTRODUCTION

1.1 Two-photon polymerization (2PP) technique

The possibility of very high localization of laser energy has led to more and more advanced laser applications. Particularly, the interaction of lasers with polymers and dielectrics is of high technological interest. The adaptation of polymers to laser characteristics and vice versa is very challenging from the scientific and technological points of view. A method which has recently attracted considerable attention is two-photon polymerization (2PP) using femtosecond lasers (Maruo, et al., 1997; Kawata, et al., 2001; Cumpston, et al., 1999; Serbin, et al., 2003; Deubel, et al., 2004), where complicated microstructures can be produced in photosensitive materials. Two-photon polymerization (2PP) is a very attractive 3D rapid microstructuring technology which provides much better structural resolution and quality than the well-known stereo-lithography (SL) technique. In spite of that, 2PP is very young technology (ten years old), and there are already several very nice and informative reviews published on this subject (Sun, et al., 2004; Blanco, et al., 2004; Prasad, 2004). In this chapter we will provide only a brief introduction to the 3D material processing and will report on our recent results obtained at the Laser Zentrum Hannover e.V. and Fraunhofer-Institut für Silicatforschung.

The very important distinction between the 2PP and SL technologies is that in the case of 2PP, near-infrared (IR) laser pulses, and in case of SL, ultraviolet (UV) laser radiation, are used for curing of photosensitive materials. From the first glance, it does not look like a big difference whether multiphoton (IR) or single-photon (UV) absorption initiate polymerization processes. Taking into account that photosensitive materials are usually transparent in the infrared and are highly absorptive in the UV range, one can initiate polymerization with IR laser pulses within the volume and fabricate 3D structures, whereas with UV laser radiation, due to single-photon absorption, polymerization also occurs at the surface. 3D structures made with tightly focused UV light usually have poor resolution. Therefore, for the fabrication of 3D structures, SL is used as a planar technology with layer-by-layer polymerization steps, whereas 2PP is a truly high-resolution 3D technology.

In our work, we apply near-infrared Ti:sapphire femtosecond laser pulses (at 800/780 nm) for 3D material processing. When tightly focused into the volume of a photosensitive material (or photoresist), they initiate 2PP process by, for example, transferring liquid into the solid state. This allows the fabrication of any computer-generated 3D structure by direct laser "recording" into the volume of photosensitive material. Because of the threshold behavior and nonlinear nature of the 2PP process, a resolution beyond the diffraction limit can be realized by controlling the laser pulse energy and number of applied pulses. Fig. 1 shows two scanning electron microscope images of 3D microstructures fabricated by 2PP technique.

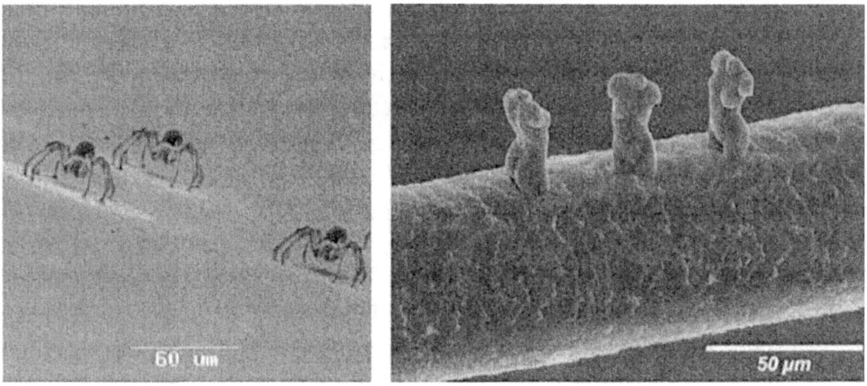

Figure 1. A scanning electron microscope (SEM) image of a microspider-array and Venus micro-models fabricated by 2PP technique.

One can see the strength and beauty of this technology and envision many potential applications. In our contribution, we will report on recent advances of this technology and future short- and long-term prospects. A comparative analysis of the structuring properties of various photosensitive materials will be presented.

Negative photoresists which can be patterned by 2PP technique can be either radically or cationically cured. For inorganic-organic hybrid polymers (ORMOCER®s, Trademark of the Fraunhofer-Gesellschaft zur Förderung der Angewandten Forschung e.V., Munich, Germany), used in our work, we applied radical initiation. In these photoresists, absorption of light generates free radicals that initiate the polymerization process. We will provide a detailed description of ORMOCER®s and their microstructuring characteristics. In cationic systems (an important example is the commercially available SU-8 photoresist, which we also apply in our work), an acid is generated upon laser irradiation. In this case, polymerization does not take place during laser irradiation (only after a postexposure bake). This is an important property of cationic photoresists, since the difference in the refractive index of exposed and unexposed area is negligible, which provides flexible irradiation strategy and allows one to combine direct laser beam writing (serial processing) and holographic exposure (parallel processing). The fact that ORMOCER®s are liquid and SU-8 is solid leads to different exposure strategies and sample configurations which are shown in Fig. 2.

2PP of photosensitive materials irradiated by femtosecond laser pulses is now considered as enabling technology for the fabrication of 3D photonic crystals (PhCs) and photonic crystal templates. In particular, 2PP allows one to introduce defects at any desired locations, which is crucial for practical applications. To realize photonic crystals with a full photonic band gap, 3D microstructuring of high refractive index materials is required. The most attractive option is to fabricate templates which are later infiltrated with a high refractive index material, followed by the removal of the original template structure. Application of most negative photoresists for the fabrication of templates is rather complicated, since the structures fabricated

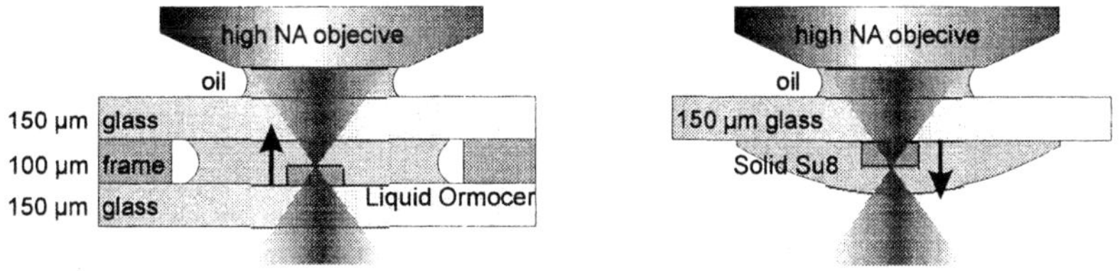

Figure 2. Different sample configurations used for the fabrication of 3D structures in ORMOCER®s (left) and SU-8 (right).

in these materials are quite stable and not simply soluble. In case of positive photoresist, the polymer is weakened and is usually more soluble in developing solutions. This is very attractive for the fabrication of 3D templates. Possibilities to use positive photoresists for the fabrication of PhC templates will be explored and discussed in this chapter.

Recently, we and some other groups in the world have studied possible applications of 2PP technique in biomedicine, looking for novel biocompatible and bioactive materials which can be structured by 2PP. Applications such as tissue engineering, drug delivery, and medical implants could greatly benefit from this approach. In particular, 2PP is a very interesting technique for the fabrication of drug delivery systems and medical implants. Fabrication of scaffolds for tissue engineering could also greatly benefit from this technique.

1.2 Two-photon activated (2PA) processing

Two or multiphoton activated laser processing represents a generalization of the 2PP technique. When negative photoresists are used, two-photon exposure results in cross-linking of polymer chains so that the unexposed resist can be washed out. For positive tone resists, light exposure leads to a chain scission into shorter units that can be dissolved and washed away in the development process. Two-photon scission of polymer chains in the positive photoresist allows one to dissolve the irradiated regions and to make 3D microstructures. The difference in processing of positive and negative photoresists is illustrated in Fig. 3. Both of these processes are induced by two-photon absorption which leads to different chemical outcomes. Using this technique, two or multiphoton activation of chemical reactions can be performed at a desired place on or within a sample.

Two-photon activated chemical and physical processes inside transparent materials allow the fabrication of complex 3D structures with high spatial resolution. The region of two-photon absorption is confined to the laser focal volume, which can be made small by using optics with a high numerical aperture (NA). All subsequent two-photon activated processes will be also confined to this volume.

In this chapter, we will also report on the fabrication of 3D microstructures in commercial positive tone solid state photoresists by 2PA. The fabrication process is not accompanied by any distortion of the unexposed regions, whereas the exposed regions can be easily removed.

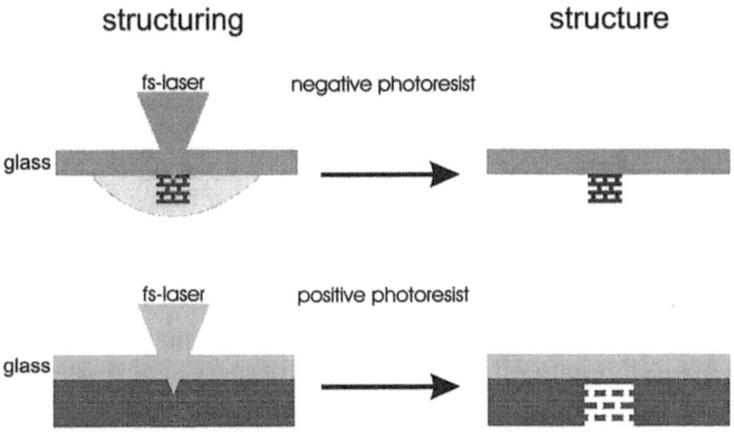

Figure 3. Difference between multiphoton structuring of negative photoresists (top) and positive photoresists (bottom).

2. MATERIALS FOR 2PP AND 2PA

There are two classes of photosensitive materials which can be structured by 2PA technique: negative and positive tone photoresists (see Fig. 4). Negative resist materials which we have used so far, can be subdivided into solid and liquid. SU-8 is a well-known commercial material.

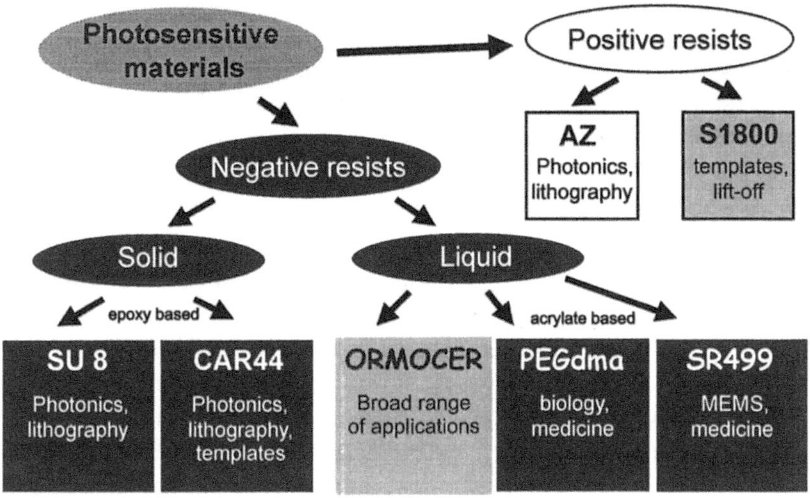

Figure 4. Photosensitive materials for 2PA studied by our group.

CAR 44 is a novel material which is more suitable for the fabrication of templates. The main advantage of this material is that the fabricated structure can be easily removed with a remover. Both materials are epoxy-based cationic photoresists. Interaction with light generates an acid in the illuminated regions. Polymerization takes place during the postbake step. Target applications of these materials are in photonics and lithography.

Liquid materials (except of ORMOCER®s) are acrylate-based materials, where the polymerization reaction is triggered by radically reacting photo-initiators during the laser irradiation. This allows online optical monitoring of the polymerization reaction. ORMOCER® is an inorganic-organic hybrid polymer which is most often used in our work and will be discussed in detail below. We investigated 2PP of PEGdma for possible applications in biology and medicine. SR499 is used for the fabrication of MEMS and Bio-MEMS.

Among positive photoresists, AZ and S1800 materials were studied. S1800 is a commercially available photoresist from Shipley and will be discussed in details below. Note that all materials shown in Fig. 4 can be structured by the 2PA technique with a resolution down to 100 - 300 nm.

2.1 Negative tone materials/inorganic-organic hybrid polymers

The materials used so far for 2PP were commercially available acrylate or epoxy-based resins, i.e. pure organic materials. Although many of these materials have good processibility, they don't fulfill major requirements if used for multi-chip module (MCM) technology; they are often chemically unstable against solvents typically applied in multilayer processing, and their thermal and mechanical stability is often quite poor.

A material class, which has attracted considerable attention for MCM applications, is represented by inorganic-organic hybrid polymers such as ORMOCER®s. Since their physical and chemical properties can be tailored, they can be employed in many devices for a large variety of applications, enabling novel properties from micro down to the nanometer scale. Due to their enhanced chemical, thermal, and mechanical stability and – simultaneously – very good processibility, they can overcome the restrictions of purely organic polymers for most applications.

For more than one decade, silicate-based inorganic-organic hybrid polymers played an important role due to their physical and chemical properties, resulting from their hybrid nature. The concept of ORMOCER®s is that they combine properties of organic polymers (toughness, functionalization, and processing at low temperatures) with those of glass-like materials (hardness, chemical and thermal stability, and transparency)

(Haas, 2000). This allows one to achieve material properties which are not accessible with composite or polymer materials.

ORMOCER®s are synthesized by sol-gel processing (Brinker, et al., 1990), where inorganic-oxidic units are connected to organic moieties on a molecular level (Kahlenberg, et al., 2005). The synthesis offers a tremendous flexibility by variation of the catalysts, temperature, and alkoxysilane scaffold. This flexibility enables one to tailor material properties as recommended by the application (Amberg-Scwab, et al., 1991; Rose, et al., 1995; Wolter, et al., 1994; www.voco.com; Streppel, et al., 2002; Bräuer, et al., 2001; Houbertz, et al., 2003a; Robertsson, et al., 1998; Fröhlich, et al., 2002; Houbertz, et al., 2001, 2003; Popall, et al., 1998; Depre, et al., 2001). For example, several abrasion-resistant coatings for different substrates (Amberg-Scwab, et al., 1991; Rose, et al., 1995) as well as series of dental materials (Wolter, et al., 1994; www.voco.com), based on ORMOCER®s, have already been introduced into the market. Specially synthesized ORMOCER®s are applied in optical interconnects or waveguides (Streppel, et al., 2002), in microoptics (Bräuer, et al., 2001; Houbertz, et al., 2003), in electro/optical applications (Robertsson, et al., 1998), as dielectric layers (Fröhlich, et al., 2002), as passivation materials for the encapsulation of microelectronic devices and components (Houbertz, et al., 2001, 2003a), in Li ion conductors for energy supply (Popall, et al., 1998), or as membrane materials for fuel cell applications (Depre, et al., 2001).

Fig. 5 shows a schematic sketch of the multifunctional precursors for the hybrid polymer synthesis as well as a brief overview about the variation of material properties. The processing of ORMOCER®s for microsystems typically consists of two steps: (1) an Si-O-Si network is established via hydrolysis/polycondensation reactions of alkoxysilanes which yields organically modified nanoscaled inorganic-oxidic units, present in a prepolymer sol. Their size (typically 1 to 10 nm) is dependent on the choice of alkoxysilane precursors and the catalytically controlled reaction conditions such as catalyst, concentration, or reaction temperature. As organic moieties, (oligo-)methacryl, acryl, styryl, or epoxy functionalities are often used in order to account for the organic cross-linking of these polycondensed alkoxysilanes. (2) An organic cross-linking is performed either photochemically and/or thermally.

Dependent on the synthesis conditions, i.e. the kind and amount of alkoxysilane precursors and catalyst, the temperature, or the solvents used for synthesis, the inorganic network formation can be influenced. The Young's modulus, as well as the mechanical and thermal stability, can be increased by increasing the inorganic content in the hybrid network.

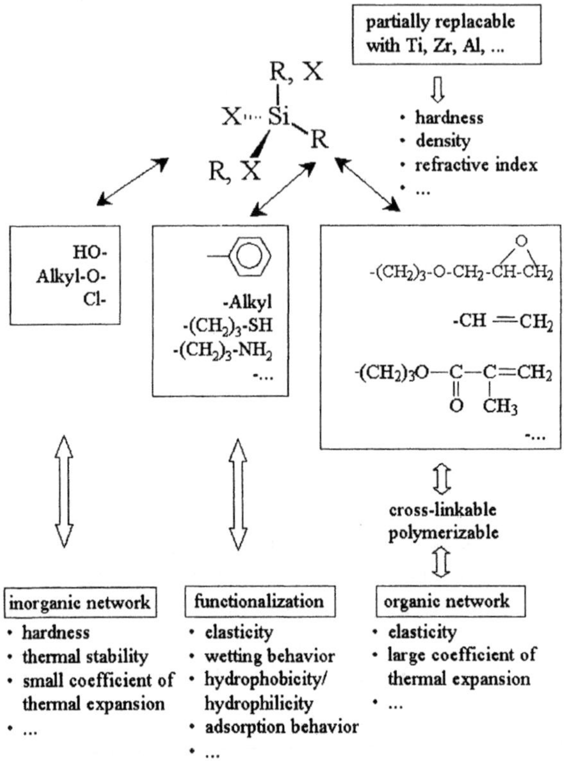

Figure 5. Schematic sketch of multifunctional precursors for hybrid polymers and the variation of properties.

In this way, the coefficient of thermal expansion (CTE) and the optical losses in the near-infrared (NIR) regime are reduced as well. The reason for the latter is that the content of Si-O-Si, which is known to have a low optical loss, is increased, while the amount of organic species is diluted. Beside influencing the Young's modulus and the CTE by the inorganic content in the ORMOCER® material, these quantities are also influenced by a modification of the connecting unit. In order to achieve highest transparency in the near-infrared (NIR) regime, these connecting units can be partially fluorinated or replaced by perfluoroaryl groups, respectively (Brinker, et al., 1990; Buestrich, et al., 2001). The organic polymerizable units are chosen depending on the requirements of the application. If UV laser lithography or 2PP is applied for patterning, (oligo-) methacryl or styryl moieties are typically chosen. For thermal cross-linking, epoxy moieties are preferred. Non-reactive groups such as, e.g., alkyl or aryl groups, which are connected to Si, also influence the material properties. Increasing their amount within the hybrid polymer will reduce the degree of polymerization due to sterical reasons, thus resulting in a reduced density within the coated layers. This, for example, has direct impact on the optical or dielectric properties such as the refractive index or permittivity.

Upon hydrolysis/polycondensation, solvents are generally present in the resin, either applied in the sol-gel process as an inorganic network influencing agent or as volatile products resulting from the chemical reaction of the different precursors. The solvents are usually removed from the material to a certain under-reduced pressure in order to achieve highest processing flexibility. For thin-film applications, other solvents can be added to adjust the viscosity of the material. Most of the ORMOCER® resins have a long storage stability without initiators at room temperature which was tested for some systems to be longer than two years. If the material is cooled down to -18 °C, the storage can be performed for even a couple of years.

A special synthesis was carried out in order to achieve a suitable inorganic-organic hybrid polymer for wafer-scale telecom applications (henceforth referred to as ORMOCER®I), where the inorganic backbone is solely established via a polycondensation reaction in order to avoid O-H groups which would significantly increase the optical absorption around 1550 nm (Rouesseau, et al., 1992). As silane precursors, diphenylsilanediol (DPD) and 3-methacryloxypropyltrimethoxysilane (MEMO) were used. The synthesis as well as further chemical characterization of the material is described elsewhere in more detail (Buestrich, et al., 2001).

The DPD is specially synthesized in order to guarantee highest purity for the resulting ORMOCER® materials and to control residual O-H groups within the final hybrid polymer. The purity is routinely characterized by NMR spectroscopy, whereas for telecom applications it is extremely important to use materials free of water.

In Fig. 6, a schematic sketch of the reaction for ORMOCER®I is shown which is, for example, applied as core material for optical waveguides (Streppel, et al., 2002, Houbertz, et al., 2003b). Characteristic for the polysiloxane network is the lack of silanol groups. An akoxylation reaction is favored by the sterical demand of the phenyl groups which, in addition, hinders a self-condensation of the diol component. According to this, a constant stoichiometric relation between the two silane precursors MEMO and DPD (1:1) is guaranteed during the synthesis. The synthesis is carried out at elevated temperatures between 60 and 100 °C with a basic catalyst. This results in short reaction times and also fosters an alkoxylation. By-products which are generated upon synthesis such as, for example, solvents like methanol, are typically removed from the ORMOCER® materials under reduced pressure.

Dependent on the application, the processing of ORMOCER®s varies significantly. For example, for dental applications such as cavity fillers, the materials have to be highly filled with particles, resulting in a paste which can be cured by blue light (www.voco.com). On the other hand, ORMOCER®s used for (large-scale) coatings have to be applicable on

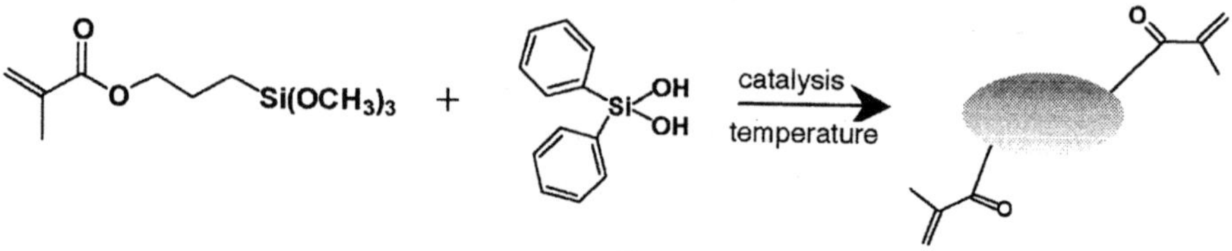

Figure 6. .Schematic sketch of the ORMOCER®I synthesis.

different surfaces such as glass or polymer surfaces as in the case of magnification lenses (Schottner, 2001).

Contrary to the above mentioned applications, ORMOCER®s for microsystems have to fulfill quite different requirements with respect to their intrinsically generated material properties by specific synthesis routes, as well as with respect to the processing technologies which have to be applied. For example, many different application and patterning methods such as spin, dip, or spray coating, and photolithography or printing methods are used for generating functional microsystems. Thus, the materials are addressing basic requirements such as:

• adaptable optical or electrical properties, for example, optical absorption, refractive index, transparency, dielectric permittivity, or dielectric loss,

• perfect surface planarization and no reaction of the processed materials with solvents (like esters, ketones, or alcohols) used for subsequent processing steps, for example in the case of multilayer applications,

• no parasitic polymerization out of the exposure area, and

• long-term environmental stability in reliability testing under various conditions.

The hybrid structure combines the properties of inorganic structures (e.g., an increased mechanical and temperature stability as compared to purely organic polymers) and those of conventional polymers (e.g., easy processing at low temperatures). In the following, the processing of ORMOCER®s for microsystems will be described in more detail.

The organic cross-linking can be either performed photochemically by conventional photolithography (see, for example Streppel, et al., 2002; Bräuer, et al., 2001; Robertsson, et al., 1998; Fröhlich, et al., 2002; Houbertz, et al., 2001), by a combination of UV lithography/replication (Houbertz, et al., 2003a), by laser-direct writing, or by using femtosecond laser pulses in order to initiate two-photon polymerization (2PP) (Serbin, et al., 2003). Besides, the cross-linking can also be initiated thermally. The photochemical patterning methods result in polymerized (solid) structures where the non-polymerized parts are removed by standard developers, whereas arbitrary 3D (nano-)structures can be written with 2PP. This allows one to apply these material class for many different applications in microsystems technology, ranging from optics via micro-or polymer electronics to energy storage or supply.

For the photochemical patterning, radical initiation is mostly used to organically cross-link the organic moieties. However, cationic initiation is also possible, but it has to be mentioned that ions might be present within the ORMOCER® layers/structures. These might influence the properties of the materials negatively for some applications, for example in the area of micro or polymer electronics. The processing conditions, as well as the amount of initiator introduced into the ORMOCER® resin, influences the degree of organic cross-linking (polymerization), and thus the physical properties (Houbertz, et al., 2004).

A general processing flow chart for the production of ORMOCER® structures with either conventional UV lithography or with 2PP is shown in Fig. 7. Before applying the ORMOCER®, a UV initiator has to be introduced into to the resin in order to allow initiation of the organic polymerization reaction upon light absorption. In order to allow one to produce micropatterned layers of different thickness by conventional UV lithography, the ORMOCER® resins are typically diluted with suited solvents such as, for example, propylacetate (PAC) or 1, 2-propanediol-monomethyletheracetate (PGMEA). For the production of nano or microstructures by using femtosecond lasers to initiate 2PP, this is not necessary since the structures are produced within the volume of the resin, thus allowing any arbitrary structures with undercuts, different thicknesses, or varying aspect ratios.

After preparation of the resins/lacquers, these are applied, for example by spin coating or dispensing for conventional UV lithography or 2PP. The patterning typically follows a sequence of soft-bake, UV exposure, post-exposure bake, and a development step. Pre and postexposure bakes are used to remove solvents and to increase adhesion. The UV patterning can be either carried out with or without inert gas flow, whereas the latter results in the formation of an inhibition layer in the case of radical initiation. Subsequently, after the development the structures are finally thermally cured at elevated temperatures (between 100 and 185 °C) at elapsed time (15 min to 3 hrs.), yielding reliable surface films/structures for further investigations. For 2PP, the processing sequence can be kept even simpler: after application of the hybrid resin (no additional solvents added), a femtosecond laser is used to initiate 2PP, followed by a development step. Optionally, a final photochemical or thermal curing can be applied.

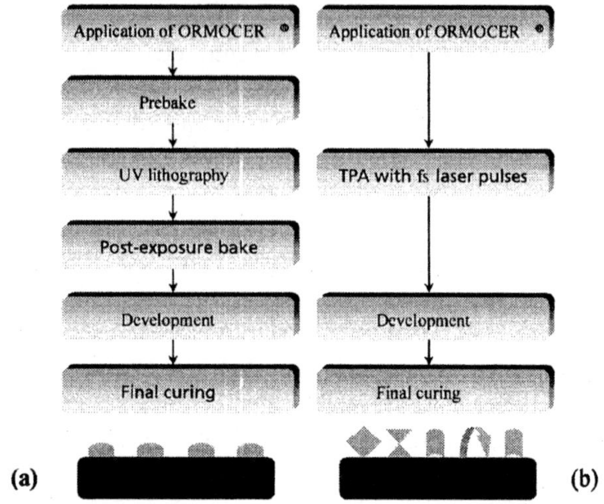

Figure 7. Processing flow charts for the production of (a) wafer-scale ORMOCER® structures using conventional UV lithography, and (b) ORMOCER® nano and microstructuring using 2PP with femtosecond lasers.

Figure 8. Examples of micro and nanostructures produced with (a) conventional UV lithography (Streppel, et al., 2002), (b) combined UV lithography/replication (Houbertz, et al., 2003a), (c) laser-direct writing, and (d) 2PP initiated by femtosecond laser pulses (Serbin, et al., 2003).

As already mentioned above, the material properties of ORMOCER®s can be adjusted over a wide range with respect to their application. An overview about the properties is given in (Haas, K.-H., 2000, Schottner, G., 2001). In Fig. 9, an absorption spectrum of ORMOCER®I resin with an added UV initiator is shown. As designed for UV processing, the material exhibits a high absorption in the UV range. This is mainly caused by the phenyl groups which, on the other hand, are also responsible for the low optical losses at 1310 nm (0.22 dB/cm) and 1550 nm (around 0.55 dB/cm), respectively.

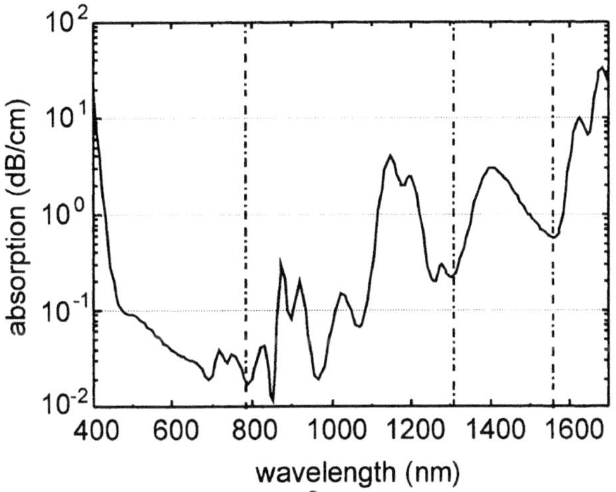

Figure 9. Absorption spectrum of ORMOCER®I resin with UV initiator (Irgacure 369) (after (Houbertz. R., 2003b)).

The refractive index of this system can be adjusted by simply mixing it with another ORMOCER® system which, for example, has a lower refractive index. The precise adjustment of the refractive index step Δn is very important for single and multimode waveguide applications. In order to precisely adjust the index, two different ORMOCER® systems can be simply mixed in certain ratios (see Fig. 10).

Besides the good optical properties, the material shows very good planarization behavior which is important for rough substrates like printed circuit boards, for example. When applied in layers or structures, the material's rms roughness is between 2 and 4 nm. These values are below λ/10, which gives the applicability for highly sophisticated (optical) devices. A typical topography is displayed in Fig. 11(a), while in Fig. 11(b) a structure generated by 2PP in ORMOCER®I, with a wall thickness of about 1 μm, is shown. For comparison, a structure fabricated by 2PP in a commercial acrylate system is shown in Fig. 11(c), where the surface roughness is in the several μm range.

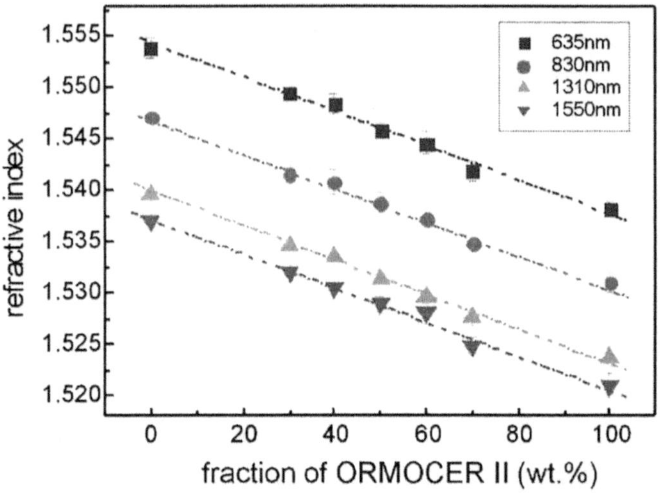

Figure 10. Index tuning for various ORMOCER® I and ORMOCER® II mixtures (after (Houbertz, R., 2003b)).

Figure 11. (a) AFM topography of a ORMOCER®I layer (b) ORMOCER®I and (c) a commercial acrylate system patterned by 2PP.

In conclusion, the material class of ORMOCER® with its tunable chemical and physical properties and processing characteristics enables many exciting applications in optics, micromechanical systems, biomedicine, etc.

2.2 Positive tone materials

In positive tone photoresists, 2PA processing induces a chain scission of polymer units in the irradiated region inside the material. After that, the irradiated regions can be dissolved. This allows one to fabricate three-dimensional hollow structures.

One class of positive photoresists, which has been used extensively in our work, is the Shipley S1800 photoresist. This resist has been developed for the integrated circuit device fabrication in the microelectronic industry. The resist can be dissolved in a propylene glycol monomethyl ether acetate and is optimized for the G-line absorption of a mercury lamp. This resist is photosensitive at wavelengths below 450 nm, as can be seen in the transmission spectrum shown in Fig. 12.

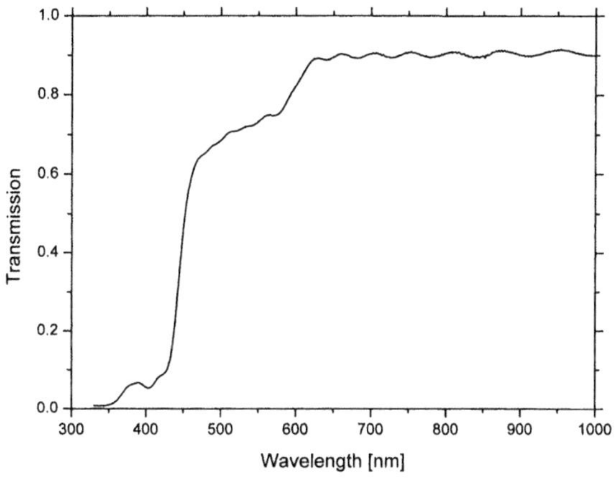

Figure 12. Transmission spectrum of the S1800-positive photoresist.

This absorption range makes S1800 a good material for multiphoton processing with Ti:Sapphire laser radiation at 800/780 nm wavelength. The S1800 resist can be spin coated in different thicknesses ranging from around 2.5 μm down to a few hundred nanometers depending on the dilution with a solvent and the spin coating speed. One can also use S1800 droplets with a thickness of up to 50 μm. The refractive index of the S1800 photoresist (below we use a specific modification of this material S1813) is about 1.63 at around 800 nm.

In thick S1813 photoresist droplets, it is possible to write hollow photonic crystal structures (Chichkov, et al., 2005). In Fig. 13, two examples of photonic crystals written in S1813 are shown.

The main reason for choosing positive photoresists instead of negative photoresists for the fabrication of photonic crystals lies in the simplicity of fabrication of replicas. Up to now, there are no negative photoresists with a high refractive index larger than 2. Therefore, it is not possible to produce three-dimensional photonic crystals with a complete photonic band gap in a single step by illuminating negative photoresists.

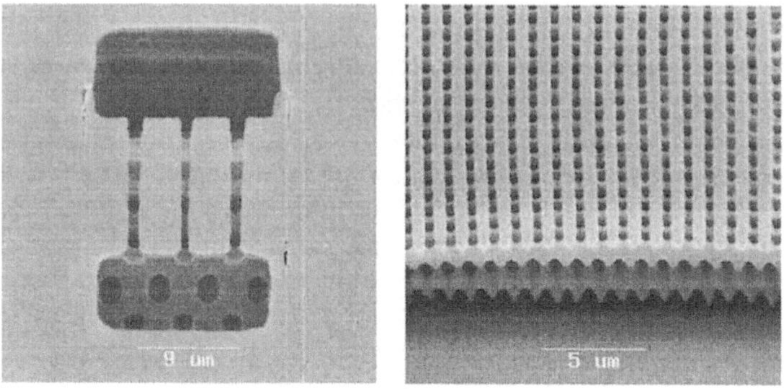

Figure 13. Hollow photonic crystal structures written in S1813 photoresist.

Thus, it is necessary to fabricate replicas from these crystals using materials with a higher refractive index. Application of most negative photoresists for the fabrication of templates is rather complicated, since the structures fabricated in these materials are quite stable and not simply soluble. In case of positive photoresists, the polymer is weakened and is more soluble in developing solutions. This is a very attractive feature for the fabrication of 3D templates.

In Fig. 14, a hollow photonic crystal structure written in S1813 (left) and a first replica of this structure (right) are shown. For the fabrication of the replica, an acrylate monomer is infiltrated into the structure. After that, the S1813 and the acrylate monomer are exposed to UV light and put into an NaOH developer. This procedure dissolves the S1813 structure completely, while the acrylate monomer remains a solid and stable polymer. The result of this procedure is a free-standing replica as shown in Fig. 14 (right side). In the first proof-of-principle experiments, only the replicas in the acrylate monomer were tested.

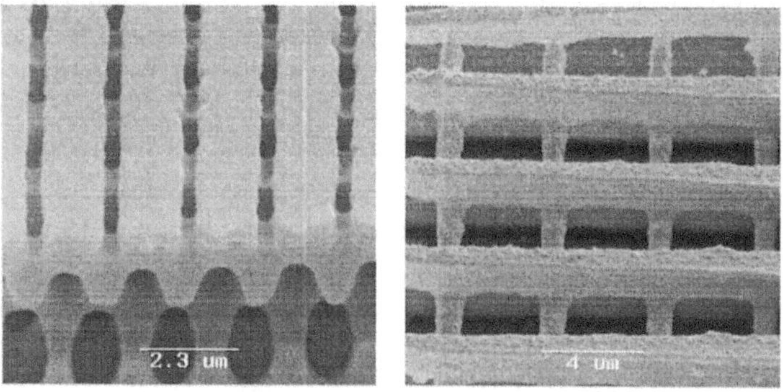

Figure 14. Hollow photonic crystal structures written in S1813 photoresist (left) and its acrylate replica (right) (Chichkov, et al., 2005).

The next step will be the infiltration with a higher refractive index material (n >2).

3. RESOLUTION LIMITS

There are several factors determining the resolution limits of 3D multiphoton material treatment with femtosecond lasers. One is set by the material composition. For example, in ORMOCER® the size of an individual organic-inorganic unit is between 2 and 10 nm, depending on the material properties. Therefore, the smallest structure, which might be fabricated in ORMOCER®, is limited by this size. The second limit depends on laser and material characteristics which define how close one can approach to the 2PP (or 2PA) threshold. The third factor is determined by the order of process nonlinearity (linear or multiphoton processing). And the classical resolution limit is set by the laser wavelength and by the numerical aperture (NA) of the imaging optics. The last limit coincides with the well-known resolution limit of photolithography defined by the following expressions (Levinson, 2001; Ito, et al., 2000):

$$CD = k_1 \frac{\lambda}{NA} \quad \text{and} \quad DOF = k_2 \frac{\lambda}{NA^2}, \tag{1}$$

where CD is the critical dimension of the printed structure size, DOF is the depth-of-focus (axial structure dimension), λ is the exposure wavelength, NA is the numerical aperture of the objective ($NA \cong r/f$ where r is the radius of the lens and f is the focal length), and k_1 and k_2 are factors depending on the photoresist material, process technology, and illumination technique.

For multiphoton polymerization, the resolution limits can be rewritten as

$$CD = \frac{k_1 \lambda}{\sqrt{q}\, NA} \quad \text{and} \quad DOF = \frac{k_2 n \lambda}{\sqrt{q}\, NA^2}, \tag{2}$$

where q is the order of the multiphoton process (q=2 for 2PP), and n is the immersion oil refractive index in an ordinary used high numerical aperture immersion oil microscope objective.

Usually, the Ti:sapphire fs-laser at a wavelength around 780 nm is used for 2PP processing. For all photosensitive materials, a threshold irradiation fluence (and/or irradiation time) exists which has to be overcome to initiate polymerization. Due to this well-defined polymerization threshold, one can reach a resolution far beyond the diffraction limit as is schematically illustrated in Fig. 15 for a Gaussian laser pulse.

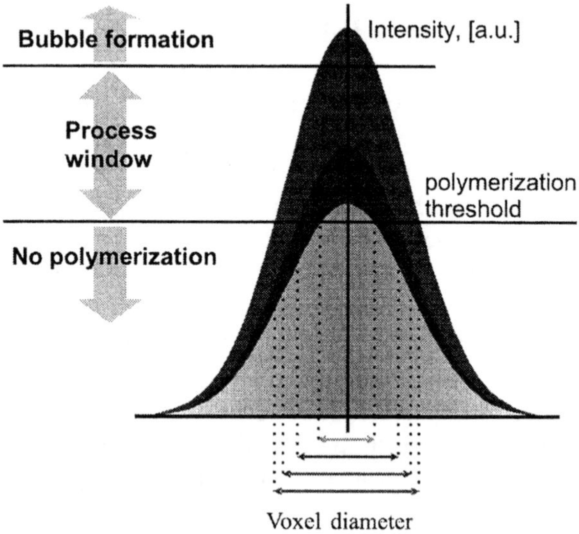

Figure 15. Light intensity distribution and voxel size (resolution) due to a well-defined polymerization threshold.

To calculate the size of the polymerized volume (volume pixel or voxel), one needs to define the polymerization threshold. For this we assume that the resin is polymerized as soon as a specific amount of initiator molecules is activated. For ORMOCER®s, this initiator threshold is roughly estimated to be approximately ρ_{th} =0.25wt%.

The density of activated initiator molecules $\rho(\boldsymbol{r},t)$ can be derived by using the following equation (see (Serbin, et al., 2003) for details)

$$\frac{\partial\rho(r,t)}{\partial t} = (\rho_0 - \rho(r,t))\sigma_2 N^2(r,t), \tag{3}$$

where σ_2 is the effective two-photon cross-section for the activation of initiator molecules, N is the photon flux, and $\rho_0(\boldsymbol{r},t)$ is the primary initiator density. We approximate the light distribution at the focal plane (z=0) by a Gaussian distribution assuming that the photon flux N_0 is constant during the laser pulse:

$$N(r,t) = N_0 e^{-\frac{2r^2}{r_0^2}}. \tag{4}$$

Neglecting the initiator losses between single laser pulses, we can estimate the voxel size d generated by 2PP as the region where $\rho \geq \rho_{th}$ is fulfilled. This leads to

$$d(N_0, t) = r_0 \sqrt{\ln\left(\sigma_2 N_0^2 n \tau_L / C\right)},$$
$$C = \ln(\rho_0 / (\rho_0 - \rho_{th})), \quad (5)$$

where n is the number of pulses, t is the total irradiation time, and τ_L is the laser-pulse duration.

Using the same expression for the axial light distribution at $r=0$ as for a Gaussian beam $N(z) = N_0 / \left(1 + z^2 / z_R^2\right)$, the pixel length can be determined by

$$l(N_0, t) = 2 z_R \sqrt{\sqrt{\sigma_2 N_0^2 n \tau_L / C} - 1}, \quad (6)$$

where z_R is the Rayleigh length.

To achieve a nearly spherical voxel ($l / d \approx 1$), it is important to work at low laser energy and short irradiation time, i.e. close to the polymerization threshold. This gives not only the best resolution, but also the best (closest to spherical) shape of the polymerized voxel.

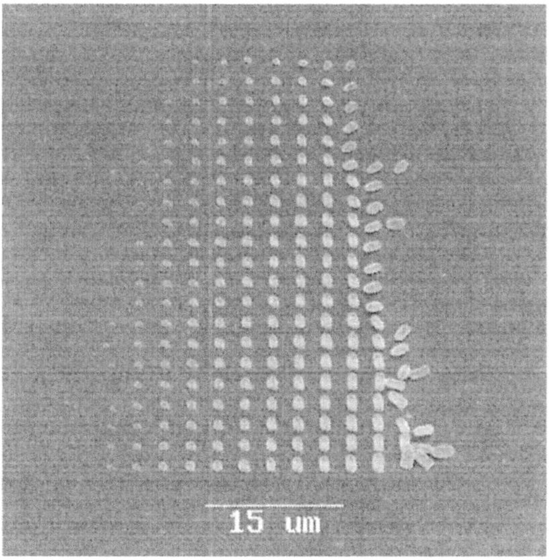

Figure 16. SEM image of a polymerized voxel array. All voxels are produced with the same irradiation time, whereas the laser pulse energy is varied and the z-position of the focus is changed.

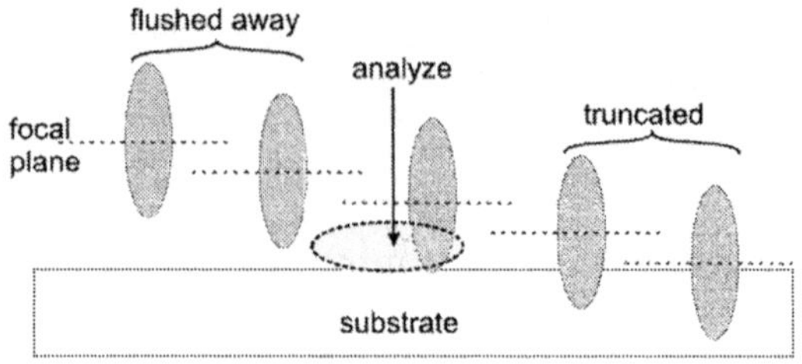

Figure 17. Single voxels in a negative photoresist, depending on the focal z-position.

To compare the theoretical predictions with the experimental results, polymerized voxels were produced in negative photoresist. An SEM-image of a voxel array used for measurements is shown in Fig. 16. The difficulty with this method is that the single voxels must stick to the glass substrate and also should not be truncated (compare with Fig. 17).

The analysis of the voxel array in Fig. 16 gives the dependence of the voxel size on the applied laser energy. Also a voxel array can be produced in which the energy is kept constant and the irradiation time is varied. This gives the size dependence on the irradiation time. These results for negative photoresists are compared with the theoretical calculations in Figs. 18 and 19.

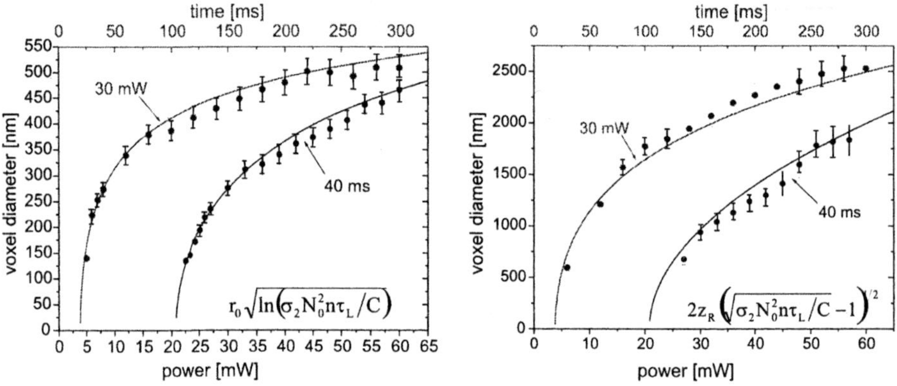

Figure 18. Voxel diameter and –length in ORMOCER®I depending on the irradiation time and laser power (theoretical and experimental results).

All curves cross the x-axis at tP^2=1.7x10^{-5}W^{2}s, in case of ORMOCER®I, and at tP^2=0.12x10^{-5}W^{2}s for SU8. Close to this threshold, one can theoretically get an infinitely small voxel. In practice, the pointing stability and power fluctuations of the laser limit the achievable resolution. The best resolution for a 2PP achieved by us today is about 100 nm.

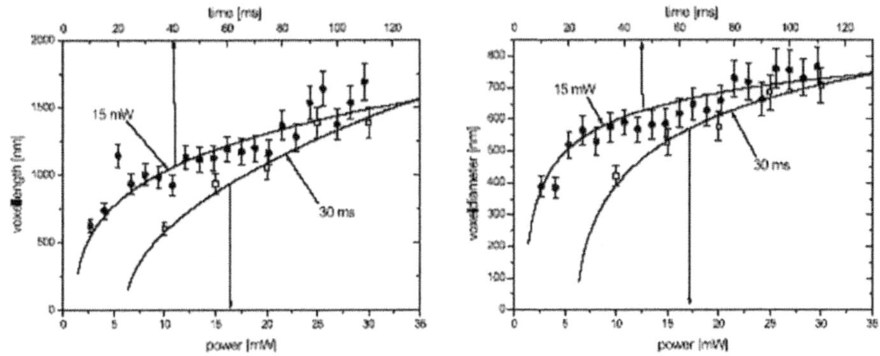

Figure 19. Voxel diameter and –length (the same as in Fig. 18) for SU8.

For the fabrication of imbedded microfluidic components such as cavities and channels, the structuring of positive photoresist with femtosecond laser pulses is a very promising technique. In this case, the resolution limits can be measured by the fabrication of microchannels into the photoresist using different laser pulse energies and scanning speed. Fig. 20 shows microchannels fabricated in S1813 photoresist. The diameter and the axial dimension of these microchannels have been measured and can be described by the same dependences as given by Eqs. 5 and 6. Experimental results and corresponding theoretical curves are shown in Fig. 21.

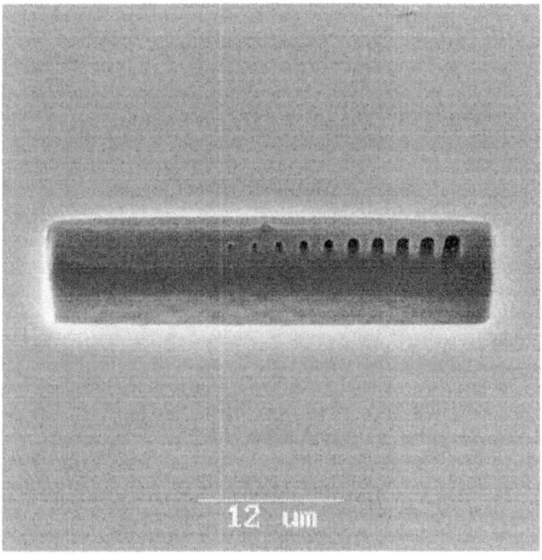

Figure 20. SEM image of microchannels fabricated in positive S1813 photoresist.

Photosensitivity of positive photoresist is much higher than that of ORMOCER®or SU8. Taking into account that structures in positive photoresists can be easily removed by solvents, these two properties make them very attractive for the fabrication of templates and 3D replicas. As shown in Figure 22, in positive photoresist it is possible to achieve resolution below 150 nm.

The nonlinear threshold character of two-photon microstructuring processes allows in all cases the fabrication of structures with a resolution in the subwavelength range, e.g. much smaller than 800 nm. Resolution measurements reported in this section show that structures down to 100 nm can be fabricated by this technique in both negative and positive tone photoresists.

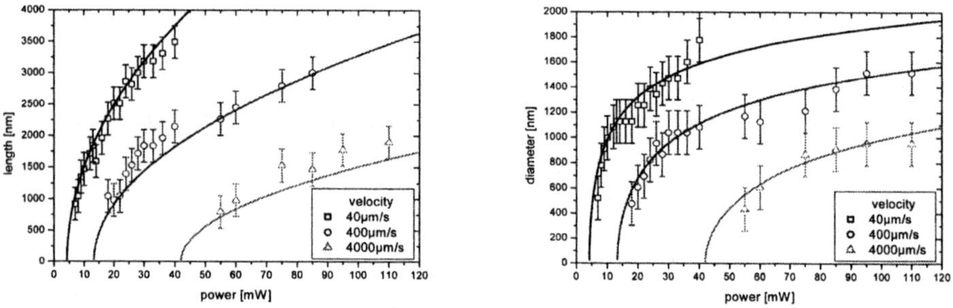

Figure 21. Diameter and length of structures in S1813 photoresist depending on writing speed and laser power (experimental and theoretical results).

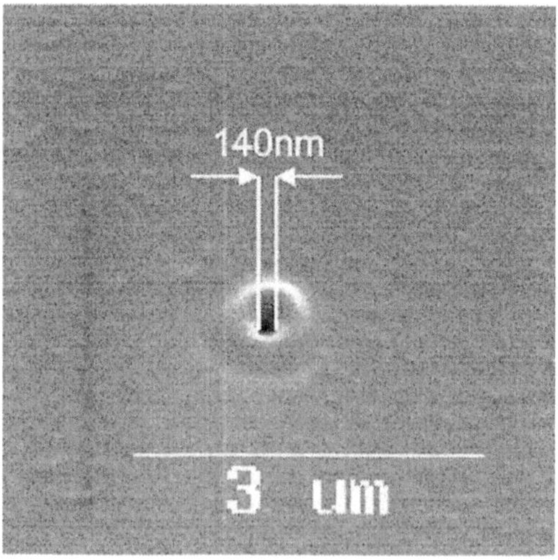

Figure 22. Smallest structure fabricated in S1813 demonstrating the achieved resolution below 150 nm.

4. DEMONSTRATED APPLICATIONS OF 2PP AND 2PA TECHNIQUES

4.1 Photonic crystals and waveguide structures

Photonic crystals (PhCs) are structures consisting of spatially alternating regions with different dielectric constants. Propagation of light inside a certain frequency range, called photonic band gap, is forbidden in such structures. This effect is similar to X-ray Bragg diffraction in regular atomic lattices. If the dielectric constant periodicity occurs in all directions, one is talking about 3D photonic crystal. Depending on the topology and dielectric constant contrast of photonic crystals, their optical properties can be tailored in a desired manner. Many fascinating applications of photonic crystals are possible: control of spontaneous emission, zero-threshold lasing, lossless guiding of light, dispersion management, nonlinear frequency conversion, etc. Futuristic prospects include applications in telecommunication as all-optical signal processing, and realization of "transistors" for light and all optical computers.

Since 1987, when a concept of 3D photonic crystal was introduced by E. Yablonovitch (Yablonovitch, 1987) and S. John (John, 1987), photonic crystals are a subject of intensive research. During the past years, sophisticated theoretical apparatus describing properties of such crystals has been developed, and many 3D PhC designs were proposed and studied theoretically. But the fabrication of photonic crystals with a full 3D (omnidirectional) band gap in the visible range is still a challenging task, mainly limited by the resolution and/or flexibility of already available technologies. For practical applications of photonic crystals, inexpensive and reliable nanofabrication techniques are required. Two most promising techniques for the fabrication of photonic crystals, holographic lithography and two-photon polymerization, are presented in details below.

4.1.1 Holographic lithography

Three-dimensional interference pattern produced by multiple laser beams is a well-studied subject because it is used for trapping of atoms. Holographic recording is a method that allows fast (literally, single shot) fabrication of up to few mm large 3D crystals with long range periodicity. The idea is to record an interference pattern created by two or more beams into a photosensitive material. If the beam parameters are chosen properly, it is possible to polymerize photosensitive resins at the high intensity regions while leaving the rest of material non-polymerized. This can be achieved if light intensities at the interference maxima are larger than the threshold value for polymerization, while intensities at the minima are lower than the polymerization threshold. Tuning the intensity inside this working window changes the filling factor and therefore, has an effect on stability of the structure. The material that is not polymerized is then washed out to reveal a solid 3D mesh in air. Various lattices can be fabricated depending on the number of beams applied and their arrangement. The period of the crystal structure depends on the wave vectors of applied laser beams. Campbell et al. (Campbell, et al., 2000) have reported four UV laser beam interference for the fabrication of periodic three-dimensional pattern in a 30 μm thick SU8 photoresist. Structures with sub-μm features were produced by a single 6 ps pulse of a UV laser. Shortly after that, Shoji and Kawata have reported the fabrication of 3D hexagonal photonic crystal lattice by a two-step process (Shoji, et al., 2000): in a first step, three-beam interference was used to obtain 2D hexagonal arrangement of rods, in the second immediate step two more interfering beams formed additional layers which are perpendicular to the rod array. In 2003, Shoji et al. also demonstrated fabrication of woodpile structure by two-step four-beam interference (Shoji, et al., 2003). Four-beam interference produced a square lattice of rod array; a woodpile is formed from two such lattices rotated by 90 degrees with respect to each other. Several groups used single-step exposure four-beam interference to fabricate FCC lattices with different filling factors (Miklyajev, et al., 2003; Wang, et al., 2003). In these papers, an umbrella-like beam configuration with a central beam symmetrically surrounded by three other beams was used. Lin et al. have recently proposed a method of holographic fabrication by using phase masks (Lin, et al., 2005). They have generated 3D woodpile structures in photosensitive SU8 in a two step process using two masks, an approach similar to that applied by Shoji et al. (Shoji, et al., 2003).

Recording into the photosensitive material is not the only way to produce 3D PhC by multiple-beam interference. Duneau et al. have proposed to use multiple-beam interference to directly grow 3D photonic crystals by chemical vapour deposition; to our knowledge a practical realization of this idea has not been demonstrated (Duneau, et al., 2004).

Holographic recording results in "perfect" structures and additional steps are needed in order to introduce defects. For materials where polymerization is driven by cationic reaction, a refractive index change appears only during the postbake. Multiphoton direct writing of defects is easier to realize in this case, since no additional effort is required in order to match refractive indices; also, final samples can be developed in a single common step. Further information on holographic recording and fabrication of photonic crystals can be found in the excellent review (Blanco, et al., 2004).

4.1.2 Two-photon polymerization

The ability of 2PP to create complex 3D structures with exceptionally high resolution (down to 100 nm) makes this technology advantageous for the fabrication of 3D photonic crystals. With this technique, one is able to introduce defects at any desired location, which is crucial for the practical applications of photonic crystals. Few groups employing 2PP have reported the fabrication of photonic crystals exhibiting photonic band gap effects. One of the first papers employed a frequency doubled femtosecond laser radiation at 400 nm to write a woodpile structure in the Nopcocure 800 resin, which has its UV absorption edge at 370nm (Sun, et al., 1999). Fabricated structures consisted of 20 layers, had in-layer spacing down to 1.2μm, and rod diameters of 1μm. Optical characterization revealed dips in transmission spectra. Attenuation of only 1.3dB per unit cell was demonstrated. Nevertheless, this work is an important achievement that revealed great potential of 2PP for the fabrication of photonic crystals. In 2001, the same group succeeded in the fabrication of a photonic crystal with a planar defect (Sun, et al., 2001). Recently, the fabrication of high-quality large-scale 3D photonic crystals with fundamental stop bands in IR, ranging from 1.3 to 1.7 μm, has been reported by another group (Deubel, et al., 2004).

A common problem with processing of negative tone photosensitive polymers is the shrinkage of the produced structure. In case of ORMOCER®I, shrinkage does not affect the structure during the 2PP polymerization, since only single voxels shrink at the position of the laser focus. After development, the fabricated structure consists from partially cross-linked voxels and polymer resin. In order to increase the structure stability, it is exposed to UV light. This final polymerization of the whole structure leads to distortions and deviations from a desired geometry.

The inset in Figure 23 illustrates the shrinkage effect on an example of a photonic crystal written in ORMOCER®I. The plot shows dependence of the shrinkage degree (ratio of the upper layer size to the size of the lower layer) on the average laser power used for 2PP. Different pre and postprocessing procedures were tested, and no considerable improvement was noticed. An upper layer can shrink by as much as 20%, compared to the layer on the substrate. The ratio of cross-linked to not cross-linked polymer in the structure depends on the exposure dose, and therefore, the more power is used for 2PP, the less shrinkage will occur during the postexposure with UV light. Increasing the laser power used for 2PP inevitably decreases the resolution of 2PP processing, which in most cases is a bad trade-off.

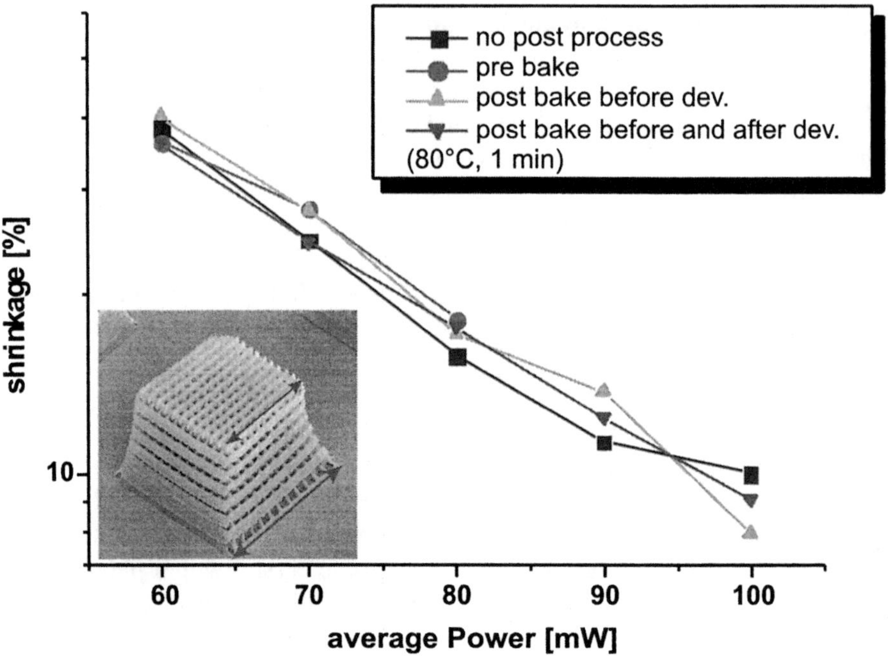

Figure 23. Shrinkage of photonic crystals fabricated from ORMOCER®s.

Shrinkage changes the size of a woodpile layer and also the in-layer distance between the rods, introducing modifications in photonic band gap characteristics. Studies of effects played by structural fluctuations on photonic band gap characteristics can be found in (Chutinan, et al., 1999).

Since the bottom layer of a crystal is attached to the substrate, it cannot shrink. Far from the substrate, the layer has more freedom to shrink. At some distance from the substrate, the saturation region occurs, where every subsequent layer shrinks by the same amount.

To solve the shrinkage problem, we have introduced a numerical compensation into the computer model for the fabrication of a woodpile structure. By measuring the shrinkage dependence on the layer number, we have empirically found a function describing this behavior. Using inverted function:

$$F_{comp} = [1 + s(1 - \exp(-0.3z)]^{-1}, \quad (7)$$

where z is a layer number (counted from the substrate), one can calculate exactly how much each single 2PP produced layer should be modified in order to obtain an undistorted woodpile structure after the UV postexposure. The dimensions of each layer in a computer model are simply multiplied by the F_{com} function, according to its number.

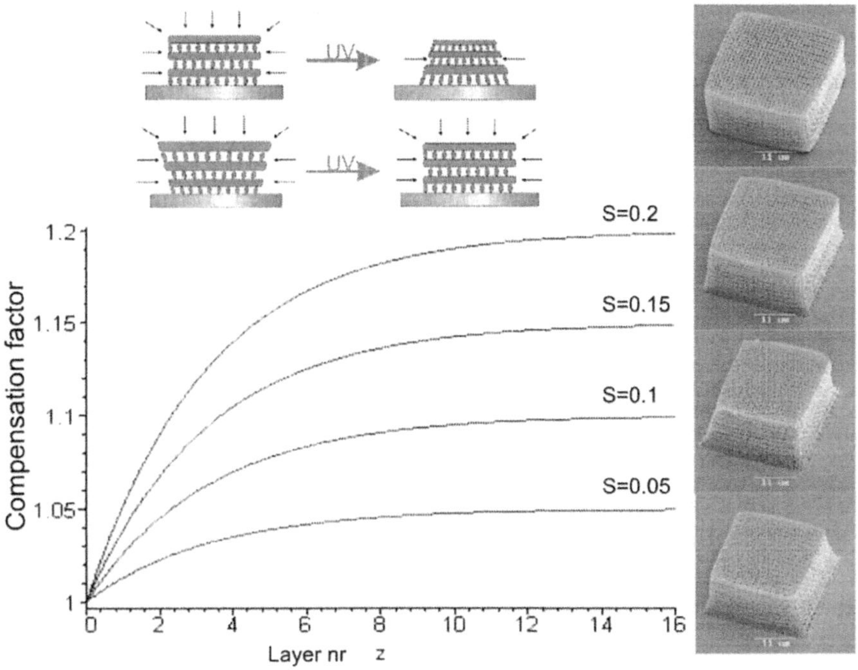

Figure 24. Numerical shrinkage compensation applied to PhCs in ORMOCER®I.

The parameter *s* allows "fine tuning", i.e. changing the slope of the function to compensate for postexposure dose effect. The plot in Figure 24 illustrates numerical shrinkage compensation for different values of *s* parameter. The inset in this figure shows scanning electron microscope (SEM) images of woodpile photonic crystals precompensated according to Eq. 7. With the increasing *s* value, the woodpile structure acquires its correct form. This simple approach allows the fabrication of undistorted photonic crystal structures.

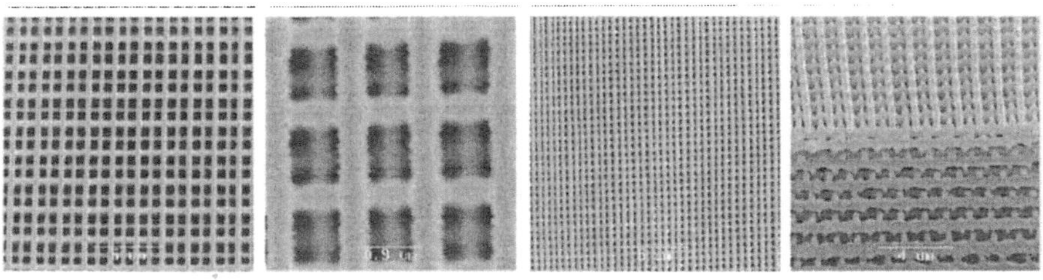

Figure 25. Photonic crystal "woodpile" structures written in ORMOCER®I.

We have fabricated woodpile photonic crystal structures in two different classes of materials, ORMOCER®I and SU8. Both materials are designed for optical applications and are highly transparent in the visible and IR range. Figure 25 shows SEM images of different photonic crystals written in ORMOCER®I. Transmission of such photonic crystals was analyzed (for details see (Serbin, et al., 2004, Chichkov, et al., 2005).

Relative width and position of a photonic band gap depend on the refractive index contrast between dielectric material and air. For woodpile structures, the refractive index contrast of at least 2.7 is required for the demonstration of full photonic bandgap. Refractive indices of most photosensitive materials processible with 2PP are below 1.8.

One possibility for the realization of photonic crystals from high-refractive index materials is to use 2PP fabricated structures as templates, and to produce inversed replicas of these structures. For this purpose, ORMOCER®s are not suitable, since they exhibit exceptional thermal and chemical stability and hence are extremely hard to remove. Structures made in SU8 can be removed by annealing at 600°C. It could be more advantageous to produce templates from positive photoresists. This topic has been discussed in Sec. 2.2.

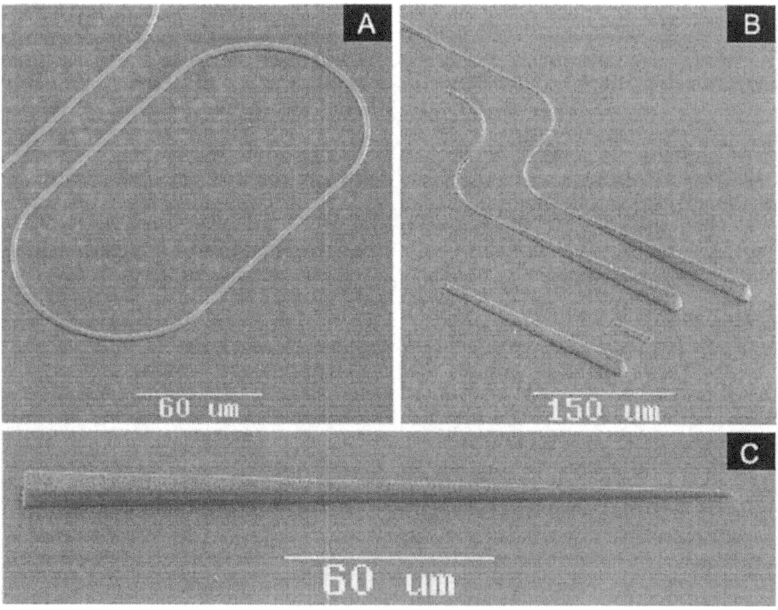

Figure 26. Waveguides fabricated by 2PP: a) ring resonator; b) and c) tapered waveguides.

Due to their excellent optical properties, ORMOCER®s are very attractive materials for the fabrication of microoptical devices (see Sec. 2.1). Using 2PP, it is straightforward to generate complicated 2D and 3D structures, like ring resonators and 3D waveguide tapers shown in Figure 26. These waveguides are suitable for guiding single-mode light at 1550 nm.

In conclusion, 3D microstructuring of photosensitive materials using two or multi-photon processing techniques has great potential for the fabrication of photonic crystals, waveguides, and microoptical components. Possible applications in medicine and biology will be highlighted in the next section.

4.2 Biomedical applications

There are several very promising bioapplications of the 2PP technique: for tissue engineering, drug delivery, medical implants, and sensorics. For tissue engineering, the ability to produce arbitrary 3D scaffolds is very appealing. Artificial fabrication of living tissue that will be able to integrate with the host tissue inside the body is a challenging task. Natural repair of a tissue at the particular site is a result of complex biological mechanisms which are the subject of intensive research. In order to encourage cells to build such a tissue, one has to create an appropriate environment exactly resembling that of a particular tissue type. Some cell types can preserve tissue-specific features in a 2D environment, others require a 3D environment. Another problem that becomes critical with increasing scaffold size is possibility of blood-vessel formation throughout the scaffold, in order to promote healing and to avoid hypoxia. 2PP in combination with the right materials allows the precise control over 3D geometry of the scaffold, and therefore, allows to model and to reproduce *in vivo* cellular microenvironment. Furthermore, high resolution of 2PP can provide control over the cell organisation inside the scaffold and consequently over the cell interactions. Another advantage of 2PP is that near-IR laser radiation, used for 2PP, is not dangerous for cells at the applied intensities and could also be used for the manipulation and encapsulation of cells.

The study of intrinsic biocompatibility of ORMOCER® is an important and necessary first step towards fabricating novel biological scaffold structures and other applications of ORMOCER® in biomedicine. We have recently studied biocompatibility of ORMOCER®I; details can be found in (Doraiswamy, et al., 2005). The obtained results have shown good adherence of different cell types to this material and a growth rate comparable to bioactive materials like ECM (extracellular matrix).

An example of a potential scaffold structure fabricated in ORMOCER®I is shown in Figure 27. In this case, amplified laser pulses were focused by a f-theta lens with a focal distance of 80 mm in order to produce long

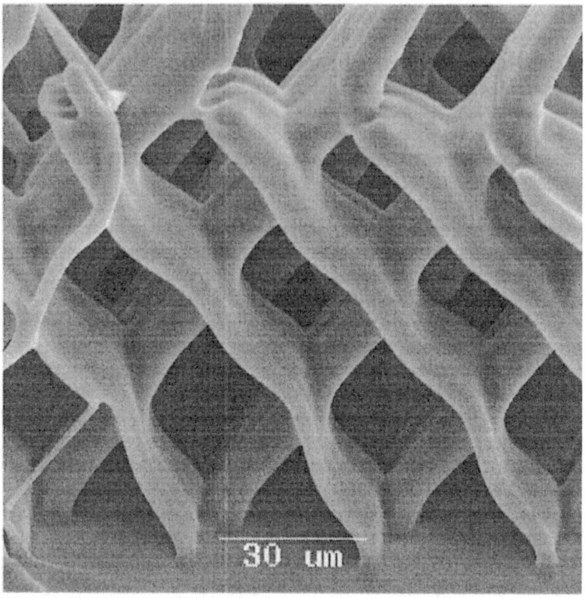

Figure 27. Scaffold structures produced by 2PP of ORMOCER®I.

filaments through the 300 μm thick ORMOCER®I layer. The structure was formed in two steps by scanning parallel lines on a tilted sample. ORMOCER®I contained 1.8% photoinitiator Irgaure 369. Since material properties can be controlled by chemical design, all of this makes ORMOCER®I a good candidate for biological applications and offers great potential for developing true *designer* 3D biological scaffolds.

For the fabrication of 3D structures with higher resolution, Zeiss 20X planachromat microscope objective was used to focus laser pulses into ORMOCER®I.

Free-standing Lego® type structures were fabricated on salt substrates. The size of the Lego® structure base is 500μm x 500μm x 40μm; it is supported by 9 50 μm in diameter and 150μm long hollow pillars, and carries 16 such pillars on top (Figure 28 (left)). After fabrication, unsolidified ORMOCER®I is washed away by 1:1 volume solution of 4-methyl-2-penthanone and isopropanol.

The shown free-standing Lego® type structures can be stacked layer-by-layer for developing heterogeneous tissue. Each layer would mimic the natural tissue construct, with layers such as epithelial, endothelial, vascular, neural, muscular, and specialized cells to a functional tissue (Narayan, et al., 2005).

Pillars of various diameters (10-150μm), and 20 μm in height were fabricated in ORMOCER® coated on glass substrates. Studies of simple pillar structures seeded with B35 cells show that cells are oriented along the pillar walls, suggesting the recognition of the structure by the cells (see Figure 28, right).

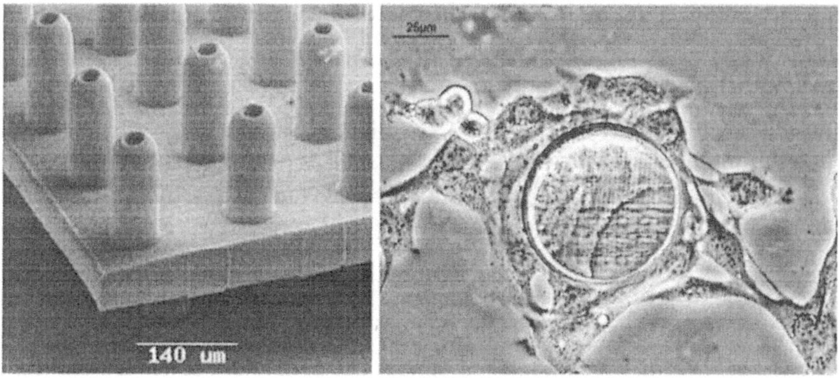

Figure 28. Free-standing Lego® type structure fabricated by 2PP (left) in ORMOCER®I, simple pillar structure seeded by cells (right).

Ideally, the scaffold should provide a temporary support for tissue regeneration and should later degrade (either during or after healing). Therefore, it is very important to develop photosensitive materials that are biodegradable and can be structured by 2PP technique. A potential class of photosensitive biodegradable materials is Poly (ethylene glycol) (PEG) based hydrogels, which are commonly used for cell encapsulation and can be photocrosslinked with UV light (Bryant, et al., 2004).

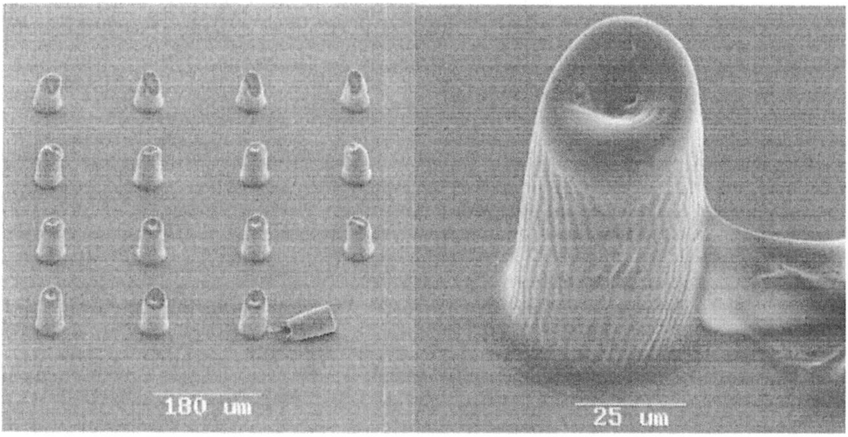

Figure 29. Cylinder structures produced by 2PP of PEGDA-based polymer.

PEG-based materials are fully synthetic and hence allow control of their physical and chemical properties, which is more difficult with natural polymers. The property that makes PEG biocompatible is its intrinsic protein resistance, which reduces protein adsorption on its surface (Albrecht, et al., 2005). In order to explore the structurability of PEG-based polymers by 2PP, we have performed experiments with Poly (ethylene glycol)-dimethacrylate, using Irgacure 784 as a photoinitiator. PEG-dimethacrylate (Mn=875) was mixed with 10% phosphate-buffered saline (PBS). 5% of Irgacure 784 solved in acetone was added by stirring. In these preliminary studies, the material is not expected to be biodegradable; it forms a quite firm constellation, resulting in well-defined structures produced by 2PP.

Figure 29 shows SEM images of pillars fabricated by 2PP of PEG-dimethacrylate-based polymer. After fabrication, the non-polymerized material was washed away with distilled water. The samples were dried at room temperature. Slight deformations of the top of the cylinders and wrinkles indicate that the produced structures are elastic, and shrink as the water is evaporating. This material can be used for the fabrication of support structures for biodegradable scaffolds made from weakly cross-linkable biodegradable hydrogels.

Microstructures created by 2PP can also be used for drug delivery. Figure 30 shows SEM images of microcapsules fabricated on the surface of a metal stent. A stent is a wire mesh tube used to prop open an artery during angioplasty, and is a common procedure applied to cure atherosclerosis. In some cases the restenosis, a renarrowing of the artery, may occur. To avoid this problem, drugs are injected into the blood, suppressing formation of new tissue at the stent site. To minimize the amount of drugs required and therefore, to reduce possible side effects, drug-eluting stents were developed. Such stents are coated with drugs that are slowly released and help to keep the blood vessel from reclosing. Flexibility of 2PP allows fabrication of microcapsules of different geometries directly on the stent surface.

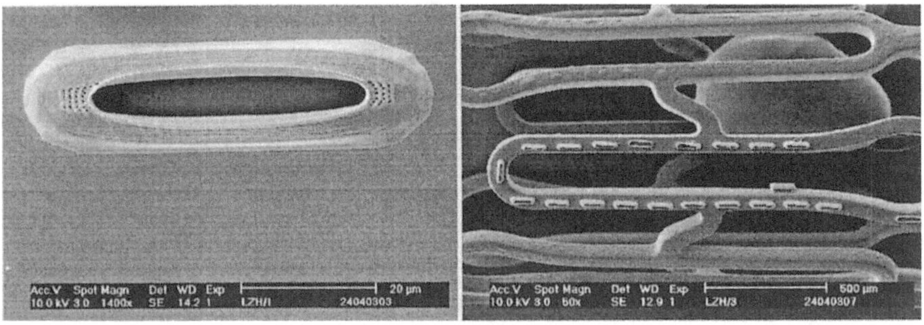

Figure 30. Microcapsules produced by 2PP of ORMOCER® on metal stents.

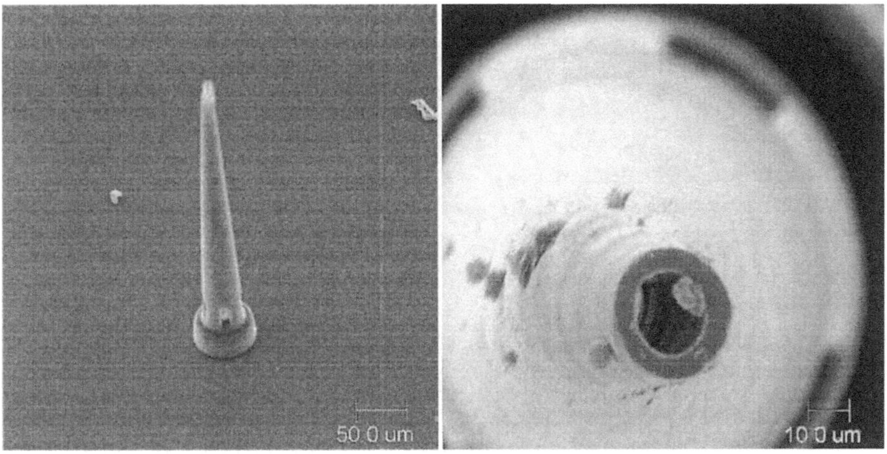

Figure 31. Microneedles for drug delivery, fabricated by 2PP of ORMOCER®.

Using 2PP, one can produce microneedles with a large range of sizes, shapes, and improved properties. In Figure 31, SEM images of microneedles fabricated from ORMOCER®are shown. Microneedle arrays were fabricated for drug delivery. Micron scale needles can increase skin permeability and improve the transdermal delivery of drugs. They can provide painless injections to people suffering from diabetes, blood clotting, or other disorders.

In conclusion, topics discussed in this section demonstrate great potential of 2PP technology for the fabrication of complex 3D structures for biomedicine.

4.3 Micromechanical and microfluidic devices

Further promising applications of two-photon activated (2PA) processing are in the field of rapid prototyping and fabrication of micromechanical and microfluidic systems. In this section, a few examples of structures fabricated by 2PA are presented. In Figure 32, a gearwheel made by 2PP of SU8 is shown.

Since the resolution limit of 2PP is at present about 100 nm, it is possible to build gearwheels with a diameter of around 1μm. Because 2PP is the real 3D processing technology, one can build any desired structure. This means not only single elements for micromechanical systems, but also complex constructions.

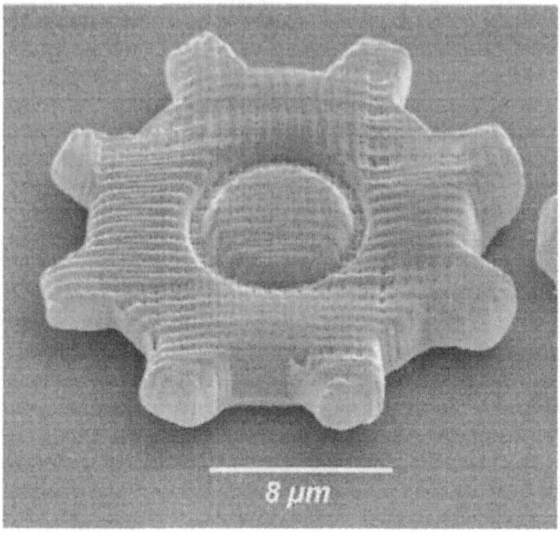

Figure 32. A gearwheel made with 2PP in SU8.

The other possible field of applications is microfluidics. With 2PA technique it is possible to build 3D microfluidic and optical devices on the same chip. A good example is the ring resonator microcavity shown in Figure 33. The cavity is fabricated in S1813 photoresist and is surrounded by 4 90^0 mirrors. The emitted light from a dye solution inside the cavity can be directed around the resonator by total internal reflection. The image shown in Figure 33 represents a horizontal cut through the device, which also can be fabricated completely buried in the resist, and the dye can be injected via microchannels.

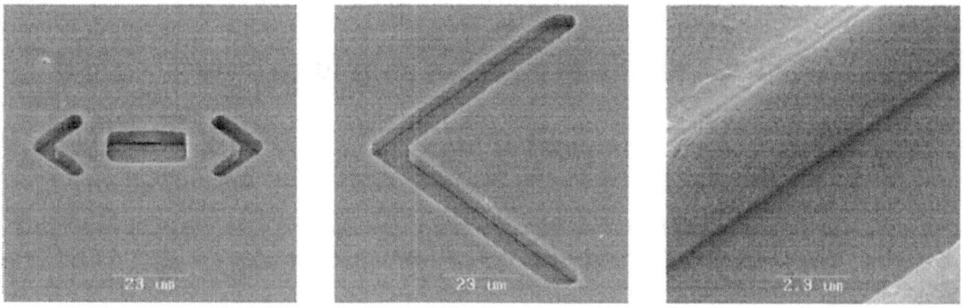

Figure 33. A microring resonator fabricated in S1813 photoresist by 2PA technique (left), close-up of the cavity mirror (two right images).

This technique provides a simple method for the fabrication of different imbedded microstructures and microfluidic devices.

5. SUMMARY AND OUTLOOK

Recent progress in three-dimensional microstructuring of photosensitive materials by femtosecond lasers has been reviewed. Due to the nonlinear nature of multiphoton laser-activated processing, application of ultrashort laser systems allows one to overcome the diffraction limit and to produce high-quality 3D microstructures with a subwavelength resolution. This is very powerful technology with many potential applications which have been briefly discussed in this chapter.

This 3D nonlinear laser processing technique is still in a rapidly developing phase and represents a very exciting field of laser physics and novel laser material processing technologies.

6. ACKNOWLEDGMENTS

The authors gratefully acknowledge very important contributions from our colleagues which have been involved in different parts of this work: T. Bauer, C. Cronauer, J. Koch, J. Li, A. Ostendorf, C. Reinhardt, B. Rubehn, and J. Serbin. Biomedical applications of 2PP technique have been studied in cooperation with A. Doraiswamy, T. Platz, R. Narayan, R. Modi, R. Auyeung, and D.B. Chrisey. This work has been supported by the DFG "Photonic crystals" research program SPP1113 and by the European Network of Excellence "Plasmo-Nano-Devices."

References

Albrecht, D.R., Tsang, V.L., Sah, R.L., and Bhatia, S.N., 2005, *Lab. Chip.***5**, 111

Amberg-Schwab, S., Arpac, E., Glaubitt, W., Rose, K., Schottner, G. and Schubert, U., 1991, *Materials Science Monographs – High Performance Ceramic Films and Coatings*, Vol. 67 (Ed: P. Vincencini), Elsevier, Amsterdam, p. 203.

Blanco, A., Busch, K., Deubel, M., Enkrich, C., Von Freymann, G., Hermatschweiler, M., Koch, W., Linden, S., Meisel, D.C. and Wegener, M., "*Three-dimensional lithograohy of Photonic-Crystals* in *Photonic Crytsals*", edited by Busch, K., Lölkes, S., Wehrspohn, R. B., Föll, H., 2004, Wiley-VCH, 153–173

Bräuer, A., Dannberg, P., Mann, G. and Popall, M., 2001, *MRS Bull.* **26**, 519

Brinker, C.J. and Scherer, W.G., 1990, *Sol-Gel Science*, Academic Press, New York

Bryant, S.J., Chowdhury, T.T., Lee, D.A., Bader, D.L., and Anseth, K.S., 2004, *Ann. Biomed. Eng.*, **22**, 407

Buestrich, R., Kahlenberg, F., Popall, M., Dannberg, P., Müller-Fiedler, R. and Rösch, O., 2001, *J. Sol-Gel Sci. Technol.* **20**, 181

Campbell, M., Sharp, D. N., Harrison, M. T., Denning, R. G. and Turberfield, A. J., 2000, *Nature London* **404**, 53

Chichkov, B. N., Koch, J., Ovsianikov, A., Passinger, S., Reinhardt, C. and Serbin, J., 2005, *Mater. Res. Soc. Symp. Proc.*, **850**, MM2.5.1

Chutinan, A. and Noda, S., 1999, *J. Opt. Soc. Am. B*, **16**, 240

Cumpston, B. H., Ananthavel, S. P., Barlow, S., Dyer, D. L., Ehrlich, J. E., Erskine, L. L., Heikal, A. A., Kuebler, S. M., Lee, I.-Y. S., McCord-Maughon, D., Qin, J., R¨ockel, H., Rumi, M., Wu, X.-L., Marder, S. R. and Perry, J. W., 1999, *Nature*, **398**, 51

Depré, L., Ingram, M., Poinsignon, Ch. and Popall, M., 2000, *Electrochim. Acta* **45**, 1377

Deubel, M., Von Freymann, G., Wegener, M., Pereira, S., Busch, K. and Soukoulis, C.M., 2004, *Nature Materials*, **3**, 444

Deubel, M., Von Freymann, G., Wegener, M., Pereira, S., Busch, K. and Soukoulis, C. M., 204, *Nature Materials* **3**, 444

Doraiswamy, A., Patz, T., Narayan1, R., Chichkov, B., Ovsianikov, A., Houbertz, R., Modi, R., Auyeung, R., Chrisey, D. B., 2005, *Mater. Res. Soc. Symp. Proc.*, Vol. 845.

Duneau, M., Delyon, F. and Audier, M., 2004, *J. Appl. Phys.*, **96**, 2428

Fröhlich, L., Houbertz, R., Jacob, S., Popall, M., Mueller-Fliedler, R., Graf, J., Munk, M. and Von Zychlinski, H., 2002, *Mater. Res. Soc. Symp. Proc.* **726**, 349

Haas, K.-H., 2000, *Adv. Eng. Mater.* **2**, 571

Houbertz, R., Domann, G., Cronauer, C., Schmitt, A., Martin, H., Park, J.-U., Fröhlich, L., Buestrich, R., Popall, M., Streppel, U., Dannberg, P., Wächter, C. and Bräuer, A., 2003, *Thin Solid Films* **442**, 194

Houbertz, R., Domann, G., Schulz, J., Olsowski, B., Fröhlich, L. and Kim, W.-S., 2004, *Appl. Phys. Lett.* **84**, 1105

Houbertz, R., Domann, G., Schulz, J., Popall, M., 2003, *Mater. Res. Soc. Symp. Proc.* **769**, 239

Houbertz, R., Fröhlich, L., Popall, M., Streppel, U., Dannberg, P., Bräuer, A., Serbin, J. and Chichkov, B.N., 2003, *Adv. Eng. Mater.* **5**, 551

Houbertz, R., Schulz, J. and Popall, M., 2001, *Mater. Res. Soc. Symp. Proc.* **665**, 321

Ito, T. and Okazaki, S., 2000, *Nature* **406**, 1027

John, S., 1987, *Phys. Rev. Lett.* **58**, 2486

Kahlenberg. F. and Popall, M., 2005, *Mater. Res. Soc. Symp.* **847**, EE14.4.1

Kawata, S., Sun, H.-B., Tanaka, T. and Takada, K., 2001, *Nature*, **412**, 697

Levinson, H.J., 2001, *Principles of Lithography*, SPIE press

Lin, Y., Herman, P.R. and Darmawikarta, K., 2005, *Appl. Phys. Lett.* **86**, 071117

Maruo, S., Nakamura, O. and Kawata, S., 1997, *Opt. Lett*, **22**,132

Miklyaev et al., 2003, *Appl. Phys. Lett.*, **82**, 1284

Narayan, R. J., Jin, C., Patz, T., Doraiswamy, A., Modi, R., Chrisey, D. B., Su, Y-Y., Lin, S. J., Ovsianikov, A. and Chichkov, B., 2005, *Advanced Materials & Processes*, 39, April

Popall, M., Andrei, M., Kappel, J., Kron, J., Olma, K. and Olsowski, B., 1998, *Electrochim. Acta* **43**, 1155

Prasad, P. N., "*Nanophotonics*", 2004, John Wiley & Sons

Robertsson, M.E., Hagel, O.J., Gustafsson, G., Dabek, A., Popall, M., Cergel, L., Wennekers, P., Kiely, P., Lebby, M. and Lindhal, T., 1998, *Proc. 48th Electron. Comp. Technol. Conf.* (Seattle, Washington, USA), Institute of Electrical and Electronics Engineers, Inc. IEEE Catalogue No. 98CH36206 (1998), p. 1413.

Rose, K., 1995, in: *Organosilicon Chemistry II*, Auner, N., and Weis, J., (eds.), VCH Weinheim, Germany, p. 649.

Rousseau, A. and Boutevin, B., 1992, *Proc. Of the Plastic Optical Fibrers Conference* (Paris) p. 33.

Schottner, G., 2001, *Chem. Mater.* **13**, 3422

Serbin, J., Egbert, A., Ostendorf, A., Chichkov, B. N., Houbertz, R., Domann, G., Schulz, J., Cronauer, J., Fröhlich, L. and Popall, M., 2003, *Opt. Lett.* **28**, 301
Serbin, J., Ovsianikov, A. and Chichkov, B., 2004, *Opt. Express*, **12,** 5221
Shoji, S. and Kawata, S., 2000, *Appl. Phys. Lett.*, **76**, 2668
Shoji, S., Sun, H.-B. and Kawata, S., 2003, *Appl. Phys. Lett.*, **83**, 608
Streppel, U., Dannberg, P., Wächter, Ch., Bräuer, A., Fröhlich, L., Houbertz, R. and Popall, M., 2002, *Opt. Mater.* **21**, 475
Sun, H-B., Kawata, S., 2004, "*Two-Photon Photopolymerization and 3D Lithographic Microfabrication*", Springer Verlag
Sun, H-B., Matsuo, S. and Misawa, H., 1999, *Appl. Phys. Lett.*, **74**, 786
Sun, H-B., Mizeikis, V., Xu, Y., Juodkazis, S., Ye, J-Y., Matsuo, S. and Misawa, H., 2001, *Appl. Phys. Lett.*, **79**, 1
Wang, X., et al., 2003, *Appl. Phys. Lett.*, **82**, 2212
Wolter, H., Storch, W. and Ott, H., 1994, *Mater. Res. Soc. Proc.* **346**, 143
www.voco.com
Yablonovitch, E., 1987, *Phys. Rev. Lett.* **58**, 2059

Chapter 7

ULTRAFAST OPTICAL MEASUREMENTS OF SHOCKED MATERIALS

David J. Funk, David S. Moore, and Shawn D. McGrane
Los Alamos National Laboratory, Los Alamos, NM 87545

1. INTRODUCTION

A molecular description of detonation has been pursued for decades (c.f. Tarver, 1997), with little or no success in obtaining high-quality data at these scales. Recently, we have setup laboratories specifically designed to address this deficiency, employing pump-probe techniques based on chirped-pulse amplified Ti:sapphire laser technology coupled with thin film targets of energetic materials (Hare, et al., 1995). Several reasons exist for these technological choices, including the relative ease of synchronicity that one can obtain using direct optical (ablative) drive followed by optical probe. In contrast to shock experiments with gas guns, in which obtaining synchronicity to a few picoseconds is extremely difficult and time-resolution occurs with fast streak cameras, optical synchronicity is simply obtained by changing the optical path lengths with translation stages and the time resolution results from the ~130 fs "frames" that are recorded as a function of delay. In addition, these timescales are required if one would like to resolve the "rise time" (e.g. ambient to peak pressure) of the shockwave, observe energy transfer and/or redistribution within the shockwave (phonons, molecular vibrations), observe shock-induced chemical reactions, or study some shock-induced phase transitions (solid-solid and solid-melt). For example, a shockwave propagating through a material at a velocity of 6 km/s causes the material to "jump" to a velocity of 1 or 2 km/s. Such a "particle velocity" translates to a time and distance scale of 2 μm/ns or

2 nm/ps. Consequently, if one is to time-resolve the processes listed above at these scales, one needs picosecond or better temporal resolution.

Furthermore, these "table-top" experiments are generated in small sample areas (0.03 mm^2) and can be conducted repetitively, achieving very high S/N ratios that cannot be readily obtained in single-shot experiments. Due to the high intensity of the amplified ultrafast laser pulses, it is quite easy to generate a variety of probe wavelengths, making the technique spectroscopically versatile; transient experiments using Raman, IR absorption, UV-Vis absorption, or CARS can easily be conducted. However, there was concern that these methodologies would not yield conditions in the shocked materials that would allow interpretable results. We have conducted a number of characterization experiments of our shockwaves using interferometric methods that have supported our choice of technology and have led to the results presented here, ultimately leading to the first observation of shock-induced chemistry in an energetic material on ps time-scales (McGrane, et al., 2004b).

2. EXPERIMENTAL METHODS

2.1 Interferometric Measurements

In the experiments described here, two separate techniques are used for interferometric characterization of the shocked material's motion: frequency domain interferometry (FDI) (Tokunga, et al., 1992; Geindre, et al., 1994; Evans, et al., 1996; Gahagan, et al., 2000) and ultrafast spatial interferometric microscopy (Takeda, et al., 1982; Gahagan, et al., 2002). Shown in Fig. 1a, is our implementation of reflection frequency domain interferometry. Here, a pair of ultrafast probe pulses is used to simultaneously measure the dynamic phase shift and reflectance during shock breakout from thin-film metal targets. For the FDI experiments, a single 800 nm, $t_p = 130$ fs, 0.7 mJ laser pulse generated by a seeded, chirped-pulse amplified Ti:sapphire laser system (Spectra Physics) was used for both shock-generation and probing. The shock generating pulse (0.2 – 0.5 mJ) was focused onto the front side of the target assembly to a spot size of d_s = 75 μm. A small portion of the main pulse (~0.04 mJ) reflected from a beam splitter was passed through an unbalanced Michelson interferometer to produce a pair of probe pulses separated in time by 4-16 ps. These s-polarized pulses were loosely focused onto the backside of the target at an angle of $\theta = 32.6°$ to a spot size of ~1000 μm to circumscribe the region of shock break-out at probe intensity less than ~5 x 10^{11} W/cm^2 (pump intensity of ~1 x 10^{14} W/cm^2).

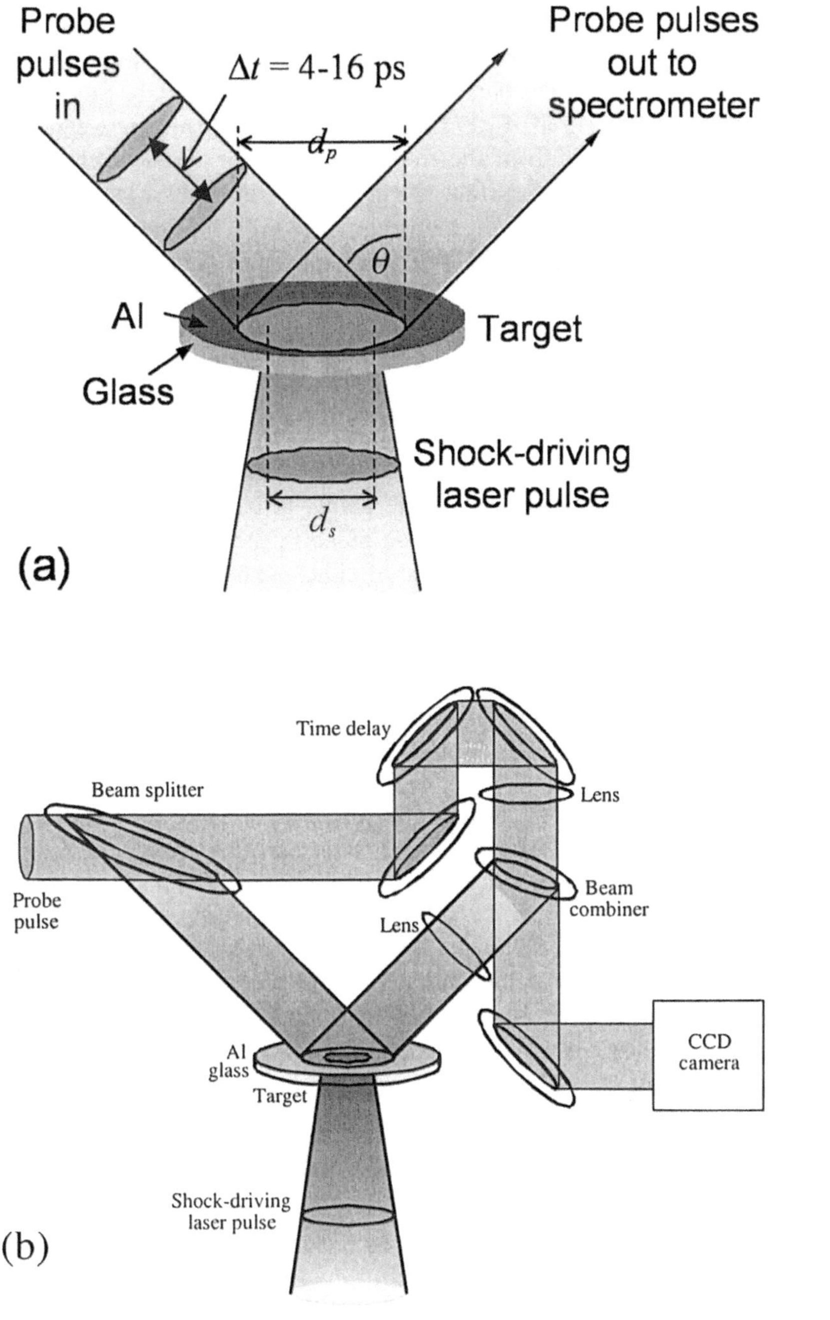

Figure 1. Schematic diagrams of the femtosecond laser-driven shock interferometry experiments: a) frequency domain interferometry b) ultrafast 2D spatial interferometric microscopy.

The reflected probe pulses were imaged at x16 magnification onto the entrance slit of a high resolution imaging spectrograph (Acton model 300i) with TE-cooled CCD detector (Photometrics model SenSys 1600). Following Geindre, et al., (1994), the relative phase shift between the probe pulses caused by motion of the free surface and/or transient changes in the optical properties of the surface during shock breakout was determined by performing an inverse Fast-Fourier transform (IFFT) on the spectral intensity interferogram recorded on the spectrograph CCD. The dynamic phase and reflectance are extracted from the analysis using baseline fitting to the unshocked region surrounding the shock breakout region.

Shown in Fig. 1b is a schematic of the set up for ultrafast spatial interferometric microscopy (Gahagan, et al., 2002). Here, a small portion of the main laser pulse (~10 μJ) is reflected from a beam splitter and passed through a modified Mach-Zehnder interferometer with the sample in one arm (the sample arm) and a variable delay, to control temporal overlap, in the other (reference) arm. The probe pulse in the sample arm was focused onto the target at an incidence angle of either 32.6° or 76.0° to a spot size of ~1000 μm to circumscribe the optically pumped region. For these studies, the probe pulse was p-polarized relative to the plane of incidence. A lens was used to image the surface (at ca. 2 pixels/μm) onto a CCD camera (same camera as above). A duplicate imaging lens was used in the reference arm. The sample and reference arms were recombined at a slight angle to produce an interference pattern on the CCD. This interferogram was transferred to and stored in a computer, and all interferograms from a time series, built up by adjusting the time delay in the shock-driving laser arm, were post processed off-line. In practice, three images were obtained at 1 Hz: (a) a "reference" interferogram, I_r, taken before the pump pulse arrives, (b) a "pump" interferogram, I_p, taken "during" the experiment and (c) a post-shot interferogram to observe the damage to the material after the experiment. Analysis is conducted using the FFT method first developed by Takeda, et al., (1982), and described in Gahagan, et al., (2002). Two 2D data maps are constructed, one containing the amplitude or reflectivity information, the other containing the phase information, which is composed of optical property and surface position data. Each time data point is obtained by averaging the 2D map values in an area that is approximately 20 microns in diameter at the center of each laser experiment.

2.2 Target Fabrication

The metal targets used in these experiments were Al and Ni films of varying thickness produced by vapor deposition onto 150 +/- 20 μm thick BK-7 microscope cover slips (Fisher Scientific). The samples were examined both with a spectroscopic ellipsometer (Sentech SE 800) and an atomic force microscope (AFM; Quesant Nomad). Best fits to the spectral ellipsometric data from a 750 nm Al sample (Ψ and Δ measured over 400-840 nm) modeled as an Al_2O_3 layer on top of the aluminum substrate yielded an Al_2O_3 layer 4.9 nm thick, consistent with literature results (Smith, et al., 1985). AFM results indicated an RMS roughness of ~ 5 nm over a 900 μm^2 area, consistent with the ellipsometric data, in which Ψ was fit less satisfactorily than Δ. Polymethylmethacrylate (PMMA), nitrocellulose (NC; 12.6% N), and polyvinyl nitrate (PVN 14.5 % N) films were spin coated from the appropriate solutions (toluene or "magic" solvent), on to 500-1000 nm thick aluminum films described above. Concentration of the solution was used to control film thickness. Thickness and surface uniformity of the films were monitored with null ellipsometry (Rudolph Research, AutoEL) at helium neon wavelength, 70^0 incidence, and using white light reflectometry spectral interference fringe analysis (Filmetrics). Both methods of film thickness measurement are sensitive to changes as small as a nanometer, and the best films produced have surface thickness variations of 1% across most of a 2 cm diameter sample. In all experiments, the targets were mounted on a computer-controlled x-y translation stage (where z was the normal to the target surface). The target was rastered at ~300 μm intervals between pump pulses so that each "experiment" involved undisturbed material.

2.3 Shock Generation using Ultrafast Lasers

All laser driven shocks begin with the absorption of part of the laser pulse impinging on a sample. The absorbed energy evolves into material motion via a variety of processes, including multiphoton and avalanche ionization, plasma formation, time-dependent plasma optical density, plasma expansion and ion heating, electron-electron relaxation, and electron-phonon coupling (von der Linde, et al., 2000; Schmitt, et al., 2004; Stuart, et al., 1996). The experiment can be designed so that a flyer is produced, where a layer of material is accelerated by the absorbed laser energy and flies across a gap where it strikes a target. Or the laser can be absorbed in a reactive molecular layer, whose reaction produces the pressure wave that travels into the target materials (c.f. Dlott, et al., 1998). Finally, the material motion directly consequent on the laser absorption can be utilized, which is the method that we use herein. Our group at Los Alamos has utilized a table-top ultrafast laser system to drive shocks into layered samples, as well as to measure the shock and particle velocities using interferometric methods. Initial experiments utilized the ~ 120 fs length pulses directly. Such a method produces very fast shock rise times, but results in triangular (very fast rise and fast fall) pressure profiles along the shock propagation direction, which is problematic for spectroscopic methods that probe along a path.

Using the short pulse, particle velocities of 0.1 to 0.4 nm/ps (km/s) were achieved, depending on the thickness of the metal layer and the metal type, with the lower particle velocities measured in thicker metal layers. This result indicates the shock is not supported, as it decreases in pressure with run distance. Methods to overcome this disadvantage are discussed below (Section 2.3.1). Also, the drive laser energy appeared to be optically limited in the substrate so that larger particle velocities were not achievable by simply increasing the drive energy (Moore, et al., 2001).

2.3.1 Temporally Shaped Shock Drive

In order to overcome the saw-tooth (triangular) pressure profile produced using ultrafast laser pulses to drive shocks, we investigated temporal pulse shaping methods (McGrane, et al., 2002). While femtosecond lasers have been successfully employed to generate ultrafast shocks, the pressure typically rises sharply over a few picoseconds then decays quickly over tens of picoseconds (Evans, et al., 1996; Funk, et al., 2001). This property makes such laser pulses extremely valuable for micromachining via photoablation (Ashkenasi, et al., 1998; Momma, et al., 1997; Evans, et al., 1996), but the highly transient nature and time-dependent pressure of femtosecond shocks makes their use in studying physics of shock compressed materials problematic.

In general, a shock-driving pulse's time-dependent intensity profile must be controlled to obtain a simple time dependent pressure profile. The actual pulse shape required depends on the details of the optical and physical processes involved in transforming the light pulse into material motion. These processes include the time dependent absorption due to multiphoton and avalanche ionization processes, plasma production, time-dependent plasma optical density, plasma expansion, energy transfer, and electron-phonon coupling (von der Linde, et al., 2000; Schmitt, et al., 2004; Stuart, et al., 1996). While the laser pulse is typically a Gaussian or $sech^2$ function of time, the desired pressure profile for studies of shock compressed materials is often a step wave, with a very sharp leading edge and a relatively long time at uniform pressure. Simultaneously obtaining both the fast pressure onset required to resolve picosecond molecular or phonon-mediated dynamics and the sustained constant pressure desired for simplifying analysis and for pressurizing material thicknesses exceeding tens of nanometers, until this development has presented a serious problem.

Here we describe a method of generating shocks that overcomes these problems and can be simply implemented in common tabletop chirped-pulse amplified lasers. Driving a shock using a chirped, amplified pulse with the temporally leading (reddest) spectral range removed results in a pressure risetime of less than 10-20 ps and a sustained constant pressure for hundreds of picoseconds (McGrane, et al., 2002). Spectrally unmodified chirped-pulses have been used to drive shocks (Benuzzi-Mounaix, et al., 1999; Hambir, et al., 2001), however this results in a complicated time-dependent pressure that can be easily avoided, as shown below. Because the spectrally modified pulses are generated by stretching a 100 fs seed pulse, a fraction of the amplified pulse can be recompressed to <200 fs for use in probing the effects of shock loading. We present results on shocks thus produced in aluminum thin-films as a function of spectral content and pulse energy, and illustrate the sustained drive by confining the shocked aluminum with a nitrocellulose thin-film.

A Ti:sapphire femtosecond laser provides the seed pulse for a chirped-pulse amplifier. The seed pulse is centered at 800 nm with a bandwidth ~7.5 nm full width at half max (FWHM) and a temporal FWHM of ~110 fs ($sech^2$). The chirped-pulse amplifier utilizes a grating based pulse stretcher and compressor, with a regenerative amplifier and two stages of further amplification, to produce up to 50 mJ per pulse at 10 Hz repetition rate, and, usually, ~110 fs pulse length. Halfway through the stretcher, where the seed pulse is spectrally dispersed, amplitude and/or phase modulation of the spectrum can be used to produce a variety of pulse shapes (Weiner, et al., 1995; Xu, et al., 2000). At this point, we simply blocked a portion of the red end of the spectrum, which corresponds to the leading temporal edge of the chirped-pulse. Since the spectral block is prior to the amplification stages, the amplified pulse energy does not change as a function of the wavelength range blocked. However, the blocked spectrum is temporally shorter and care must be taken not to exceed the damage threshold of amplifier materials. A beam splitter is placed after the amplification stages, but prior to the compressor stage, to remove 80% of the spectrally modified, chirped-pulse, while allowing the remaining fraction to be recompressed.

The effect of this spectral clipping on shock generation was studied by measuring the spectra, the time-dependent intensity profiles, and the time-dependent surface motion in shocked aluminum and nitrocellulose-coated aluminum thin-films. Spectra were measured with diffuse scattering into a fiber coupled CCD spectrograph (resolution 0.3 nm). An approximate temporal width (FWHM) of the compressed pulse was measured by autocorrelation. The temporal intensity profile of the chirped-pulse was determined by cross correlating with the compressed pulse in a 1 mm BBO crystal and measuring the intensity of the second harmonic generated as a function of delay between pulses. The spatial interferometry technique is used to measure surface motion and the temporally shaped pulse is focused to ~150 μm diameter on a 250 nm thick aluminum layer through a 170 μm thick glass substrate to provide the shock drive. The sample is translated to obtain a fresh region between shots.

We have explored the parameters of shock generation as function of spectral content at 2.5 mJ and pulse energy from 0.4 to 5 mJ with clipped spectral content. The spectrum of the unaltered amplified pulse, along with two spectra clipped at approximately 806 and 804 nm on the long wavelength side, are shown in Fig. 2(A). The portion of the pulse that was recompressed is affected by the missing spectral components and exhibited autocorrelation times of 110, 140, and 175 fs FWHM for the unaltered, 806 nm clipped, and 804 nm clipped spectra, respectively. The spectrum provides a simple measure of the time-dependent intensity of the chirped-pulse, which is examined more directly through cross correlation.

The shocks generated in 250 nm thick aluminum films by pulses of the two spectral contents shown in Fig. 2(A) were characterized using spatial interferometry. Fig. 2(C) shows the time-dependent phase shift and corresponding surface displacement in the aluminum thin-films. The one-dimensional surface displacement Δx is related to the phase shift $\Delta\phi$ geometrically by the equation $\Delta x = \Delta\phi\lambda(4\pi n\cos\theta)^{-1}$, where λ is the probe wavelength, n is the refractive index of the transparent medium (in this case, air) and θ is the angle of incidence (c.f. Gahagan, et al., 2000).

The phase shift plotted is the value determined at the peak of the spatial intensity, but note that the phase shift has a Gaussian spatial distribution with a FWHM of ~150 μm. The inset to Fig. 2(C) illustrates how clipping the spectrum can remove the initial >100 ps of pressure ramping and achieve pressure rise times of 10-20 ps (determined by fitting free surface velocity to a tanh function as; see Section 3.1).

The time-dependent phase shifts in aluminum films were measured for the spectrum clipped at 804 nm at various pulse energies. The results are shown in Fig. 3(A). Fig. 3(B) illustrates the final pressures achieved in the aluminum and the corresponding shock and particle velocities in the bulk film derived from the free surface velocity measured. The slopes of phase shift versus time, $\Delta\phi/\Delta t$, of Fig. 3(A) at times >100 ps determine the free surface velocity u_{fs}, and the aluminum particle velocity u_p in the bulk material is $\approx u_{fs}/2$ (to within several percent at these pressures). The shock speed u_s and pressure P in the aluminum can be determined from the particle velocity, the aluminum Hugoniot (experimental u_s vs. u_p relation), and the Hugoniot-Rankine equations (Zel'dovich and Raizier, 1966) that account for conservation of mass, energy, and momentum across a shock discontinuity.

For aluminum, $u_s = c_s + 1.34\, u_p$, where c_s is the speed of sound, 5.35 nm/ps (Marsh, 1980). Pressures are calculated using the relation $P = u_p u_s \rho$, where ρ is the density of unshocked aluminum, 2.7 g/cm^3.

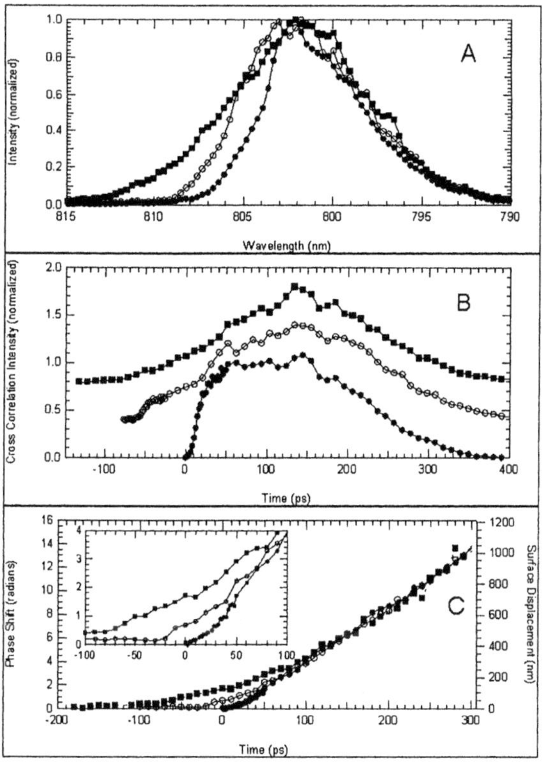

Figure 2. (A) Spectra of chirped amplified pulses; ■, unaltered; ○, spectrally clipped at 806 nm; ●, spectrally clipped at 804 nm. (B) Cross correlation (with sub-200 fs pulse) measurements of time-dependent intensity for chirped-pulse spectra of (A) (offset vertically for clarity). (C) Phase shifts and motion of aluminum free surface for shocks generated with 2.5 mJ pulse energy, using chirped-pulse spectra of (A); inset magnifies short times.

Our particular methodology limited pulse energies to 5 mJ or less, but this is not a fundamental limit and higher pressures could be obtained with higher intensities.

The 2.5 mJ pulse clipped at 804 nm was driven through the 250 nm Al layer into a 700 nm thick spin coated nitrocellulose film to examine the effect of confining the aluminum surface. The resulting time dependent phase shift at the aluminum/nitrocellulose interface is given in Figure 4, which shows that the shock achieves its terminal velocity within 20 ps and maintains a uniform velocity over hundreds of picoseconds.

For the confined reflecting surface, the slope $\Delta\phi/\Delta t$ is equal to the particle velocity u_p of the nitrocellulose at the interface, and the pressure is ~20 GPa, using the Hugoniot for cellulose acetate (Marsh, 1980).

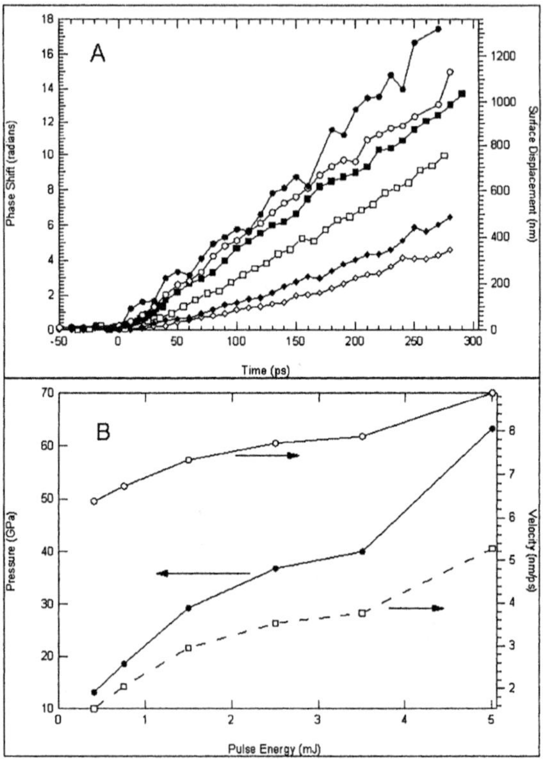

Figure 3. (A) Phase shifts and motion of aluminum free surface for chirped-pulse spectrum clipped at 804 nm and pulse energies: ●, 5.0; ○, 3.5; ■, 2.5; □, 1.5; ◆, 0.75; ◊, 0.4 mJ. (B) Calculated ●, pressure; ○, shock velocity; □, bulk particle velocity as a function of pulse energy in aluminum films, determined with the aluminum Hugoniot and the free surface velocities from the slopes in (a) at times >100 ps.

The interface measurements in the presence of the nitrocellulose support the interpretation that the pressure is maintained for hundreds of picoseconds, as the phase shift is linear with time even when the aluminum surface is confined.

2.4 Transient Infrared Spectroscopy

For these studies, in addition to the interferometric measurements, time-dependent infrared absorption measurements were also made. Tunable mid-IR output from an Optical Parametric Amplifier (OPA; Spectra Physics), pumped by a small portion (~ 1 mJ) of the main femtosecond laser beam was split into two beams, a sample and reference. The sample laser beam was directed on and focused to a ~45 μm spot that intersected the center area of the sample to be shocked. For these experiments, a spectrally modified and temporally stretched driving pulse was used to generate the shockwaves; choice of this pulse format yielded high-pressure (~200 kbar),

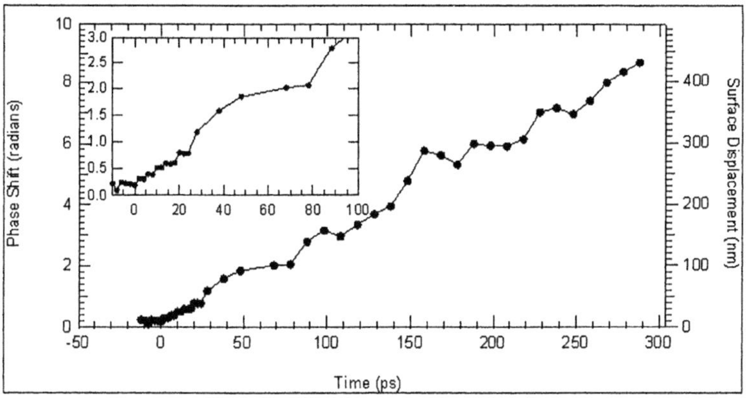

Figure 4. Phase shift and surface motion of aluminum/nitrocellulose interface for chirped-pulse spectrum clipped at 804 nm with energy 2.5 mJ. Nitrocellulose film is 700 nm thick, calculated pressure in nitrocellulose is ~20 Gpa; inset magnifies short times.

sustained (~250 ps) shockwaves in the energetic films (McGrane 2002). The reflected IR laser beam was then directed through a 0.125 m IR spectrometer (Oriel, 75 lines/mm grating blazed at 7 microns) onto the upper half of a 256 x 256 HgCdTe focal plane array (FPA) cooled to liquid N_2 temperatures. Similarly, the reference arm was also directed through the IR spectrometer and collected on the lower half of the FPA. Spectral calibration was performed through use of water vapor absorptions observed in the spectra. Time-dependent spectra were obtained as above, by delaying the pump (shock driving) beam relative to the probe.

Analysis of the infrared spectra requires accounting for thin film interference effects, which upon shock compression, change the composite reflectivity, as in the PMMA films discussed below (Section 3.3). To analyze the thin film interference effects, the infrared complex refractive index spectra for ambient samples of PVN (molecular structure $(-CH(-ONO_2)CH_2-)_n$), and the inert polymethylmethacrylate (PMMA) (Moore and McGrane, 2003), were determined. This was accomplished by recording FTIR spectra at 12 angles and 2 orthogonal polarizations for 5 film thicknesses, followed by fitting the data to the thin-film equations. This method of determining complex infrared refractive index and modeling the effect of shock propagation on the absorption spectrum is detailed elsewhere (Moore, et al., 2004); the only modification is the inclusion of the dispersive rarefaction wave that releases the pressure (McGrane 2004b).

3. RESULTS AND DISCUSSION

3.1 Shockwave Rise Time Measurements

Thin nickel and aluminum films were shocked using ~ 0.5 mJ of the output of the 800 nm laser (Gahagan, et al., 2000). FDI was used to measure the phase shift caused by the shocked material's free surface motion, yielding the acceleration and final velocity of shocked Al and Ni thin films. Fig. 5 shows the relative phase shift, $\Delta\phi$, as a function of pump delay time for three nickel films: (A) a 467 nm film probed at λ = 400 nm, Δt = 6 ps, (B) a 467 nm film probed at λ = 800 nm, Δt = 8, and (C) an 839 nm film probed at λ = 800 nm, Δt = 8 ps. The targets were each fabricated with a thin (50 nm) region to be used as a fiducial for measuring the shock breakout time relative to initiation, from which the average shock velocity through the sample was determined. Time t = 0 in Fig. 5 corresponds to the estimated arrival of the laser pulse at the nickel surface. All three samples exhibit a phase profile consistent with that expected for a sharp, but finite width shock. The phase profile of such an idealized discontinuous (zero rise time) shock front would be characterized by an *instantaneous* transition from the initial $\Delta\phi = 0$ value (when both probe pulses arrive before the shock), to a region of linear rise for the length of the time separation between the probe pulses (i.e., only the trailing probe pulse sees the surface moving at constant velocity), followed by a second *instantaneous* change to a constant $\Delta\phi$, proportional to the free surface velocity (both pulses arrive after the shock and see a constant surface velocity). The measured phase profiles exhibit similar characteristics, but with smooth rather than instantaneous transitions on the time scales of the measurement, indicative of an observable acceleration of the free surface. To obtain a particle velocity and shockwave rise time, we assume a hyperbolic tangent form for the free surface velocity, $u_{fs}(t) = u_p[1 + \tanh((t - t_0)/\tau_{fs})]$, where τ_{fs} and t_0 are fitting parameters characterizing the free surface velocity profile. The final free surface velocity is taken to be twice the final particle velocity of the shock state, $u_{fs}(t >> t_0) = 2u_p$, and assumes a reflected Hugoniot or Walsh equation of state for the rarefaction wave. Extracting a shockwave pressure profile from the measured free surface velocity is not straightforward. The principle difficulty lies in determining the relative timing between the arrival of the shockwave at the surface and the development of the rarefaction wave passing back into the material. In one limit, we may imagine that the two events occur simultaneously. That is, as the shock pressure rises incrementally, the rarefaction develops and the free surface accelerates to the velocity corresponding to isentropic relaxation from the shock state. In this case, we may approximate the time-dependent particle velocity as half the free surface velocity, $u_p(t) = u_{fs}(t)/2$. Thus, the free surface velocity, particle velocity, and shock velocity are all characterized by the single time constant τ_{fs}. Defining the shockwave rise time as the 10%-90% width of the hyperbolic tangent rise gives $\tau_{sh} = 2.3\tau_{fs}$. Values of τ_{sh} are for aluminum and nickel films and were found to be 5-6 ps.

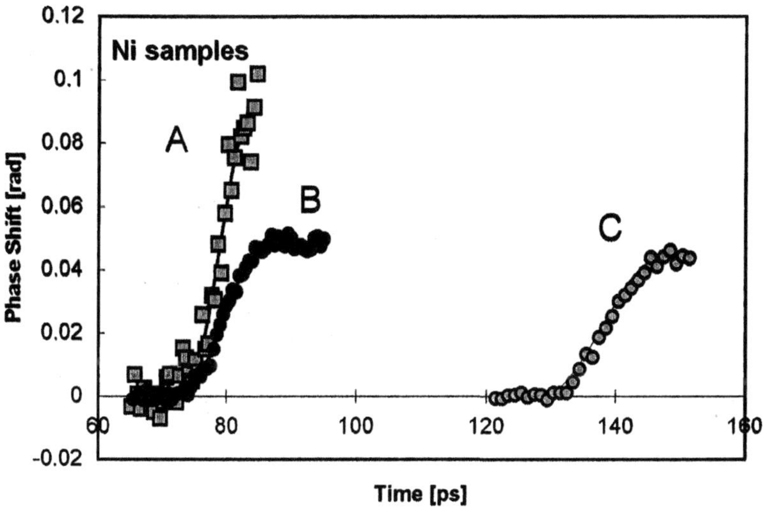

Figure 5. Plot of relative phase difference between the probe beams vs. relative delay (to the pump laser) for a shockwave breakout from three nickel films. (A) 467 nm film probed at λ = 400 nm, Δt = 6 ps, (B) 467 nm film probed at λ = 800 nm, Δt = 8, and (C) 839 nm film probed at λ = 800 nm, Δt = 8 ps.

We have also conducted experiments to investigate the effect of surface roughness by probing through the glass layer where the interface roughness is much smaller than at the free surface of a vapor deposited film. We found no measurable difference in the rise time. The short rise times we measure suggest that the thickness of these shocks is tens of lattice spacings. These results are larger than, but comparable with, molecular dynamics simulations of shocks in other metals (e.g. Kadau, 2002) and may have implications for reaction chemistry in energetic materials under similar shock loading conditions, such as direct pumping into transition states.

3.2 Shock-Induced Changes in the Complex Index

Figure 6 shows three sets of FDI measurement of the relative phase shift as a function of delay time between the shock generation pulse and the probe pulses for a 750 nm thick Al film (Funk, et al., 2001). All data were taken with an 800 nm p-polarized probe but at different angles of incidence and temporal separations (32.6°, 8 ps; 82.5°, 4 ps; 84.5°, 4 ps). The *negative* phase shift that occurs during shock breakout is very apparent and unexpected (we use the convention in our data analysis that material motion alone would yield a *positive* phase shift). We have also conducted experiments with the identical sample under similar conditions, but with a probe wavelength of 400 nm and probe pulse separation of 4 ps. No experiment done with the probe wavelength at 400 nm resulted in an observable negative phase shift during shock breakout.

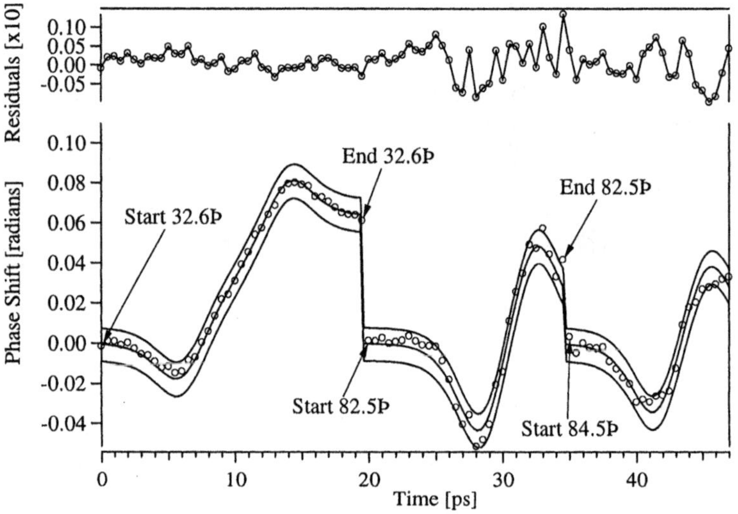

Figure 6. Plot of measured phase difference between the probe beams vs. relative delay (to the pump laser) for a shockwave breakout from a 750 nm thick Al thin film. The probes are p-polarized 800 nm light taken at three angles of incidence (which are offset in time for clarity). Note the maximum negative phase difference occurs in the data taken near the quasi-polarizing angle in aluminum (82.5°). The dashed lines are 95% confidence limit prediction bands.

Reflection from an air-metallic interface is governed by Maxwell's equations and the appropriate boundary conditions leading to the Fresnel relations for the reflection amplitudes of s and p-polarized light. Thus, upon reflection from a stationary metallic surface, the electric field undergoes a phase shift, ϕ_n, with magnitude r_n, that can be accurately calculated from knowledge of the complex index of refraction, polarization state, and the angle of incidence of the light striking the sample. Moreover, this phase shift will be influenced by any time-dependent changes in the complex index of refraction of the material. Thus, we hypothesize that the differences in the p-polarized 800 nm probe data and the 400 nm probe data are due to differences in the measured phase shifts, resulting from the pressure induced shift of the *U(200)* interband transition in aluminum. To quantify the influence of changes in the complex index, we modeled the data of Fig. 6. Measurement at incident angles near the quasi-polarizing angle (near 82.5° in aluminum) minimized the contribution of the phase signal due to surface motion, which is proportional to the cosine of the angle of incidence (cf. Eq. 1), and maximized the effect of the complex index changes. We took the changes in the complex index to be proportional to the acceleration of the surface, which is related to the pressure. To extract an *n*-effective (n_{eff}) and *k*-effective (k_{eff}), and to account for the Al_2O_3 over-layer in phase shift calculations, we treated the over-layer/aluminum as a thin film in which *only* the complex index of the aluminum is time dependent. We set the index of the Al_2O_3 layer (n=1.76, k=0) to literature values (Gervais, 1991) and we fixed the layer thickness at 4.9 nm. This choice is consistent with the fact that the index of sapphire changes very little under moderate shock conditions (< 12.0 GPa) (Barker and Hollenbach, 1972) and that the compression the sapphire undergoes due to the shockwave (less than 2% or 0.1 nm) causes a phase shift less than the noise in our method (0.7 nm RMS; Gahagan, 2000). The three data sets were fit simultaneously to the differences of the following equation for the time-dependent phase shift of each probe by varying τ, u_p, t_j, n_{eff}, and k_{eff} in equation (1),

$$\phi_j(t) = \Delta\phi_{n,\,n_{eff},\,k,\,k_{eff}} \operatorname{sech}^2\left(\frac{t - t_j + \delta t_j}{\tau}\right) + \frac{4\pi\cos(\theta_o)}{\lambda_o}\int_{t_i}^{t_f} u_p\left(1 + \tanh\left(\frac{t - t_j + \delta t_j}{\tau}\right)\right)dt \quad , \qquad (1)$$

where τ is the hyperbolic tangent time constant, u_p is the shock state particle velocity, t_j are offsets for the data from each set (time relative to the pump, which can change from day to day) and n_{eff} and k_{eff} are the effective complex index of shocked aluminum. A comparison of the n_{eff} and k_{eff} with n and k calculated from Sturm and Ashcroft's model (1981), using the modified parameters introduced by Dandrea and Ashcroft (1985), and affected by the shift in position of *U(200)*, indicated that the absolute magnitudes of the changes in the complex index are different, but the trends are in the correct direction. Thus, we conclude that our data are being influenced by the pressure dependent shift of the *U(200)* interband transition in aluminum, and its effect on the complex index (Funk, 2001). Finally, we have also taken data with nickel substrates at three angles of incidence (Funk, et al., 2001). In the case of nickel, we found the observed optical contributions are of the opposite sign as those for aluminum, indicating that the nickel data taken at 32.6° contain an optical phase contribution in the same direction as surface motion. However, in contrast to aluminum, this may be interpretable in terms of a Drude response, as nickel has no interband transition near 800 nm. We note that the absolute position of the surface derived from these measurements *would be in error* if the optical effects were not taken into account.

3.3 Subpicosecond Interferometry of Shocked PMMA Films

Shown in Fig. 7 are the measured (using ultrafast spatial interferometric microscopy) time dependent phase shifts of a 625 nm PMMA film acquired using *s* and *p* polarization at both 32.6^0 and 76.3^0 angles of incidence (McGrane, 2003). The solid lines are fits to the surface motion, u_p =2.45 km/s, and the dashed and dotted lines represent fits for *s* and *p* polarization, respectively, of the phase shift including the interference effects (see below). The shock velocity, U_s=6.5 km/s was also found from the fit, and using the PMMA density of 1.186 g/cm^3 (Marsh, 1980), we calculate the pressure in the PMMA to be 19 GPa. Note that all data shown are recorded prior to the shock completely traversing the film to avoid complications of the PMMA surface release.

To determine the effect of the weakly reflective shockwave and allow modeling of the data, the expected behavior of the thin-film was constructed as shown in Fig. 8, consisting of the following layers: 1) 500 nm aluminum, 2) 4 nm Al_2O_3, 3) time-dependent layer of shocked PMMA [d(t)= (U_s-u_p)*t], and 4) unshocked PMMA [l(t)= d_o- U_s*t]. The Fresnel coefficient for reflection from the thin-film was calculated using standard matrix formalism (Born and Wolfe, 1970). This coefficient was then written in the form as $r=|r|\exp(i\ \phi)$, where *r* is the reflection amplitude and ϕ is the phase shift, and the change in phase calculated as the difference in the phase shift expected from the ambient film and that from shocked film for the proper angle of incidence and polarization. A best fit was obtained to all four data sets by varying u_p, U_s and $n_{shocked}$ and invoking the following assumptions: a constant u_p, and negligible contribution to phase shift from shock heated

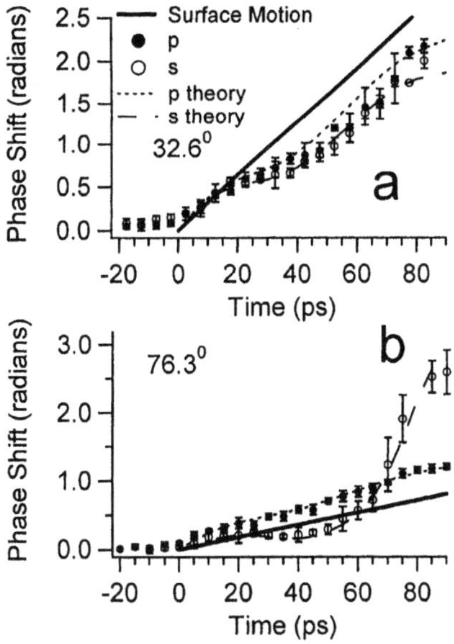

Figure 7. Phase shifts for 625 nm PMMA on Al during shock. Solid lines are fits to surface motion phase shifts, determining u_p =2.45 km/s, U_s=6.5 km/s(P=19 GPa), and $n_{shocked}$ =1.77. Dotted (p-polarization) and dashed (s-polarization) lines are calculated including thin film interference. Experimental points are, ○, p and ●, s polarization at (a) 32.6^0 and (b) 76.3^0.

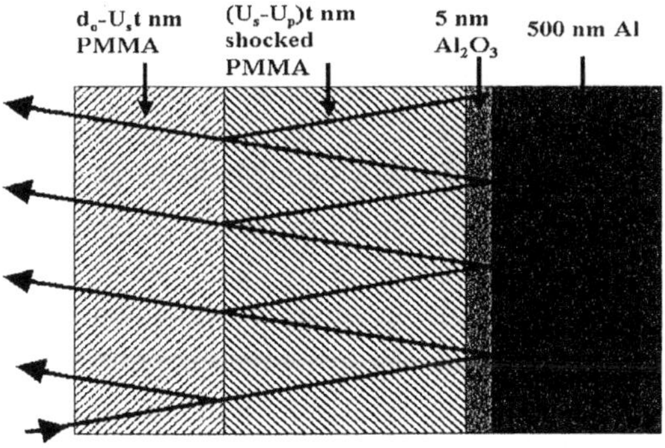

Figure 8. Diagram of thin-film structure and time dependent thicknesses as shock transits sample from right to left. Shock velocity is U_s and Al interface velocity is u_p. Arrows indicate path of light partially reflected off interfaces, leading to thin-film interference.

Al and PMMA. The fit obtained with these assumptions matches the data within the experimental error (as shown in Fig. 7), implying the assumptions made are applicable in this regime. The fitted $n_{shocked}$=1.77 compares favorably with that calculated using the Gladstone-Dale approximation, $n_{shocked}$=1.79 (Wise and Chhabildas, 1985; Chhabildas and Asay, 1979), which has been shown to be valid in this pressure regime. The plot of "true" surface motion extracted from the fit and plotted in Fig. 7 indicates that shock interferometry experiments with thin-film transparent materials cannot measure the motion of the reflective surface directly because of thin-film interference effects. These interference effects arise from multiple reflections of the interferometry probe laser pulse within the shocked and unshocked thin-film layers due to their different indices of refraction. The interference effects change as each layer thickness changes with shock transit through the film. Therefore, thin-film shock interferometry experiments must deconvolve the surface motion from the thin-film interference, which can be accomplished by repeating measurements at two angles of incidence and two polarizations. A simple model of the shock process coupled with implementation of thin-film interference equations allowed calculation of these effects. PMMA films shocked to 19 GPa matched the theoretical predictions within experimental error, justifying a number of assumptions made in modeling the data: that the particle velocity produced by our shocks is nearly constant, that the Gladstone-Dale model for shocked refractive index is valid here, that heating of the sample introduces negligible phase shift, and that the thin-film Hugoniot is essentially equal to that of the bulk material. These results support the relevance of short time, small length scale measurements for increasing understanding of larger scale shock phenomenon.

3.4 Shock-Induced Chemical Reaction of Energetic Polymers

Laser shocked polyvinyl nitrate films have been studied interferometrically (McGrane, et al., 2004a) and with infrared absorption, and have shown clear evidence of ultrafast shock-induced chemical reaction (McGrane, et al., 2004b). The effects of shock loading on the infrared absorption resonance (measured in reflectance) associated with the 1270 cm^{-1} symmetric nitro group stretch (Moore and McGrane, 2003), $\nu_s(NO_2)$, of 940 nm thick PVN films are shown as a function of shock pressure in Fig. 9 (McGrane, et al., 2004b). The spectral data are distributed on the x-axis, and the y-axis defines the time since the shockwave has entered the polymer film, determined interferometrically. The calculated spectra include only thin-film effects (Moore and McGrane, 2003; Moore, et al., 2004), and do not account for pressure and temperature shifts or chemical reaction. The areas of disagreement between the calculated and observed shocked spectra are thus the areas of most interest. In Fig. 9, spectra at 7 and 9 GPa exhibit loss of absorption followed by recovery, as expected for thin-film interference. In contrast, data at 17 and 18 GPa show a permanent loss of absorption, as expected for chemical reaction. The 9 GPa peak reflectance

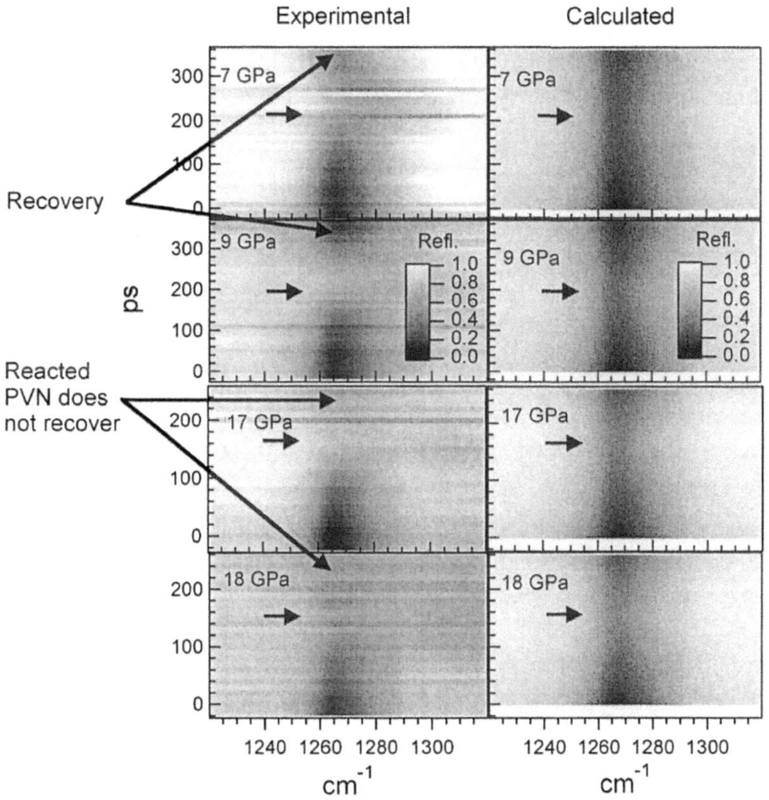

Figure 9. Time resolved infrared absorption (reflectance) spectra of the $\nu_s(NO_2)$ band in 940 nm thick PVN films during shock loading. Calculated plots include only thin film interference effects, excluding pressure and temperature shifts and chemical reaction, making the differences between the experimental and calculated spectra the primary objects of interest. Shock pressures were determined by interferometry. Arrows show the time at which the shock has fully traversed the film and rarefaction begins.

increases from 0.14 to 0.73, then recovers to 0.20 as the rarefaction wave releases the sample pressure. In contrast, when the sample is shocked to 18 GPa, the reflectance changes to 0.79, and then recovers slightly to 0.71. The

7 and 9 GPa data show slightly greater reflectance than the fully shocked calculated spectra, which we attribute to temperature and pressure effects on the index of refraction that are unaccounted for in the calculation. Recovery of the reflectance upon rarefaction (at 7 and 9 GPa) indicates that the origin of the spectral changes is predominantly thin film interference, whereas lack of recovery upon rarefaction (at 17 and 18 GPa) indicates chemical reaction (McGrane, et al., 2004b).

The effects of shock loading on the 1625 cm^{-1} antisymmetric nitro group stretch band, $\nu_{as}(NO_2)$, of 940 nm PVN films are shown in Fig. 10. The increase in reflectance at the $\nu_{as}(NO_2)$ resonance is less complete when fully shocked, than observed in the symmetric stretch. There is no clear evidence

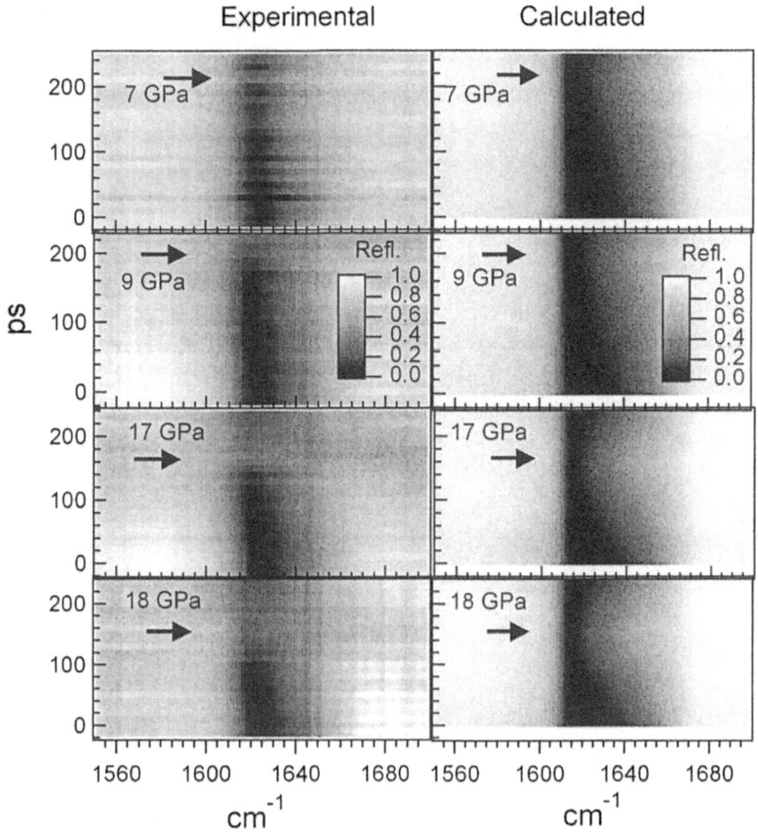

Figure 10. As Fig. 9, but the ν_{as} (NO_2) spectral region of a 940 nm PVN film.

of recovery; the 7 and 9 GPa data were not acquired at times far enough into the rarefaction to show recovery. In all cases, the fully shocked PVN exhibits greater increases in reflectance than the thin-film interference calculations predict, and the loss of absorption persists into the rarefaction. To more clearly understand the $\nu_{as}(NO_2)$ mode behavior, especially during the rarefaction, we repeated the 18 GPa experiment with a 700 nm PVN film; the infrared absorption spectra are shown in Fig. 11. The loss of the 700 nm PVN $\nu_{as}(NO_2)$ absorption is far more obvious, and clearly does not recover during rarefaction. The differences seen in Figs. 10 and 11 illustrate that changing film thickness affects the thin-film interference effects as well as the time available for reaction.

The above evidence for chemical reaction, i.e., loss of $\nu_{as}(NO_2)$ and $\nu_s(NO_2)$ mode absorption strength, occurred at times when the rarefaction is traversing back through the shocked sample. This fact raised the possibility that the rarefaction plays an essential role in the loss of absorption. To address this possibility, a multilayer film experiment was performed. A 440 nm window of inert PMMA was spin coated onto 575 nm of PVN. The PMMA layer maintains the pressure in the PVN after the shockwave has passed.

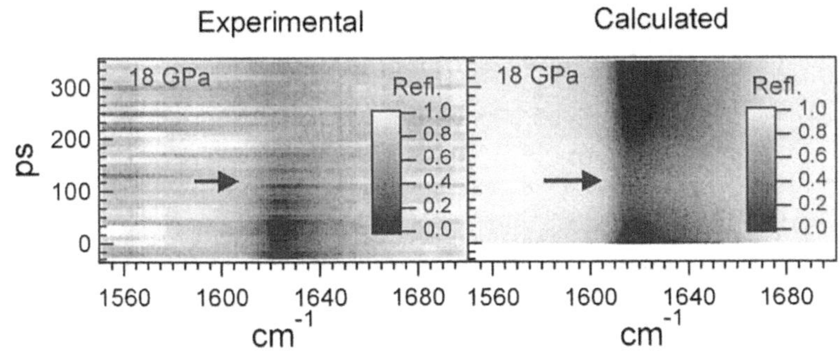

Figure 11. As Fig. 10, but for a 700 nm thick PVN film shocked to 18 GPa.

Figure 12 shows the time-dependent IR absorption spectra (measured in reflectance) in the $\nu_{as}(NO_2)$ mode region during shock and rarefaction of the stacked film. The caption provides the times that the shock or rarefaction reaches various interfaces. Only half of the loss of PVN absorption occurs by 100 ps, when all the PVN is shocked. The rest of the absorption loss occurs while the shock is traversing the PMMA. At the same time, there is a larger increase in absorption spread over higher frequencies than was apparent in the single PVN layer data. Again, there is no recovery of the initial PVN absorption spectrum after rarefaction. The PMMA absorption has changed slightly, which is likely caused by both residual temperature broadening and by inaccurate accounting of thin-film effects due to the significant refractive index changes in the reacted PVN.

The similar permanent loss of absorption in the $\nu_{as}(NO_2)$ and $\nu_s(NO_2)$ modes on shock loading indicates chemical reaction affecting the nitro group. The fact that the absorptions do not completely disappear may be due either to partial reaction quenched by the rarefaction or to formation of products that have absorptions in the same spectral regions. Indeed, both nitrous acid (Ogilvie and Duvall, 1983), a possible product from bimolecular

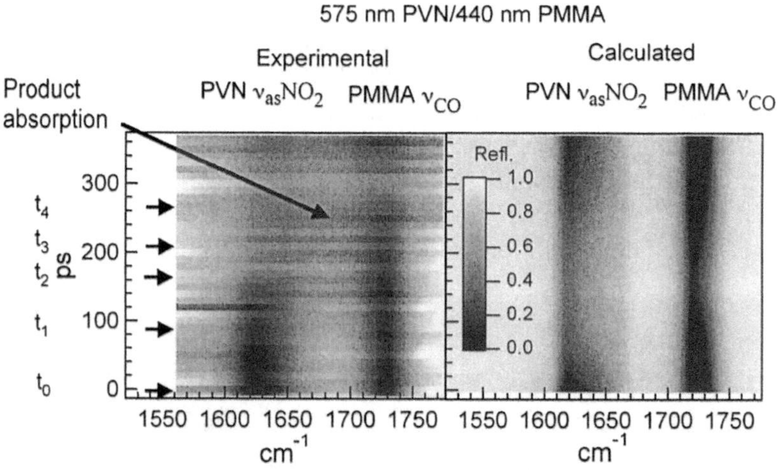

Figure 12. Time-dependent IR absorption (reflectance) spectra for a 575 nm PVN + 440 nm PMMA stacked film shocked to 18 GPa. The 1624 cm-1 peak is PVN $\nu_{as}(NO_2)$ and the 1728 cm-1 peak is the PMMA carbonyl stretch. Arrows denote timings: t_0, shock enters PVN; t_1, shock has full transited PVN; t_2, shock has transited PMMA and rarefaction begins; t_3, head of rarefaction fan reaches PVN/PMMA interface; t_4, head of rarefaction fan reaches Al.

reaction, and nitrogen dioxide (Duvall, et al., 1982), a possible product from unimolecular decomposition, are expected to have absorptions in both the $\nu_{as}(NO_2)$ and $\nu_s(NO_2)$ mode spectral regions. The spectra of such likely products are not yet known at these pressures and temperatures, and further experiments and calculations are required to quantitatively explain the reaction kinetics and identify the contributing reactions.

The observation that the reaction requires an induction time of tens of picoseconds can be used to differentiate between proposed mechanisms of how shockwave energy causes chemical reaction. An induction time of this magnitude is expected for reaction mechanisms that involve vibrational energy transfer, such as multiphonon up-pumping (Dlott and Fayer, 1980), where the shockwave excites low frequency phonons that multiply annihilate to excite the higher frequency modes. The induction time is also consistent with electronic excitation relaxing into highly excited vibrational states before dissociation (Kuklja, 2003), and experiments are underway to search for electronic excitations. On the other hand, prompt mechanisms, such as direct high-frequency vibrational excitation by the shockwave, or direct electronic excitation (Kuklja, 2003) and prompt excited-state dissociation, should occur on subpicosecond time scales, in contrast to the data presented here.

The infrared absorption results presented above demonstrate that it is possible to spectroscopically monitor shock-induced chemical reactions on picosecond time scales at the beginning of the reaction zone. This demonstration opens the door to further probing of such events with the myriad of ultrafast laser-based spectroscopic tools now available, promising to provide more insight into the effects of extreme pressure and temperature jumps at the molecular scale.

4. CONCLUSIONS

Much of the research discussed above was conducted in preparation for the studies that examined shock-induced chemistry in energetic materials. The knowledge we obtained regarding the influence of direct laser-drive upon the target on which the energetic polymer was coated, has led to our confidence that the reaction chemistry studies provide relevant information regarding the first decomposition steps in shock-loaded energetic materials. For example, the PMMA interferometric data were well-modeled using the bulk Hugoniots of both the PMMA and the aluminum; thus, the use of direct-drive did not perturb the thermodynamic state significantly from the Hugoniot, allowing us to make the assumption that on these timescales (hundreds of picoseconds), the states are well-approximated as the same relevant state obtained with gas-gun systems. We have also shown that caution must be exercised when attempting to use the interferometric data for the characterization of material motion; changes in the material properties can influence the interferometric data and must be deconvolved to yield the "true" surface motion. However, this also offers the possibility of using the changes in material properties as a measure of the shocked material's thermodynamic state; discontinuities will exist in the index of refraction as the material crosses a phase boundary. These changes may then be used as a characterization tool if they are accurately measured using the interferometric techniques to measure phase boundaries under shock-loaded conditions. Finally, the culmination of these studies involved the first observation of reaction under shock loading conditions of an energetic material. We have shown that when an energetic polymer, PVN, is shocked to ~200 kbar, we observe a disappearance of the NO_2 vibration infrared absorption(s), indicating chemical reaction as the shockwave traverses the film.

This work was performed at Los Alamos National Laboratory by the University of California under the auspices of the Department of Energy. under contract W-7405-ENG.

5. REFERENCES

Ashcroft, N.W., and Sturm, K., 1981, Interband absorption and the optical properties of polyvalent metals, *Phys. Rev. B* **24**:2315.

Ashkenasi, D., Varel, H., Rosenfeld, A., Henz, S., Herrmann, J., Campbell, E.E.B., 1998, Application of self-focusing of ps laser pulses for three-dimensional microstructuring of transparent materials, *Appl. Phys. Lett.* **72**:1442-4.

Barker, L.M., and Hollenbach, R.E., 1970, Shock-wave studies of PMMA, fused silica, and sapphire *J. Appl. Phys.* **41**(10):4208.

Benuzzi-Mounaix, A., Koenig, M., Boudenne, J.M., Hall, T.A., Batani, D., Scianitti, F., Masini, A., Di Santo, D., 1999, Chirped pulse reflectivity and frequency domain interferometry in laser driven shock experiments, *Phys. Rev. E* **60**:R2488-2491.

Born, M. and Wolf, E., 1970, *Principles of Optics* 4th ed., Pergamon Press, New York.

Chhabildas, L.C., and Asay, J.R., 1979, Rise-time measurements of shock transitions in aluminum, copper, and steel, *J. Appl. Phys.* **50**:2749.

Dandrea, R.G., and Ashcroft, N.W., 1985, High pressure as a probe of electron structure: aluminium, *Phys. Rev. B,* **32**(10):6936.

Dlott, D.D., and Fayer, M.D., 1990, Shocked molecular solids: vibrational up pumping, defect hot spot formation, and the onset of chemistry, *J. Chem. Phys.*, **92**:3798-12.

Dlott, D.D., Hambir, S., Franken, J., 1998, The new wave in shockwaves, *J. Phys. Chem. B* **102**:2121-2130.

Duvall, G.E., Ogilvie, K.M., Wilson, R., Bellamy, P.M., Wei, P.S.P., 1982, Optical spectroscopy in a shocked liquid, *Nature* **296**:846-847.

Evans, R., Badger, A.D., Faillìès, F., Mahdieh, M., Hall, T.A., Audebert, P., Geindre, J.-P., Gauthier, J.-C., Mysyrowicz, A., Grillon, G., and Antonetti, A., 1996, Time- and space-resolved optical probing of femtosecond-laser-driven shockwaves in aluminum, *Phys. Rev. Lett.* **77**:3359.

Funk, D.J., Moore, D.S., Gahagan, K.T., Buelow, S.J., Reho, J.H., Fisher, G.L., and Rabie, R.L., 2001, Ultrafast measurement of the optical properties of aluminum during shock-wave breakout, *Phys. Rev. B* **64**(11):115114.

Gahagan, K.T., Moore, D.S., Funk, D. J., Rabie, R.L., Buelow, S. J., and Nicholson, J.N., 2000, Measurement of shockwave rise times in metal thin films, *Phys. Rev. Lett.* **85**(15):3205-3208.

Gahagan, K.T., Moore, D.S., Funk, D.J., Reho, J. H., Rabie, R.L., 2002, Ultrafast interferometric microscopy for laser-driven shockwave characterization, *J. Appl. Phys.* **92**(7):3679-3682.

Geindre, J. P., Audebert, P., Rousse, A., Fallìès, F., Gauthier, J.C., Mysyrowicz, A., Dos Santos, A., Hamoniaux, G., and Antonetti, A., 1994, Frequency-domain interferometer for measuring the phase and amplitude of a femtosecond pulse probing a laser-produced plasma, *Opt. Lett.* **19**:1997-9.

Gervais, F., 1991, in: *Handbook of Optical Constants of Solids II*, E. D. Palik, ed., Academic Press, San Diego, p. 761.

Hambir, S.A., Kim, H., Dlott, D.D., and Frey, R.B., 2001, Real time ultrafast spectroscopy of shock front pore collapse, *J. App. Phys.* **90**:5139.

Hare, D.E., Franken, J., Dlott, D.D., 1995, A new method for studying picosecond dynamics of shocked solids: application to crystalline energetic materials, *Chem. Phys. Lett.,* **244**(3-4):224-230.

Kadau, K., Germann, T.C., Lomdahl, P.S., and Holian, B.L., 2002, Microscopic view of structural phase transitions induced by shockwaves, *Science* **296**:1681.

Kuklja, M.M., 2003, On the initiation of chemical reactions by electronic excitations in molecular solids, *Appl. Phys. A* **76**:359.

Marsh, S.P., 1980, *LASL Shock Hugoniot Data*, University of California, Berkeley.

McGrane, S.D., Moore, D.S., Funk, D.J., and Rabie, R.L., 2002, Spectrally modified chirped pulse generation of sustained shockwaves, *Appl. Phys. Lett.* **80**:3919.

McGrane, S.D., Moore, D.S., Funk, D.J., 2003, Sub-picosecond shock interferometry of transparent thin films, *J. Appl. Phys.* **93**(9):5063-5068.

McGrane, S.D., Moore, D.S., Funk, D.J., 2004a, in: *Shock Compression of Condensed Matter-2003*, M.D. Furnish, Y.M. Gupta, J.W. Forbes, eds, AIP Proceedings Vol 706, Melville, NY.

McGrane, S.D., Moore, D.S., and Funk, D.J., 2004b, Shock induced reaction observed via ultrafast infrared absorption in poly(vinyl nitrate) films, *J. Phys. Chem.* A **108**:9342-9347.

Momma, C., Nolte, S., Chichkov, B.N., Alvenslebeen, F.v., and Tünnermann, 1997, Precise laser ablation with ultrashort pulses, *A., Appl. Surf. Sci.* **109/110**:15.

Moore, D.S. Gahagan, K.T., Reho, J.H., Funk, D.J. Buelow, S.J., Rabie, R.L., and Lippert, T., 2001, Ultrafast nonlinear optical method for generation of planar shocks, *Appl. Phys. Lett.* **78**:40-42.

Moore, D.S. and McGrane, S.D., 2003, Comparative infrared and Raman spectroscopy of energetic polymers, *J. Mol. Struct.* **661**:561-566.

Moore, D.S., McGrane, D.J. Funk, 2004, Infrared complex refractive index measurements and simulated reflection mode infrared absorption spectroscopy of shock-compressed polymer thin films, *Appl. Spectrosc.* **58**:491-498.

Ogilvie, K.M. and Duvall, G.E., 1983, Shock-induced changes in the electronic spectra of liquid CS_2, *J. Chem. Phys.* **78**:1077-1087.

Schmitt, M.J., Kopp, R.A., Moore, D.S., and McGrane, S.D., 2004, Analysis of laser-driven shocks in confined in unconfined geometries, in: *Shock Compression of Condensed Matter – 2003*, M.D. Furnish, et al., eds, AIP, Melville, NY, pp. 1409-1412.

Smith, D.Y., Shiles, E., and Inokuti, M., 1985, in: Handbook of Optical Constants of Solids, E. D. Palik, ed., Academic Press, San Diego, p. 374.

Stuart, B.C., Freit, M.D., Herman, S., Rubenchik, A.M., Shore, B.W., and Perry, M.D., 1996, Nanosecond-to-femtosecond laser-induced breakdown in dielectrics, *Phys. Rev. B* **53**:1749.

Takeda, M., Ina, H., Kobayashi, S., 1982, Fourier-transform method of fringe-pattern analysis for computer-based topography and inteferometry, *J. Opt. Soc. Am.* **72:**156-160.

Tarver, C.M., 1997, Multiple roles of highly vibrationally excited molecules in the reaction zones of detonation waves, *J. Phys. Chem.* **101**(27):4845.

Tokunaga, E., Terasaki, A., and Kobayashi, T., 1992, Frequency-domain interferometer for femtosecond time-resolved phase spectroscopy, *Opt. Lett.* **17**:1131.

von der Linde, D., Sokolowski-Tinten, K., 2000, The physical mechanisms of short-pulse laser ablation, *Appl. Surf. Sci.* **154/155**:1.

Weiner, A.M., 1995, Femtosecond optical pulse shaping and processing, *Prog. Quant. Electr.* **19**:161.

Wise, J.L., and Chhabildas, L.C., 1985, in: *Shockwaves in Condensed Matter*, Y. M. Gupta, ed., Plenum Press, New York.

Xu, L., Li, L., Nakagawa, N., Morita, R., Yamashita, M., 2000, Application of a spatial light modulator for programmable optical pulse compression to the sub-6-fs regime, *IEEE Phot. Tech. Lett.* **12**:1540.

Zel'dovich, Y.B., Raiser, Y.P., 1966, in: *Physics of Shockwaves and High Temperature Hydrodynamic Phenomena*, Academic Press, New York.

Chapter 8

TIME AND SPACE-RESOLVED SPECTROSCOPY

Spatial, temporal and spectral resolution in laser-materials processing and spectroscopic analysis

Richard F. Haglund, Jr.
Department of Physics and Astronomy and W. M. Keck Foundation Free-Electron Laser Center, Vanderbilt University, Nashville TN 37235

1. INTRODUCTION

A distinguishing feature of laser-materials interactions is their *hierarchical* character (Stoneham, et al. 1999). The initial photon interaction is with *microscale* sites — atoms, molecular complexes, clusters, defect sites, grain boundaries — in the material; this interaction is properly described by atomic or molecular-scale models. Moreover, since the flux of laser photons is generally high, the sum of these individual interactions produces a *mesoscale* effect that ultimately leads to local thermodynamic equilibrium. Useful modifications to materials properties are generally described as occurring at the *macroscale*, that is, on micro to millimeter length scales. Similar considerations apply to time scales, which range from femtosecond electron-electron interactions to very long-time relaxation processes that lead, for example, to fatigue or long-term degradation of materials properties in laser-surface modification, or to long time-scale chemical reactions in a laser ablation plume. And, increasingly, the interest in coherent diagnostics for and control of the laser-materials interaction requires attending to the spectral characteristics both of the laser pulse and of the materials being modified. Thus, an indispensable key to understanding and controlling laser modifications of materials is characterizing the spatial, temporal, and spectral characteristics of the laser interaction with materials over the multiscale parameter space in distance, time, and frequency (or wavelength).

New experimental developments in laser ablation are being matched by calculations of rapidly increasing precision and scope, calculations that are indispensable to developing confidence in our understanding of the physical mechanisms. Because laser ablation and similar laser-materials interactions have this inherently multiscale character, the computational challenges likewise require ranging over large reaches in time, space, and spectral width. Thus both the experimental and theoretical/computational research projects require enhanced understanding of the spatial, temporal, and spectral characteristics of laser-materials interactions.

In this chapter, we consider examples of techniques that access both mechanisms of ultrafast laser-materials modification processes in materials and the results of the modification in the ablated or remanent material. The topics covered include ways of monitoring

- The temporal and spatial characteristics of the laser pulses themselves;
- The temporal and spatial evolution of the laser-modified material; and
- The temporal and spatial evolution of material removed by the laser.

What follows is an eclectic, even idiosyncratic, selection drawn from an already large and burgeoning literature. However, there is an underlying theme in all the examples: the use of spatial, temporal, or spectral control — or combinations thereof — either to modify materials by moving them along a particular potential energy surface, or to follow the process in a way that yields an accurate multiscale picture of the physical or chemical dynamics.

2. LASER-MATERIALS INTERACTION PHYSICS

Detailed analysis of the ultrafast laser-materials interaction are presented elsewhere in this volume; this section mainly highlights those spatial, temporal, and spectral dimensions that are of particular interest for laser interactions with materials or for the analysis of those interactions.

Laser ablation or materials modification by picosecond or femtosecond pulses is critically dependent on the spatial and temporal *density of excitation*, that is, on the number of quanta deposited into the material per unit volume and per unit time. The combination of density of excitation and the strength of electron-lattice coupling largely determine the outcome of any ultrafast laser process.

Materials processing with femtosecond Ti:sapphire and excimer lasers begins with single-photon or multiphoton electronic excitations that relax by a series of complex energy-transfer processes, first among electrons in the conduction band and at later times with the lattice atoms. (Chichkov, et al. 1996; Itoh and Stoneham 2001; Stuart, et al. 1995). In this case, the initial near-IR excitation produces spatially and temporally dense *electronic* excitation that destabilizes the local bonds and leads to bond-breaking, ablation, and irreversible phase transformations (Sokolowski-Tinten, et al. 1998). Irradiation by picosecond-pulse, tunable lasers in the midinfrared produces dense *vibrational* excitation, once again followed by nuclear motion, bond-breaking, and other structural modifications, often while remaining in the electronic ground state (Bubb, et al. 2002). In resonant IR excitation, the IR photons excite anharmonic vibrations that couple on time scales of a few picoseconds to the harmonic vibrations that constitute the phonon bath. Thus, nuclear motion and bond-breaking may begin if the density of excitation is sufficiently high, before the energy leaks out of the excited mode.

Thus, materials modification by picosecond and femtosecond lasers may involve outcomes of many different processes on multiple time and length scales: photon absorption in electronic or vibrational transitions followed by nuclear motion (displacement, amorphization or recrystallization, ablation); changes in local electronic structure; and changes in composition induced either by ejection of a component of the material, by adsorption or binding of exogenous atoms or molecules.

At the shortest time scales, interest often centers on measuring the rate at which particular processes are occurring; such is the case, for example, in conventional pump-probe measurements with femtosecond pulse lasers. The rate at which the measured process occurs is determined set by the laser intensity I and the multiphoton cross-sections $\sigma_{(k)}$ for the process as follows,

$$\frac{dN^*}{dt} = \eta N_0 \sigma_{(k)} \left(I/\hbar\omega\right)^k , \qquad (1)$$

where N^+ is the number density of atoms or molecules taken from the initial to the final state, η a quantum efficiency; N^0 is the number of atoms or molecules in the laser-irradiated volume; $\sigma_{(k)}$ the kth order cross-section; and $\Phi \equiv \left(I/\hbar\omega\right)$ is the photon flux, the number of photons per unit time per unit area. For nanosecond and longer pulses, or in the case where one wishes to measure the yield of a particular process, the total integrated effect is proportional to the specific energy deposited in a volume V and fluence F_L:

$$Yield \propto (E/V) \cong F_L \alpha(\omega, I) \cong I_0 \tau_L \left[\alpha_0(\omega) + \beta I + \ldots\right] , \tag{2}$$

where $\alpha_0(\omega)$ is the linear absorption coefficient, β is the third-order nonlinear absorption coefficient, and τ_L is the laser pulse duration.

The ellipsis in Eq. (2) refers to higher-order or multiphoton absorption processes, quite likely in picosecond and subpicosecond laser interactions with materials. An estimate of the probability of a k-photon excitation can be made by calculating the probability P_k for k photons to be simultaneously in the volume L^3, occupied by a PTFE molecule and its nearest neighbors when the average number of photons per unit volume is m (Andrews 1985):

$$P_k = \frac{m^k}{k!}, \quad m = \frac{I}{hc/\lambda} \cdot \frac{L}{c/n} L^2 = \frac{n\lambda I L^3}{hc^2} , \tag{3}$$

where n is the index of refraction and h is the Planck constant. For conditions typical of ultrashort-pulse lasers, there is a non-negligible probability that a given unit cell or molecular unit will experience a multi-quantum vibrational transition during a laser pulse, generating strong localized lattice relaxation and nuclear motion leading to intermolecular bond-breaking and ejection from the surface.

A necessary (though not sufficient) condition to take advantage of the unique characteristics of ultrafast laser excitation is that the absorbed energy be localized in the laser absorption volume on a time scale short compared to thermal diffusion times; otherwise, energy will dissipate out of the absorption zone before it is able to begin moving along the desired configuration coordinate. These conditions are derived from optical and acoustic materials properties; they are (Zhigilei and Garrison 2000)

$$\tau_p << \tau_{thermal} \approx L_p^2 / D_{thermal}, \quad \tau_p \leq \tau_s \approx L_p / C_s , \tag{4}$$

where τ_p is the laser pulse duration, and $\tau_{thermal}$ and τ_s are, respectively, the thermal and stress confinement times. L_p is the optical penetration depth, $D_{thermal}$ is the characteristic diffusion constant, and C_s is the speed of sound in the material. Since sound speeds are of order 10^3 m·s^{-1} in solid materials, and with $D_{thermal}$ ranging from 0.1 to 10 cm^{2}s^{-1}, Equation (4) dictates that pulse durations of 100 ps or less will be both thermally and mechanically confined even in metals.

For nonmetals, on the other hand, penetration depths are much greater (of order 0.1-10 μm, depending on wavelength), and stress confinement may not be guaranteed even when thermal confinement is.

These constraints on thermal and mechanical confinement of the laser-induced modification depend critically on the strength of the electron-lattice or electron-phonon coupling, and determine the ultimate temporal and length scales associated with the laser-induced materials modification.

3. TEMPORAL MEASUREMENT AND CONTROL

Temporal measurements are central to understanding the dynamics of laser-materials interactions. Ultrafast lasers have made it possible to deposit energy impulsively into materials, on time scales shorter than virtually all the characteristic relaxation times of interest. In this section, we briefly describe autocorrelation measurements of ultrafast pulse durations, and give four examples of critical temporal measurement schemes that have been or can be applied as ablation diagnostics: a multi time scale pump-probe system; ultrafast microscopy of ablated surfaces; broadband measurements of dielectric function dynamics; and ultrafast pulse sequencing.

3.1 Measurement of ultrafast pulse duration

The first indispensable requirement for temporal ablation diagnostics is a measurement of the laser pulse duration. The standard pump-probe technique (Demtröder 2002) measures either the interferometric (electric field) or intensity autocorrelation functions, $G^{(1)}(\tau)$ or $G^{(2)}(\tau)$, given by

$$G^{(1)}(\tau)=\frac{\langle E(t)\cdot E(t+\tau)\rangle}{\langle E^2(t)\rangle} \quad G^{(2)}(\tau)=\frac{\langle I(t)\cdot I(t+\tau)\rangle}{\langle I^2(t)\rangle} \quad . \tag{5}$$

Figure 1 shows a typical configuration for measuring the field correlation function $G^{(1)}(\tau)$. The intensity autocorrelation function $G^{(2)}(\tau)$ is usually measured in a background-free Michelson interferometer configuration such as that shown in Fig. 1. Similar autocorrelation techniques can be used to extract complete pulse duration and phase information (Section 5.2).

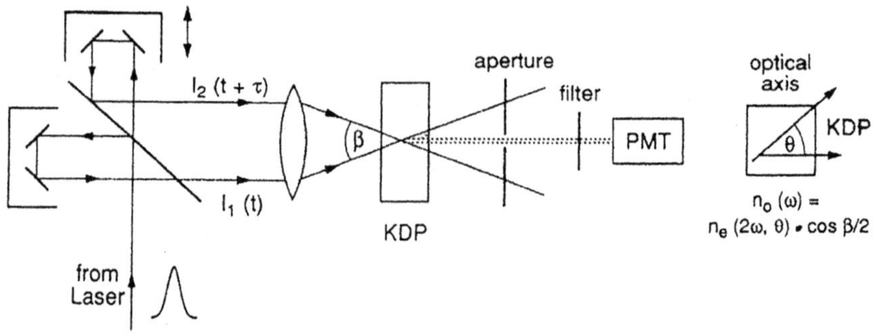

Figure 1. Interferometric configuration for background-free measurement of ultrafast pulse duration from the second-harmonic signal produced by intensity correlations of the first-harmonic laser signal.

3.2 Multiscale measurement of ablation dynamics

Typical pump-probe measurements of ablation dynamics are carried out at time scales less than 10s of ns, owing to challenges of maintaining alignment in optical delay lines that are longer than a few meters. Yet one of the key problems one would like to study is to follow the complete course of a materials modification experiment, from fs to ms, in keeping with the multiscale character of these processes. New technology for electronically synchronizing two lasers makes it possible to have this temporal dynamic range, as shown in some detail (Yu, et al. 2005) in Fig. 2.

The Coherent, Inc. master and slave oscillators are pumped by a single 10W frequency-doubled Nd:YAG laser; the oscillator pulses are amplified in a pair of regenerative amplifiers pumped by a 20W Ar ion laser. The key to the synchronization of the two systems is a commercial synchronization system that can lock the two laser pulses with a temporal jitter of less than 10 fs. Similar laser systems are available from other manufacturers and have been described elsewhere. These kinds of systems make it possible to track processes such as laser micromachining of brittle solids, in which relaxation processes may take place with time scales as long as seconds or minutes.

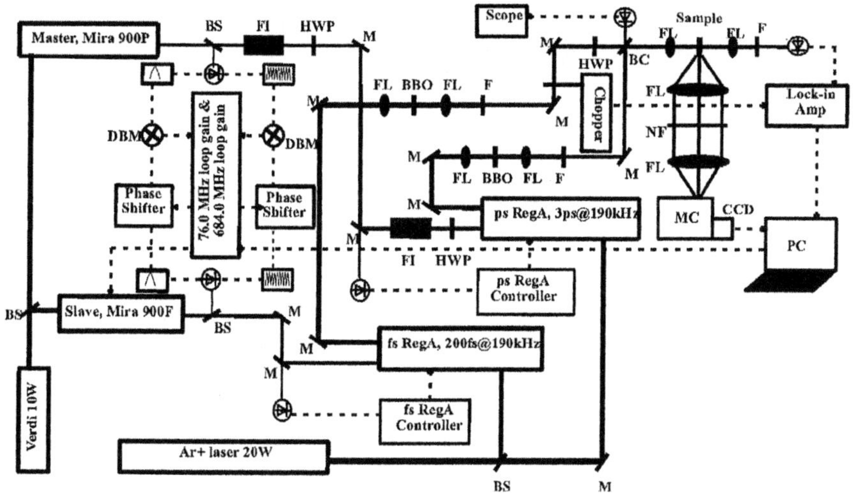

Figure 2. Dual-laser, ps-fs pump-probe laser system for transient absorption and Raman spectroscopy. The lasers are locked together by a commercial synchronization system, and run from a microcomputer. Illustration from (Yu, et al. 2005).

3.3 Ultrafast pump-probe microscopy

Time-resolved microscopy can be accomplished in a pump-probe configuration, with time scales limited only by delay-line technology. In pioneering studies of surface damage to semiconductors and metals (Sokolowski-Tinten, et al. 1998), a 120-fs pulse from an amplified dye laser (620 nm) provided a pump fluence of order 0.5 $J \cdot cm^{-2}$ to an irradiated area. The pump focal spot was then illuminated by a time-delayed probe beam, and the reflected signal viewed through a microscope objective from a direction normal to the surface. Fig. 3 shows a Si wafer irradiated above the threshold for ablation but below that for plasma formation. In the first picosecond after laser irradiation, the surface is highly reflective, characteristic of a metallic phase with electron densities in excess of 10^{22} cm^{-3}. As a rarefaction wave carries into the bulk, alternating layers of dense and rarified regions form and hydrodynamic motion begins. Since the rarefaction wave propagates at sound speeds (~ 10^3 m/s), one infers that material

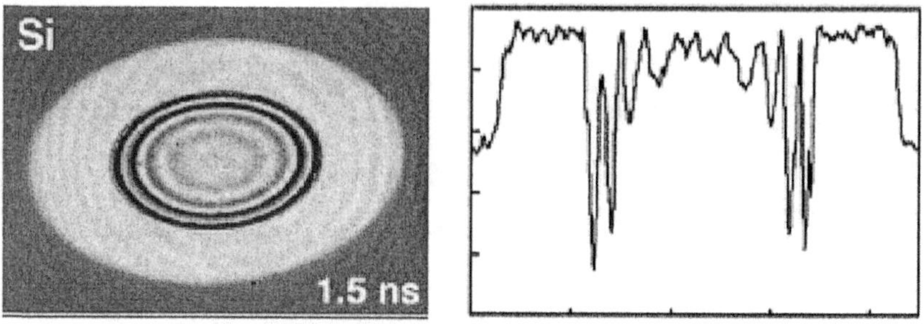

Figure 3. (Left) Ultrafast reflection micrograph of a Si sample irradiated by a 620 nm pump pulse and probed 1.5 ns later. (Right) Interferogram of the irradiated area showing the formation of the signature Newton's rings that track hydrodynamic motion of molten material with alternating regions of high and low density (Sokolowski-Tinten, et al. 1998).

motion at the surface begins 0.1-1 ns after the pump pulse (Rethfeld, et al. 2004).

Approximately 1 ns after the pump pulse, a series of Newton's rings also appears, due to the interference of parallel interfaces between regions of high and low refractive index as the strongly heated, pressurized Si is ablated and simultaneously begins the much slower process of equilibration with the cold region above and outside the absorption zone. The right panel in Fig. 3 shows the reflectivity profile of this interference pattern created by reflections from the high and low-index components of the expanding laser plume. This behavior is almost certainly a general characteristic of all solids with moderate thermal conductivities; however, the situation is substantially more complex for insulators because of longer-lived relaxation processes induced by electron-phonon coupling found in insulators.

3.4 Dynamics of dielectric function during ablation

In nonmetallic materials, the dielectric function is one of the key indicators of the electronic structure, and as such, can also be a microscopic indicator at femtosecond time scales. An illustrative example of ultrafast melting in a direct-gap semiconductor is furnished by experiments on GaAs using ultrafast time-resolved ellipsometry (Callan, et al. 2000). Cr-doped bulk GaAs (100) samples were irradiated by a 70-fs pump pulse, and probed after a variable time delay by a weak broadband pulse (1.5-3.5 eV) generated focusing the probe beam into a 2 mm thick CaF_2 window. By measuring the spectral reflectivity of the broadband probe at two angles of incidence and numerically inverting the Fresnel reflectivity, the real and imaginary parts of the complex dielectric function ε(ω) could be extracted as a function of pump-probe delay time.

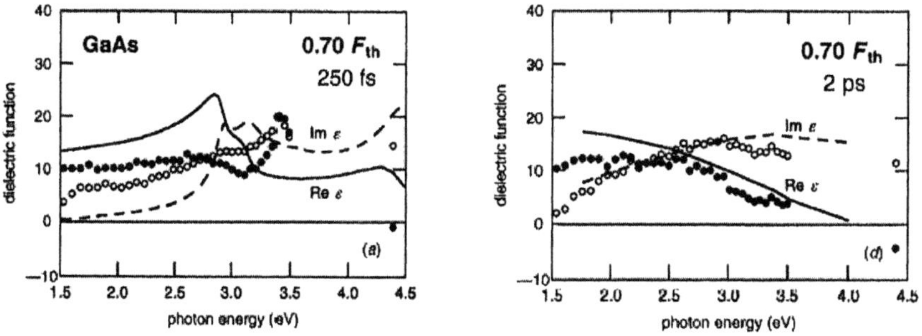

Figure 4. Dielectric functions of GaAs irradiated at $0.7F_{th}$ at 250 fs and 2 ps pump-probe delay. $\varepsilon(\omega)$ evolves from semiconducting at 250 fs to metallic by 2 ps, following arrival of the pump pulse (Callan, et al. 2000).

The snapshots shown in Fig. 4 illustrate this evolution, for a fluence of $0.7 \cdot F_{th}$, where F_{th} is the threshold for microscopically observable single-shot damage. Only 250 fs after the pump pulse, the dielectric function already deviates significantly from its room temperature shape (compare the experimental points with the solid and dashed curves). This evolution continues at 500 fs, until at 2 ps pump-probe delay, the dielectric function resembles much more nearly those for amorphous GaAs at room temperature; by 16 ps after irradiation, the dielectric functions have evolved even beyond the amorphous GaAs into a permanently altered structure. Indeed, at $1.6 \cdot F_{th}$, the dielectric function at longer times more nearly resembles that of a Drude metal than the original semiconductor.

3.5 Ultrafast pulse sequencing

The multiscale character of ultrafast laser-induced modifications to materials leads naturally to the thought that one might wish to tailor the arrival of multiple pulses in order to introduce energy into the system at specific times in order to "steer" the modification. This idea has been implemented by several groups (Stoian, et al. 2002; Stoian, et al. 2005).

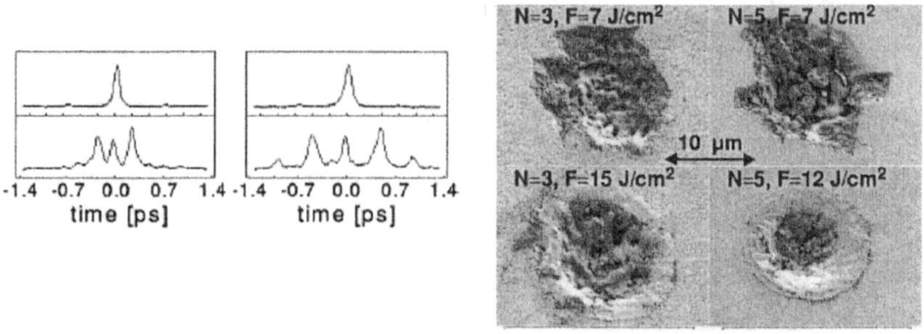

Figure 5. (Left) Laser pulse shapes and sequences for four different experiments on laser ablation of calcium fluoride. (Right) Laser ablation signatures of the four corresponding pulse structures shown at left. From (Stoian, et al. 2002; Stoian, et al. 2005).

The power of this technique for altering even as crude a materials modification as the shape of a laser ablation crater in an insulator is illustrated in Fig. 5, where the morphology of a laser ablation crater in CaF_2 is seen by electron microscopy as a function of several different pulse sequences. Despite the fact that sub-ps pulses were used — much shorter than the electron-phonon relaxation time — the use of the modulated pulse sequence leads to a much smoother ablation crater. CaF_2 has a very fast *e-ph* relaxation time, of order 1 ps, but even then the deposition of energy can be modulated on such a time scale to produce a better outcome than that achievable by a single pulse. In the case of CaF_2, a brittle material with very efficient carrier trapping properties, the effects of electronic excitation by a fast laser pulse include the creation of localized lattice deformation, atomic displacements and lattice heating. By using modulated pulse sequences, it is possible to time the energy deposition profile such that the kinetic energy of the constituent atoms is enhanced, leading to efficient ablation, while coupling a minimal amount of energy to the surrounding lattice and thus reducing the problem of collateral thermal damage. Studies in Al_2O_3 and SiO_2 show that the optimal sequencing and amplitude of an ultrafast ablation pulse depends on the properties of the materials. So far no experiments combining real-time microscopy with pulse sequencing have appeared in the literature.

4. SPATIAL MEASUREMENT AND CONTROL

The length scales of interest in laser ablation range from atomic scale to the mesoscale and macroscale. The properties of ultrafast laser pulses can be used to achieve high spatial resolution in studies of laser-materials interactions by a variety of techniques, including adapting conventional microscopies to pump-probe configurations; making use of the high intensity of ultrafast pump pulses to generate images at higher frequencies than the pump laser; using multiphoton excitation to see into structures that obscure or scatter visible photons; and combining diffraction techniques with ultrafast excitation of materials to observe structural phase transitions.

4.1 Ultrafast multiphoton microscopy

Multiphoton microscopy first developed as a way of improving the performance of time-resolved confocal microscopy in biological specimens, which suffer both from laser-induced damage when laser light is strongly absorbed, as well as from background fluorescence induced by visible light signals (Denk, et al. 1990; Williams, et al. 1994). Another applications area that quickly adopted two-photon microscopy was optical data storage (Strickler and Webb 1991). Multiphoton microscopy, using the third harmonic signal generated by nonlinear conversion of 800 nm light from a fs Ti:sapphire laser, has also been used to image microscale structures in many materials, such as cellular structure in plants (Oron, et al. 2003; Squier, et al. 1998); this is possible because virtually all materials have a non-negligible third-order dielectric susceptibility $\chi^{(3)}$ that can be exploited to generate the third harmonic. The same technique has also been used to image regions of dielectric breakdown in glass (Squier and Muller 1999). By using the third harmonic, it is often possible to find a region of transparency that is opaque to the fundamental laser frequency.

Of more immediate interest in the area of materials modification and proc-

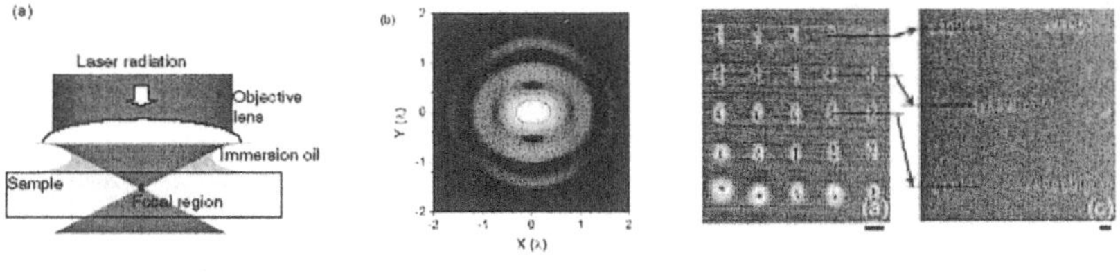

Figure 6. (a) Immersion microscope for two-photon laser microlithography. (b) Calculated intensity distribution for the geometry in (a). (Right) Nanorods fabricated lithographically in polystyrene film viewed from the top and the side. From (Juodkazis, et al. 2004).

essing is the possibility of making functional materials — such as polymers (Albota, et al. 1998) or photochromic glasses (Masuda, et al. 2003) — specifically designed to be processed by ultrafast two-photon excitation. The spatial resolution gained from TPM arises from: (1) the steep intensity dependence of two-photon excitation cross sections, and (2) discrimination against background processes excited by a single photon. In addition, where the solid is transparent to single-photon absorption, depth resolution can be further enhanced by using short focal-length microscope lenses.

These properties have recently been used to fabricate nanorods in polymers (Juodkazis, et al. 2005) and in materials that undergo a glass transition (Juodkazis, et al. 2004) at a critical temperature. Fig. 6 (a) shows the set-up of the immersion microscope, while (b) displays the calculated intensity distribution at the focal plane for a laser beam polarized in the *x*-direction. The neighboring figure shows nanorods with a diameter around 50 nm fabricated by the two-photon microlithography in polystyrene.

4.2 Ultrafast diffraction techniques

Pump-probe techniques can also be used to provide atomic-scale information about laser-induced modifications to crystal structure by diffraction. In early experiments the probe was focused on a metal tape to produce a beam of incoherent line radiation from a plasma-driven X-ray source (*e.g.*, Cu K_{α}) that

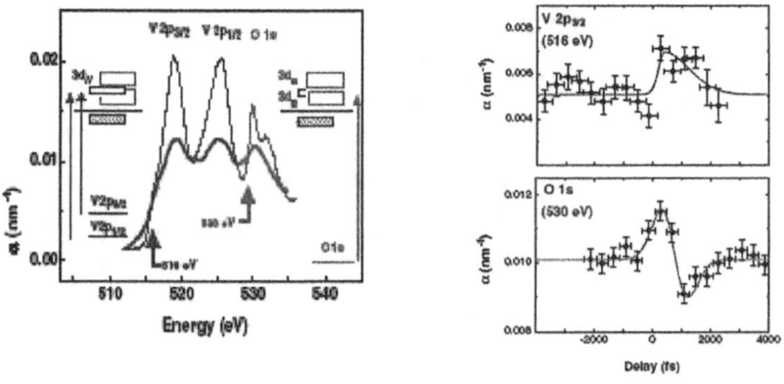

Figure 7. (Left) Static X-ray absorption fine structure such as would be found in a conventional XAFS experiment on VO_2. (Right) Dynamical evolution of the V and O atomic levels involved in the metal-insulator transition of VO_2 excited by a fs laser. Source: (Cavalleri, et al. 2005).

functions as the probe. The technique was used, for example, to elucidate the femtosecond dynamics of melting in semiconductors (Cavalleri, et al. 2000), of the structural change accompanying solid-solid metal-insulator transition in VO_2 (Cavalleri, et al. 2001).

Other experiments use a fast X-ray probe pulse from a synchrotron source (typically in the picosecond range), synchronized to an ultrafast laser pulse used as the pump (Chen 2005). The advantage this brings comes from the higher photon flux of the synchrotron beams and the opportunity for employing the many sophisticated X-ray beam-manipulation and scattering techniques available. For example, this femtosecond-pump/picosecond X-ray probe technique was employed to unravel the complex dynamics of the ferroelectric transition in the tetrathiafulvalene-*p*-chloranil (TTF-CA, $C_6H_4S_4$-$C_6Cl_4O_2$), an organic charge-transfer crystal (Collet, et al. 2003).

For example, the experiments done on the structural phase transition in VO_2 with the plasma line source were able to confirm only the time scale of the transition; more recent experiments using the Advanced Light Source have located the lower bound for the transition speed (Cavalleri, et al. 2004) and exhibited the details of the band-filling dynamics during the phase transition by using near-edge X-ray absorption fine-structure spectroscopy (NEXAFS) (Cavalleri, et al. 2005). A fs X-ray source, created by laser modulation of the electron-bunch energy in an insertion device, produced off-axis fs X radiation that was captured and focused using a toroidal mirror, yielding a resolution of 4 eV over the range from 100 to 800 eV.

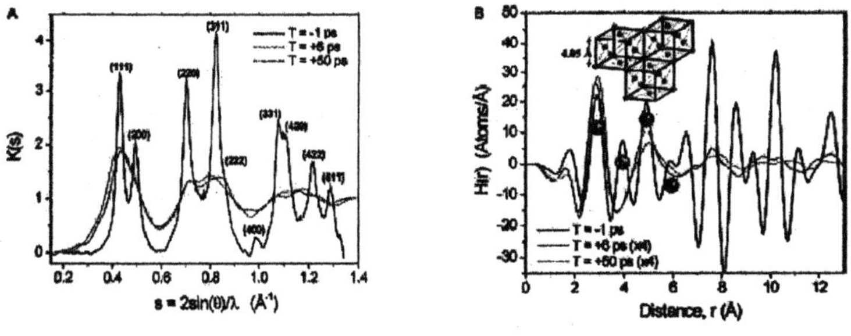

Figure 8. (A) Snapshots at various times of the pair correlation function for Al irradiated by a Ti:sapphire fs laser pulse above the melting threshold. (B) Time-dependent radial distribution function calculated from the data in (A)m for various atomic locations. From (Siwick, et al. 2004).

A different view of phase transitions and ultrafast coherent phonons in solids has been enabled by the evolution of femtosecond electron-diffraction (ED) sources (Siwick, et al. 2004). A fs pump laser is used to drive a phase transition, while a probe beam simultaneously ejects a fs burst of photoelectrons from a photocathode. ED benefits from the high cross-section of electron-electron interactions compared to X-rays, while short electron mean free paths in condensed phases yield high time resolution as measured by cross-correlation (Siwick, et al. 2005). Figure 7 shows the radial electron-density function for a melting transition initiated by launching coherent phonons into an Al film. The darkest curve, taken before the laser pulse strikes the sample, shows distinct oscillations characteristic long-range correlations of the fcc Al lattice. By 6 ps after the coherent phonons are launched, longer-range correlations have largely disappeared, although the liquid phase is not fully equilibrated even at 50 ps.

4.3 Coherent ultrafast Raman microscopy

Raman spectroscopy is widely used for identifying molecules in gas, liquid or solid phases. However, in ultrafast spectroscopy, conventional Raman scattering is disadvantageous because of low cross sections. Coherent anti-Stokes Raman scattering (CARS), on the other hand, overcomes this disadvantage and yields microscopic information about laser ablation processes in the small focal volumes characteristic of laser ablation or materials-modification processes. For example, an early picosecond CARS measurement of laser ablation of thin PMMA films showed a time-resolved snapshot of the volume expansion and a photon-induced chemical reaction observed by a time-resolved spectral line shift (Hare and Dlott 1994).

CARS is particularly useful in resolving the effects of rapid mechanical deformation on thin organic layers (Hambir, et al. 1997), a situation frequently encountered in tribology. This effect was simulated (Patterson, et al. 2005) by driving a shockwave into an alkane self-assembled monolayer (SAM) using the geometry shown in Fig. 9. An ultrafast planar shock is generated (Moore, et al. 2001) by absorption of a fs pulse in a thin Au film, accelerating the Au surface to a velocity of 0.5 nm/ps. The SAM response was probed simultaneously by sum-frequency generation combining probe pulses from broadband and narrowband signals derived from a Ti:sapphire laser. The purely elastic compression in a SAM of pentadecane thiol (PDT) indicates a complete recovery of the SAM structure. However, in octadecane thiol (ODT), the SFG signal returns only to a fraction of its initial value; above a normalized volume change DV~0.07, a *trans*-to-*gauche* isomerization of the ODT, fingerprinted by the CARS signal, produces *gauche* defects in the ODT and alters the tribological response of the SAM.

4.4 Ultrafast near-field spatial mapping

There has been much interest recently in micro and nanoparticles as masks for subwavelength scale lithography by near-field exposure of a resist or other substrate (Piglmayer, et al. 2002). In patterning schemes using dielectric spheres, the micro or nanosphere acts as a lens to focus the light from an incident laser beam on a substrate that is either ablated or exposed for subsequent wet-chemistry development. For micron-size spheres of polystyrene, for example, with diameters comparable to or larger than the wavelength of incident light, the near-field focal plane is approximately *R*/4 behind the surface of the sphere (Münzer, et al. 2001). For an ablatable substrate, near-field irradiation produces a circular damage spot with a diameter of order *R*/2, as predicted by diffraction theory. For particles smaller than the wavelength of light, however, the incident laser light undergoes Mie scattering, and the interaction with the substrate is quite different.

Femtosecond laser pulses can also be used to image the near-field intensity distributions for nanoscale dielectric and metal particles by atomic-force microscopy of the ablation patterns left in the substrate (Leiderer, et al. 2004). The targets for laser irradiation were variously PS nanoparticles of diameter 320 nm,

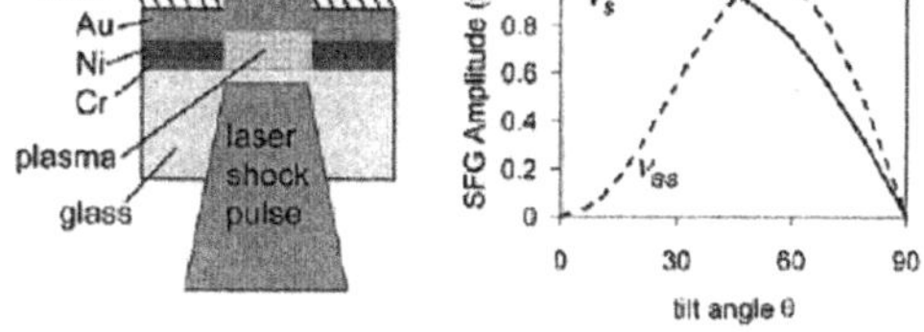

Figure 9. (a) Experimental configuration for measuring molecular effects from a shock-wave initiated by a laser on a self-assembled monolayer. (b) Orientation of the SAM molecules on the film. (c) Amplitude of the sum-frequency signal from the shock front interface.

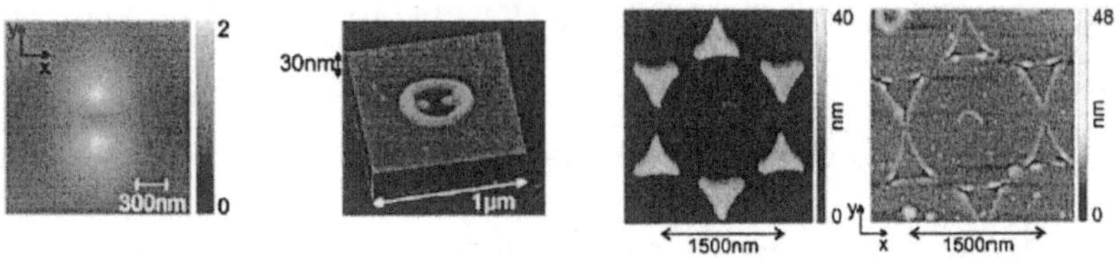

Figure 10. (Left) calculation of the near field of a dielectric sphere corresponding to the damage spot produced by a fs laser pulse on Si. (Right) Au nanotriangles produced by colloid-mask lithography and the damage spots remaining after irradiation by a fs laser pulse with the electric field polarized along the y-axis. From (Leiderer, et al. 2004).

arranged as a two-dimensional close-packed array on a Si substrate, or triangular Au nanoparticles (450 nm on a side, 25 nm high) produced by physical vapor deposition of Au on a PS array and subsequent lift-off of the PS spheres. A single pulse from a Ti:sapphire laser (pulse duration 150 fs, pulse energies up to 10 mJ) was allowed to strike the sample; following irradiation, an atomic-force microscope scanned the laser-irradiated area to determine where material had been ablated.

As shown in Fig. 10, the near-field ablation pattern underneath the PS nanoparticle resembles that of a dipole, oriented parallel to the polarization direction of the incident light beam. Interestingly, in the case of the Au nanotriangles, the maximum field enhancement appears to be the largest for tips oriented perpendicular to the polarization direction of the incident electric field. This is not consistent with earlier theories, and will apparently require rethinking the boundary conditions of the incident electric field parallel and perpendicular to the boundary surface.

4.5 Simultaneous spatio-temporal pulse mapping

It is tempting to ask if there is a way of simultaneously monitoring both temporal and spatial characteristics of a laser pulse. And indeed a recent paper proposes just such an opportunity (Balistreri, et al. 2001).

In the experiments, as shown in Fig. 11 (left), the fiber tip of a scanning photon-tunneling microscope (SPTM) is scanned along a Si_3N_4 mesa waveguide in which a fs pulse is propagating; the pulse intensity in this case is sufficiently low that self-phase modulation and group velocity dispersion are absent. The evanescent field of the laser pulse is sensed by the fiber probe, and once picked

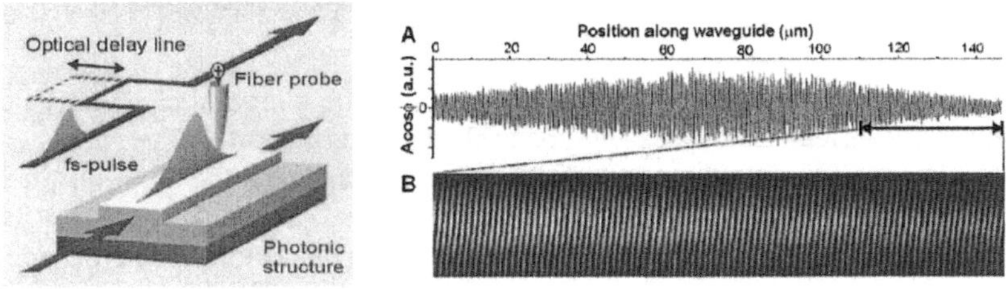

Figure 11. (Left) Experimental configuration for simultaneous measurement of pulse duration and spatial position of a fs laser pulse in a photonic structure. (Right) False-color image and interferogram for the same laser pulse, showing the pulse shape and position (top trace) and the frequency interferogram or chirp (bottom trace). Source: (Balistreri, et al. 2001).

up in the SPTM, a far-field optical pulse is generated that is correlated with the probe signal. By varying the length of the optical delay line while the SPTM is scanning, the transit of the fs laser pulse is monitored directly in the waveguide. At the same time, the frequency chirp in the pulse, and hence its changing spectral content as a function of time, can be monitored by interfering the SPTM pulse with the light from the fs reference pulse propagating in the optical delay line. The wavefronts are shown to be planar in this case; the wavelength can be read off from the plot of the phase of the laser pulse $A\cos\phi$ in the upper trace.

An oft unanswered question in laser ablation and materials processing is how the input laser pulse characteristics change as a function of penetration depth in the material in which it is propagating (Mollenauer, et al. 1989) or which is to be modified. Historically, this question has been answered by measuring the pulse after it has passed through a given length of material. This fs pulse-tracking technique offers the real prospect of being able to fully characterize the pulse, including phase distortion and phase singularities (Balistreri, et al. 2000), as a function of penetration depth.

5. SPECTRAL MEASUREMENT AND CONTROL

The broad frequency spread of ultrashort laser pulses can be used to advantage in characterizing or controlling the laser-materials interaction. Here we present examples that indicate how this unique property of ultrafast pulses can be used in examples potentially useful in laser ablation studies.

5.1 Frequency-resolved optical gating

Closely related to diagnostics and control of the temporal characteristics of the laser modification process is spectral characterization. The most widely used technique for monitoring the frequency or phase content of an ultrafast laser pulse is frequency-resolved optical gating (FROG) and various modifications thereof, developed by Trebino and coworkers somewhat more than a decade ago (Kane, et al. 1994; Kane and Trebino 1993a, 1993b). The basic principle follows that of the standard pump-probe measurement, with a critical difference; In FROG, the laser pulse is divided, one pulse delayed by an interval τ with respect to the other, and then mixed in a nonlinear optical medium. The critical difference comes at this stage: one pulse is used as a gate that lets through a temporal slice of the other pulse, producing a signal field that is proportional to

$$\vec{E}_s(t) \propto \vec{E}_{gate}(t)\left|\vec{E}_{probe}(t-\tau)\right|^2 . \tag{6}$$

The spectrogram compiled from the complete record of the signal field as the gate is swept through the pulse is called the FROG trace, and it contains the complete information about the temporal and spectral content of the pulse. The reconstruction of the data is usually based on complex fast-Fourier transform techniques that are now easily implemented on a personal computer, and have become the laboratory standard for characterizing pulses significantly shorter than 1 ps in duration.

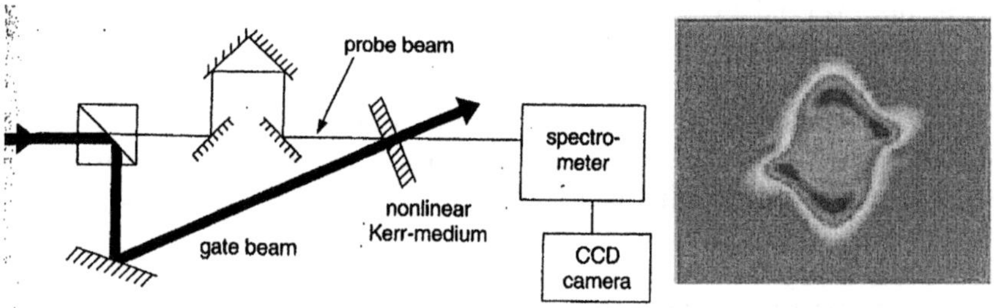

Figure 12. (Left) Experimental geometry for a FROG measurement. (Right) Third-harmonic FROG trace for a Ti:sapphire laser pulse. The vertical axis is time, and the horizontal axis is frequency (Trebino, et al. 1997).

5.2 Coherent control of materials modification

Manipulation of the frequency content of an ultrashort pulse opens the door to one of the most powerful techniques for materials modification, still in its infancy as applied to materials modification: coherent control. Coherent control techniques combine the selective excitation of quantum states with control of the spectral content of ultrashort laser pulses (Rabitz and Zhu 2000; Rabitz, et al. 2004; Warren, et al. 1993). The burgeoning literature on coherent control shows such diverse examples as altering the angular momentum distribution of the products of multiphoton excitation (Dudovich, et al. 2004); controlling energy flow in light harvesting (Herek, et al. 2002); and spatiotemporal regulation of phonons in solids (Feurer, et al. 2003). Here we present a single example of coherent control of materials processing involving the relative yields of a liquid-phase chemical reaction.

Coherent control experiments rely generically on methods for shaping the spectral content, most commonly by passing the fs laser pulse through a spatial light modulator (Weiner 2000; Weiner, et al. 1990). In the experiment (Brixner, et al. 2001), a mixture of DCM and $[Rb(dpb)_3]^{2+}$ dissolved in methanol is subjected to irradiation by a 80 fs, 1 mJ pulses from a Ti:sapphire laser. The excited molecules evolve into charge-separated states whose emission rates yield the ratio of DCM to $[Rb(dpb)_3]^{2+}$; these rates are monitored by the photomultiplier tubes (PMTs) in the figure. Then, a genetic algorithm is used to modify the

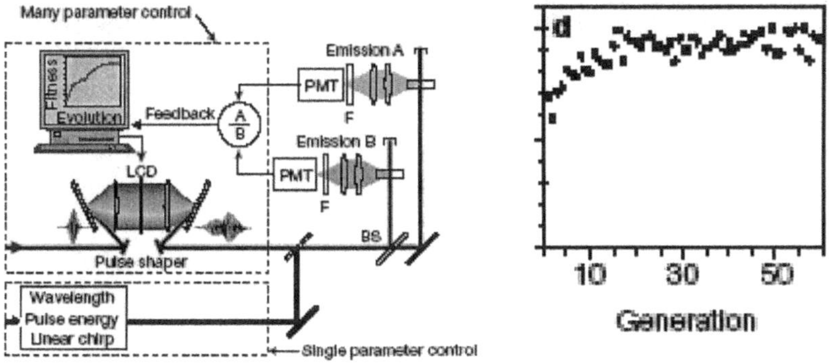

Figure 13. (Left) Experimental configuration for using a genetic algorithm together with fs pulse shaping to control the outcome of a chemical reaction. (Right) Yield of the $[Rb(dpb)_3]^{2+}$ ions as a function of the generation number in the algorithm. Source: (Brixner, et al. 2001).

laser parameters in an attempt to achieve control over the relative probability for reaching these charge-separated excited states. As shown in Fig. 13, two control regimes are used: single-parameter control for the laser and multiparameter control of the pulse shaper and the feedback loop. As the photomultipliers monitor the ion yield of the chemical reaction, the genetic algorithm varies the phase pattern superimposed on the Ti:sapphire laser pulse and seeks to optimize the yield as specified by the experimenter.

5.3 Spectral control of desorption and ablation

Laser-induced desorption and ablation, as well as bulk materials modification, from and on insulators has been a topic of intense debate since chirped-pulse amplified Ti:sapphire lasers first became available. These studies gave us the first clear evidence that there are significant changes in laser interactions with insulators as one moves to pulse durations shorter than a few picoseconds (Du, et al. 1994; Perry, et al. 1999; Stuart, et al. 1996; Stuart, et al. 1995). However, these studies were based on identification of bulk laser-induced damage phenomenology on a scale of microns; the models of multiphoton excitation followed by electron avalanche generation were able to reproduce some of the relevant features. Nevertheless, important aspects of the microscopic mechanisms could not be discerned from these relatively large spatial-scale measurements, and refinements to the models are still under development (Rethfeld 2004).

It has long been known that energy localization at the atomic scale plays a critical role in laser-induced modifications to insulators; nowhere is this more evident than in laser-induced changes in the insulator surface. Here enhanced metallization and defect formation have been shown to play significant roles. It is useful to consider recent experiments (Hess, et al. 2005) demonstrating that it is possible to use frequency-selective excitation to drive desorption in specific directions and to alter the velocity distributions of the desorbing particles (Henyk, et al. 2003). These experiments rely on the use of specific frequencies, pulse durations and pulse sequences (Hess, et al. 2002) to generate specific surface excitations. As a byproduct, the experiments show that specific surface states play an important role in the desorption process; indeed, it is possible not only to identify the precursor surface states for desorption, but also to excite those states and produce desorption (Joly, et al. 2003). An example is shown in Fig. 14, in which narrowband laser excitation at various energies is used to excite band-to-band transitions (at higher photon energy) and surface-state excitations (at slightly lower photon energies, since the surface states are below the band edge). Hyperthermal Cl (Br) atoms are ejected in both cases, but the higher photon-energy excitation yields both surface and bulk excitations producing the two-component velocity distributions observed in the figure. Similar effects have been reported on adsorbed alkali atoms on noble-metal substrates, (Petek and Ogawa 2002), although the mechanism of energy localization is not nearly as well understood.

Excitation of specific *vibrational* excitation at the picosecond time scale has been shown by a number of experiments using a broadly tunable free-electron laser irradiating molecular solids. Figure 15 (a) shows resonant infrared wavelengths of poly(ethylene glycol) (PEG), molecular weight distribution centered at 1500 amu, corresponding to the backbone mode (C=C stretch, 8.96 μm), C-H sidechain mode (C-H stretch, 3.40, 3.45 μm), and end-group excitation (C-O stretch, 2.94 μm); the *relative* absorption coefficients favor the C=C excitation by a significant factor as indicated in the figure caption (Bubb, et al. 2002). Nevertheless, the ablation yield due to the 2.94 μm excitation is substantially greater than for either of the other two vibrational excitations; moreover, there is less fragmentation of the PEG when irradiated at the O-H end-group mode.

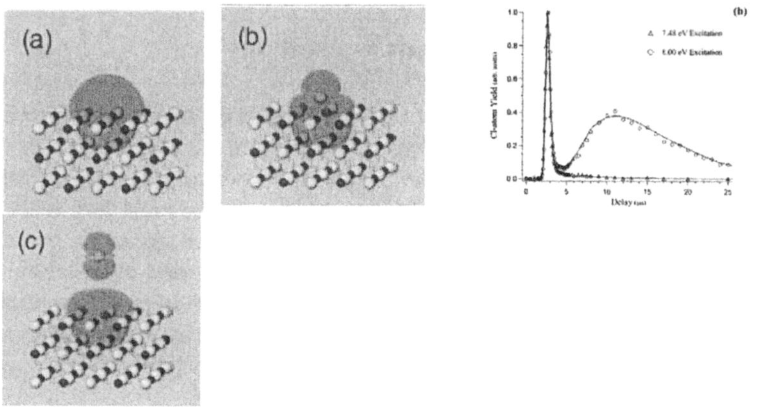

Figure 14. (Left) Calculated evolution of the self-trapped exciton in an electronically excited alkali halide crystal. (Right) Two-component velocity distributions of halogen ions desorbed from the surfaces of alkali halide crystals by narrow-band state-selective laser pulses.

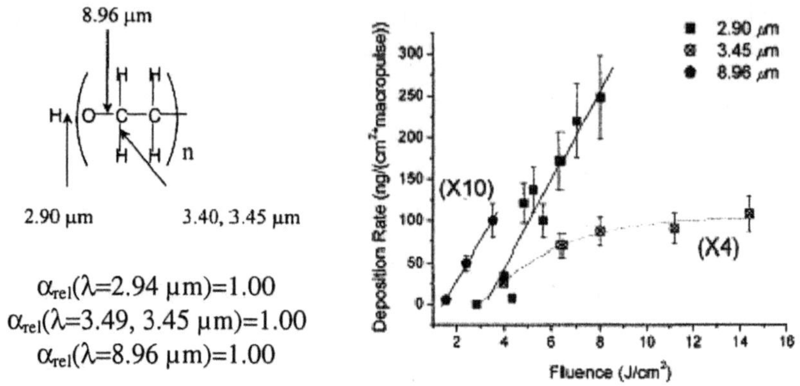

Figure 15. (Left) Molecular structure of the PEG monomer, showing resonant vibrational modes and relative absorption coefficients. (Right) Ablation yields for irradiation at the specified wavelengths by a train of picosecond pulses from a free-electron laser. Source: (Bubb, et al. 2002).

5.4 Coherent-control Raman microscopy

Raman spectroscopy, a third-order nonlinear optical process, is a powerful tool for identifying molecular fingerprints. Inherently low Raman cross-sections have driven the development of coherent tools. The development of ultrashort pulse lasers has led to increasing use of this technique in complex, real-time spectromicroscopy; the most common form is coherent anti-Stokes Raman scattering (CARS) spectroscopy, as shown in Fig. 16a. Picosecond CARS spectroscopy has been successfully used, for example, in ultrafast studies of shock compression in polymers and biological materials (Hambir, et al. 1997; Zumbusch, et al. 1999). However, these schemes have typically used two synchronized, narrow-band ps laser pulses.

Recently, however, a single-pulse CARS microscope has been developed that capitalizes on the *spectral* characteristics of ultrashort pulses (Dudovich, et al. 2002). The experimental technique is shown schematically in Fig. 16. An ultrafast pulse is sent through a computer-controlled spatial light modulator (SLM) that controls the phase function of the output pulse, with a spectral resolution of order 0.5 nm, yielding a pulse train modulated by the SLM. This pulse train is focused onto the sample, where it populates vibrational energy levels at frequencies in proportion to

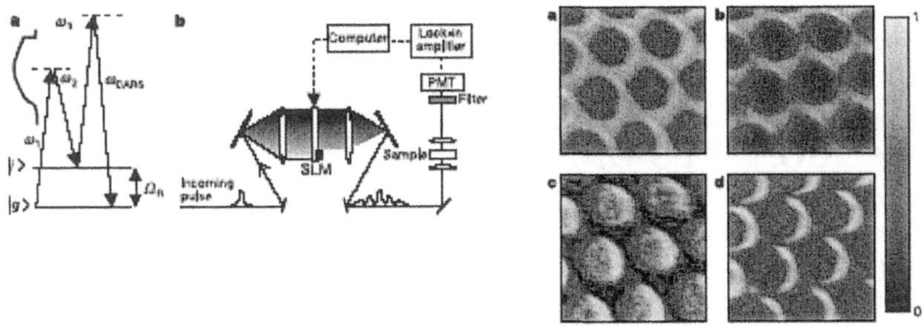

Figure 16. (a) CARS excitation scheme, showing the CARS signal at Ω_R. (b) Single-beam CARS spectroscopy experiment, showing the role of the spatial light modulator in creating the multiple interference paths that lead to a constructive CARS signal. (c) False-color images from single-beam CARS depth-resolved microscopy of CH_2Br_2 contained in 10-μm diameter wells in a glass capillary plate. The lower right-hand image is the CARS resonant image. From (Dudovich, et al. 2002).

$$n(\Omega_R) \propto \left| \int d\omega \cdot E(\omega) E^*(\omega - \Omega_R) \right|^2, \; E(\omega) = |E(\omega)| e^{i\Phi(\omega)} \, , \qquad (7)$$

where $E(\omega)$ is the complex amplitude of the applied field. The relative phase $\Delta\Phi(\omega,\Omega_R) = \Phi(\omega) - \Phi(\omega - \Omega_R)$ determines whether constructive or destructive interference occurs along the multiple possible routes to the population of the level with energy $\hbar\Omega_R$. By modulating the spectral phase with a period, all levels for which are populated by constructive interference. The non-resonant background is suppressed by a periodic complex modulation of the phase function, leading to a nearly hundred-fold reduction in the noise from non-resonant transitions. If the spatial light modulator is used to modulate the spectral phase at a frequency Ω such that $\Omega_R = N\Omega$ (where N is an integer), the condition for populating the level with resonant frequency Ω_R is satisfied and the CARS condition is reached.

By combining easily accessible imaging software now available in microcomputers, and subtracting non-resonant from resonant signals, it is possible to provide spatially-resolved, and even time-resolved, images of particular molecular or solid-state complexes. Because this requires only a single beam, this

CARS technique is much easier to implement in laser processing experiments than multiple-beam techniques previously used.

6. CONCLUSIONS

Because ultrafast lasers have pulse durations shorter than most characteristic relaxation times of condensed phases, it has become more important than ever to characterize their temporal, spatial, and spectral content in detail. Of increasing importance are the broad spectral bandwidth, the enhanced probability of multiphoton electronic excitations, and the possibility of creating extremely high spatio-temporal densities of electronic (or, in the case of picosecond infrared free-electron lasers, vibrational) excitation. Narrow-band tunable laser sources continue to have an important place here, because they permit state-selective excitations. Because ultrafast lasers can be used both to control the direction of laser-induced materials modification and to follow the temporal and spatial evolution of those modifications, the kinds of techniques described here are likely to be much more frequently used in the future. The most advanced techniques for doing this include:

- Temporal characterization based on autocorrelation and pump-probe techniques, coupled to microscopy;
- The spatial evolution of the laser-modified material using X-ray and electron diffraction methods; and
- Monitoring the temporal and spatial evolution of material removed by the laser using nonlinear time-resolved spectroscopy, such as CARS.

REFERENCES

Albota, M., Beljonne, D., Bredas, J. L., Ehrlich, J. E., Fu, J. Y., Heikal, A. A., Hess, S. E., Kogej, T., Levin, M. D., Marder, S. R., McCord-Maughon, D., Perry, J. W., Rockel, H., Rumi, M., Subramaniam, C., Webb, W. W., Wu, X. L. and Xu, C., 1998, Design of organic molecules with large two-photon absorption cross sections, *Science* 281(5383):1653-1656.

Andrews, D. L., 1985, A Simple Statistical Treatment Of Multiphoton Absorption, *American Journal Of Physics* 53(10):1001-1002.

Balistreri, M. L. M., Gersen, H., Korterik, J. P., Kuipers, L. and van Hulst, N. F., 2001, Tracking femtosecond laser pulses in space and time, *Science* 294(5544):1080-1082.

Balistreri, M. L. M., Korterik, J. P., Kuipers, L. and van Hulst, N. F., 2000, Local observations of phase singularities in optical fields in waveguide structures, *Physical Review Letters* 85(2):294-297.

Brixner, T., Damrauer, N. H., Niklaus, P. and Gerber, G., 2001, Photoselective adaptive femtosecond quantum control in the liquid phase, *Nature* 414(6859):57-60.

Bubb, D. M., Papantonakis, M. R., Toftmann, B., Horwitz, J. S., McGill, R. A., Chrisey, D. B. and Haglund, R. F., 2002, Effect of ablation parameters on infrared pulsed laser deposition of poly(ethylene glycol) films, *Journal of Applied Physics* 91(12):9809-9814.

Callan, J. P., Kim, A. M. T., Huang, L. and Mazur, E., 2000, Ultrafast electron and lattice dynamics in semiconductors at high excited carrier densities, *Chemical Physics* 251(1-3):167-179.

Cavalleri, A., Dekorsy, T., Chong, H. H. W., Kieffer, J. C. and Schoenlein, R. W., 2004, Evidence for a structurally-driven insulator-to-metal transition in VO2: A view from the ultrafast timescale, *Physical Review B* 70(16):

Cavalleri, A., Rini, M., Chong, H. H. W., Fourmaux, S., Glover, T. E., Heimann, P. A., Kieffer, J. C. and Schoenlein, R. W., 2005, Band-selective measurements of electron dynamics in VO2 using femtosecond near-edge x-ray absorption, *Physical Review Letters* 95(6):

Cavalleri, A., Siders, C. W., Brown, F. L. H., Leitner, D. M., Toth, C., Squier, J. A., Barty, C. P. J., Wilson, K. R., Sokolowski-Tinten, K., von Hoegen, M. H., von der Linde, D. and Kammler, M., 2000, Anharmonic lattice dynamics in germanium measured with ultrafast x-ray diffraction, *Physical Review Letters* 85(3):586-589.

Cavalleri, A., Toth, C., Siders, C. W., Squier, J. A., Raksi, F., Forget, P. and Kieffer, J. C., 2001, Femtosecond structural dynamics in VO2 during an ultrafast solid-solid phase transition, *Physical Review Letters* 8723(23):

Chen, L. X., 2005, Probing transient molecular structures in photochemical processes using laser-initiated time-resolved X-ray absorption spectroscopy, *Annual Review Of Physical Chemistry* 56(221-254.

Chichkov, B. N., Momma, C., Nolte, S., vonAlvensleben, F. and Tunnermann, A., 1996, Femtosecond, picosecond and nanosecond laser ablation of solids, *Applied Physics a-Materials Science & Processing* 63(2):109-115.

Collet, E., Lemee-Cailleau, M. H., Buron-Le Cointe, M., Cailleau, H., Wulff, M., Luty, T., Koshihara, S. Y., Meyer, M., Toupet, L., Rabiller, P. and Techert, S., 2003, Laser-induced ferroelectric structural order in an organic charge-transfer crystal, *Science* 300(5619):612-615.

Demtröder, W., 2002, *Laser Spectroscopy: Basic Concepts and Instrumentation* (Berlin, Springer Verlag)

Denk, W., Strickler, J. H. and Webb, W. W., 1990, 2-Photon Laser Scanning Fluorescence Microscopy, *Science* 248(4951):73-76.

Du, D., Liu, X., Korn, G., Squier, J. and Mourou, G., 1994, Laser-induced breakdown by impact ionization in SiO2 with pulse widths from 7 ns to 150 fs, *Appl. Phys. Lett.* 64(23):3071-3073.

Dudovich, N., Oron, D. and Silberberg, Y., 2002, Single-pulse coherently controlled nonlinear Raman spectroscopy and microscopy, *Nature* 418(6897):512-514.

Dudovich, N., Oron, D. and Silberberg, Y., 2004, Quantum control of the angular momentum distribution in multiphoton absorption processes, *Physical Review Letters* 92(10):

Feurer, T., Vaughan, J. C. and Nelson, K. A., 2003, Spatiotemporal coherent control of lattice vibrational waves, *Science* 299(5605):374-377.

Hambir, S. A., Franken, J., Hare, D. E., Chronister, E. L., Baer, B. J. and Dlott, D. D., 1997, Ultrahigh time-resolution vibrational spectroscopy of shocked molecular solids, *Journal Of Applied Physics* 81(5):2157-2166.

Hare, D. E. and Dlott, D. D., 1994, Picosecond Coherent Raman-Study Of Solid-State Chemical-Reactions During Laser Polymer Ablation, *Applied Physics Letters* 64(6):715-717.

Henyk, M., Joly, A. G., Beck, K. M. and Hess, W. P., 2003, Photon stimulated desorption from KI: Laser control of I-atom velocity distributions, *Surface Science* 528(1-3):219-223.

Herek, J. L., Wohlleben, W., Cogdell, R. J., Zeidler, D. and Motzkus, M., 2002, Quantum control of energy flow in light harvesting, *Nature* 417(6888):533-535.

Hess, W. P., Joly, A. G., Beck, K. M., Henyk, M., Sushko, P. V., Trevisanutto, P. E. and Shluger, A. L., 2005, Laser control of desorption through selective surface excitation, *Journal Of Physical Chemistry B* 109(42):19563-19578.

Hess, W. P., Joly, A. G., Gerrity, D. P., Beck, K. M., Sushko, P. V. and Shluger, A. L., 2002, Control of laser desorption using tunable single pulses and pulse pairs, *Journal Of Chemical Physics* 116(18):8144-8151.

Itoh, N. and Stoneham, A. M., 2001, *Materials Processing by Electronic Excitation* (Oxford, Oxford University Press)

Joly, A. G., Beck, K. M., Henyk, M., Hess, W. P., Sushko, P. V. and Shluger, A. L., 2003, Surface electronic spectra detected by atomic desorption, *Surface Science* 544(1):L683-L688.

Juodkazis, S., Mizeikis, V., Seet, K. K., Miwa, M. and Misawa, H., 2005, Two-photon lithography of nanorods in SU-8 photoresist, *Nanotechnology* 16(6):846-849.

Juodkazis, S., Yamasaki, K., Mizeikis, V., Matsuo, S. and Misawa, H., 2004, Formation of embedded patterns in glasses using femtosecond irradiation, *Applied Physics a-Materials Science & Processing* 79(4-6):1549-1553.

Kane, D. J., Taylor, A. J., Trebino, R. and Delong, K. W., 1994, Single-Shot Measurement Of The Intensity And Phase Of A Femtosecond Uv Laser-Pulse With Frequency-Resolved Optical Gating, *Optics Letters* 19(14):1061-1063.

Kane, D. J. and Trebino, R., 1993a, Characterization Of Arbitrary Femtosecond Pulses Using Frequency-Resolved Optical Gating, *Ieee Journal Of Quantum Electronics* 29(2):571-579.

Kane, D. J. and Trebino, R., 1993b, Single-Shot Measurement Of The Intensity And Phase Of An Arbitrary Ultrashort Pulse By Using Frequency-Resolved Optical Gating, *Optics Letters* 18(10):823-825.

Leiderer, P., Bartels, C., Konig-Birk, J., Mosbacher, M. and Boneberg, J., 2004, Imaging optical near-fields of nanostructures, *Applied Physics Letters* 85(22):5370-5372.

Masuda, M., Sugioka, K., Cheng, Y., Aoki, N., Kawachi, M., Shihoyama, K., Toyoda, K., Helvajian, H. and Midorikawa, K., 2003, 3-D microstructuring inside photosensitive glass by femtosecond laser excitation, *Applied Physics A-Materials Science & Processing* 76(5):857-860.

Mollenauer, L. F., Smith, K., Gordon, J. P. and Menyuk, C. R., 1989, Resistance Of Solitons To The Effects Of Polarization Dispersion In Optical Fibers, *Optics Letters* 14(21):1219-1221.

Moore, D. S., Gahagan, K. T., Reho, J. H., Funk, D. J., Buelow, S. J., Rabie, R. L. and Lippert, T., 2001, Ultrafast nonlinear optical method for generation of planar shocks, *Appl. Phys. Lett.* 78(1):40-42.

Münzer, H. J., Mosbacher, M., Bertsch, M., Zimmermann, J., Leiderer, P. and Boneberg, J., 2001, Local held enhancement effects for nanostructuring of surfaces, *Journal Of Microscopy-Oxford* 202(129-135.

Oron, D., Tal, E. and Silberberg, Y., 2003, Depth-resolved multiphoton polarization microscopy by third-harmonic generation, *Optics Letters* 28(23):2315-2317.

Patterson, J. E., Lagutchev, A., Huang, W. and Dlott, D. D., 2005, Ultrafast dynamics of shock compression of molecular monolayers, *Physical Review Letters* 94(1):

Perry, M. D., Stuart, B. C., Banks, P. S., Feit, M. D., Yanovsky, V. and Rubenchik, A. M., 1999, Ultrashort-pulse laser machining of dielectric materials, *Journal of Applied Physics* 85(9):6803-6810.

Petek, H. and Ogawa, S., 2002, Surface femtochemistry: Observation and quantum control of frustrated desorption of alkali atoms from noble metals, *Annual Review Of Physical Chemistry* 53(507-531.

Piglmayer, K., Denk, R. and Bauerle, D., 2002, Laser-induced surface patterning by means of microspheres, *Applied Physics Letters* 80(25):4693-4695.

Rabitz, H. and Zhu, W. S., 2000, Optimal control of molecular motion: Design, implementation, and inversion, *Accounts Of Chemical Research* 33(8):572-578.

Rabitz, H. A., Hsieh, M. M. and Rosenthal, C. M., 2004, Quantum optimally controlled transition landscapes, *Science* 303(5666):1998-2001.

Rethfeld, B., 2004, Unified model for the free-electron avalanche in laser-irradiated dielectrics, *Physical Review Letters* 92(18):

Rethfeld, B., Sokolowski-Tinten, K., von der Linde, D. and Anisimov, S. I., 2004, Timescales in the response of materials to femtosecond laser excitation, *Applied Physics A-Materials Science & Processing* 79(4-6):767-769.

Siwick, B. J., Dwyer, J. R., Jordan, R. E. and Miller, R. J. D., 2004, Femtosecond electron diffraction studies of strongly driven structural phase transitions, *Chemical Physics* 299(2-3):285-305.

Siwick, B. J., Green, A. A., Hebeisen, C. T. and Miller, R. J. D., 2005, Characterization of ultrashort electron pulses by electron-laser pulse cross correlation, *Optics Letters* 30(9):1057-1059.

Sokolowski-Tinten, K., Bialkowski, J., Cavalleri, A., von der Linde, D., Oparin, A., Meyer-ter-Vehn, J. and Anisimov, S. I., 1998, Transient states of matter during short pulse laser ablation, *Physical Review Letters* 81(1):224-227.

Squier, J. A. and Muller, M., 1999, Third-harmonic generation imaging of laser-induced breakdown in glass, *Applied Optics* 38(27):5789-5794.

Squier, J. A., Muller, M., Brakenhoff, G. J. and Wilson, K. R., 1998, Third harmonic generation microscopy, *Optics Express* 3(9):315-324.

Stoian, R., Boyle, M., Thoss, A., Rosenfeld, A., Korn, G., Hertel, I. V. and Campbell, E. E. B., 2002, Laser ablation of dielectrics with temporally shaped femtosecond pulses, *Applied Physics Letters* 80(3):353-355.

Stoian, R., Mermillod-Blondin, A., Winkler, S. W., Rosenfeld, A., Hertel, I. V., Spyridaki, M., Koudoumas, E., Tzanetakis, P., Fotakis, C., Burakov, I. M. and Bulgakova, N. M., 2005, Temporal pulse manipulation and consequences for ultrafast laser processing of materials, *Optical Engineering* 44(5):

Stoneham, A. M., Ramos, M. M. D. and Ribeiro, R. M., 1999, The mesoscopic modeling of laser ablation, *Appl. Phys. A* 69(S81-S86.

Strickler, J. H. and Webb, W. W., 1991, 3-Dimensional Optical-Data Storage In Refractive Media By 2-Photon Point Excitation, *Optics Letters* 16(22):1780-1782.

Stuart, B. C., Feit, M. D., Herman, S., Rubenchik, A. M., Shore, B. W. and Perry, M. D., 1996, Nanosecond-to-femtosecond laser-induced breakdown in dielectrics, *Physical Review B* 53(4):1749-1761.

Stuart, B. C., Feit, M. D., Rubenchik, A. M., Shore, B. W. and Perry, M. D., 1995, Laser-Induced Damage in Dielectrics with Nanosecond to Subpicosecond Pulses, *Physical Review Letters* 74(12):2248-2251.

Trebino, R., DeLong, K. W., Fittinghoff, D. N., Sweetser, J. N., Krumbugel, M. A., Richman, B. A. and Kane, D. J., 1997, Measuring ultrashort laser pulses in the time-frequency domain using frequency-resolved optical gating, *Review Of Scientific Instruments* 68(9):3277-3295.

Warren, W. S., Rabitz, H. and Dahleh, M., 1993, Coherent Control Of Quantum Dynamics - The Dream Is Alive, *Science* 259(5101):1581-1589.

Weiner, A. M., 2000, Femtosecond pulse shaping using spatial light modulators, *Review Of Scientific Instruments* 71(5):1929-1960.

Weiner, A. M., Leaird, D. E., Wiederrecht, G. P. and Nelson, K. A., 1990, Femtosecond Pulse Sequences Used For Optical Manipulation Of Molecular-Motion, *Science* 247(4948):1317-1319.

Williams, R. M., Piston, D. W. and Webb, W. W., 1994, 2-Photon Molecular-Excitation Provides Intrinsic 3-Dimensional Resolution For Laser-Based Microscopy And Microphotochemistry, *Faseb Journal* 8(11):804-813.

Yu, A. C., Ye, X., Ionascu, D., Cao, W. X. and Champion, P. M., 2005, Two-color pump-probe laser spectroscopy instrument with picosecond time-resolved electronic delay and extended scan range, *Review Of Scientific Instruments* 76(11):

Zhigilei, L. V. and Garrison, B. J., 2000, Microscopic mechanisms of laser ablation of organic solids in the thermal and stress confinement irradiation regimes, *Journal Of Applied Physics* 88(3):1281-1298.

Zumbusch, A., Holtom, G. R. and Xie, X. S., 1999, Three-dimensional vibrational imaging by coherent anti-Stokes Raman scattering, *Physical Review Letters* 82(20):4142-4145.

Chapter 9

PHYSICAL CHEMISTRY OF ULTRAFAST LASER INTERACTIONS WITH SOLIDS

Wolfgang Kautek
University of Vienna, Department of Physical Chemistry
Waehringer Strasse 42, A-1090 Vienna, Austria

1. INTRODUCTION

Intensive research in femtosecond pulse laser micromachining of inorganic and biological samples has been initiated in few European laboratories more than a decade ago (Küper and Stuke 1987; Kautek and Krüger 1994; Kautek, et al. 1994; Krüger and Kautek 1999; Bäuerle 2000; Krüger and Kautek 2004). The interaction of femtosecond laser pulses results in ultimate high-precision processing not accessible with pulses in the picosecond, nanosecond or even longer duration region.

These pioneering studies focused on polymers (Küper and Stuke 1987; Krüger and Kautek 1999, 2004), metallic materials (Kautek and Krüger 1994; Krüger and Kautek 1999, 2004), dielectrics (Kautek and Krüger 1994; Krüger and Kautek 1999), human soft tissue (Kautek, et al. 1994), and various other biological materials (Kautek and Krüger 1994; Krüger and Kautek 1999, 2004).

Sub-picosecond pulse lasers allow an approach to the physical limits of material science and material processing engineering. The material interaction of near infrared ultrashort laser pulses down to durations of 5 fs have been investigated and their processing potentials were established (Kautek, et al. 1996; Lenzner, et al. 1998, 2000).

Top-down structuring strategies for functional microstructures were always accompanied by regular ripple structures with submicrometer periodicities not only on dielectrics (Krüger and Kautek 1996; Daminelli, et al. 2002; Costache, et al. 2002; Reif, et al. 2002; Costatche, et al. 2003), but also on silicon (Bonse, et al. 1999, 2000, 2002a; Jeschke, et al. 2002; Bonse, et al. 2004; Costache, et al. 2004), indium phosphide (Bonse, et al. 2001, 2002b), and titanium nitride (Bonse, et al.1999, 2000a, 2000b).

Recent investigations on the femtosecond pulse laser interaction with high-performance ceramics like silicon carbide, aluminium nitride, and a composite compound SiC-TiC-TiB_2, indicated that a direct correlation between chemical composition and ripple character exists (Rudolph, et al. 2003; Rudolph and Kautek 2004). Moreover, femtosecond laser interaction with the silicon-water interface showed that a contacting condensed phase also has a strong physicochemical influence on these phenomena (Daminelli, et al. 2004).

Femtosecond laser pulses are close to an industrial use. Therefore, laser protection and safety equipment has been investigated with respect to its resistance and protection performance for femtosecond laser illumination. Recently, systematic studies of the beam-material interactions were performed to specify safety equipment as e.g. protective eyewear and protective shields (Krüger, et al. 2003; Martin, et al. 2003a, 2003b; Hertwig, et al. 2004a, 2004b; Mero, et al. 2005; Hertwig, et al. 2004c).

In this context, near-ablation threshold phenomena, such as the generation of nanostructures, and correlations of their morphology with the composition of the substrate and the nature of the interface are reviewed on the basis of the physicochemical concepts such as thermodynamics and kinetics (Kautek, et al. 2004, 2005). Thus surface energy changes and time-resolved molecular dynamic modelling of non-thermal melting processes in the femtosecond time regime were considered in the context of literature data of self-assembled surfaces rearrangements (Jeschke, et al. 2002).

2. CERAMICS

The surface of irradiated TiN areas shows highly directed ripple structures perpendicular to the electric field vector of the incident laser pulses with two completely different periodicities in dependence on the number of applied laser pulses N and the laser fluence (Bonse, et al. 2000, Fig. 1). For $N = 100$ pulses and a fluence of 0.3 Jcm^{-2}, which is slightly above the surface damage threshold, the ripples show nanoscale periodicities of 170 nm. At an increased laser fluence (3.8 Jcm^{-2}, $N = 10$) the ripple period rises to ~590 nm. Fig. 2 shows the development of these coarse ripples at constant fluence with increasing number of pulses indicating the necessity of repeated pulsing in order to observe such self-assembled structures (Turchanin 2002).

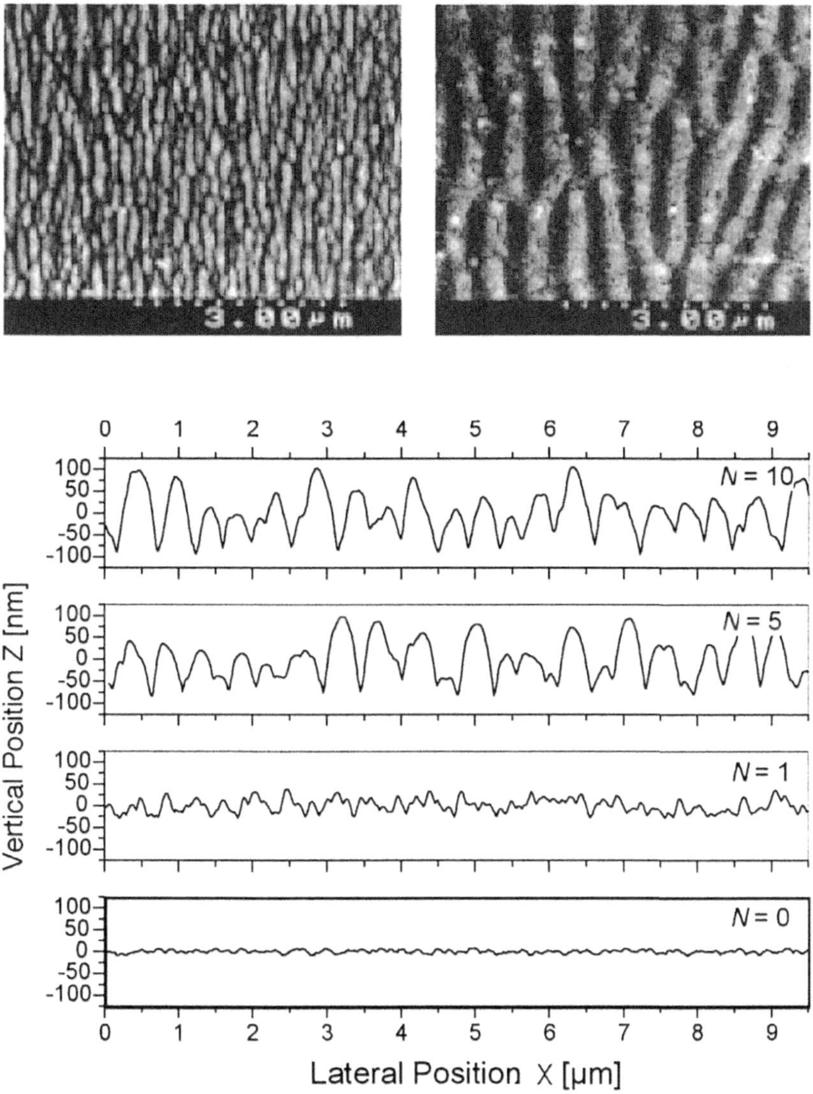

Figure 2. Atomic Force Traces of TiN irradiated with 130 fs, 3.8 J/cm^2, varied pulse number $N = 0$, 1, 5, and 10 pulses.

Femtosecond laser irradiation in the multipulse regime of other high-tech compound ceramic materials such as e.g. AlN and SiC, also led to periodic nanostructures (Rudolph, et al. 2003; Rudolph and Kautek 2004). Again, two periodicities could be discriminated, near the ablation threshold 200–300 nm, and in the ablation regime 610 nm. This phenomenon has also been described on barium borosilicate glass (Krüger and Kautek 1996), fused silica (Krüger and Kautek 1996), wide bandgap insulators (BaF_2 and CaF_2) (Costache, et al. 2002; Reif, et al. 2002; Costache, et al. 2003) and on silicon (Costache, et al. 2004) calling the classical interpretation for ripples formation in relation to interference processes (Bäuerle 2000) into question. Actually, only these two types of structures were observed and were correlated with the electric field of the incident beam. No transient index gratings were created as a result of the intensity variation.

Silicon carbide plays an important role in many industrial applications because of its hardness, high melting temperature, and chemical and thermal resistance, combined with its low weight. A composite ceramic compound SiC–TiC–TiB_2 was designed for tribological applications. Addition of TiC and TiB_2 to the SiC matrix reduces the wear rate. Mechanical machining of these materials is difficult due to their hardness up to 22 GPa. Short and ultra-short pulse laser machining was investigated as a solution to avoid mechanical tool wear and minimize thermal and mechanical stress (Rudolph, et al. 2003). μ-Raman measurements after fs-treatment of SiC–TiC–TiB_2 showed rutile (TiO_2) coverage of the Ti-containing grains, whereas SiC was covered by unoriented graphite and a reduced Si species, but negligible oxide. The melted regions were converted to resolidified nanocrystalline and amorphous phases. Also XPS analyses on SiC proved the absence of silicon oxide in contrast to nanosecond treatment, which resulted in an oxide coverage.

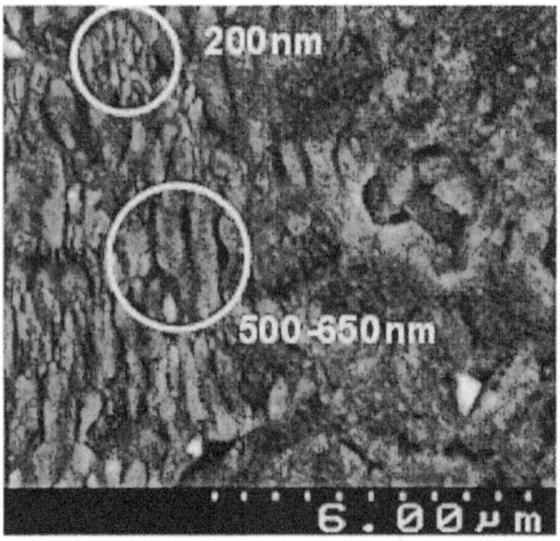

Figure 3. SiC–TiC–TiB_2. Detail near crater edge, fluence near ablation threshold. 800 nm, 130 fs, $N = 100$.

These investigations on the ceramic alloy SiC–TiC–TiB_2 yielded the first indication that the periodic nanostructure formation is primarily controlled by the microscopic chemical composition of the substrate surface. Ten pulses with 0.9 Jcm^{-2} (ablation threshold 0.24 Jcm^{-2}) caused widespread ripples with a periodicity of ~500 nm, a finer type with 200 nm arranged in clusters, and a third disk-like feature with a period of 600–650 nm (Rudolph and Kautek 2004, Fig. 3). Energy dispersive X-ray analysis (EDX) showed that the cluster-like ripple structures, with a periodicity of 200 nm, consisted of SiC. The widespread ripple structures contained mainly TiC with a periodicity of 500 nm. The disks consisted of TiB or TiB_2 with dimensions of 600–650 nm. That means that the ripples with the largest periodicity were the one with the highest share of TiC. The other two components, SiC and TiB_2, were localized only in small regions.

Generally, there are two ripple types as a function of fluence. At higher fluence, structures with a periodicity of 600-650 nm near to the laser wavelength of 800 nm were observed on all investigated ceramic compounds. At lower values near the ablation threshold, periods of 200–500 nm occurred depending on the chemical composition. In the latter case, material-dependent parameters determine the ripple formation process. Far above the ablation threshold fluence, the nature of the material becomes less important and laser parameters control the morphology. The 600-650 nm structures can be explained in the classical ripple model, which corresponds to a nonuniform energy input into the sample, modulated by the interference between the incident wave and an induced surface wave (Bäuerle 2000).

Laser ablation can also lead to the formation of non-coherent structures as e.g. at fluences of 0.20 Jcm^{-2} near the ablation threshold. Such features may be due to spatio-temporal ordering requiring at least two degrees of freedom, as e.g. the temperature and the melt film thickness. This behaviour can be understood from the behaviour of zero isoclines of a particular variable (e.g. melt thickness) where the time-derivative is zero.

On the other hand, transitions of fluid-vapour interfaces on heating and cooling through the binodal and spinodal regions can involve oscillatory instabilities that may result in self-assembled surface features. Actually, interfacial oscillatory instabilities have also been observed in classically heated fluid systems such as a binary metallic fluid, under the influence of two competing forces with different time scales (Turchanin and Freyland 2004). Such phenomena are based on wetting-dewetting transitions at the interface driven by spinodal decomposition. Surface freezing and wetting transitions may parallel thermal melting. Surface freezing is an interfacial phase transition where an ordered solid-like film forms at the liquid-vapour interface at temperatures above bulk freezing. It may occur by nucleation of a strongly undercooled liquid wetting film. In any case, the driving force of wetting and freezing films formation is the lowering of surface free energy (Turchanin, et al. 2002). Regular coherent interconnected structures (labyrinth patterns) have also been observed recently with the electrochemical metal phase formation in the underpotential and overpotential deposition region typical for spinodal decomposition mechanisms (Dogel, et al. 2003; Cahn 1965). The influence of chemistry, i.e. interatomic interactions, on surface phase transitions such as surface freezing is a widely open question so far. The regular spinodal structures described above are formed near the thermodynamic equilibrium on long time scales (up to minutes) dissimilar to the present phase changes in the femtosecond and picosecond time regime (Debenedetti 1996). A spinodal decomposition upon laser heating of a melt phase, however, involves a massive pressure rise. The whole volume undergoes a so-called phase explosion under these conditions. It involves the rapid spontaneous growth of small density fluctuations extending over large spatial scales, i.e. the entire liquid volume (Ruf and Dausinger 2004), and nanoscale ripples cannot be formed.

In conclusion of the so-far described experimental results, the phenomenon of the fs-laser-induced generation of regular nanostructures can be correlated with physicochemical processes, i.e. surface energy changes triggered upon femtosecond pulse laser irradiation. Processes far from equilibrium states, such as strong electron heating coupled with lattice destabilization, may play a major role in changing the kinetic paths from the initial to final thermodynamic state of the surface (see below) (Jeschke, et al. 2002). The high-intensity laser-triggered destabilization thus only affects the kinetic conditions without affecting the thermodynamic final state.

The equilibrium shape of a crystalline solid's surface is correlated to the tendency to attain a minimum surface free energy (Kautek, et al. 2005; Herring 1951; Adamson 1976). This has to take into account that the surface free energy for different faces is usually different. In a good approximation the total surface enthalpy, H_s and the surface energy, E_s are not distinguished:

$$E_s \sim H_s = G_s + TS_s \ , \qquad (1)$$

with G_s the surface free enthalpy and S_s the surface entropy.

In this context, chemical influences can be rationalized. The surface energy, E_s, of a covalently bonded crystal, in the simplest assumption, is one-half of the energy to rupture bonds passing through the unit area (Harkins 1949, Harkins 1950)

$$E_s = 1/2\ E_{\text{cohesion}} \qquad (2)$$

This implies that higher bonding strengths in a solid, i.e. a higher E_{cohesion}, lead to higher surface energy, E_s. An analogous proportionality between E_s and lattice energy of ionic crystals exists as well (Pampuch 1991).

It could be shown that the equilibrium surface for any given plane is not smooth (Temperley 1952). Also the total surface energy, E_S, is a minimum for a given apparent plane area, which requires an improbably ordered arrangement, and the minimum surface free energy, $G_{s,\text{relax}}$, is attained for a saw-toothed surface with teeth and waves as much as several tens of nanometers in height, representing an optimum energy entropy balance. This simplified "saw-tooth" structure can well be correlated with the observed ripple structures of less than 200 nm periodicity. It also has been pointed out that this resulting equilibrium roughness may involve a surface-melting step (Burton and Cabrera 1949). The total surface energy, E_s, is generally larger than the surface free energy G_s. If rearrangement and relaxation is allowed e.g. by non-thermal surface melting due to strong electronic excitation after repetitive femtosecond laser pulses (Jeschke, et al. 2002), saw-tooth-analogous or wavy features may be probable.

Accordingly, the probability and the rate of forming relaxed nanostructured surfaces should depend on the value of the surface free energy gain, ΔG, which always has a negative sign

$$\Delta G = G_{s,\,relax} - G_s \quad (3)$$

This thermodynamic relation holds independently from the mechanistic path, even when the kinetics is modified by far-off equilibrium states such as hot electrons which can change the transition state of the lattice on the way to the end product.

Actually, the tendency to relax and form self-assembled nanostructures decreased in the order SiC, TiC, TiB_2 (Rudolph and Kautek 2004). Assuming a correlation between tensile strength, lattice energy, and E_s, the relatively stronger tendency of the SiC surface to relax to a nanostructured surface, in contrast to the TiC and TiB_x surfaces exhibiting less surface energy, could thus be rationalized. This strong chemical influence can also be supported by a greater experimental E_s value of SiC than of TiC.

It is well established that the removal of atoms and clusters from an ordered surface leading to various surface defects, rather causes the surface energy to decrease, so that the tendency to reassemble (ΔG) would be lessened. The idea that ripples may arise from relaxation more than from interference effects as in the classical model (Costache, et al. 2000; Reif, et al. 2002; Costache, et al. 2003, 2004) agrees with the energy reduction picture (Eq. 3). The concept that surface (coulomb) explosion may be the cause of instability and therefore lead to self-assembly (Costache, et al. 2000, Reif, et al. 2002, Costache, et al. 2003, Costache, et al. 2004) would contradict the above surface energy approach.

It should be further noted that bonding of foreign ligands e.g. in a surface conversion reaction like e.g. oxidation of the TiC or TiB_x grains (see above) (Rudolph, et al. 2003; Rudolph and Kautek 2004), or a ligand addition/exchange to the surface atoms in fluid contact (see below) may affect this mechanism drastically.

3. SEMICONDUCTORS

3.1 Silicon

A single pulse on silicon results in the formation of circular substructures (holes) within the cavities (Bonse, et al. 2002a, Fig. 4). These holes vanish or are obscured by other morphological features when the same spot is illuminated with subsequent pulses. Locally enhanced carrier densities

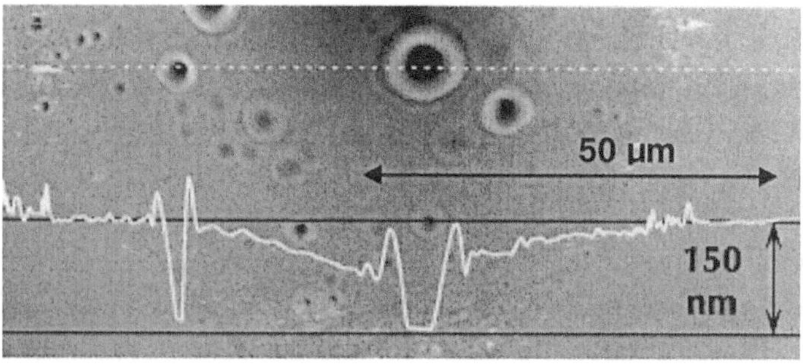

Figure 4. AFM of silicon generated with a single laser pulse (780 nm, 5 fs, 7.7 J/cm^2). Line-scan along the dotted white line.

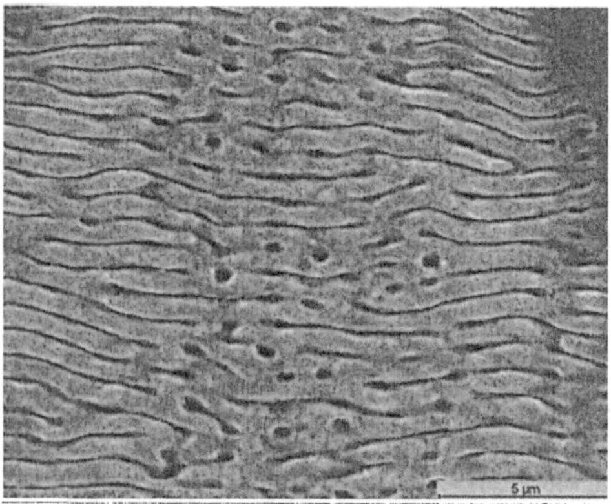

Figure 5. Silicon irradiated in air. 0.20 Jcm^{-2}, N = 100.

generated either by an inhomogeneous laser beam profile or by locally enhanced absorption (scratches, crystal defects, dust) could be ruled out. An indirect two-photon absorption with a coefficient of only 1 cm/GW (Reintjes and McGroddy 1973) as the dominant carrier-generating mechanism may be considered, which is strongly saturated. Enhancement of absorption in the depth of the semiconductor (where the light intensity already dropped one or more orders of magnitude) could account for an evolving inhomogeneous energy deposition. Consequently, after the strongly saturated and overheated surface layer is removed by phase explosion, normal boiling, including inhomogeneous nucleation of bubbles, occurs in the remaining liquid layer (Kelly and Miotello 1999). This scenario is supported by the fact that larger bubbles are formed in regions of higher fluences, i.e. regions of higher temperature (and therefore slower cooling).

Near-ablation threshold irradiation with repetitive pulsing then masks these structures, and regular ripples with periodicities of ~700 nm occur near the edges of the illuminated area (right side of Fig. 5).

3.2 Silicon-Fluid Interface

Material laser processing in the presence of water has attracted interest due to many motivations. Higher plasma pressure and longer duration of the shockwaves may be advantageous for laser shock processing, where changes in the material structure and stress state result in improved surface hardness, fatigue strength, and corrosion resistance of the material. Generation of bubbles as well as water explosion have been employed in steam laser cleaning for the removal of particles from surfaces (Oltra, et al. 2000). Water convection and bubble motion contribute to the removal of debris redeposition. The high heat capacity of water provides a better heat sink, effectively cooling heat sensitive substrates and the ejected material. Formation of nanoparticles by laser ablation of solids in liquids has been achieved thanks to the confinement effects on vaporized material within the liquid layer. A water layer during material laser processing may also allow the coupling of electrochemical techniques for in-situ monitoring of laser machining on differently conducting multilayers (Cortona and Kautek 2001; Noack and Vogel 1999).

Femtosecond laser ablation of silicon in water contact has been studied only recently (Daminelli, et al. 2004). Its behaviour in dry femtosecond laser ablation is well known (Bonse, et al. 2002a). The plasma breakdown threshold of water is 0.58 –1.11 Jcm^{-2} for pulse duration of 100 fs at 580 nm (Noack and Vogel 1999), somewhat above the ablation threshold of silicon of 0.26 Jcm^{-2} (800 nm, 130 fs). Under water confinement, near the ablation threshold, and well below the water breakdown fluence, no regular ripples but chaotic nanostructures occurred under fluid contact, whereas well-oriented ripples were formed without the water phase (Daminelli, et al. 2004). In the case of maximum fluences above the threshold (near-Gauss-shaped beam profile), one can also observe the 100 nm ripples at a maximum fluence leading to the "large" ripples which are formed not in correspondence with the chemical composition of the solid and the interface.

This suggests, according to the findings on ceramic samples, that different chemical surface conditions lead to deviating relaxation features. One can expect that fluid contact decreases the surface energy and surface tension, respectively, so that the driving force of the surface near regions to relax and self-organize (ΔG, Eq. 3) is drastically reduced. Actually, the wet surface did not show regular self-assembly structures in contrast to the dry surface near the ablation threshold.

The effect of the fluid contact on the laser-treated surface bases on the physicochemical changes due to ligand exchange versus the dry surface. If a clean solid surface is immersed in a liquid, there generally is liberation of heat, q_{imm} (Debenedetti 1996),

$$q_{\text{imm}} = E_{\text{sl}} - E_{\text{s}} \tag{4}$$

and the surface energy attains a smaller value E_{sl}. In water contact, this energy reduction is practically independent of the nature of the solid (Debenedetti 1996). Actually, water causes an extremely high surface energy reduction due to its high dipole moment. Water contact, therefore, releases the need of surface rearrangement so that no self-assembled regular ripples occur at near-threshold radiant exposures in contrast to the dry surface at comparable fluence (Fig. 5). At higher maximum fluences above the threshold, ablation processes do not allow interfacial chemical conditions to control the surface relaxation characteristics so that even there self-assembled regular ripples may occur (not shown here). This aspect is not yet fully understood, and needs further investigation.

Single fs-pulse could yield melted silicon surfaces either generating Si bubbles which left behind round holes after resolidification (above) (Bonse, et al. 2002a), or heating adjacent water which then formed water bubbles (Katayama, et al. 2003). These could oscillate and emit an acoustic wave imposing a wavy circular ripple structure on the surface melt layer. Those single-pulse features, however, contrasted from the observed multipulse ripples in their size and orientation.

4. FEMTOSECOND PERTURBATION OF BONDS

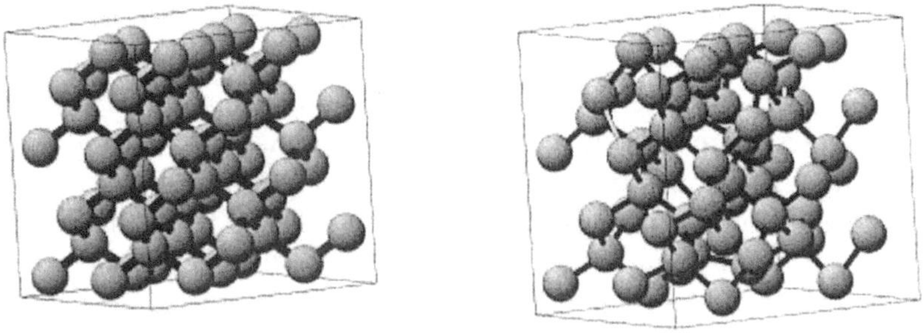

Fig. 6. Ultrashort pulse laser interaction with silicon after 100 fs (right) vs. original status (left). Pulse duration 20 fs.

The small periodic features strongly depend on the composition and the interfacial chemistry. A possible explanation may be a self-organization structure formation during the relaxation of a highly non-equilibrium surface including non-thermal melting processes. Support for this mechanism comes from a theoretical simulation of femtosecond ablation (Jeschke, et al. 2002). Without ablation, i.e. without bond-breaking, rapid excitation of electrons can cause massive instability in the crystal lattice within a few 10 fs due to a perturbation of the interatomic bonds. Below threshold fluences, however above a modification threshold (compare study on Si modification thresholds Bonse, et al. 2002a), this instability, then, relaxes on a several 100 fs time scale by surface reorganization in regular, periodic structures. A typical signature of such structures is bifurcations. Molecular dynamics (MD) simulations on the basis of an electronic tight-binding Hamiltonian in real-space took into account all atomic degrees of freedom. Non-equilibrium occupation numbers for the energy levels of the system which, being time-dependent, lead to lattice dynamics on time-dependent potential energy surfaces, were calculated. Thus, a theoretical framework is provided for the treatment of strong nonequilibrium situations in materials where atomic and electronic degrees of freedom play an equally important role. Figure 6 shows a snapshot of the lattice dynamics due to excitation with a laser pulse of duration 20 fs. During the first 100 fs after the peak maximum a moderate expansion of the system occurs. If sufficient energy above the ablation threshold is delivered, strong bond-breaking and ablation starts to take place. This may also be followed by classical thermal melting when the electron energy has been dissipated to the lattice leading to the well-known wavelength-scale ripples independent of the chemistry and the bond strength. If, however, the non-thermal dilatation of the lattice is repeated, thermodynamically driven enthalpy changes may lead to nanoscale self-organization features in finite time periods as observed in this context.

5. POLYMERS

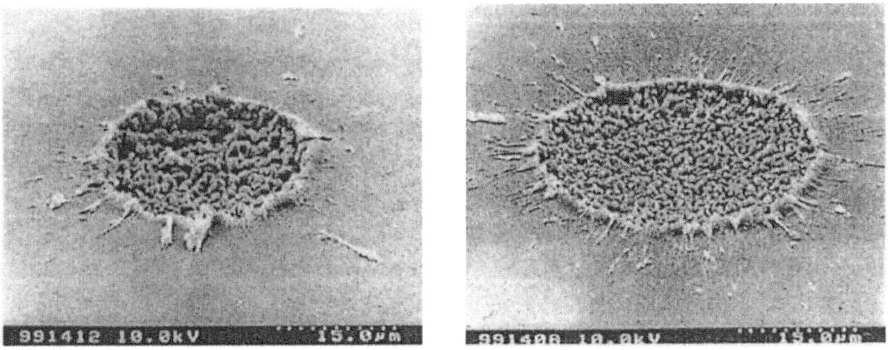

Fig. 7. Femtosecond laser ablation of PMMA (left) and PC (right). 150 fs, 3.0 Jcm^{-2}, $N = 5$, 800 nm.

Fs-laser-induced cavities could be generated on polyimide (PI), polycarbonate (PC), polyethylenterephthalate (PET), polytetrafluoroethylene (PTFE), and polymethylmethacrylate (PMMA) with smaller diameters than the quasi-Gaussian $1/e^2$ diameter of the beam (Bonse, et al. 2000; Baudach, et al, 2001). Diameters changed significantly during the first 50-60 pulses. The ablation rates per pulse were of the order of < 1μm. The single pulse threshold increased from 1.0 Jcm^{-2} of PI to 2.6 cm^{-2} of PMMA. This trend goes along with the increase of the optical bandgap, suggesting a multi-photon absorption mechanism.

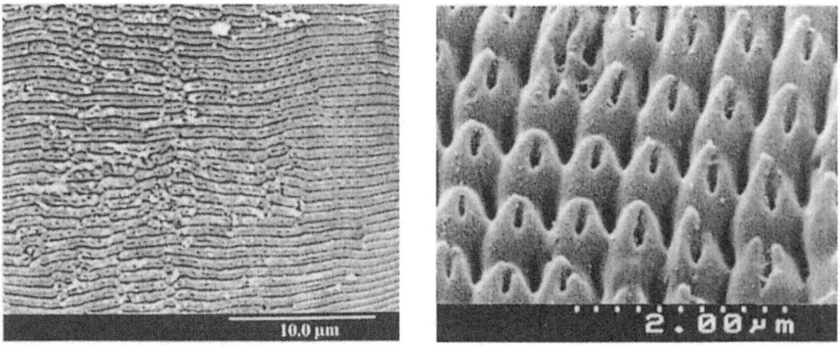

Figure 8. Femtosecond laser ablation of polyimide (PI). (Left) Linear polarization. (Right) Circular polarization. 1.3 Jcm^{-2}, $N = 50$, 150 fs.

Incubation effects were observed for all polymers, i.e. thresholds decreased with the number of pulses. A stronger incubation was observed for PET, PC, and PMMA in contrast to the "inert" polymers PI and PTFE (Bonse, et al. 2000; Baudach, et al. 2001). The ester bonds in PC, PET, and PMMA show much less stability than e.g. the CO-N or ether bridges in PI towards repeated laser pulsing. All polymers except PI showed melting together with the generation of volatile substances, yielding small bubble holes and swelling (Fig. 7). PI does not melt and swell because it only sublimates and/or degrades. The remaining material forms periodical surface structures (Baudach, et al. 1999, Fig. 8). Linear polarization of repeated laser pulses caused ripples parallel to the electric field vector. Circular polarization resulted in nanostructure arrays of the order of ~ 100nm.

6. CONCLUSIONS

Controlled chemical and morphological surface conversions may be attractive for surface technological and nanotechnological purposes. The remaining roughnesses are self-assembled, being not only detrimental to micromachining applications but also opening a range of nanotechnological applications.

Orientation of the coarse type of ripples perpendicular or parallel (polyimide) to the E-vector of the laser beam may suggest their relation to surface electromagnetic waves induced by the incident laser beam in the solid, and their interference with the incident laser beam within the absorption layer.

Such surface electromagnetic waves exist only during the femtosecond laser pulse, before electron-phonon relaxation occurs. Thus, complex dielectric functions may differ from equilibrium, leading to transient sub-picosecond interference parameters resulting in various (coarse, ~600nm) ripple periodicities.

A ripple type with periodicities around 200 nm and less could be generated near the ablation threshold on various ceramics and semiconductors in air contact. This phenomenon can be correlated with surface energy changes triggered by femtosecond-laser-induced non-thermal melting processes on the 100 fs scale. A clear dependence on the chemical composition of the bulk and/or the conversion layer could be detected. Materials with high surface energy (SiC, Si) exhibited a strong tendency to relax to the nanoripples, whereas materials with low surface energies (TiC and TiB_2) did not show this process. At higher fluences, these chemical differences were masked and classical coarse ripples with periodicities approaching the laser wavelength (800 nm) occurred.

Bonding of foreign ligands e.g. in a surface conversion reaction like oxidation of the TiC or TiB_2 grains or a ligand addition/exchange to the surface atoms in fluid contact, may affect this mechanism drastically. Water contact reduces the surface energy practically independent of the nature of the solid and releases the need of surface rearrangement so that no self-assembled regular nanoripples occur at threshold fluences.

ACKNOWLEDGMENTS

Partial financial support was provided by the European Community, in the BRITE-EURAM III project BRP-CT96-0265, "Deployment of In-Situ Optical Monitoring Techniques for Tayloring Thin Film Properties for Specific Advanced Industrial Applications", BRPR-CT96-0294," Novel Rubber Steelcord Adhesion Technology", the TMR Project "Modeling and Diagnostic of Pulsed Laser–Solid Interaction, Applications to Laser Cleaning", FMRX-CT98-0188, by the German Ministry for Research and Technology (BMBF) in the framework of LASER 2000 (Laserinduzierte Fertigungsverfahren, Verbundprojekt ABLATE, #13N 7048/7), and the project "Safety for Applications of Femtosecond Laser Technology" - SAFEST (BMBF-Projektverband Femtosekundentechnologie).

REFERENCES

Adamson, A.W. 1976, *Physical Chemistry of Surfaces*, John Wiley & Sons New York
Baudach, S., Bonse, J. and Kautek, W. 1999, *Appl. Phys. A* **69**, pp. S395-S398
Baudach, S., Bonse, J., Krüger, J. and W. Kautek, 2000, *Appl. Surf. Sci.* **154-155** 555 - 560.
Baudach, S., Krüger, J. and Kautek, W. *The Review of Laser Engineering* **29** (2001), 705-709.
Bäuerle, D. 2000, *Laser Processing and Chemistry*, Springer Verlag Berlin Heidelberg New York
Bonse, J., Baudach, S., Krüger, J., Kautek, W. and Lenzner, M. 2002a, *Appl. Phys. A* **74**, 19
Bonse, J., Wrobel, J., M. Brzezinka, K. -W., Esser, N. and Kautek, W. 2002b, *Appl. Surf. Sci.* **202**, 272
Bonse, J., Brzezinka, K.-W. and Meixner, A. J. 2004, *Appl. Surf. Sci.* **221**, 215
Bonse, J., Geuß, M., Baudach, S., Sturm, H. and Kautek, W. 1999, *Appl. Phys. A* **69**, 399
Bonse, J., Rudolph, P., Krüger, J., Baudach, S. and Kautek, W. 2000a, *Appl. Surf. Sci.* **154-155**, 659
Bonse, J., Sturm, H., Schmidt, D. and Kautek, W. 2000b, *Appl. Phys. A* **71**, 657
Bonse, J., Wrobel, J. ,M. Krüger, J. and Kautek, W. 2001, *Appl. Phys. A* **72**, 89
Burton, W.K. and Cabrera, N. 1949, N. *Discuss. Faraday Soc.* **5**, 33
Cahn, J.W. 1965, *J. Chem. Phys.* **42**, 93
Cortona, A. and Kautek, W. 2001, *Phys. Chem. Chem. Phys.* **3**, 5283
Costache, F., Henyk, M. and Reif, J. 2002, *Appl. Surf. Sci.* **186**, 352
Costache, F., Henyk, M. and Reif, J. 2003, *Appl. Surf. Sci.* **208**, 486
Costache, F., Kouteva-Arguirova, S. and Reif, J. 2004, *Sol. St. Phen.* **95-96**, 635
Daminelli, G., Krüger, J. and Kautek, W. 2004, *Thin Solid Films* **467**, 334
Daminelli, G., Meja, P., Cortona, A., Krüger, J., Autric, M. and Kautek, W. 2002, *SPIE Proceedings* **4760**, 239
Debenedetti, P. 1996, *Metastable liquids,Concepts and Principles*, Princeton University Press Princeton NJ
Dogel, J. and Fryland, W. 2003, *Phys. Chem. Chem. Phys.* **5**, 2484
Harkins, W.D. 1949, *J.Chem. Soc. (London)* **A 62**, 167
Harkins, W.D. 1950, *J.Chem. Soc. (London)* **A 63**, 444
Herring, C., 1951, "Some theorems on the free energy of crystal surfaces," *Phys. Rev.*, **82**(1) 87-93.
Hertwig, A., Martin , S., Krüger J. and Kautek, W. 2004a, *Thin Solid Films* **453-454**, 527
Hertwig, A., Martin, S., Krüger, J. and Kautek, W. 2004b, *Appl. Phys. A* **79**, 1075
Hertwig, A., Martin, S., Krueger, J., Spielmann, C., Lenner, M. and Kautek, W. 2004c, in: *Femtosecond Technology for Technical and Medical Applications, Topics Appl. Phys.* **96**, 287
Jeschke, H.O., Garcia, M.E., Lenzner, M., Bonse, J., Krüger, J. and Kautek, W. 2002, *Appl. Surf. Sci.* **197-198**, 839
Katayama, K., Yonecubo, H. and Sawada, T. 2003 *Appl. Phys. Lett.* **82**, 4244
Kautek, W. and Daminelli, G. 2003, *Electrochim. Acta* **48**, 3249
Kautek, W. and Krüger, J. 1994, *SPIE Proceedings* **2207**, 600
Kautek, W., Krüger, J. Lenzner, M., Sartania, S., Spielmann, C. and Krausz, F. 1996, *Appl. Phys. Lett.* **69**, 3146

Kautek, W., Mitterer, S., Krüger, J., Husinsky, W. and Grabner, G. 1994, *Appl. Phys. A* **58**, 513

Kautek, W., Rudolph, P., Daminelli, G. and Krüger, J. 2005, *Appl. Phys. A* **81**, 65

Kautek, W., Rudolph, P., Daminelli, G., Hertwig, A., Martin, S., Bonse, J. and Krüger, J. 2004, *SPIE Proceedings* **5448**, 213

Kelly, R. and Miotello, A. 1999, *Phys. Rev.* **E60**, 2616

Krüger, J. and Kautek, W. 1996, *Appl. Surf. Sci* **96-98**, 430

Krüger, J. and Kautek, W. 1999, *Laser Physics* **9**, 30

Krüger, J. and Kautek, W. 2004, *Advances in Polymer Science*, Vol. 168 (Springer Verlag Heidelberg p. 247.

Krüger, J., Lenzner, M., Martin, S., Lenner, M., Spielmann, C., Fiedler A. and Kautek, W. 2003, *Appl. Surf. Sci.* **208-209**, 233

Krüger, J., Meja, P., Autric, M. and Kautek, W. 2002, *Appl. Surf. Sci.* **186**, 374

Küper, S. and Stuke, M. 1987, *Appl. Phys. B* **44**, 199

Lenzner, M., Krausz, F., Krüger, J. and Kautek, W. 2000, *Appl. Surf. Sci.* **154-155**, 11

Lenzner, M., Krüger, J., Sartania, S., Cheng, Z., Spielmann, C., Mourou, G., Kautek, W. and Krausz, F. 1998, *Phys. Rev. Lett.* **80**, 4076

Martin, S., Krüger, J., Hertwig, A., Fiedler, A. and Kautek, W. 2003a, *Appl. Surf. Sci.* **208-209**, 333

Martin, S., Hertwig, A., Lenzner, M., Krüger, J. and Kautek, W. 2003b, *Appl. Phys. A* **77**, 883

Mero, M., Clapp, B., Jasapara, J., Rudolph, W., Ristau, D., Starke, K., Krueger, J., Martin, S. and Kautek, W. 2005, *Opt. Eng.* **44/5**, 051107

Noack, J. and Vogel, A. 1999, *IEEE J. Quant. Electron.* **35**, 1156

Oltra, R. ,Arenholz, E., Leiderer, P., Kautek, W., Fotakis, C., Autric, M., Afonso, C. and Wazen, P. 2000, *SPIE Proceedings* **3885**, 499

Pampuch, R. 1991, *Constitution and Properties of Ceramic Materials* Elsevier, Amsterdam Oxford New York Tokyo

Reif, J., Costache, F., Henyk, M. and Pandelov, S.V. 2002, *Appl. Surf. Sci.* **197-198**, 891

Reintjes, J.F. and McGroddy, J.C. 1973, *Phys. Rev. Lett.* **30**, 901

Rudolph, P. and Kautek, W. 2004, *Thin Solid Films* **453-454**, 537

Rudolph, P., Brzezinka, K.-W., Wäsche, R. and Kautek, W. 2003, *Appl. Surf. Sci.* **208-209**, 285

Ruf, A. and Dausinger, F. 2004 in *Femtosecond Technology for Technical and Medical Applications*, Topics Appl. Phys. **96**, 105

Temperley, H.N.V. 1952, *Proc. Cambridge Phil. Soc.* **48**, 683

Turchanin, A. and Freyland, W. 2004, *Chem. Phys. Lett.* 387, 106

Turchanin, A., Freyland, W. and Nattland, D. 2002, *Phys. Chem. Chem. Phys.* **4**, 647

Chapter 10

FEMTOSECOND PLASMA-MEDIATED NANOSURGERY OF CELLS AND TISSUES

Alfred Vogel[1], Joachim Noack[1], Gereon Hüttman[1] and Günther Paltauf [2]
[1]Institut für Biomedizinische Optik, Universität zu Lübeck, Peter-Monnik Weg 4, D-23562 Lübeck, Germany; [2]Institut für Physik, Karl-Franzens-Universität Graz, Universitätsplatz 5, A-8010 Graz, Austria

1. INTRODUCTION

1.1 Cell surgery

Nonlinear absorption of short and ultrashort laser pulses focused through microscope objectives of high numerical aperture (NA) can be used to achieve very fine and highly localized laser effects inside of biological media that are transparent at low irradiance (Shen, 1984; Vogel and Venugopalan, 2003; Venugopalan, et al., 2002; König, et al., 1999; Vogel, et al., 2005) as well as in the bulk of photonic materials (Shaffer, et al., 2001; Minoshima, et al., 2001).

With moderate NAs and nanosecond laser pulses, this possibility has already been utilized already in the 1980s for intraocular surgery (Steinert and Puliafito, 1986; Vogel, et al., 1986). After the advent of femtosecond lasers, it was also employed for corneal intrastromal refractive surgery (Ratkay-Traub, et al., 2003; Heisterkamp, et al., 2003) and for the creation of corneal flaps in excimer laser refractive surgery (LASIK) (Juhasz, et al., 1999; Ratkay-Traub, et al., 2003; Heisterkamp, et al., 2003; Han, et al., 2004). However, with moderate NAs, the spatial distribution of the deposited energy is influenced by nonlinear self-focusing, normal group velocity dispersion, and plasma-defocusing leading to filamentation and streak formation in the biological material (Shen, 1984; Vogel, et al., 1996a;

Heisterkamp, et al., 2002; Liu, et al., 2003; Kasparian, et al., 2004; Kolesik, et al., 2004; Arnold, et al., 2005). The nonlinear propagation effects become ever more important when the laser pulse duration is reduced and a larger laser power is required to produce optical breakdown. Therefore, it is not possible to achieve highly localized energy deposition when femtosecond pulses are focused into the bulk of transparent media at low NA. With increasing numerical aperture the spot size becomes smaller, and thus the power that is necessary to overcome the threshold irradiance decreases. Beyond a certain numerical aperture, the breakdown power is smaller than the critical power for self-focusing, and localized energy deposition on a submicrometer scale can be achieved. For femtosecond optical breakdown in water and glass this was found to be the case for NA ≥ 0.9 (Schaffer, et al., 2001).

Recent years have seen a continuous rise of interest in micro and nanosurgery on a cellular and subcellular level. One important application is the separation of individual cells or other small amounts of biomaterial from heterogeneous tissue samples for subsequent genomic or proteomic analysis. Sensitive analytical techniques such as polymerase chain reaction (PCR) enable the analysis of very small amounts of materials, which allows for even more specific investigations of cell constituents and their functions. Key technologies for sample preparation are laser microdissection (LMD) (Meier-Ruge, et al., 1976), and subsequent laser pressure catapulting (LPC) of the dissected specimens into a vial for further analysis (Schütze and Lahr, 1998; Schütze, et al., 1998; Niyaz and Sägmüller, 2005). A related technique is laser-induced cell lysis and catapulting of the cell content into a micropipet for time-resolved capillary electrophoresis (Sims, et al., 1998). Laser microbeams have also been applied to dissect chromosomes (Berns, et al., 1981; Liang, et al., 1993; Greulich, 1999; König, et al., 2001), and fuse cells (Schütze and Clement-Sengwald, 1994). Laser-induced transient permeabilisation of the cell membrane is of great interest for a gentle transfection of genes and transfer of other substances into specific cell types (Tsukakoshi, et al., 1984; Tao, et al., 1987; Krasieva, et al., 1998; Soughayer, et al., 2000; Tirlapur and König, 2002; Zeira, et al., 2003; Paterson, et al., 2005.

Laser-generated inactivation of specific proteins or cell organelles, together with an analysis of the induced deviations from the normal development, provides information about the function of the respective proteins and organelles and can be utilized to study cell proliferation, embryonal development, or stress-induced reaction pathways. Two complementary strategies for functional studies have been followed. In the 'systemic' approach, specific proteins or DNA sequences are targeted by means of antibodies attached to metallic nanoparticles or chromophores (Huettmann and Birngruber, 1999; Jay and Sakurai, 1999; Pitsillides, et al., 2003; Yao, et al., 2005; Garwe, et al., 2005). When the antibody-absorber conjugates have bound to the target protein(s), the entire cell or group of cells is exposed to a short-pulsed laser beam. Protein inactivation occurs through linear absorption of the laser irradiation in the nanoparticles or chromophores, respectively, resulting in thermomechanical or photochemical destruction of the target proteins regardless of their localisation within the cell. Alternatively, in the 'local' approach, which is investigated in the present paper, one or a few specific target structures are irradiated by a tightly focused laser beam. As the laser energy is deposited via non-linear absorption, surgery can be performed at any desired location within a cell or a small organism, regardless of its linear absorption properties.

1.2 Historical development

Historically, light inactivation of cells or cell organelles was first attempted in 1912 by Tschachotin using 280 nm irradiation from a magnesium spark imaged by a microscope objective on a 5 μm wide spot on the cell (Tschachotin, 1912). This type of apparatus was highly refined in the 1950s by Bessis and Nomarski (1960), and the resolution increased into the sub-micrometer regime. However, these instruments required very long exposure times. After the advent of the laser, a high-brightness light source was available that enabled reduction of the exposure time into the microsecond range (Bessis, et al., 1962). First experiments on mitochondrial inactivation were performed using free-running ruby laser pulses with about 500 μs duration that were focused into a 5 μm spot (Amy and Storb, 1965). Later, chromosomal dissection was demonstrated using argon laser irradiation with 20-30 μs duration (Berns, et al., 1969, 1971). Owing to the good quality of the argon laser beam and the shorter wavelength, it could be focused into a much smaller spot than the multimode emission of the initial ruby lasers. It is important to note that microsecond pulses are still 'long' in the context of cell surgery because during pulses longer than about 10 μs, a stationary temperature distribution similar to that produced by continuous wave (cw) irradiation evolves around the laser focus (Vogel, et al., 2005).

Soon researchers began to also use short-pulsed laser irradiation, mostly with wavelengths in the UV region of the optical spectrum and with durations of a few nanoseconds (Bessis, 1971; Meier-Ruge, et al., 1976; Berns, et al., 1981; Schütze and Clement-Sengwald, 1994; Krasieva, et al., 1998; Greulich, 1999; Colombelli, et al., 2004, 2005a, 2005b). It was found that short laser pulses enable localized energy deposition at arbitrary locations without external sensitizing agents, even though the ablation threshold can still be lowered by staining of the target structures. With nanosecond pulses, energies between 0.25 μJ and 250 μJ were required to produce the desired ablative effect, depending on the laser wavelength, beam profile, numerical aperture, and the quality of the optical scheme used for coupling the laser beam into the microscope. Use of UV wavelengths that are well-absorbed by biomolecules yielded lower ablation thresholds than the use of visible or near IR irradiation under similar focusing conditions. Recently, it was demonstrated that pulsed laser microdissection relies on plasma formation supported by linear absorption, and that this is associated with violent mechanical effects (shockwave emission and cavitation bubble formation) reaching well beyond the region of energy deposition (Venugopalan, et al., 2003). Pulse energies in the microjoule range typical for nanosecond laser microbeams can therefore severely affect the cell viability.

In search for finer effects, researchers employed first picosecond pulses that could produce intracellular dissections with energies of 70-140 nJ (Liang, et al., 1993), and later femtosecond pulses that enabled to lower the ablation threshold to an energy range between 0.4 nJ and a few nanojoules (König, et al., 1999, Yanik, et al., 2004). Due to the low energy threshold for plasma formation (Vogel, et al., 1999; Noack and Vogel, 1999), femtosecond pulses can create very fine effects with a spatial extent below the optical diffraction limit. This has been demonstrated in chromosomes (König, et al., 1999; König, et al., 2001), various other cell organelles (Meldrum, et al., 2003; Watanabe, et al., 2004; Heisterkamp, et al., 2005; Sacconi, et al., 2005; Shen, et al., 2005), small organisms (Yanik, et al., 2004; Supatto, et al., 2005a,b; Chung, et al., 2005), and tissues (König, et al., 2002; Zeira, et al., 2003; Riemann, et al., 2005). Subdiffraction limited resolution can be achieved because the nonlinear absorption diminishes the volume into which the laser energy is deposited. While for nanosecond pulses the optical breakdown threshold depends strongly on the linear absorption at the laser focus, femtosecond optical breakdown exhibits a much weaker dependence on the absorption coefficient of the target material (Oraevsky, et al., 1996).

This facilitates the targeting of arbitrary cellular structures. Because the wavelength dependence of femtosecond breakdown is weak (Vogel and Noack, 2001), IR wavelengths that can penetrate deeply into the tissue can be used without compromising the precision of tissue effects as observed with ns pulses (Krasieva, et al., 1998; Venugopalan, et al., 2002). Moreover, when pulses from a femtosecond oscillator are used, it becomes possible to combine nonlinear material modification with nonlinear imaging techniques based on 2-photon fluorescence excitation or second harmonic generation (König, et al., 2002; Tirlapur and König, 2002; Yanik, et al., 2004; König, et al., 2005; Saccioni et al., 2005; Supatto, et al., 2005a,b). Additional progress was possible through the use of modern gene fusion products such as green fluorescent proteins (GFP) which permit the visualization and ablation of cellular structures that are below the resolution of a light microscope (Botvinick, et al., 2004; Yanik, et al., 2004; Supatto, et al., 2005a,b). The above advances allow for an unprecedented precision of aiming, surgery, and of the analysis of the created immediate and long-term effects. This potential of fs and ps pulses has been utilized in a variety of functional studies to elucidate the mechanisms of chromosome separation during cell division (Liang, et al., 1993; Grill, et al., 2003), induce highly localized DNA damage (Meldrum, et al., 2003), measure the biophysical properties of the cytoskeleton and mitochondria (Shen, et al., 2005; Colombelli, et al., 2005b; Maxwell, et al., 2005), stimulate calcium waves in living cells (Smith, et al., 2001), demonstrate nerve regeneration after axotomy within a living C. elegans worm (Yanik, et al., 2004), map thermosensation in C. elegans (Chungs, et al., 2005), and to shed light on morphogenetic movements in embryonal development (Berns, et al., 1981; Supatto, et al., 2005a,b).

1.3 Objectives of the present study

The high precision of the femtosecond laser effects is certainly related to the fact that the energy threshold for femtosecond optical breakdown is very low. The low breakdown threshold is, however, not sufficient to explain the fineness of the laser effects because laser-induced breakdown is generally associated with mechanical effects such as shockwave emission and bubble formation that extend beyond the focal region (Vogel, et al., 1996b; Venugopalan, et al., 2002). We found in previous theoretical studies that plasmas with a large free electron density are produced in a fairly large irradiance range below the breakdown threshold that was defined by a critical free electron density $\rho_{cr} = 10^{21}$ cm^{-3} (Vogel and Noack, 2001; Vogel, et al., 2002). To understand the full potential of femtosecond pulses for highly localized material processing and modification of biological media, one therefore needs to include the irradiance range *below* the optical breakdown threshold. Moreover, one needs to elucidate why the conversion of absorbed laser light into mechanical energy above the breakdown threshold is much smaller than for longer pulse durations (Vogel, et al., 1999; Vogel and Venugopalan, 2003).

The present study investigates the chemical, thermal, and thermomechanical effects arising from low-density plasmas to explain the mechanisms underlying femtosecond-laser nanosurgery of cells and biological tissues. One technique for nanosurgery uses long series of pulses from fs oscillators with repetition rates in the order of 80 MHz, and pulse energies well below the optical breakdown threshold that do not much exceed the energies used for nonlinear imaging (König, et al., 1999; König, et al., 2001; Tirlapur and König, 2002; Zeira, et al., 2003; Saccioni, et al., 2005; Supatto, et al., 2005a,b; König, et al., 2005). The other approach uses amplified pulse series at 1 kHz repetition rate with pulse energies slightly above the threshold for transient bubble formation (Yanik, et al., 2004; Watanabe, et al., 2004; Heisterkamp, et al., 2005; Shen, et al., 2005). To cover both parameter regimes, we investigate plasma formation and plasma-induced effects for an irradiance range reaching from the values used for nonlinear imaging to those producing bubble formation. We consider repetition rates in the kilohertz range where the mechanical and thermal events induced by subsequent pulses are largely independent, and in the megahertz range where accumulative effects are likely to occur.

We use a rate equation model considering multiphoton ionization, tunnel ionization, and avalanche ionization to numerically simulate plasma formation. The value of the energy density created by each laser pulse is then used to calculate the temperature distribution in the focal region after application of a single laser pulse and of series of pulses. The results of the temperature calculations yield, finally, the starting point for calculations of the thermoelastic stresses that are generated during the formation of the low-density plasmas, and of stress-induced bubble formation. All calculations are performed for a numerical aperture of NA = 1.3 and the wavelength of the titanium sapphire laser (λ = 800 nm). Where possible, the findings of the numerical simulations are compared to experimental results.

2. MODELING OF PLASMA FORMATION

The process of plasma formation through laser-induced breakdown in transparent biological media is schematically depicted in Fig. 1. It essentially consists of the formation of quasi-free electrons by an interplay of photo-ionization and avalanche ionization.

It has been shown experimentally that the optical breakdown threshold in water is very similar to that in ocular and other biological media (Docchio, et al., 1986). For convenience, we shall therefore focus attention on plasma formation in pure water. Whereas the optical breakdown in gases leads to the generation of free electrons and ions, in condensed matter electrons are either bound to a particular molecule or they are "quasi-free" if they have sufficient kinetic energy to be able to move without being captured by local potential energy barriers. Transitions between bound and quasi-free states are the equivalent of ionization of molecules in gases. To describe the breakdown process in water, Sacchi (1991) has proposed that water should be treated as an amorphous semiconductor and the excitation energy Δ regarded as the energy required for a transition from the molecular $1b_1$ orbital into an excitation band (band gap 6.5 eV) (Grand, et al., 1970; Nikogosyan, et al., 1983). We follow this approach. For simplicity, we use the terms "free electrons" and "ionization" as abbreviations for "quasi-free electrons" and

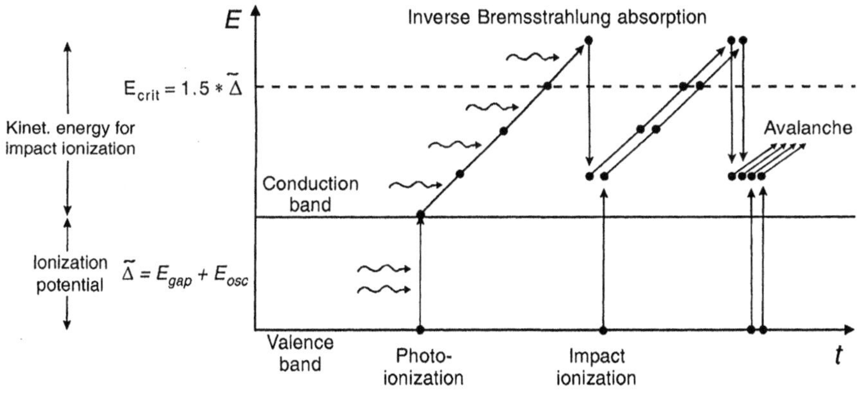

Figure 1. Interplay of photoionization, inverse Bremsstrahlung absorption, and impact ionization in the process of plasma formation. Recurring sequences of inverse Bremsstrahlung absorption events and impact ionization lead to an avalanche growth in the number of free electrons. The consequences of the conservation laws for energy and momentum on the energetics of impact ionization are discussed in the text.

"excitation into the conduction band". Nonlinear absorption processes of liquid water actually do not only consist of ionization, but also include dissociation of the water molecules (Nikogosyan, et al., 1983). However, in our model dissociation is neglected to reduce the complexity of the numerical code.

The excitation energy into the conduction band can be provided either by photoionization (multiphoton ionization or tunneling (Keldysh, 1965; Ammosov, et al., 1986), or by impact ionization (Shen, 1984; Thornber, 1981; Arnold and Cartier, 1992; Ridley, 1999). In previous breakdown models, it was often assumed that a free electron could be produced as soon as the band gap Δ was exceeded either by the sum of the simultaneously absorbed photons, or by the kinetic energy of an impacting free electron (Kennedy, 1995; Feng, et al., 1997; Noack and Vogel, 1999; Tien, et al., 1999). However, for very short laser pulses where breakdown occurs at large irradiance values, the band gap energy has to be replaced by the effective ionization potential to account for the oscillation energy of the electron due to the electric laser field. The ionization potential of individual atoms is (Keldysh, 1965)

$$\tilde{\Delta} = \Delta + e^2 F^2 / (4m\omega^2), \tag{1}$$

where ω and F denote the circular frequency and amplitude of the electric laser field, e is the electron charge, and $1/m = 1/m_c + 1/m_v$ is the exciton reduced mass that is given by the effective masses m_c of the quasi-free electron in the conduction band, and m_v of the hole in the valence band. The second term in equation (1) can be neglected in nanosecond optical breakdown but must be considered in femtosecond optical breakdown where F is orders of magnitude larger. For condensed matter, the description of the ionization potential is more complex than Eq. (1) (Keldysh, 1965 ; Vogel, et al., 2005).

Multiphoton ionization (MPI) and tunneling are the mechanisms governing photoionization for different field strengths and frequencies of the electromagnetic field. In his classical paper, Keldysh (1965) introduced a parameter $\gamma = \omega / \omega_t$ to distinguish tunneling and MPI regimes. Here, $1/\omega_t$ stands for the tunneling time through the atomic potential barrier which is inversely proportional to the strength of the electromagnetic field. For values $\gamma << 1$, as obtained with low frequencies and large field strengths, tunneling is responsible for ionization, while for values $\gamma >> 1$ typical for optical frequencies and moderate field strengths, the probability of MPI is much higher than that of tunneling. However, femtosecond optical breakdown requires very high field strengths for which the tunneling time through the atomic potential barrier is extremely short, leading to values $\gamma < 1$ of the Keldysh parameter even for optical frequencies (Tien, et al., 1999). Approximations of the Keldysh theory considering only multiphoton ionization that were used in previous breakdown models (Kennedy, 1995; Feng, et al., 1997; Nock and Vogel, 1999) are thus inappropriate for the modeling of femtosecond breakdown, especially for pulse durations ≤ 100 fs.

Once a free electron is produced in the medium, it can absorb photons in a non-resonant process called "inverse Bremsstrahlung" in the course of collisions with heavy charged particles (ions or atomic nuclei). A third particle (ion/atom) is necessary for energy and momentum to be conserved during absorption, as they cannot both be conserved if only an electron and a photon interact. The electron gains kinetic energy during the absorption of the photon. After a sequence of several inverse Bremsstrahlung absorption events, the kinetic energy is sufficiently large to produce another free electron through impact ionization (Thornber, 1981; Arnold and Cartier, 1992; Ridley, 1999; Kaiser, et al., 2000). Two free electrons with low kinetic energies are now available which can gain energy through inverse Bremsstrahlung absorption (Fig. 1). The recurring sequence of inverse Bremsstrahlung absorption events and impact ionization leads to an avalanche growth in the number of free electrons if the irradiance is high enough to overcome the losses of free electrons through diffusion out of the focal volume, and through recombination. The energy gain through inverse Bremsstrahlung must, moreover, be more rapid than the energy loss by collisions with heavy particles occurring without simultaneous absorption of a photon (the fraction of energy lost is proportional to the ratio of the electron and ion masses). The whole process is called "avalanche ionization", or "cascade ionization".

For impact ionization to occur, the kinetic energy of the impacting electron must be larger than the effective ionization potential $\tilde{\Delta}$ to satisfy the conservation laws for energy and momentum (Keldysh, 1960; Ridley, 1999). According to Ridley (1999), the critical energy for bands with parabolic energy dispersion is

$$E_{crit} = \left[(1+2\mu)/(1+\mu)\right]\tilde{\Delta}, \quad \text{with } \mu = m_c / m_v . \tag{2}$$

The value of μ depends on the band structure; it is 1 for a symmetric band structure with the Fermi level at the center of the bandgap, but smaller for semiconductors (Ridley, 1999). Kaiser, et al., (2000) assumed $\mu = 1$ for α-SiO_2, and since we did not find information on the value of μ for water, we follow their assumption. This implies that a kinetic energy of $E_{crit} = 1.5\tilde{\Delta}$ is required for impact ionization (Kaiser, et al., 2000; Rethfeld, 2004).

The excess energy of $0.5\tilde{\Delta}$ that remains after impact ionization is distributed among the collision partners. Thus, each quasi-free electron produced by impact ionization has to gain less energy than $1.5\tilde{\Delta}$ to reach the critical energy. However, the average energy leading to an impact ionization event is larger than E_{crit} because the impact ionization rate increases with kinetic energy (Keldysh, 1960; Arnold and Cartier, 1992; Kaiser, et al., 2000; Rethfeld, 2004). To consider both factors, we assume that the average energy gain required for a free electron to cause impact ionization is $1.5\tilde{\Delta}$, as illustrated in Fig. 1.

While strong-field ionization is almost "instantaneous," there are time constraints on cascade ionization because several consecutive inverse Bremsstrahlung absorption events are necessary for a free electron to pick up the critical energy for impact ionization. For a bandgap of 6.5 eV in water and a Keldysh parameter $\gamma = 2$, the effective ionization potential is $\tilde{\Delta} \approx 7.3$ eV, and the average gain in kinetic energy required to enable impact ionization is $(3/2)\tilde{\Delta} \approx 10.95$ eV. When laser irradiation of $\lambda = 800$ nm wavelength with a photon energy of 1.55 eV is used to produce optical breakdown, an electron must undergo at least $n = 8$ inverse Bremsstrahlung absorption events before impact ionization can occur. As mentioned above, inverse Bremsstrahlung absorption can only occur during collisions of the electrons with heavy particles. In condensed matter, the time τ between collisions was estimated to be roughly 1 fs (Bloembergen, 1984). Recent experimental investigations yielded a value of $\tau = 1.7$ fs for fused silica (Sun, et al., 2005). Based on this value, the minimum time for one doubling sequence of the number of free electrons by cascade ionization is $\tau_{ion} = \tau n =$ 13.6 fs, even if every collision involves absorption of a photon. A detailed analysis of the time constraints in cascade ionization was presented by Kaiser, et al., (2001) and Rethfeld (2004). They come to the conclusion that cascade ionization plays only a minor role in femtosecond breakdown compared to multiphoton effects – in striking contrast to Joglekar, et al., (2004), who present some experimental evidence for the opposite statement.

In our study, we shall combine the complete Keldysh model for strong-field ionization (Eq. (38) in Keldsh's paper (1965)) with the description of avalanche ionization used by Shen (1984), Kennedy (1995), and Stuart, et al., (1996), which is based on the Drude model. Since the numerical model used by Kaiser, et al., (2000) and Rethfeld, (2004) is very complex, we consider the time constraints in cascade ionization in a simpler way by evaluating the contribution of cascade ionization at time *t* using the electron density created at the retarded time $t_{ret} = t - t_{ion}$ (Vogel, et al., 2005).

In most theoretical investigations, the electron density

$$\rho_{cr}' = \omega^2 (m_c \varepsilon_0 / e^2) \tag{3}$$

above which the plasma becomes both strongly reflective and absorbing is used as breakdown criterion (Stuart, et al., 1996; Lenzner, et al., 1998; Tien, et al., 1999; Kaiser, et al., 2000; Mao, et al., 2004). Here ε_0 denotes the vacuum dielectric permittivity. We use a free electron density of $\rho_{cr} = 10^{21}$ cm^{-3} as breakdown criterion, which is close to ρ_{cr} for λ = 1064 nm. The experimental threshold criterion is bubble formation. Plasma luminescence, which is often used as threshold criterion for nanosecond breakdown, is not (or only with great difficulties) detectable for ultrashort laser pulses (Hammer, et al., 1996; Noack and Vogel, 1999).

Since all calculations are performed for a numerical aperture of NA = 1.3, nonlinear propagation effects in the biologic medium can be neglected even for pulse durations as short as 100 fs because Schaffer, et al., (2001) showed that these nonlinear effects influence the breakdown threshold only for NA < 0.9. Self-focusing and filamentation may play a role well above the breakdown threshold, but are not relevant for the pulse energies used in nanosurgery on cells.

The time evolution of the electron density ρ_c in the conduction band under the influence of a laser pulse with Gaussian temporal shape was calculated using a rate equation of the generic form (Noack and Vogel, 1999)

$$d\rho_c / dt = \eta_{\text{photo}} + \eta_{\text{casc}} \rho_c - \eta_{\text{diff}} \rho_c - \eta_{\text{rec}} \rho_c^2 . \tag{4}$$

The first term represents the production of free electrons mediated by the strong electric field in the laser focus (photoionization via multiphoton and tunneling ionization), the second term represents the contribution of cascade ionization, and the last two terms describe the losses through diffusion of electrons out of the focal volume, and recombination. The cascade ionization rate η_{casc} and the diffusion loss rate η_{diff} are proportional to the number of already produced free electrons, while the recombination rate η_{rec} is proportional to ρ_c^2, as it involves an interaction between two charged particles (an electron-hole pair). Even though diffusion and recombination do not play a significant role during femtosecond laser pulses, they were included to enable a comparison to plasma formation by nanosecond pulses. The explicit form of the individual terms in Eq. (4) has been described in detail by Vogel, et al., (2005).

The temporal evolution of the electron density, $\rho(t)$, was calculated for laser pulses focused into pure water at a numerical aperture of NA = 1.3. At least one free "seed" electron produced by photoionization is required for the start of the cascade. Therefore, the term for cascade ionization is only

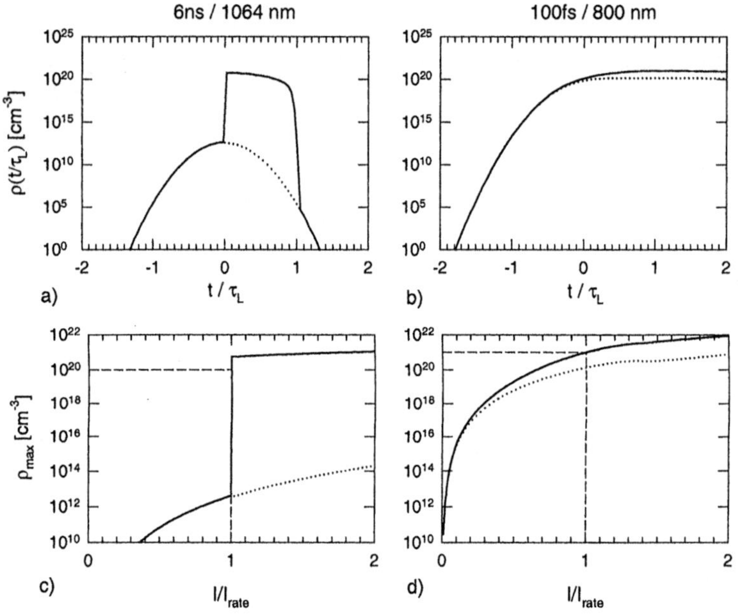

Figure 2. Top row: Evolution of the free-electron density during the laser pulse at the optical breakdown threshold for 6 ns, 1064 nm pulses and for 100 fs, 800 nm pulses. The time t is normalized with respect to the laser pulse duration τ_L. The contribution of multiphoton ionization to the total free-electron density is plotted as a dotted line. Bottom row: Maximum free electron density ρ_{max} achieved during the laser pulse as a function of irradiance, for the same laser parameters. The irradiance I is normalized with respect to the threshold irradiance I_{rate}. The threshold I_{rate} and the corresponding value of ρ_{max} are marked by dotted lines.

The dynamics of plasma formation is extremely different for nanosecond and femtosecond pulses. With nanosecond pulses, no free electrons are formed for irradiance values below the optical breakdown threshold because the irradiance is too low to provide seed electrons by means of multiphoton ionization (Fig. 2c). Once the irradiance is high enough to provide a seed electron, the ionization cascade can start. It proceeds very rapidly, owing to the high irradiance (Fig. 2a). The electron density shoots up by 9 orders of magnitude within a small fraction of the laser pulse duration until its rise is stopped by recombination which is proportional to ρ_c^2. The breakdown threshold is, hence, extremely sharp - either a highly ionized plasma is produced, or no plasma at all. These numerical predictions are supported by the experimental observation that at the threshold of nanosecond optical breakdown with IR laser pulses the transmission of the focal volume drops abruptly to less than 50% of the value without plasma formation (Nahen and Vogel, 1996; Noack, 1998c). The transmission loss for shorter pulse durations is much less abrupt (Nahen and Vogel, 1996 ; Noack, et al., 1998b ; Noack, 1998c; Vogel, et al., 1999).

With femtosecond pulses, a much higher irradiance is necessary for optical breakdown to be completed during the laser pulse duration than with nanosecond pulses. This favors the generation of free electrons through photoionization because multiphoton ionization exhibits a strong irradiance dependence $\propto I^k$ (k representing the number of photons required for crossing the ionization potential) as opposed to $\propto I$ for the cascade ionization rate (Vogel, et al., 2005). While with nanosecond pulses the total number of free electrons generated through avalanche ionization is 10^9 times larger than the number generated through multiphoton ionization (Fig. 2a), it is only 12 times larger with 100 fs pulses at 800 nm (Fig. 2b). As a consequence of the increasing importance of multiphoton ionization with shorter pulse durations, there is never a lack of seed electrons for avalanche ionization. An avalanche is initiated at irradiance values considerably lower than the breakdown threshold. The free-electron density reached at the end of the avalanche depends on irradiance in a much smoother way (Fig. 2d) than for ns pulses (Fig. 2c). Therefore, one can generate any desired free-electron density by selecting an appropriate irradiance value.

Figure 3 presents threshold values for irradiance, I_{rate}, and radiant exposure, $F_{rate} = I_{rate} \times \tau_L$, required to reach a critical free electron density of

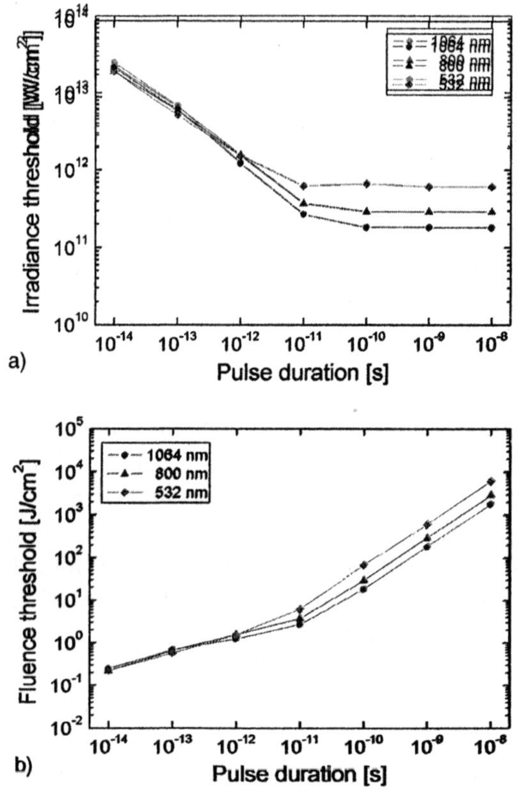

Figure 3. Calculated optical breakdown thresholds ($\rho_{cr} = 10^{21}$ cm^{-3}) as a function of laser pulse duration for various wavelengths; (a) irradiance threshold, (b) radiant exposure threshold.

$\rho_{cr} = 10^{21}$ cm^{-3}. The thresholds were calculated for various wavelengths and pulse durations ranging from 10 fs to 10 ns. Two regimes can be distinguished; for $\tau_L < 10$ ps, the threshold radiant exposure F_{rate} exhibits only a weak dependence on pulse duration. This reflects the fact that recombination plays only a minor role during ultrashort laser pulses. Therefore, only one set of free electrons is produced that corresponds to an approximately constant energy density within the focal volume. This is in accordance with the experimental threshold criterion of bubble formation that requires a specific energy density, which varies little with laser parameters. By contrast, for longer pulses more than one set of free electrons is produced and recombines during the laser pulse. Here it is the threshold irradiance I_{rate} that remains approximately constant, because a minimum irradiance is required to provide the seed electrons for the ionization cascade by multiphoton ionization and to drive the cascade sufficiently fast to reach the critical free-electron density within the laser pulse duration. As a consequence, the radiant exposure threshold and plasma energy density increase steeply with increasing pulse duration.

The predicted form of the F_{rate} (τ_L) dependence qualitatively matches experimental observations on the pulse duration dependence of single shot damage thresholds at surfaces of transparent large bandgap dielectrics (Du, et al., 1996, Tien, et al., 1999) and ablation thresholds of corneal tissue (Du, et al., 1994).

2.4 Low-density plasmas in bulk media

Figure 2d indicates that femtosecond pulses focused into bulk transparent media can create low-density plasmas in which the energy density remains below the level that leads to cavity formation in the medium. Experimental evidence for the existence of such low-density plasmas was provided by Mao, et al., (2004) through measurements of the free-electron density in MgO and SiO_2. Free electrons are produced in a fairly large irradiance range below the optical breakdown threshold, with a deterministic relationship between free-electron density and irradiance. Low-density plasmas thus offer the possibility to deliberately produce chemical changes, heating, and thermomechanical effects by varying the irradiance.

For larger irradiances, plasmas in bulk media grow beyond the region of the beam waist, which is not possible for plasma formation at surfaces. At surfaces, the energy deposition becomes confined to a thin layer of less than 100 nm thickness once the free electron density reaches the critical density because the superficial plasma layer is highly absorbing and reflecting (Stuart, et al., 1996; von der Linde and Schüler, 1996; Joglekar, et al., 2004; Feit, et al., 2004). By contrast, in bulk media there is no restriction for the region of optical breakdown to spread towards the incoming laser beam with increasing irradiance. At large irradiances, breakdown already starts to occur before the femtosecond pulse reaches the beam waist, and both irradiance and beam propagation are influenced by the plasma generation (Hammer, et al., 1997, Arnold, et al., 2005). These effects shield the focal region, enlarge the size of the breakdown region, and limit the free electron density and energy density reached in the entire breakdown volume (Fan, et al., 2002a, 2002b; Arnold, et al., 2005; Rayner, et al., 2005). Low density plasmas can, therefore, easily be produced in bulk media while at surfaces the self-induced confinement of plasma formation to a thin layer leads to a rapid rise of free-electron density with irradiance, and the irradiance range in which low-density plasmas can be formed is very small (Stuart et al, 1996; Joglekar, et al., 2004).

3. FOCAL IRRADIANCE AND FREE-ELECTRON DISTRIBUTION

The temperature and stress distribution in the focal region depend on the distribution of quasi-free electrons produced during femtosecond optical breakdown. Therefore, we must explore the shape of the irradiance and free-electron density distributions within the focal volume before we can investigate the resulting temperature and stress effects. The irradiance distribution in the focal volume of a diffraction limited microscope objective used to focus a plane wave has an approximately ellipsoidal shape (Born and Wolf, 1970; Ditlbacher, et al., 2004; Vogel, et al., 2005). For our numerical simulations, the focal volume will therefore be approximated by an ellipsoid with short axis d and long axis l. The short axis d of the ellipsoid is identified with the diameter of the central maximum of the Airy pattern in the focal plane

$$d = 1.22\,\lambda / NA\,. \tag{5}$$

The symbol λ refers to the vacuum wavelength of light. The refractive index of the medium is contained in the value of the numerical aperture (NA) of the microscope objective. The ratio l/d of the long and short axes is

$$l/d = (3 - 2\cos\alpha - \cos 2\alpha)^{1/2}/(1\text{-}\cos\alpha) \tag{6}$$

for optical setups with very large solid angles (Grill and Stelzer, 1999). Here α is the half angle of the light cone such as used in the definition of the numerical aperture $NA = n_0 \sin\alpha$. For NA = 1.3, which in water corresponds to an angle of α = 77.8°, we find l/d = 2.4. For λ = 800 nm, the above considerations yield focal dimensions of d = 750 nm, and l = 1800 nm.

The mathematical form of the diffraction-limited irradiance distribution in the Fraunhofer diffraction pattern of a microscope objective (Born and Wolf, 1970) is too complex for convenient computation of the temperature and stress evolution induced by optical breakdown. We approximate the ellipsoidal region of high irradiance in the center of the focal region by a Gaussian function

$$\rho_{max}(r,z) = \rho_{max}[I(0,0)]\exp[-2(r^2/a^2 + z^2/b^2)] \;, \qquad (7)$$

where r and z are the coordinates in radial and axial direction, respectively, and $a = d/2$ and $b = l/2$ denote the short and long axis of the ellipsoid. The boundaries of the ellipsoid correspond to the $1/e^2$ values of the Gaussian irradiance distribution.

To derive the free-electron distribution ρ_{max} (r,z) from the irradiance distribution I (r,z), we assume that for femtosecond pulses the free-electron density at the end of the laser pulse is approximately proportional to i^k, where k is the number of photons required for multiphoton ionization. This simplifying assumption corresponds to the low-intensity approximation of the Keldysh theory and neglects the weaker irradiance dependence of avalanche

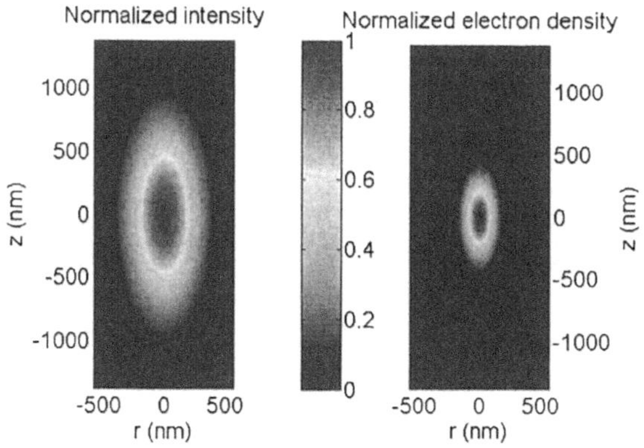

Figure 4. Normalized irradiance distribution (a) and electron density distribution (b) in the focal region for NA = 1.3 and $\lambda = 800$ nm that are assumed for the numerical calculations of the temperature and stress evolution induced by femtosecond optical breakdown.

Ionization that usually dominates plasma formation during the second half of a laser pulse (Fig. 2b). For $\rho_{max} \leq 5\times10^{20}$ cm^{-3}, the proportionality $\rho_{max} \propto I^{k}$ has been confirmed by the experimental results of Mao, et al., (2004). The spatial distribution of the free-electron density can thus be expressed as

$$\rho_{max}(r,z) = \rho_{max}[I(0,0)]\exp[-2k\,(r^2/a^2 + z^2/b^2)]\ . \qquad (8)$$

Figure 4 shows the irradiance and electron density distribution in the focal region according to Eqs. (7) and (8) for NA = 1.3 and $\lambda = 800$ nm, for which $k = 5$. Due to the nonlinear absorption process underlying optical breakdown, the free-electron distribution is much narrower than the irradiance distribution. For $\lambda = 800$ nm and breakdown in water, it is narrower by a factor of $\sqrt{5} = 2.24$, which corresponds to a reduction of the affected volume by a factor of 11.2 below the diffraction limited focal volume. The diameter of the free-electron distribution at the $1/e^2$- values amounts to 336 nm, the length to 806 nm. Femtosecond-laser nanoprocessing can achieve a 2-3 fold better precision than cell surgery using cw irradiation, and enables manipulation at arbitrary locations.

4. CHEMICAL EFFECTS

Plasma-mediated chemical effects of low-density plasmas in biological media can be classified into two groups: 1). Changes of the water molecules by which reactive oxygen species (ROS) are created that affect organic molecules, and 2). Direct changes of the organic molecules in resonant electron-molecule scattering.

1. The creation of ROS such as OH* and H_2O_2, through various pathways following ionization and dissociation of water molecules has been investigated by Nikogosyan, et al., (1983) and recently reviewed by Garret, et al., (2005). Both oxygen species are known to cause cell damage (Tirlapur, et al., 2001). Heisterkamp, et al., (2002) confirmed the dissociation of water molecules during femtosecond laser-induced plasma formation by chemical analysis of the gas content of the bubbles.

2. Capture of electrons into an antibonding molecular orbital can initiate fragmentation of biomolecules (Boudaiffa, et al., 2000; Hotop, 2001; Gohlke and Illenberger, 2002; Huels, et al., 2003; Garret, et al., 2005). Such capture

can occur when the electron possesses a "resonant" energy for which there is sufficient overlap between the nuclear wave functions of the initial ground state and the final anion state. For a molecule XY this process corresponds to $e^- + XY \rightarrow XY^{*-}$, where the XY^{*-} has a repulsive potential along the X-Y bond coordinate. After a time of 10^{-15} to 10^{-11} s, the transient molecular anion state decays either by electron autodetachment leaving a vibrationally excited molecule (VE), or by dissociation along one, or several specific bonds such as $XY^{*-} \rightarrow X^{\bullet} + Y^{-}$ (DA). Various authors describe resonant formation of DNA strand breaking induced by low-energy electrons (3-20 eV) (Boudaiffa, et al., 2000, Gohlke and Illenberger, 2002, Huels, et al., 2003). Boudaiffa, et al., (2000) found that the maximum single-strand break (SSB) and double-strand break (DSB) yields per incident electron are roughly one or two orders of magnitude larger than those for 10-25 eV photons. It is conceivable that accumulative effects of this kind can lead to a dissociation/dissection of biological structures that are exposed to femtosecond-laser-generated low-density plasmas.

The irradiance threshold for chemical changes by low-density plasmas can be assessed using the plot of free-electron density versus irradiance presented in Fig. 2d. At NA = 1.3 and 800 nm wavelength, one free electron per focal volume corresponds to a density of $\rho = 2.1\times10^{13}$ cm^{-3}. Our calculations yield the result that this value is reached at an irradiance of $I = 0.26\times10^{12}$ W cm^{-2}, which is 0.04 times the irradiance threshold for breakdown defined as $\rho_c = \rho_{cr} = 10^{21}$ cm^{-3}. Tirlapur, et al., (2001) experimentally observed membrane dysfunction and DNA strand breaks leading to apoptosis-like cell death after scanning irradiation of PtK2 cells with a peak irradiance of $I \approx 0.44\times10^{12}$ W/cm^{-2} in the focal region, or 0.067 times the calculated breakdown threshold. The observed damage pattern of membrane dysfunction and DNA strand breaks matched the effects expected from ROS and free electrons. The damage resembled the type of injury otherwise associated with single-photon aborption of UV radiation (Tirlapur, et al., 2001). However, in Tirlapur's experiments it arose through nonlinear absorption of NIR irradiation and the exposure of cells to low-density plasmas.

The irradiance producing lethal changes when laser pulse series are scanned over entire cells ($0.067\times I_{rate}$) is slightly higher than the model prediction for the irradiance producing one free electron per pulse in the focal volume ($0.04\times I_{rate}$). According to our model, about 10 free electrons in the focal volume are produced by each laser pulse when lethal changes occur. Considering that the cell is exposed to thousands of pulses during the scanning irradiation, cumulative chemical damage may easily arise from the free electrons. By contrast, when locally confined irradiation is used to achieve knockout of individual cell organelles or intracellular dissection, the irradiance threshold for cell death is considerably higher, and dissection can be performed without affecting cell viability.

5. TEMPERATURE EVOLUTION

5.1 Calculation of temperature distribution

The deposition of laser energy into the medium is mediated by the generation and subsequent acceleration of free electrons. The energy carried by the free electrons is transferred to the heavy particles in the interaction volume through collisions, electron hydration, and nonradiative recombination processes resulting in a heating of the atomic, molecular, and ionic plasma constituents. To assess the time needed to establish an equilibrium temperature, we need to look at the characteristic time for electron cooling (the transfer of kinetic electron energy during collisions) and at the time scale for recombination which in water progresses through hydration of the free electrons. The time constant for electron cooling is in the order of only a few picoseconds (Nolte, et al., 1997), and the time constant for hydration of free electrons in water is even shorter, about 300 fs (Nikogosyan, et al., 1983). However, the hydrated states possess a relatively long lifetime of up to 300 ns (Nikogosyan, et al., 1983). In the framework of our model, the different steps are treated as one recombination process. As the frequency of recombination events is proportional to ρ_c^2, the recombination time depends on the free-electron density. It takes about 40 ps until the free electron density decreases by one order of magnitude from a peak value of $\rho_c = 10^{20}$ cm^{-3}, and about 20 ps for a peak value of $\rho_c = 10^{21}$ cm^{-3} (Noack and Vogel, 1999; Vogel and Noack, 2001). For low-density plasmas it will thus take between a few picoseconds and tens of picoseconds until a "thermodynamic" temperature is established (Garret, et al., 2005).

The temperature rise can be determined by calculating the volumetric energy density gained by the plasma during the laser pulse. This calculation is particularly easy for femtosecond pulses because the pulse duration is considerably shorter than the electron cooling and recombination times.

Therefore, hardly any energy is transferred during the laser pulse, and the energy density deposited into the interaction volume is simply given by the total number density ρ_{max} of the free electrons produced during the pulse, multiplied by the mean energy gain of each electron. The mean energy gain of an electron is given by the sum of ionization potential $\tilde{\Delta}$ and average kinetic energy, the latter of which is $(5/4)\tilde{\Delta}$ for free electrons produced by cascade ionization (see Section 2 and Fig. 1). This yields the following simple relation for the plasma energy density ε at the end of the laser pulse:

$$\varepsilon = \rho_{max}\,(9/4)\,\tilde{\Delta}\ . \tag{9}$$

The temperature rise in the interaction volume after a single laser pulse can then be calculated by $\Delta T = \varepsilon / (\rho_0 C_p)$, where C_p is the heat capacity and ρ_0 the mass density of the medium. The evolution of the temperature distribution after single 100 fs pulse ($\lambda = 800$ nm) during application of series of 100 fs pulses emitted at various repetition rates was calculated by solving the differential equation for heat diffusion as described by Vogel, et al., (2005).

5.2 Evolution of the temperature distribution

The spatial temperature distribution at the end of a single fs-laser pulse, before heat diffusion sets in, reproduces the shape of the free-electron distribution of Fig. 4. Hence, the diameter of the initial temperature distribution ($1/e^2$- values) amounts to 336 nm, and the length to 806 nm. Figure 5 shows the calculated temperature evolution at the center of the laser focus when series of 800 nm, 100 fs pulses are focused into water at different repetition rates (80 MHz and 1 MHz) and numerical apertures (NA = 1.3 and NA = 0.6). It was assumed that with each pulse an energy density of 1 J cm^{-3} at the center of the initial temperature distribution is deposited. For other values of the volumetric energy density, the shape of the temperature vs. time curve will be the same but the absolute values of the temperature varies proportional to the peak density of absorbed power. For comparison, we also calculated the temperature evolutions during cw irradiation with the same average power as for the pulsed irradiation. For 80 MHz repetition rate, pulsed and continuous energy deposition differ significantly only during the first 100 ns.

The calculations in Fig. 5a for tightly focused irradiation with 80 MHz repetition rate reveal that the temperature is only 6.8 times larger after a few microseconds than the temperature increase caused by a single pulse. This implies that only a moderate heat accumulation occurs during plasma-

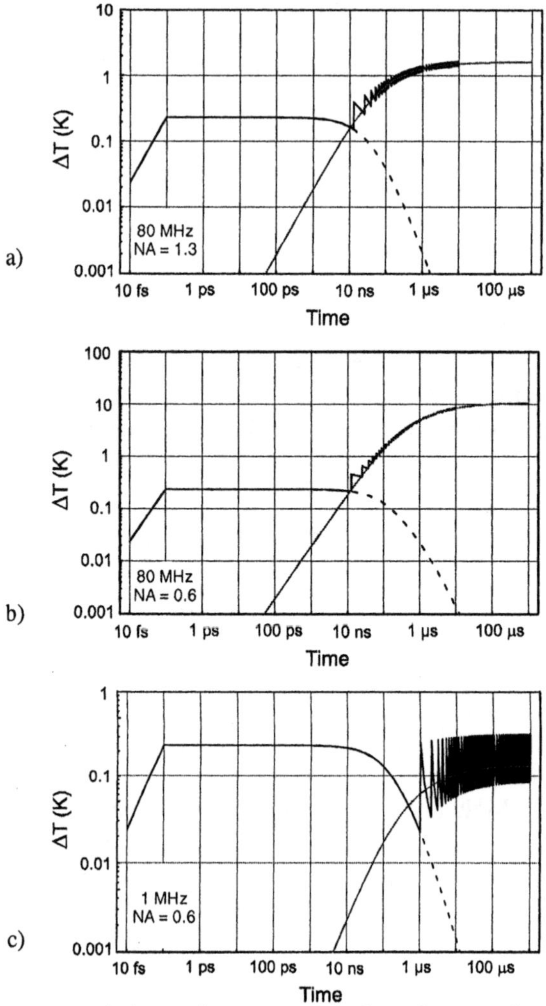

Figure 5. Temperature evolution at the center of the laser focus produced by a series of 800 nm, 100 fs pulses focused into water. a) 80 MHz repetition rate, NA = 1.3; b) 80 MHz repetition rate, NA = 0.6; c) 1 MHz repetition rate, NA = 0.6. The volumetric energy density deposited per pulse is always 1 Jcm^{-3} at the focus center. The dashed lines represent the temperature decay after a single pulse. For comparison, the temperature evolution during cw irradiation with the same average power as for the pulsed irradiation is also shown.

energy deposition and heat diffusion has been established. This is related to the small size of the focal volume which allows for rapid heat diffusion in all directions. For NA = 0.6, the temperature distribution is significantly broader (radial FWHM ≈ 1.5 μm, axial FWHM ≈ 2.900 μm).

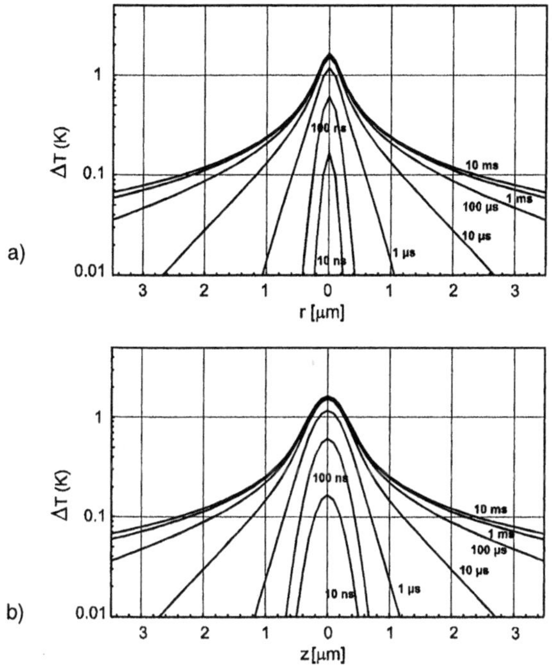

Figure 6. Temperature distribution in radial direction and axial direction produced by series of 800 nm, 100 fs pulses focused into water at numerical apertures of NA = 1.3 at a pulse repetition rate of 80 MHz. The volumetric energy density deposited at the focus center was 1 J cm^{-3} for each pulse.

At first sight, the results of our temperature calculations might suggest that an irradiance range below the optical breakdown threshold exists where predominantly thermal effects in biological media can be produced. However, one needs to consider that about 10^6 free electrons per pulse are generated in the focal volume at the irradiance, which creates a temperature difference of 11.8°C per pulse and a peak temperature of 100°C after a pulse series of several microseconds (for NA = 1.3). Any thermal denaturation of biomolecules will thus always be mixed with free electron-induced chemical effects, and the latter will probably dominate.

6. THERMOELASTIC STRESS GENERATION AND STRESS-INDUCED BUBBLE FORMATION

6.1 Calculation of stress distribution and bubble formation

The temperature rise in the focal volume occurs during thermalization of the energy carried by the free electrons, i. e. within a few picoseconds to tens of picoseconds (see Section 5.1). This time interval is much shorter than the acoustic transit time from the center of the focus to its periphery. Therefore, no acoustic relaxation is possible during the thermalization time, and the thermoelastic stresses caused by the temperature rise stay confined in the focal volume, leading to a maximum pressure rise (Paltauf and Schmidt-Kloiber, 1999; Paltauf and Dyer, 2003, Vogel and Venugopalan, 2003). Conservation of momentum requires that the stress wave emitted from a finite volume within an extended medium must contain both compressive and tensile components such that the integral of the stress over time vanishes (Sigrist and Kneubühl, 1978; Paltauf and Schmidt-Kloiber, 1999). In water, the tensile stress will cause the formation of a cavitation bubble when the strength of the liquid is exceeded. For cell surgery, the threshold for bubble formation defines the onset of disruptive mechanisms contributing to dissection.

To determine the evolution of the thermoelastic stress distribution in the vicinity of the laser focus, we solved the three-dimensional thermoelastic wave equation. A starting point for the calculation of the thermoelastic stress wave propagation is the temperature distribution at the end of a single femtosecond laser pulse, which reproduces the free-electron distribution described by Eq. (8). In the following calculations, this temperature distribution is characterized by T_{max}, the temperature in °C in the center of the focal volume. From this temperature distribution, the initial thermoelastic pressure, before the acoustic wave has started to propagate, was calculated using

$$p(\vec{r}) = \int_{T_1}^{T_2(\vec{r})} \frac{\beta(T)}{K(T)} dT , \qquad (10)$$

where $T_1 = 20°C$ is the temperature before the laser pulse, and $T_2(\vec{r})$ is the temperature of the plasma after the laser pulse, which depends on the location within the focal volume. The temperature dependence of the thermal expansion coefficient β and the compressibility K was taken into account, using values for metastable water reported by Skripov, et al., (1988). The time and space-dependent pressure distribution $p(\vec{r},t)$, due to the relaxation of the initial thermoelastic pressure, was calculated using a k-space (spatial frequency) domain propagation model (Köstli, et al., 2001; Cox and Beard, 2005).

Because the heated volume is very small (≈0.07 μm^3) and the region subjected to large tensile stress amplitudes is even smaller (see Fig. 14, below), the presence of inhomogeneous nuclei that could facilitate bubble formation is unlikely. Therefore, we have to consider the tensile strength of pure water to estimate the bubble formation threshold in femtosecond optical breakdown. We use the crossing of the "kinetic spinodal" as defined by Kiselev (1999) as threshold criterion for bubble formation. In the thermodynamic theory of phase transitions, the locus of states of infinite compressibility $(\delta p / \delta V)_T = 0$, the spinodal, is considered as a boundary of fluid metastable (superheated) states. Physically, however, the metastable state becomes short-lived due to statistical fluctuations well before the spinodal is reached (Skripov, et al., 1988; Debenedetti, 1996). The "kinetic spinodal" is the locus in the phase diagram where the lifetime of metastable states becomes shorter than a relaxation time to local equilibrium. If the surface tension is known, the physical boundary of metastable states in this

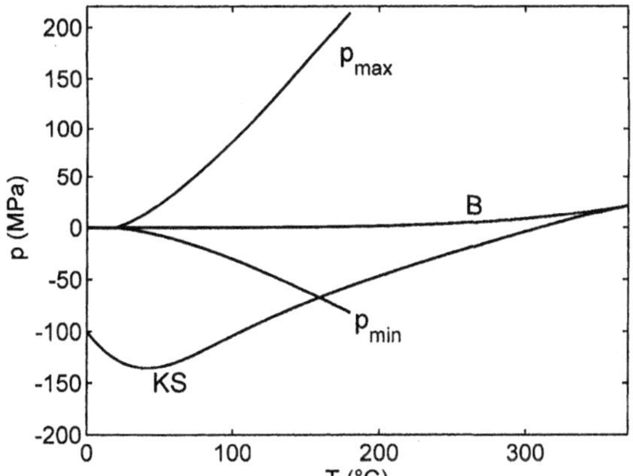

Figure 7. Peak compressive and tensile thermoelastic stresses in the focus center produced by a 800 nm, 100 fs pulse focused into water at NA = 1.3, plotted as a function of temperature together with the binodal (B) and the kinetic spinodal (KS) of water. The kinetic spinodal was calculated by Kiselev (1999) using the analytic equation of state of Saul and Wagner (1989).

approach is completely determined by the equation of state only, i.e. by the equilibrium properties of the system (Kiselev, 1999).

In Fig. 7, the kinetic spinodal is plotted together with the peak compressive and tensile thermoelastic stresses in the focus center that is produced when an 800 nm, 100 fs pulse is focused into water at NA = 1.3. The temperature at which the tensile stress curve reaches the kinetic spinodal is defined as bubble formation threshold. For larger laser pulse energies, the kinetic spinodal will be reached in an increasingly large part of the focal region.

To calculate the dynamics of the cavitation bubble produced after crossing the kinetic spinodal, first the size of the bubble nucleus was determined. It was identified with the extent of the region in which the negative pressure exceeds the kinetic spinodal limit $p(\vec{r},t) < p_{ks}(\vec{r})$. The initial radius of a spherical bubble with the same volume was taken as the starting nucleus for the cavitation bubble. The heated and stretched material within the nucleus commences to expand instantaneously (within less than 1 ps) once the kinetic spinodal is reached (Garrison, et al, 2003). As the driving force for the expansion only the negative part of the time-dependent stress in the center of the focal volume was considered, because the nucleus does not exist before the tensile stress arrives.

After the passage of the tensile stress transient, the vapor pressure p_v inside the bubble continues to drive the bubble expansion. The initial vapor pressure is calculated for a temperature averaged over all volume elements within the nucleus. During bubble growth, it will drop due to the cooling of the expanding bubble content. This cooling is counteracted by heat diffusion into the bubble from the liquid surrounding the bubble. The temperature of this liquid, on the other hand, drops because of heat diffusion out of the focal volume. To quantify the temporal evolution of the driving pressure, we consider two limiting cases defined by (1) isothermal, and (2) adiabatic conditions for the bubble content with respect to the surrounding liquid (Vogel, et al., 2005). In case 2, the vapor pressure in the bubble drops considerably faster than in case 1. In both cases, the ongoing phase transition in the bubble was neglected to obtain tractable expressions for $p_v(t)$. This simplification enabled us to use the Gilmore model to describe the cavitation bubble dynamics (Gilmore, 1952; Knapp, et al., 1971; Paltauf and Schmidt-Kloiber, 1996). To obtain a correct description of the bubble dynamics in a heated and stretched liquid, we considered the temperature-dependence of the surface tension at the bubble wall (NIST, 2005). This is a refinement of our previous study (Vogel, et al., 2005) in which we assumed a constant value (surface tension at room temperature) in all calculations.

6.2 Evolution of the stress distribution

The thermalization time of the energy carried by the free electrons was assumed to be 10 ps. For NA = 1.3, and λ = 800 nm, and a sound velocity in water of c_0 = 1500 m/s, the acoustic transit time to the periphery of the heated region with 168 nm radius is 112 ps. Thus the dimensionless thermalization time (thermalization time divided by acoustic relaxation time) is $t_p^* = 0.09$,

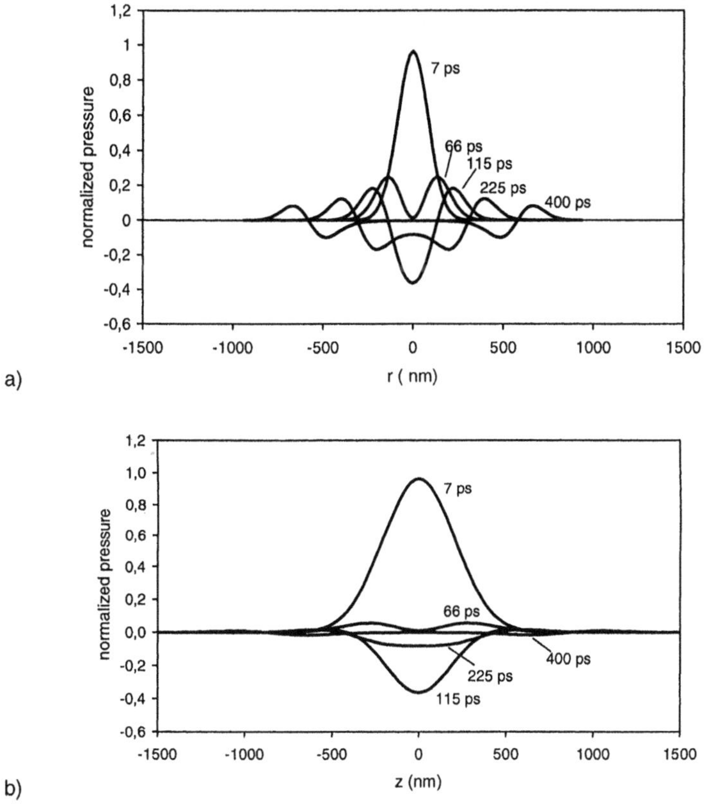

Figure 8. Stress distribution produced by a single femtosecond pulse of 800 nm wavelength focused into water (NA = 1.3), for various times after the release of the laser pulse; (a) in radial direction, (b) in axial direction. The pressure amplitudes are normalized to the peak compressive stress created in the focal volume.

which corresponds to a very high degree of stress confinement. The "thermalization pulse" used to calculate the temperature and pressure rise by means of Eq. (10) was assumed to have a Gaussian temporal shape, with peak at $t = 0$. Figure 8 shows the spatial stress distribution in radial and axial direction for various points in time after the release of the laser pulse, normalized to the peak compressive stress. The compressive stress generates a stress wave traveling into the surrounding medium. When the thermal expansion comes to a rest, inertial forces lead to the generation of a relaxation wave that propagates from the periphery of the focal volume towards its center and is focused in the center of symmetry. Because of the geometrical focusing, it turns into a tensile stress wave that achieves maximum amplitude at the center of symmetry. The duration of the entire stress wave is ≈ 200 ps. The stress wave amplitudes outside the focal region have a bipolar shape as expected for thermoelastic waves. Because of the elongated shape of the focal volume, they are considerably larger in radial than in axial direction.

Measurements of the stress waves produced by femtosecond optical breakdown at large NA and close to the breakdown threshold are very challenging because of the submicrometer size of the breakdown volume, and the sub-nanosecond duration of the stress transients. Therefore, we performed measurements at smaller numerical aperture (NA = 0.2) to assess the stress amplitudes arising during femtosecond optical breakdown. Investigations for irradiances well times above the breakdown threshold were done by means of streak photography and subsequent digital image analysis of the streak recordings (Noack and Vogel, 1998a; Noack, et al., 1998b). Differentiation of the stress wave propagation curves $r(t)$ obtained from the streak recordings yields the stress wave velocity that is related to the pressure amplitude by the known Rankine-Hugoniot-relationship for water (Rice and Walsh, 1957). The analysis provides the entire pressure vs. distance curve perpendicular to the optical axis in the vicinity of the breakdown region as shown in Fig. 9.

The determination of the shockwave pressure becomes inaccurate for pressure amplitudes below 100 MPa where the deviation of the propagation velocity from the sonic velocity becomes too small to be measured accurately with the streak technique (Noack and Vogel, 1998a). Therefore, the streak technique could only be applied for shock wave measurements at energies 15-150 times above the breakdown threshold. Stress wave amplitudes closer to the optical breakdown threshold were determined indirectly by hydrophone measurements at 6 mm distance from the focus (Noack, 1998c), and extrapolation of these data to the plasma rim. Measurement results for energies from close to the breakdown threshold up to 80 times threshold are shown in Fig. 10.

The results of far-field hydrophone measurements can be extrapolated to the boundary of the focal region if the decay constant n of the pressure decay $p \propto r^n$, with increasing propagation distance r is known. The decay constant was determined to n = 1.13 by comparing pressure values at the plasma rim and in the

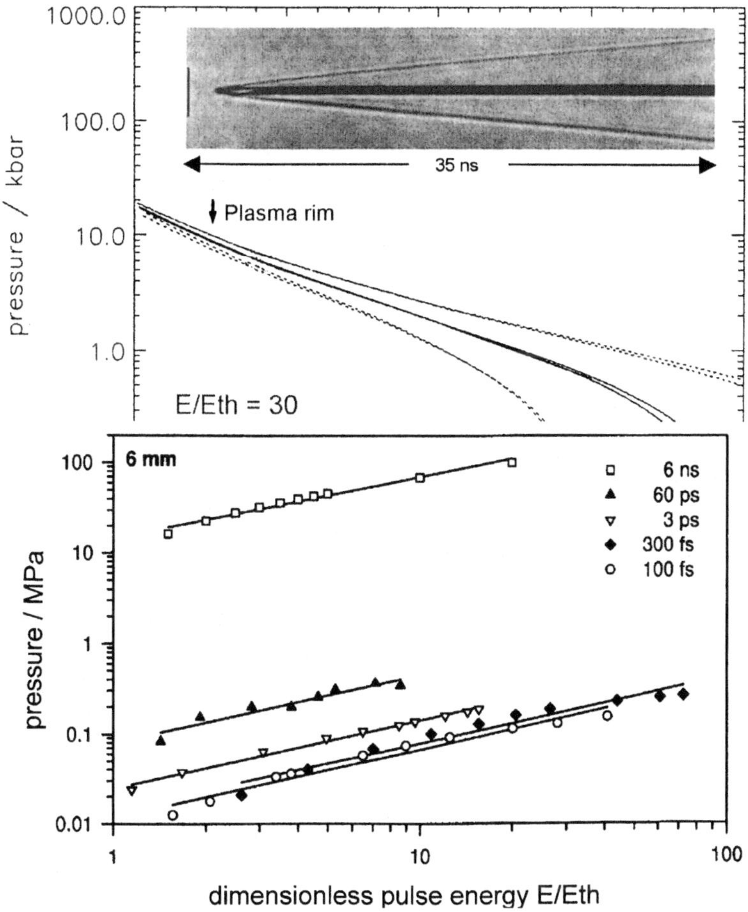

Figure 10. Stress wave amplitude for 100 fs pulses and longer pulse durations as a function of the dimensionless laser pulse energy E/E_{th}. The pressure amplitudes were measured by means of a hydrophone at 6 mm distance from the laser focus. Extrapolation of the data for 100 fs pulses to $E/E_{th} = 1$ yields a pressure value of ≈ 0.008 MPa.

far-field that were measured at larger laser pulse energies. From the data in Fig. 10, a pressure of ≈0.008 MPa at 6 mm distance is deduced for the threshold energy $E = E_{th}$. This corresponds to a pressure value of 61 MPa at the plasma rim when a decay constant n = 1.13 is assumed. The plasma radius at E_{th} is identified with the focal radius of 2.2 μm that was measured using a knife edge technique (Noack, 1998c).

Our calculations of the thermoelastic stress generation predict a peak pressure of 181 MPa at the bubble formation threshold (see Section 6.3 below). According to Fig. 8a, the stress transient that leaves the heated region in radial direction has a peak pressure of ≈25% of the maximum compressive amplitude within the focal volume. We thus obtain a theoretical prediction of 45 MPa for the amplitude of the thermoelastic stress wave at the plasma rim. Considering the uncertainties in the location of the plasma rim and the differences in numerical aperture between experiment and calculation, the agreement between experimental results (61 MPa) and calculated data (45 MPa) is very good.

Both experiments and calculations reveal that stress confinement in femtosecond optical breakdown results in the generation of high pressure values even though the temperature rise is only relatively small. In a purely thermal process starting from room temperature, a temperature rise of, for example, 180°C would produce a saturation vapor pressure of 1.6 MPa. The compressive pressure transient produced by the same temperature rise under stress confinement conditions has an amplitude of 220 MPa, which is more than two orders of magnitude larger than the vapor pressure.

The situation is different for optical breakdown at longer pulse durations where the stress confinement condition is not fulfilled. Here, high pressures are always associated with high temperatures and plasma energy densities. For pulses longer than the thermalization time of the free-electron energy, a dynamic equilibrium between generation of free electrons and thermalization of their energy is established during the laser pulse (Noack and Vogel, 1999). This leads to high value of the plasma energy density in the order of 30-40 $kJcm^{-3}$ for ns-pulses (Vogel, et al., 1996b, 1999) and temperatures of several thousand degrees Kelvin (Stolarski, et al., 1995). The duration of the resulting shock wave is determined by the time it takes for the high pressure within the plasma to decrease during the plasma expansion (Vogel, et al., 1996b). For NA = 0.9, it was found to be about 25-40 ns (Venugopalan, et al., 2002). By contrast, the duration of the thermoelastic stress transients is determined by the geometric dimensions of the breakdown volume which are in the submicrometer range. This leads to a duration of the stress transients of less than 300 ps.

6.3 Threshold for stress-induced bubble formation

The tensile stress produced during femtosecond optical breakdown makes it possible that a cavitation bubble can be generated by a relatively small temperature rise in the liquid. The threshold for bubble formation is defined by the temperature rise leading to a crossing of the kinetic spinodal, as shown in Fig. 7. For λ = 800 nm, NA = 1.3, and a room temperature of 20°C, the critical temperature rise and the corresponding critical tensile stress are ΔT = 139°C, and p = -67.5 MPa, respectively. The corresponding compressive pressure is 181 MPa.

The temperature rise of 139°C at the threshold for bubble formation corresponds to an increase in energy density of 582 J cm^{-3} which, according to Eq. (9), is produced by a free-electron density of $\rho_c = 0.249\times10^{21}$ cm^{-3}. This electron density is less than the breakdown criterion of $\rho_{cr} = 10^{21}$ cm^{-3} assumed in our numerical calculations and in most other theoretical studies of plasma formation. The discrepancy between the threshold values relying on different breakdown criteria needs to be kept in mind when comparing the results of experimental studies, where bubble formation serves as breakdown criterion, with those of numerical simulations.

The fact that femtosecond optical breakdown is associated with only a relatively small temperature rise explains why plasma luminescence is no longer visible for pulse durations shorter than about 10 ps (Kennedy, et al., 1997; Noack, et al., 1998b). For pulse durations longer than the thermalization time, large amounts of energy are transferred from the free electrons to the heavy particles during the laser pulse (Noack and Vogel, 1999), resulting in a temperature of several thousand degrees Kelvin, bubble formation, and a bright plasma luminescence (Barnes and Rieckhoff, 1968; Stolarski, et al., 1995; Chapyak, et al., 1997). By contrast, a peak temperature of 159°C reached at the threshold for bubble formation with 100 fs pulses is too low to produce blackbody radiation in the visible range of the optical spectrum. Moreover, the recombination radiation of femtosecond-laser-produced plasmas is weak because only one "set" of free electrons is produced that recombines after the end of the laser pulse. Therefore, bubble formation is a more practical breakdown criterion for ultrashort laser pulses than plasma luminescence.

A comparison between experimental threshold data from various researchers and threshold values predicted by our model (Fig. 3) has been compiled by Vogel, et al., (2005). Our numerical predictions lie within the range of experimental data for all pulse durations. However, the experimental data scatter within a range of one order of magnitude for femtosecond and nanosecond pulses, and only slightly less for picosecond pulses. These large variations reflect the difficulty of performing precise threshold measurements in the bulk of water. The measurements were influenced either by aberrations in the focusing optics, mode beating of longitudinal resonator modes resulting in picosecond intensity peaks during nanosecond pulses, or nonlinear beam propagation. A numerical aperture NA ≥ 0.9 is required for a pulse duration of 100 fs excluding a diminution of the spot size by self-focusing and the corresponding apparent reduction of the breakdown threshold (Schaffer, et al., 2001). Future measurements with aberration-free temporally Gaussian laser pulses focused at large NA will have to provide a reliable database.

In addition, a better adjustment of the numerical breakdown criterion to the experimental criterion of bubble formation is needed to enable a meaningful comparison of experimental data with model predictions. While bubble formation requires an approximately constant energy density within the focal volume for all laser pulse durations and wavelengths, the energy density associated with a fixed value of the free electron density, such as $\rho_{cr} = 10^{21}$ cm^{-3}, varies considerably with pulse duration. Thus the assumption of a constant free-electron density as breakdown criterion is quite arbitrary, especially for cases where the threshold is smooth, i.e. where ρ_{max} increases continuously with irradiance. In these cases it seems more reasonable to relate the critical free-electron density to the energy density within the medium that leads to bubble formation. Equation (9) provides the required link between electron and energy density, and an analysis of ρ_{max} (I/I_{rate}) curves such as in Fig. 2, then yields the corresponding threshold irradiance.

6.4 Cavitation bubble dynamics

Figure 11 shows a 2D representation of the evolution of the thermoelastic stress wave and of the region in which the kinetic spinodal is surpassed (bubble nucleus) for a peak temperature of 200°C, slightly above the threshold for bubble formation. The subsequent bubble dynamics are shown in Fig. 12, and the dependence between maximum bubble radius and peak

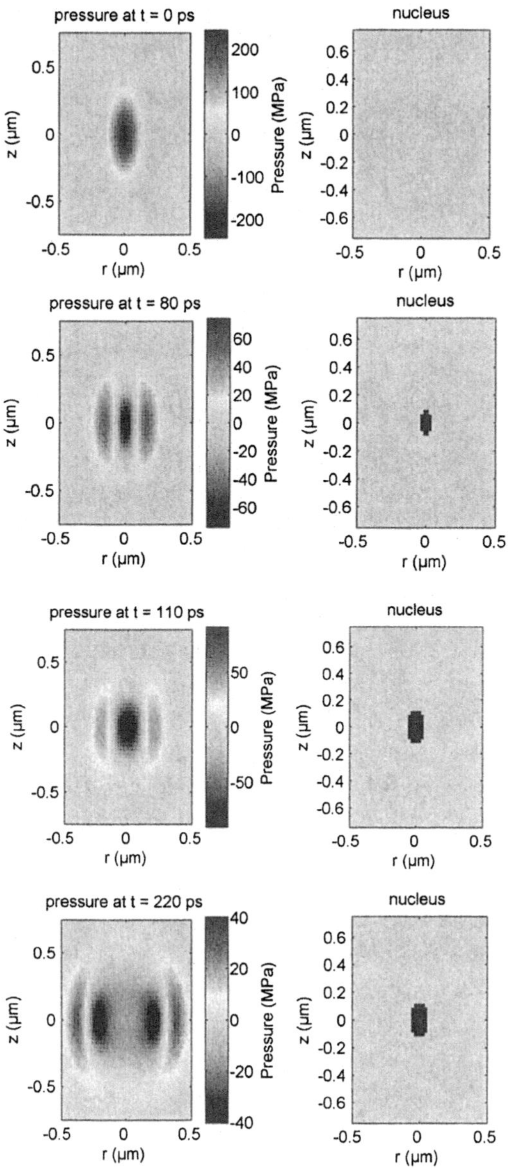

Figure 11. 2D Plots of thermoelastic stress evolution (left), and of the region in which the kinetic spinodal limit is exceeded (right). This region demarcates the size of the bubble nucleus that is then expanded by the thermoelastic tensile stress wave. The calculations were performed for a peak temperature of 200°C.

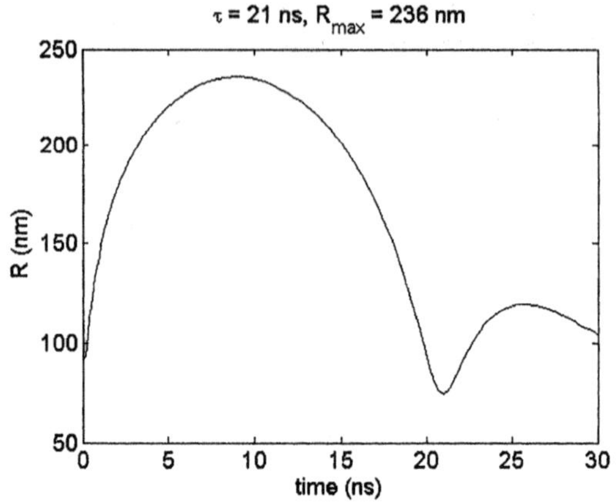

Figure 12. Radius time curve of the cavitation bubble produced by a single femtosecond laser pulse focused at NA = 1.3 that leads to a peak temperature of T_{max} = 200 °C at the focus center. For the calculations, isothermal conditions for the bubble content with respect to the surrounding liquid were assumed.

temperature is presented in Fig. 13. We see that very similar bubble sizes are predicted for cases 1 and 2, respectively (isothermal and adiabatic conditions for the bubble content with respect to the surrounding liquid). Moreover, the bubble motion in case 2 (adiabatic conditions) is almost identical to that in a third case where the vapor pressure is not at all taken into account and only the negative pressure pulse drives the bubble expansion (not shown). This implies that the thermoelastic stress is more important for driving the bubble expansion than the vapor pressure.

The most prominent feature of the transient bubbles produced close to the threshold of femtosecond optical breakdown is their small size and short lifetime. The bubble radius amounts to only about 200 nm in water, and will be even smaller in a viscoelastic medium such as the cytoplasm. This makes a dissection mechanism associated with bubble formation compatible with intracellular nanosurgery, in contrast to nanosecond optical breakdown (6 ns, 1064 nm) where the smallest bubble radius in water observed detected for NA = 0.9 was R_{max} = 45 μm (Venugopalan, et al., 2002). The small bubble size corresponds to a small energy

$$E_B = (4\pi/3)\,[(p_\infty - p_v) + p_\sigma]\,R_{max}^3 \tag{11}$$

of the expanded bubble. Here $p_\sigma = \sigma/2R$ denotes the pressure exerted by the surface tension at the bubble wall. For the sake of simplicity, we neglect the time dependence of the bubble radius and use the value $p_\sigma(R_{max})$. For the case presented in Fig. 12, E_B amounts to 1.1×10^{-14} J (11 femtojoule). The smallness of the bubble energy is partly explained by the small energy content of the stress transient creating the bubble. This energy equals the energy stored in thermoelastic compression of the heated fluid volume (Paltauf and Dyer, 2003),

$$E_{TE} = 1/(2\rho_0 c_0^2) \int p_0^2 \, dV \,, \tag{12}$$

where p_0 is the thermoelastic pressure and the integration encompasses the entire volume heated by the laser pulse. For a temperature rise in the center of the focal volume of 180°C (Fig. 12) that leads to a maximum pressure p_0 of 221 MPa, E_{TE} amounts to 7.8×10^{-14} J. The total heat content of the plasma is

$$E_{tot} = \rho_0 C_p \int \Delta T \, dV \,, \tag{13}$$

giving $E_{tot} = 1.66 \times 10^{-11}$ J under the same conditions. The energetic conversion efficiency from heat into the thermoelastic wave is thus $E_{TE}/E_{tot} = 0.46\%$, and the conversion from thermoelastic energy into bubble energy is 14.1%.

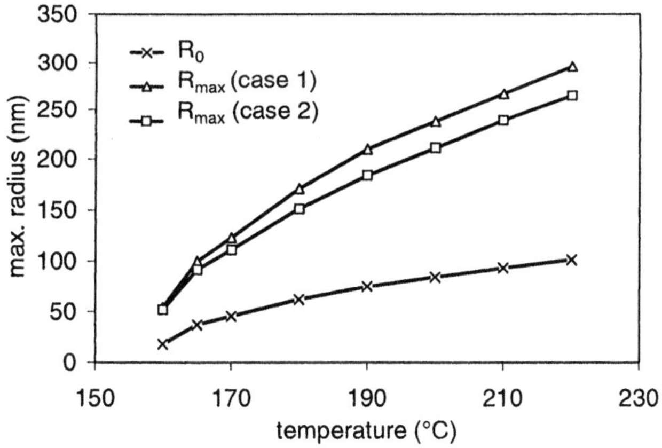

Figure 13. Maximum bubble radius R_{max} as a function of the peak temperature in the center of the focal volume, together with the radius of the nucleus, R_0. Cases 1 and 2 stand for isothermal and adiabatic conditions for the bubble content with respect to the surrounding liquid.

This result of a low conversion efficiency from absorbed laser energy into mechanical energy for femtosecond pulses (0.065%) is supported by experimental results (Noack, et al., 1998b; Vogel, et al., 1999). By contrast, for nanosecond optical breakdown with NA = 0.9, the conversion efficiency was experimentally found to be 12.7 % for λ = 1064 nm, and 3.3% for λ = 532 nm at the breakdown threshold, and it reached values of 53% and 33.5 %, respectively, for energies 10-fold above threshold (Venugopalan, et al., 2002).

6.4.1 Experimental data on cavitation bubble dynamics

Results of time-resolved investigations of the effects of transient femtosecond laser-induced bubbles on cells are not yet available. However, Dayton, et al., (2001) investigated the oscillations of 1.5-μm radius bubbles that were phagocytosed by leukocytes and stimulated by a rarefaction-first one-cycle acoustic pulse with 440 ns duration. By means of streak photography and high-speed photography with 100 Mill. f/s, they observed that phagocytosed bubbles expanded about 20-45% less than free microbubbles in response to a single acoustic pulse of the same intensity. The difference is due to the viscosity of the cytoplasm and the elastic modulus of the cytoskeleton. Bubbles subjected to a tensile stress amplitude of 0.9 MPa expanded to a radius of 6 μm without rupturing the cell membrane. Larger oscillations caused immediate cell lysis. The viability of the non-lysed cells after insonation was not tested, but it is evident that the bubble oscillations strain the cell membrane and deform or even rupture the cytoskeleton. In the case of femtosecond optical breakdown, the radius of the bubble nucleus is much smaller ($\approx$ 90 nm compared to 1.5 μm), and the tensile stress transient acting on the bubbles is much shorter than in the case investigated by Dayton ($\approx$ 100 ps compared to 220 ns). Therefore, the resulting bubbles cause little structural damage within a cell and do not affect cell viability.

Lin, et al., (1999) and Leszczynski, et al., (2001) investigated the thresholds for cell death produced by cavitation induced around absorbing microparticles irradiated by nanosecond laser pulses. They observed that an energy of 3 nJ absorbed by a single particle of 1 μm diameter produced sufficiently strong cavitation to kill a trabecular meshwork cell after irradiation with a single laser pulse. Pulses with 1 nJ absorbed energy produced lethality after several exposures (Lin, et al., 1999). Viability was lost even when no morphologic damage was apparent immediately after the collapse of the transient bubble with less than 1 μs lifetime, corresponding to $R_{max} \leq 5.5$ μm). Nuclear staining of nonviable cells by ethidium bromide confirmed that cell death was associated with membrane damage. According to Neumann and Brinkmann (2005b), a bubble radius of 3 μm within a cell of 7.5 μm radius is sufficient to cause an enlargement of the membrane by 4% that will result in membrane rupture (Needham and Nunn, 1990; Boal, 2002). The results of our calculations in Fig. 20 demonstrate that the radius of fs-laser-produced transient bubbles remains well below this damage threshold. This applies even for laser pulse energies of a few nanojoule because for $\rho_{cr} = 10^{21}$ cm^{-3} and 1 μm plasma length about 99% of the incident energy is transmitted through the focal region (Noack and Vogel, 1999). The heated volume is much smaller than the volume of the microparticles investigated by Lin, et al., (1999), and the deposited heat energy corresponding to a peak temperature of T_{max} = 200°C or 300 °C is only 16.6 or 25.8 pJ, respectively, much less than in Lin's case.

Bubbles around gold nanoparticles are of interest in the context of nanoparticle cell surgery (Section 1.1). When particles with 4.5 nm radius were irradiated by 400 nm, 50 fs pulses, bubbles of up to 20 nm radius were observed by X-ray scattering (Kotaidis and Plech, 2005). The small size of these bubbles, which is one order of magnitude less than for those produced by focused fs pulses, is consistent with the fact that the collective action of a large number of nanoparticles is required to produce the desired surgical effect.

The transient bubbles produced by single laser pulses can only be detected by very fast measurement schemes. However, during high-repetition rate pulse series accumulative thermal effects and chemical dissociation of biomolecules come into play (Sections 4 and 5.2) that can produce long-lasting bubbles, which are easily observable under the microscope (König, et al., 2002; Supatto, et al., 2005a,b; Riemann, et al., 2005). Even though thermoelastic forces may still support the bubble growth, it is mainly driven by chemical or thermal decomposition of biomolecules into small volatile fragments and by boiling of cell water. After the end of the fs pulse train, the vapor will rapidly condense, but the volatile decomposition products will disappear only by dissolution into the surrounding liquid and thus form a longer-lasting bubble.

7. IMPLICATIONS FOR LASER EFFECTS ON BIOLOGICAL CELLS AND TISSUES

Two parameter regimes have been established for femtosecond laser nanosurgery. One technique uses long pulse series from fs oscillators with repetition rates in the order of 80 MHz and pulse energies well below the optical breakdown threshold (König, et al., 1999, 2001; Smith, et al., 2001; Tirlapur and König, 2002; Zeira, et al., 2003; Saccioni, et al., 2005; Supatto, et al., 2005a,b; Riemann, et al., 2005; König, et al., 2005). From 40000 pulses (König, et al., 1999) to several million pulses (Tirlapur, et al., 2002; Saccioni, et al., 2005) have been applied at one specific location to achieve the desired dissection or membrane permeabilisation. The other approach uses amplified pulse series at 1 kHz repetition rate with pulse energies slightly above the threshold for transient bubble formation (Yanik, et al., 2004; Watanabe, et al., 2004; Watanabe, et al., 2005; Heisterkamp, et al., 2005; Maxwell, et al., 2005). Here the number of pulses applied at one location varied between thirty (Watanabe, et al., 2004) and several hundred (Yanik, et al., 2004, Heisterkamp, et al., 2005, Maxwell, et al., 2005).

Based on the discussion of the physical effects associated with femtosecond laser-induced plasma formation in the previous sections, we now proceed to explain the working mechanisms of both modalities for cell surgery. For this purpose, the different low-density plasma effects and physical breakdown phenomena are summarized in Fig. 14, together with experimental damage, transfection and dissection thresholds on cells. The different effects are scaled by the corresponding values of free-electron density and irradiance.

Chemical cell damage (2) refers to membrane dysfunction and DNA strand breaks leading to apoptosis-like cell death observed after scanning irradiation of PtK2 cells with 800 nm pulses at 80 MHz repetition rate (Tirlapur, et al., 2001). Chromosome dissection (3) relates to the intranuclear chromosome dissection (König, et al., 1999), and (4) to cell transfection by transient membrane permeabilisation (König and Tirlapur, 2002), both performed using 80 MHz pulse trains from a femtosecond oscillator. Mitochondrion ablation (8) refers to the ablation of a single mitochondrion in a living cell using 1 kHz pulse trains (Shen, et al., 2005), and axon dissection (9) applies to axotomy in life C-elegans worms carried out with sequences of pulses emitted at 1 kHz repetition rate from a regenerative amplifier (Yanik, et al., 2004). Points (1), (5), (6) and (7) stand for physical events or threshold criteria.

7.1 Pulse trains at MHz repetition rates with energies below the threshold for bubble formation

The irradiance threshold (2) for cell death induced by laser pulse series of 80 MHz repetition rate scanned over the entire cell volume ($0.067 \times I_{rate}$) is lower than the irradiance threshold for intracellular dissection (3). However, this does not imply that intracellular dissection with 80 MHz pulse series must lead to severe cell damage, because locally confined irradiation does not affect cell viability in the same way as scanning irradiation.

The threshold for intranuclear chromosome dissection with 80 MHz pulse series (3) is almost 4 times as large as the irradiance (1) producing one free electron per pulse in the focal volume ($0.15 \times I_{rate}$ vs. $0.04 \times I_{rate}$). In fact, about 1000 free electrons per pulse are produced with the parameters used for

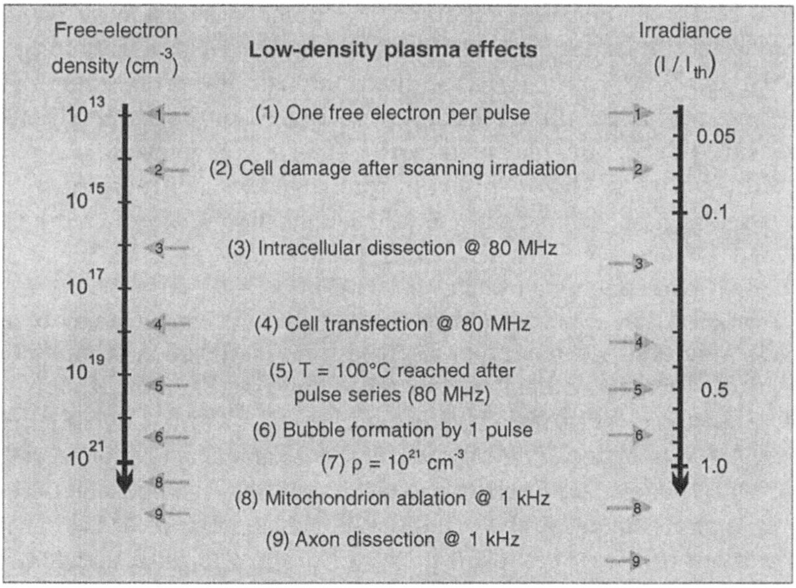

Figure 14. Overall view of physical breakdown phenomena induced by femtosecond laser pulses, together with experimental damage, transfection, and dissection thresholds on cells, scaled by free-electron density and irradiance. The irradiance values are normalized to the optical breakdown threshold I_{th} defined by $\rho_{cr} = 10^{21}$ cm^{-3}. All data refer to plasma formation in water with femtosecond pulses of about 100 fs pulses and 800 nm wavelength.

dissection. Therefore, it is very likely that the intracellular ablation produced by long trains of femtosecond pulses in the low-density plasma regime relies on cumulative free electron-mediated chemical effects. This hypothesis is supported by the facts that the individual pulses produce a thermoelastic tensile stress of only ≈ 0.014 MPa, and a pulse series of 100 μs duration results in a temperature rise of only ≈ 0.076°C. These values for tensile stress and temperature rise are by far too small to cause any cutting effect or other types of cell injury. The breaking of chemical bonds, as described in Section 4, may first lead to a disintegration of the structural integrity of biomolecules and finally to a dissection of subcellular structures. Bond-breaking may be initiated both by resonant interactions with low-energy electrons, and by multiphoton processes of lower order that do not yet create free electrons (König, et al., 1999; Koester, et al., 1999; Hopt, et al., 2001; Eggeling, et al., 2005).

Interestingly, transient membrane permeabilisation for gene transfer (4) requires a considerably larger laser dose than chromosome dissection. Not only is the irradiance larger, but the number of applied pulses ($\approx 10^6$) also exceeds by far the quantity necessary for chromosome dissection ($\approx 4\times10^4$). Chromosome dissection may be facilitated by the DNA absorption around 260 nm enabling nonlinear absorption through lower-order multiphoton processes. Moreover, while breakage of relatively few bonds is sufficient for chromosome dissection, the creation of a relatively large opening is required for diffusion of a DNA plasmid through the cell membrane. The corresponding laser parameters are still within the regime of free electron-mediated chemical effects but already quite close to the range where cumulative heat effects start to play a role (5).

At larger laser powers, bubbles with a lifetime in the order of a few seconds were observed that probably arise from dissociation of biomolecules into volatile, non-condensable fragments (König, et al., 2002; Masters, et al., 2004; Supatto, et al., 2005a,b; Riemann, et al., 2005). As discussed in Section 6.4, this dissociation of relatively large amounts of biomaterials can be attributed both to free electron-chemical and photochemical bond-breaking as well as to accumulative thermal effects. The appearance of the bubbles is an indication of severe cell damage or cell death within the targeted region and defines an upper limit for the laser power suitable for nanosurgery. A criterion for successful intratissue dissection at lower energy levels is the appearance of intense autofluorescence in perinuclear cell regions (König, et al., 2002; Supatto, et al., 2005a,b) that is likely due to the destruction of mitochondria at the rim of the laser cut (Oehring, et al., 2000).

7.2 Pulses at kHz repetition rates with energies above the bubble formation threshold

When pulse trains of 1 kHz repetition rate are employed for nanosurgery, pulse energies ranging from 2-40 nJ are used (Yanik, et al., 2004; Watanabe, et al., 2004; Shen, et al., 2005; Heisterkamp, et al., 2005). Examples are the ablation of single mitochondria (8) by several hundred 2-nJ pulses (Shen, et al., 2005), and the severing of axons in a live C-elegans worm (9) with a similar number of 10-nJ pulses (Yanik, et al., 2004). These energies are above the threshold for thermoelastically induced formation of minute transient cavitation bubbles (6), and are thus associated with mechanical disruption effects. The tensile thermoelastic stress waves enable dissection of cellular structures at low volumetric energy densities, and the small size of the heated volume (Fig. 8) correlates with a radius of the expanded bubble in the order of only 120-300 nm (Fig. 20). This explains why fs-laser-induced bubble formation does not necessarily lead to cell damage whereas ns-laser-induced bubble generation is usually associated with cell death (Lin, et al., 1999; Leszczynski, et al., 2001; Pitsillides, et al., 2003; Roegener, et al., 2004; Neumann and Brinkmann, 2005b).

Due to the contribution of mechanical effects to dissection, the total energy required for nanosurgery with kHz pulse series is less than necessary with MHz pulse trains. For example, ablation of a mitochondrion using 1 kHz pulses required a total energy of less than 1 μJ (Shen, et al., 2005), while for intranuclear chromosome dissection with 80 MHz pulses an energy of 15 μJ was needed (König, et al., 1999).

For sufficiently large pulse energies, bubble expansion and shockwave pressure can cause effects far beyond the focal volume which lead to cell death (Watanabe, et al., 2004; Watanabe, 2005; Zohdy, et al., 2005). To avoid unwanted side effects, irradiances should be used that are only slightly above the bubble formation threshold. Useful techniques for an online monitoring of the ablation threshold during laser surgery are to detect the onset of photobleaching, or of light scattering by bubbles generated at the laser focus. Heisterkamp, et al., (2005) found that the threshold for photobleaching is just below the ablation threshold. Neumann and

Brinkmann (2005a) described a light scattering technique for an online detection of micrometer-sized bubbles produced by pulsed laser irradiation.

We conclude that, depending on the repetition rate of the fs-laser pulses, nanosurgery relies on two very different mechanisms. With MHz repetition rates, dissection is due to accumulative chemical effects in low-density plasmas. In this regime, no transient bubbles with submicrosecond lifetime are produced, and the formation of long-lived bubbles by accumulative chemical and thermal effects must be avoided. With pulse trains at kHz repetition rate, the accumulative creation of chemical effects would take too long to be practical. Therefore, pulse energies are raised to a level where the thermoelastic generation of minute transient bubbles enables nanosurgery. Due to their short lifetime of less than 100 ns and the long time intervals between the laser pulses, no cumulative bubble growth occurs as long as pulse energies close to the bubble formation threshold are used. Both modes of femtosecond-laser nanoprocessing can achieve a 2-3 fold better precision than cell surgery using cw irradiation, and enable manipulation at arbitrary locations.

Acknowledgement

This work was sponsored by the US Air Force Office of Scientific Research through its European Office of Aerospace Research and Development under grant number FA8655-02-1-3047, and, in parts, by the German Bundesministerium für Bildung und Forschung, under grant number 13N8461.

REFERENCES

Ammosov, M.V., Delone, N. B., and Krainov, V. P., 1986, Tunnel ionization of complex atoms and of atomic ions in an alternating electromagnetic field, *Sov. Phys. JETP* **64**, 1191-1194.

Amy, R. L., and Storb, R., 1965, Selective mitochondrial damage by a ruby laser microbeam: an electron microscopic study, *Science* **150**, 756-758.

Arnold, C. L., Heisterkamp, A., Ertmer, W. and Lubatschowski, H., 2005, Streak formation as side effect od optical breakdown during processing the bulk of transparent Kerr media with ultrashort laser pulses, *Appl. Phys. B* **80**, 247-253.

Arnold, D., and Cartier, E., 1992, Theory of laser-induced free-electron heating and impact ionization in wide-band-gap solids, *Phys. Rev. B* **46**, 15102-15115.

Barnes, P. A., and Rieckhoff, K. E., 1968, Laser induced underwater sparks. *Appl. Phys. Lett.* **13**, 282-284.

Berns, M. W., Olson, R. S., and Rounds, D. E., 1969, In vitro production of chromosomal lesions with an argon laser microbeam, *Nature* **221**, 74-75.

Berns, M. W., Cheng, W. K., Floyd, A. D. and Ohnuki, Y., 1971, Chromosome lesions produced with an argon laser microbeam without dye sensitization, *Science* **171**, 903-905.

Berns, M. W., Aist, J., Edwards, J., Strahs, K., Girton, J., McNeil, P., Kitzes, J. B., Hammer-Wilson, M., Liaw, L.-H., Siemens, A., Koonce, M., Peterson, S., Brenner, S., Burt, J., Walter, R., Bryant, P. J., van Dyk, D., Coulombe, J., Cahill, T. and Berns, G. S., 1981, Laser Microsurgery in cell and developmental biology, *Science* **213**, 505-513.

Bessis, M., and Nomarski, G., 1960, Irradiation ultra-violette des organites cellulaires avec observation continue en contraste en phase, *J. Biophys. Biochem. Cytol.* **8**, 777-792.

Bessis, M., Gires, F., Mayer, G., and Nomarski, G., 1962, Irradiation des organites cellulaires à l'aide d'un laser à rubis, *C. R. Acad. Sci.III - Vie* **255**, 1010.

Bessis, M., 1971, Selective destruction of cell organelles by laser beam. Theory and practical applications, *Adv. Biol. Med. Phys.* **13**, 209-213.

Bloembergen, N., 1974, Laser-induced electric breakdown in solids, *IEEE J. Quantum Electr.* **QE-10**, 375-386.

Boal, D., 2002, *Mechanics of the cell*, Cambridge Univ. Press, Cambridge, UK.

Born, M., and Wolf, E., 1970, *Principles of Optics*, Pergamon Press, Oxford, 1970.

Botvinick, E. L., Venugopalan, V., Shah, J. V., Liaw, L. H., and Berns, M., 2004, Controlled ablation of microtubules using a picosecond laser, *Biophys. J.* **87**, 4203-4212.

Boudaiffa, B., Cloutier, P., Hunting, D., Huels, M. A., and Sanche, L., 2000, Resonant formation of DNA strand breaks by low-energy (3 to 20 eV) electrons, *Science* **287**, 1658-1660.

Chapyak, E. J., Godwin, R. P., and Vogel, A., 1997, A comparison of numerical simulations and laboratory studies of shock waves and cavitation bubble growth produced by optical breakdown in water, *Proc. SPIE* **2975**, 335-342.

Chung, S. H., Clark, D. A., Gabel, C. V., Mazur, E., and Samuel, A. D. T., 2005, Mapping thermosensation to a dendrite in C. elegans using femtosecond laser dissetion, *J. Neurosci.* **25** (at press)

Colombelli, J., Grill, S. W., and Stelzer, E. H. K., 2004, Ultraviolett diffraction limited nanosurgery of live biological tissues, *Rev. Sci. Instrum.* **75**, 472-478.

Colombelli, J., Reynaud, and Stelzer, E. H. K., 2005a, Subcellular nanosurgery with a pulsed subnanosecond UV-A laser, *Med. Laser Appl.* **20,** 217-222.

Colombelli, J., Reynaud, E. G., Rietdorf, J., Pepperkork, R., and Stelzer, E. H. K., 2005b, Pulsed UV laser nanosurgery: retrieving the cytoskeleton dynamics in vivo, *Traffic* **6,** 1093-1102.

Cox, B. T., and Beard, P. C., 2005, Fast calculation of pulsed photoacoustic fields in fluids using k-space methods, *J. Acoust. Soc. Am.* **117**, 3616-3627.

Dayton, P. A., Chomas, J. E., Lunn, A. F. H., Allen, J. S., Lindner, J. R., Simon, S. I., and Ferrara, K. W., 2001, Optical and acoustical dynamics of microbubble contrast agents inside neutrophils, *Biophys. J.* **80**, 1547-1556.

Debenedetti, P. G., 1996, *Metastable Liquids: Concepts and Principles*, Princeton University Press, Princeton.

Ditlbacher, H., Krenn, J. R., Leitner, A., and Aussenegg, F. R., 2004, Surface plasmon polariton based optical beam profiler, *Opt. Lett.* **29**, 1408-1410.

Docchio, F., Sachhi, C. A., and Marshall, J., 1986, Experimental investigation of optical breakdown thresholds in ocular media under single pulse irradiation with different pulse durations, *Lasers Ophthalmol.* **1**, 83-93.

Du, D., Squier, J., Kurtz, R., Elner, V., Liu, X., Güttmann, G., and Mourou, G., 1994, Damage threshold as a function of pulse duration in biological tissue, in P. F. Barbara, W. H. Knox, G. A. Mourou, and A. H. Zewail *Ultrafast Phenomena IX,* Springer, New York, pp. 254-255.

Du, D., Liu, X., and Mourou, G., 1996, Reduction of multi-photon ionization in dielectrics due to collisions, *Appl. Phys. B* **63**, 617-621.

Eggeling, C., Volkmer, A., and Seidel, C. A. M., 2005, Molecular photobleaching kinetics of Rhodamine 6G by one- and two-photon induced confocal fluorescence microscopy, *Chem. Phys. Chem.* **6**, 791-804.

Fan, C. H., Sun, J., and Longtin, J. P., 2002a, Breakdown threshold and localized electron density in water induced by ultrashort laser pulses, *J. Appl. Phys.* **91**, 2530-2536.

Fan, C. H., Sun, J., and Longtin, J. P., 2002b, Plasma absorption of femtosecond laser pulses in dielectrics, *J. Heat Transf. – T. ASME* **124**, 275-283.

Feit, M. D., Komashko, A. M., and Rubenchik, A. M., 2004, Ultra-short pulse laser interaction with transparent dielectrics, *Appl. Phys. A* **79**, 1657-1661.

Feng, Q., Moloney, J. V., Newell, A. C., Wright, E. M., Cook, K., Kennedy, P. K., Hammer, D. X., Rockwell, B. A., and Thompson, C. R., 1997, Theory and simulation on the threshold of water breakdown induced by focused ultrashort laser pulses, *IEEE J. Quantum Electron.* **33**, 127-137.

Garret, B. C., et al., 2005, Role of water in electron-initiated processes and radical chemistry; issues and scientific advances, *Chem Rev.* **105**, 355-389.

Garrison, B., Itina, T. E., and Zhigilei, L. V., 2003, Limit of overheating and the threshold behavior in laser ablation. *Phys. Rev. E* **68**, 041501.

Garwe, F., Csáki, A., Maubach, G., Steinbrück, A., Weise, A., König, K., Fritsche, W., 2005, Laser pulse energy conversion on sequence-specifically bound metal nanoparticles and its application for DNA manipulation, *Med. Laser Appl.* **20**, 201-206.

Gilmore, F. R., 1952, Calif. Inst. Techn. Rep. **26-4**.

Gohlke, S., and Illenberger, E., 2002, Probing biomolecules: Gas phase experiments and biological relevance, *Europhys. News* **33**, 207-209.

Grand, D., Bernas, A., and Amouyal, E., 1979, Photoionization of aqueous indole; conduction band edge and energy gap in liquid water, *Chem. Phys.* **44**, 73-79.

Greulich, K. O., 1999, *Micromanipulation by Light in Biology and Medicine,* Birkhäuser, Basel, Boston, Berlin, 300pp.

Grill, S. W., Howard, J., Schäffer, E., Stelzer, E. H. K. and Hyman, A. A., 2003, The distribution of active force generators controls mitotic spindle position, *Science* **301**, 518-521.

Grill, S., and Stelzer, E. H. K., 1999, Method to calculate lateral and axial gain factors of optical setups with a large solid angle, *J. Opt. Soc. Am. A* **16**, 2658-2665.

Hammer, D. X., Thomas, R. J., Noojin, G. D., Rockwell, B. A., Kennedy, P. A., and Roach, W. P., 1996, Experimental investigation of ultrashort pulse laser-induced breakdown thresholds in aqueous media, *IEEE J. Quantum Electr.* **QE-3**, 670-678.

Hammer, D. X., Jansen, E. D., Frenz, M., Noojin, G. D., Thomas, R. J., Noack, J., Vogel, A., Rockwell, B. A., and Welch, A. J., 1997, Shielding properties of laser-induced breakdown in water for pulse durations from 5 ns to 125 fs, *Appl. Opt.* **36**, 5630-5640.

Han, M., Zickler, L., Giese, G., Walter, M., Loesel, F. H., and Bille, J. F., 2004, Second-harmonic imaging of cornea after intrastromal femtosecond laser ablation, *J. Biomed. Opt.* **9**, 760-766.

Heisterkamp, A., Ripken, T., Lubatschowski, H., Mamom, T., Drommer, W., Welling, H., and Ertmer, W., 2002, Nonlinear side-effects of fs-pulses inside corneal tissue during photodisruption, *Appl. Phys. B* **74**, 419-425.

Heisterkamp, A., Mamom, T., Kermani, O., Drommer, W., Welling, H., Ertmer, W., and Lubatschowski, H., 2003, Intrastromal refractive surgery with ultrashort laser pulses: in vivo study on the rabbit eye, *Graefes Arch. Clin. Exp. Ophthalmol.* **241**, 511-517.

Heisterkamp, A., Maxwell, I. Z., Mazur, E., Underwood, J. M., Nickerson, J. A., Kumar, S. and Ingber, D. E., 2005, Pulse energy dependence of subcellular dissection by femtosecond laser pulses, *Opt. Expr.* **13**, 3690-3696.

Hopt, A., and Neher, E., 2001, Highly nonlinear photodamage in two-photon fluorescence microscopy, *Biophys. J.* **80**, 2029-2036.

Hotop, H., 2001, Dynamics of low energy electron collisions with molecules and clusters, in: L.G. Christophorou, J.K. Olthoff (eds.): *Proc. Int. Symp. on Gaseous Dielectrics IX 22-25 May 2001, Ellicott City, MD, USA*, Kluwer Academic/Plenum Press, New York, pp. 3-14.

Huels, M. A., Boudaiffa, B., Cloutier, P., Hunting, D., and Sanche, L., 2003, Single, double, and multiple double strand breaks induced in DNA by 3-100 eV electrons, *J. Am. Chem. Soc.* **125**, 4467-4477.

Huettmann, G. and Birngruber, R., 1999, On the possibility of high-precision photothermal microeffects and the measurement of fast thermal denaturation of proteins, *IEEE J. Sel. Topics Quantum Electron.* **5**, 954-962.

Jay, D. G., and Sakurai, T., 1999, Chromophore-assisted laser inactivation (CALI) to elucidate cellular mechanisms of cancer, *Biochim. Biophys. Acta* **1424**, M39-M48.

Joglekar, A. P., Liu, H., Meyhöfer, E., Mourou, G., and L. Hunt, A., 2004, Optics at critical intensity. Applications to nanomorphing, *Proc. Nat. Acad. Sci.* **101**, 5856-5861.

Juhasz, T., Loesel, F. H., Kurtz, R. M., Horvath, C., Bille, J. F., and Mourou, G., 1999, Corneal refractive surgery with femtosecond lasers, *IEEE J. Sel. Topics Quantum Electron.* **5**, 902-910.

Kaiser, A., Rethfeld, B., Vicanek, M., and Simon, G., 2000, Microscopic processes in dielectrics under irradiation by subpicosecond pulses, *Phys. Rev. B* **61**, 11437-11450.

Kasparian, J., Solle, J., Richard, M., and Wolf, J.-P., 2004, Ray-tracing simulation of ionization-free filamentation, *Appl. Phys. B.* **79**, 947-951.

Keldysh, L. V., 1960, Kinetic theory of impact ionization in semiconductors, *Sov. Phys. JETP* **11**, 509-518.

Keldysh, L. V., 1965, Ionization in the field of a strong electromagnetic wave, *Sov. Phys. JETP* **20**, 1307-1314.

Kennedy, P. K., 1995, A first-order model for computation of laser-induced breakdown thresholds in ocular and aqueous media: Part I – Theory, *IEEE J. Quantum Electron.* **QE-31**, 2241-2249.

Kennedy, P. K., Hammer, D. X., and Rockwell, B. A., 1997, Laser-induced breakdown in aqueous media, *Progr. Quantum Electron.* **21**, 155-248.

Kiselev, S. B., 1999, Kinetic boundary of metastable states in superheated and stretched liquids, *Physica A* **269**, 252-268.

Knapp, R. T., Daily, J. W., and Hammitt, F. G., 1971, *Cavitation*, McGraw-Hill, New York, 1971, pp. 117-131.

Koester, H. J., Baur, D., Uhl, R., and Hell, S. W., 1999, Ca^{2+} fluorescence imaging with pico- and femtosecond two-photon excitation: signal and photodamage, *Biophys. J.* **77**, 2226-2236.

Kolesik, M., Wright, E. M., and Moloney, J. V., 2004, Dynamic nonlinear X-waves for femtosecond pulse propagation in water, *Phys. Rev. Lett.* **92**, 253901.

König, K., Riemann, I., Fischer, P., and Halbhuber, K., 1999, Intracellular nanosurgery with near infrared femtosecond laser pulses, *Cell. Mol. Biol.* **45**, 195-201.

König, K., Riemann, I., and Fritsche, W., 2001, Nanodissection of human chromosomes with near-infrared femtosecond laser pulses, *Opt. Lett.* **26**, 819-821.

König, K., Krauss, O., and Riemann, I., 2002, Intratissue surgery with 80 MHz nanojoule femtosecond laser pulses in the near infrared, *Opt. Express* **10**, 171-176.

König, K., Riemann, I., Stracke, F., and Le Harzic, R., 2005, Nanoprocessing with nanojoule near-infrared femtosecond laser pulses, *Med. Laser Appl.* **20**, 169-184.

Köstli, K. P., Frenz, M., Bebie, H., and Weber, H. P., 2001, Temporal Backward Projection of Optoacoustic Pressure Transients Using Fourier Transform Methods, *Physics in Medicine and Biology* **46,** 1863-1872.

Kotaidis, V., and Plech, A., 2005, Cavitation dynamics on the nanoscale, *Appl. Phys. Lett.* **87**, 213102.

Krasieva, T. B., Chapman, C. F., LaMorte, V. J., Venugopalan, V., Berns, M. W. and Tromberg, B. J., 1998, Cell permeabilization and molecular transport by laser microirradiation, *Proc. SPIE* **3260**, 38-44.

Lenzner, M., Krüger, J., Sartania, S., Cheng, Z., Spielmann, Ch., Mourou, G., Kautek, W., and Krausz, F., 1998, Femtosecond optical breakdown in dielectrica, *Phys. Rev. Lett.* **80**, 4076-4079.

Leszczynski, D., Pitsillides, C. M., Pastila, R. K., Anderson, R. R., and Lin, C. P., 2001, Laser-beam triggered microcavitation : a novel method for selective cell destruction, *Radiat. Res.* **156**, 399-407.

Liang, H., Wright, W. H., Cheng, S., He, W., and Berns, M. W., 1993, Micromanipulation of chromosomes in PTK2 cells using laser microsurgery (optical scalpel) in combination with laser-induced optical force (optical tweezers), *Exp. Cell Res.* **204**, 110-120.

Lin, C. P., Kelly, M. W., Sibayan, S. A. B., Latina, M. A., and Anderson, R. R., 1999, Selective cell killing by microparticle absorption of pulsed laser radiation, *IEEE J. Sel. Top. Quantum Electron.* **5**, 963-968.

Liu, W., Kosareva, O., Golubtsov, I. S., Iwasaki, A., Becker, A., Kandidov, V. P. and Chin, S. L., 2003, Femtosecond laser pulse fiolamentation versus optical breakdown in H_2O, *Appl. Phys. B* **76**, 215-229.

Mao, S. S., Quéré, F., Guizard, S., Mao, X. Russo, R. E., Petite, G., and Martin, P., 2004, Dynamics of femtosecond laser interactions with dielectrics, *Appl. Phys. A* **79**, 1695-1709.

Masters, B. R., So, P. T. C., Buehler, C., Barry, N., Sutin, J. D., Mantulin, W. W., and Gratton, E., 2004, Mitigating thermal mechanical damage potential during two-photon dermal imaging, *J. Biomed. Opt.* **9**, 1265-1270.

Maxwell, I., Chung, S., and Mazur, E., 2005, Nanoprocessing of subcellular targets using femtosecond laser pulses, *Med. Laser Appl.* **20**, 193-200.

Meier-Ruge, W., Bielser, W., Remy, E., Hillenkamp, F., Nitsche, R., and Unsöld, R., 1976, The laser in the Lowry technique for microdissection of freeze-dried tissue slices, *Histochem J.* **8**, 387–401.

Meldrum, R. A., Botchway, S. W., Wharton, C. W., and Hirst, G. J., 2003, Nanoscale spatial induction of ultraviolet photoproducts in cellular DNA by three-photon near-infrared absorption, *EMBO Rep.* **12**, 1144-1149.

Minoshima, K., Kowalevicz, A. M., Hartl, I., Ippen, E., and Fujimoto, J. G., 2001, Photonic device fabrication in glass by use of nonlinear materials processing with a femtosecond laser oscillator, *Opt. Lett.* **26,** 1516-1518.

Nahen, K., and Vogel, A., 1996, Plasma formation in water by picosecond and nanosecond Nd:YAG laser pulses - Part II: Transmission, scattering, and reflection, *IEEE J. Selected Topics Quantum Electron.* **2,** 861-871.

Needham, D., and Nunn, R. S., 1990, Elastic deformation and failure of liquid bilayer membranes containing cholesterol, *Biophys. J.* **58**, 997-1009.

Neumann, J., and Brinkmann, R., 2005a, Boiling nucleation on melanosomes and microbeads transiently heated by nanosecond and microsecond laser pulses, *J. Biomed. Opt.* **10**, 024001.

Neumann, J. and Brinkmann, R., 2005b, Nucleation and dynamics of bubbles forming around laser heated microabsorbers, *Proc. SPIE* **5863**, 586307, 1-9.

NIST, 2005, National Institute of Standards Chemistry Web Book, Thermophysical Properties of Fluid Systems, http://webbook.nist.gov/chemistry/fluid/

Niyaz, Y., and Sägmüller, B., 2005, Non-contact laser microdissection and pressure catapulting: Automation via object-oriented image processing, *Med. Laser Appl.* **20**, 223-232.

Nikogosyan, D. N., Oraevsky, A. A., and Rupasov, V., 1983, Two-photon ionization and dissociation of liquid water by powerful laser UV radiation, *Chem. Phys.* **77**, 131-143.

Noack, J., and Vogel, A., 1998a, Single shot spatially resolved characterization of laser-induced shock waves in water, *Appl. Opt.* **37**, 4092-4099.

Noack, J., Hammer, D. X., Noojin, G. D., Rockwell, B. A., and Vogel, A., 1998b, Influence of pulse duration on mechanical effects after laser-induced breakdown in water, *J. Appl. Phys.* **83**, 7488-7495.

Noack, J., 1998c, *Optischer Durchbruch in Wasser mit Laserpulsen zwischen 100 ns und 100 fs*, PhD Dissertation, University of Lübeck, Lübeck, 237 pp..

Noack, J., and Vogel, A., 1999, Laser-induced plasma formation in water at nanosecond to femtosecond time scales: Calculation of thresholds, absorption coefficients, and energy density, *IEEE J. Quantum Electron.* **35**, 1156-1167.

Nolte, S., Momma, C., Jacobs, H., Tünnermann, A., Chikov, B. N., Wellegehausen, B., and Welling, H., 1997, Ablation of metals by ultrashort laser pulses, *J. Opt. Soc. Am. B* **14**, 2716-2722.

Oehring, H., Riemann, I., Fischer, P., Halbhuber, K. J., and König, K., 2000, Ultrastructure and reproduction behavior of single CHO-K1 cells exposed to near-infrared femtosecond laser pulses, *Scanning* **22**, 263-270.

Oraevsky, A. A., Da Silva, L. B., Rubenchik, A. M., Feit, M. D., Glinsky, M. E., Perry, M. D. Mammini, B. M., Small IV, W., and Stuart, B., 1996, Plasma mediated ablation of biological tissues with nanosecond-to-femtosecond laser pulses: relative role of linear and nonlinear absorption, *IEEE J. Sel. Top. Quantum Electron.* **2**, 801-809.

Paltauf, G., and Schmidt-Kloiber, H., 1996, Microcavity dynamics during laser-induced spallation of liquids and gels, *Appl. Phys. A* **62**, 303-311.

Paltauf, G. and Schmidt-Kloiber, H., 1999, Photoacoustic cavitation in spherical and cylindrical absorbers, *Appl. Phys. A* **68**, 525-531.

Paltauf, G. and Dyer, P., 2003, Photomechanical processes and effects in ablation, *Chem Rev.* **103**, 487-518.

Paterson, L., Agate, B., Comrie, M., Ferguson, R., Lake, T. K., Morris, J. E., Carruthers, A. E., Brown, C. T. A., Sibbett, W., Bryant, P. E., Gunn-Moore, F., Riches, A. C. and Dholakia, K., 2005, Photoporation and cell transfection using a violet diode laser, *Opt. Express* **13**, 595-600.

Pitsillides, C. M., Joe, E. K., Wei, X., Anderson, R. R., and Lin, C. P., 2003, Selective cell targeting with light absorbing microparticles and nanoparticles, *Biophys. J.* **84**, 4023-4032.

Ratkay-Traub, I., Ferincz, I. E., Juhasz, T., Kurtz, R. M., and Krueger, R. R., 2003, First clinical results with the femtosecond neodymium-glass laser in refractive surgery, *J. Refract. Surg.* **19**, 94-103.

Rayner, D. M., Naumov, A., and Corkum, P. B., 2005, Ultrashort pulse non-linear optical absorption in transparent media, *Opt. Expr.* **13**, 3208-3217.

Rethfeld, B., 2004, Unified model for the free-electron avalanche in laser-irradiated dielectrics, *Phys. Rev. Lett.* **92**, 187401.

Rice, M. H., and Walsh, J. M., 1957, Equation of state of water to 250 kilobars, *J. Chem. Phys.* **26**, 824-830.

Ridley, B. K., 1999, *Quantum Processes in Semiconductors*, Oxford University Press, Oxford, 436 pp..

Riemann, I., Anhut, T., Stracke, F., Le Harzic, R., and König, K., 2005, Multiphoton nanosurgery in cells and tissues. *Proc. SPIE* **5695**, 216-224.

Roegener, J., Brinkmann, R., and Lin, C. P., 2004, Pump-probe detection of laser-induced microbubble formation in retinal pigment epithelium cells, *J. Biomed. Opt.* **9**, 367-371.

Sacchi, C. A., 1991, Laser-induced electric breakdown in water, *J. Opt. Soc. Am. B.* **8**, 337-345.

Sacconi, L., Tolic-Norrelyke, I M., Antolini, R., and F. S. Pavone, 2005, Combined intracellular three-dimensional imaging and selective nanosurgery by a nonlinear microscope, *J. Biomed. Opt.* **10**, 014002.

Saul, A., and Wagner, W., 1989, A fundamental equation for water covering the range from the melting line to 1273 K at pressures up to 25 000 MPa, *J. Phys. Chem. Ref. Data* **18**, 1537-1564.

Schaffer, C. B., Brodeur, A., García, J. F., and Mazur, E., 2001, Micromachining bulk glass by use of femtosecond laser pulses with nanojoule energy, *Opt. Lett.* **26**, 93-95.

Schütze, K., and Clement-Sengewald, A., 1994, Catch and move – cut or fuse, *Nature* **368**, 667-668.

Schütze, K., and Lahr, G., 1998, Identification of expressed genes by laser-mediated manipulation of single cells, *Nat Biotechnol.* **16**, 737–742.

Schütze, K., Pösl, H., and Lahr, G., 1998, Laser micromanipulation systems as universal tools in molecular biology and medicine, *Cell. Mol. Biol.* **44**, 735-746.

Shen, N., Datta, D., Schaffer, C. B., LeDuc, P., Ingber, D. E., and Mazur, E., 2005, Ablation of cytoskeletal filaments and mitochondria in live cells using a femtosecond laser nanocissor, *MCB Mech. Chem. Biosystems* **2**, 17-25.

Shen, Y. R., 1984, *The Principles of Nonlinear Optics,* Wiley, New York, 563 pp.

Sigrist, M. W., and Kneubühl, F. K., 1978, Laser-generated stress waves in liquids, *J. Acoust. Soc. Am.* **64**, 1652-1663.

Sims, C. E., Meredith, G. D., Krasieva, T. B., Berns, M. W., Tromberg, B. J., and Allbritton, N. L., 1998, Laser-micropipet combination for single-cell analysis, *Anal. Chem.* **700**, 4570–4577.

Skripov, V. P., Sinitsin, E. N., Pavlov, P. A., Ermakov, G. V., Muratov, G. N., Bulanov N. V., and Baidakov, V. G., 1988, *Thermophysical properties of liquids in the metastable (superheated) state*, Gordon and Breach, New York, 1988.

Smith, N. I., Fujita, K., Kaneko, T., Katoh, K., Nakamura, O., Kawata, S., and S.Takamatsu, S. Generation of calcium waves in living cells by pulsed laser-induced photodisruption, *Appl. Phys. Lett.* **79**, 1208-1210.

Soughayer, J. S., Krasieva, T., Jacobson, S. C., Ramsey, J. M., Tromberg, B. C. and Albritton, N. L., 2000, Caracterization of cellular optoporation with distance, *Anal. Chem.* **72**, 1342-1347.

Steinert, R. F., and Puliafito, C. A., 1986, *The Nd:YAG Laser in Ophthalmology*, W. B. Saunders, Philadelphia, 154 pp.

Stolarski, J., Hardman, J., Bramlette, C. G., Noojin, G. D., Thomas, R. J., Rockwell, B. A., and Roach, W. P.,1995, Integrated light spectroscopy of laser-induced breakdown in aqueous media, *Proc. SPIE* **2391**, 100-109.

Stuart, B. C., Feit, M. D., Hermann, S., Rubenchik, A. M., Shore, B. W., and Perry, M. D., 1996, Nanosecond to femtosecond laser-induced breakdown in dielectrics, *Phys. Rev. B* **53**, 1749-1761.

Sun, Q., Jiang, H., Liu, Y., Wu, Z., Yang, H., and Gong, Q., 2005, Measurement of the collision time of dense electronic plasma induced by a femtosecond laser in fused silica, *Opt. Lett.* **30**, 320-322.

Supatto, W., Dèbarre, D., Moulia, B., Brouzés, E., Martin, J.-L., Farge, E., and Beaurepaire, E., 2005a, In vivo modulation of morphogenetic movements in Drosophila embryos with femtosecond laser pulses, *Proc. Nat. Acad. Sci. USA* **102**, 1047-1052.

Supatto, W., Dèbarre, D., Farge, E., and Beaurepaire, E., 2005b, Femtosecond pulse-induced microprocessing of live *Drosophila* embryos, *Med. Laser Appl.* **20**, 207-216.

Tao, W., Wilkinson, J., Stanbridge, E., and Berns, M. W., 1987, Direct gene transfer into human cultured cells facilitated by laser micropuncture of the cell membrane, *Proc. Natl. Acad. Sci. USA* **84**, 4180-4184.

Thornber, K. K., 1981, Applications of scaling to problems in high-field electronic transport, *J. Appl. Phys.* **52**, 279-290.

Tien, A. C., Backus, S., Kapteyn, H., Murnane, M., and Mourou, G., 1999, Short-pulse laser damage in transparent materials as a function of pulse duration, *Phys. Rev. Lett.* **82**, 3883-3886.

Tirlapur, U. K., König, K., Peuckert, C., Krieg, R., and Halbhuber, K.-J., 2001, Femtosecond near-infrared laser pulses elicit generation of reactive oxygen species in mammalian cells leading to apoptosis-like death, *Exp. Cell Res.* **263**, 88-97.

Tirlapur, U. K., and König, K., 2002, Targeted transfection by femtosecond laser, *Nature* **418**, 290-291.

Tschachotin, S., 1912, Die mikroskopische Strahlenstichmethode, eine Zelloperationsmethode, *Biol. Zentralbl.* **32**, 623-630.

Tsukakoshi, M., Kurata, S., Nomiya, Y., Ikawa, Y., and Kasuya, T., 1984, A novel method of DNA tranfection by laser-microbeam cell surgery, *Appl. Phys. B* **35**, 135-140.

Venugopalan, V., Guerra, A., Nahen, K., and A. Vogel, 2002, The role of laser-induced plasma formation in pulsed cellular microsurgery and micromanipulation, *Phys. Rev. Lett.* **88**, 078103, 1-4.

Vogel, A., Hentschel, W., Holzfuss, J., and W. Lauterborn, 1986, Cavitation buble dynamics and acoustic transient generation in ocular surgery with pulsed neodymium:YAG lasers, *Ophthalmology* **93**, 1259-1269.

Vogel, A., Nahen, K., and Theisen, D., 1996a, Plasma formation in water by picosecond and nanosecond Nd:YAG laser pulses - Part I: Optical breakdown at threshold and superthreshold irradiance, *IEEE J. Selected Topics Quantum Electron.* **2**, 847-860.

Vogel, A., Busch, S., and Parlitz, U., 1996b, Shock wave emission and cavitation bubble generation by picosecond and nanosecond optical breakdown in water, *J. Acoust. Soc. Am.* **100**, 148-165.

Vogel, A., Noack, J., Nahen, K., Theisen, D., Busch, S., Parlitz, U., Hammer, D. X., Nojin, G. D., Rockwell, B. A., and Birngruber, R., 1999, Energy balance of optical breakdown in water at nanosecond to femtosecond time scales. *Appl. Phys. B* **68**, 271-280.

Vogel, A., and Noack, J., 2001, Numerical simulation of optical breakdown for cellular surgery at nanosecond to femtosecond time scales, *Proc. SPIE* **4260**, 83-93.

Vogel, A., Noack, J., Hüttmann, G., and Paltauf, G., 2002, Femtosecond-laser-produced low-density plasmas in transparent biological media: A tool for the creation of chemical, thermal and thermomechanical effects below the optical breakdown threshold, *Proc. SPIE* **4633**, 23-37.

Vogel, A., and Venugopalan V, 2003, Mechanisms of pulsed laser ablation of biological tissues, *Chem Rev.* **103**, 577-644.

Vogel, A., Noack, J., Hüttmann, G., and Paltauf, G., 2005, Mechanisms of femtosecond laser nanosurgery of cells and tissues, *Appl. Phys. B* **81**, 1015-1046.

von der Linde, D., and Schüler, H., 1996, Breakdown threshold and plasma formation in femtosecond laser-solid interaction, *J. Opt. Soc. Am. B* **13**, 216-222.

Watanabe, W., Arakawa, N., Matsunaga, S., Higashi, T., Fukui, K., Isobe, K. and Itoh, K., 2004, Femtosecond laser disruption of subcellular organelles in a living cell, *Opt. Express* **12**, 4203-4213.

Watanabe, W., 2005, Femtosecond laser disruption of mitochondria in living cells, *Med. Laser Appl.***30**, 185-192.

Yanik, M. F., Cinar, H., Cinar, H. N., Chisholm, A. D., Jin, Y., and Ben-Yakar, A., 2004, Functional regeneration after laser axotomy, *Nature* **432**, 822.

Yao, C. P., Rahmanzadeh, R., Endl, E., Zhang, Z., Gerdes, J., and Hüttmann, G., 2005, Elevation of plasma membrane permeability by laser irradiation of selectively bound nanoparticles, *J. Biomed. Optics* **10**, 064012.

Zeira, E., Manevitch, A., Khatchatouriants, A., Pappo, O., Hyam, E., Darash-Yahana, M., Tavor, E., Honigman, A., Lewis, A., and Galun, E., 2003, Femtosecond infrared laser – an afficient and safe in vivo gene delivery system for prolonged expression, *Mol. Ther.* **8**, 342-350.

Zohdy, M. J., Tse, C., Ye, J. Y., and O'Donnell, M., 2005, Optical and acoustic detection of laser-generated microbubbles in single cells, *IEEE Trans. Ultrason. Ferr.* **52** (at press).

Chapter 11

DESIGNED POLYMERS FOR ABLATION

Lukas Urech & Thomas Lippert
General Energy Research Department, Paul Scherrer Institut, 5232 Villigen PSI, Switzerland, thomas.lippert@psi.ch

1. INTRODUCTION

Since laser ablation of polymers was first reported by Srinivasan, et al., (Srinivasan and Mayne-Banton 1982) and Kawamura, et al., (Kawamura, et al., 1982) in 1982, numerous reviews on laser ablation of a large variety of polymers and the according ablation mechanisms have been published (Lazare and Granier, 1989; Srinivasan and Braren, 1989; Bäuerle, 2000; Lippert and Dickinson, 2003; Lippert, 2004). There is still ongoing discussion about the ablation mechanism, e.g. whether it is dominated by photothermal or photochemical processes. In many studies, a photothermal process was favored for commercially available polymers such as Kapton™ (Arnold and Bityurin, 1999; Bityurin, et al., 2003). However, polymers that show a photochemical ablation behavior at the irradiation wavelength are preferable for structuring, as damage to the surrounding material via thermal processes is minimized. In addition, a conversion of the polymer into gaseous product is advantageous, as minimal amounts of ablation products are redeposited on the structured surface, and additional cleaning procedures are not necessary. One approach to test for photochemical features in the ablation process is the design of polymers that contain photochemically active groups. An irradiation with the appropriate wavelength can then lead to photochemical decomposition or bond-breaking in the polymer.

Even though polymer ablation has been a research field for over 20 years, its potential has not yet been fully explored. One possible reason for this is because commercially available polymers such as polyimide (PI), polymethylmethacrylate (PMMA), and polycarbonate (PC), etc. (Suzuki, et al., 1997) that are applied in many ablation studies have several characteristics which make them unsuitable for high quality structuring. These include low sensitivity, carbonization upon irradiation, and redeposition of ablation products on the polymer surface (Lippert, et al., 2003). Therefore, novel photopolymers for laser ablation have been designed. The most important criteria for these polymers are:

1. High absorption coefficients ($\geq$ 20000 cm^{-1}) at the irradiation wavelength.
2. Exothermic decomposition at well-defined positions of the polymer backbone.
3. Decomposition of the polymer into gaseous products, which do not contaminate the polymer surface (Bennett, et al., 1996; Lippert, et al., 1999a).

The applications for polymers in laser ablation can be separated into two main fields. In the first one, a structure is produced in the polymer, whereas in the second field, it is the ablation products, which are of specific interest. Structuring of polymers today is used industrially for the production of nozzles for inkjet printers (Aoki, 1998) and to prepare via holes in multichip modules through polyimide by IBM (Patel and Wassick, 1997), and for many other applications, e.g. fabrication of microoptical devices.

Examples for the second group are polymers as fuel in the micro laser plasma thruster (μ-LPT), pulsed laser deposition (PLD) of polymers, or matrix assisted pulsed laser evaporation (MAPLE), which is a deposition technique that can be used to deposit highly uniform thin-films (Pique, et al., 2002).

POLYMER ABLATION

In the context of comparing ablation data it is important to clearly establish the method by which the usual ablation parameters (ablation rate, threshold fluence, and effective absorption coefficient) are defined as these values can vary considerably depending on the approach taken.

The ablation process is often described by the following equation (Andrew, et al., 1983; Srinivasan and Braren, 1984):

$$d(F) = \frac{1}{\alpha_{eff}} \ln\left(\frac{F}{F_{th}}\right), \quad (1)$$

where $d(F)$ represents the ablation rate per pulse, α_{eff} is the effective absorption coefficient, and F is the irradiation fluence. F_{th} is the ablation threshold fluence and is defined as the minimum fluence where the onset of ablation can be observed.

It is important to specify how the ablation rate is defined, either as the ablation depth after one pulse at a given fluence, or as the slope of a plot of the ablation depth versus the pulse number for a given fluence. These two different analytical methods can result in very different ablation rates, especially in the case of polymers where ablation does not start with the first pulse, but after multiple pulses. This phenomena is known as "incubation" and is related to a chemical or physical modification of the polymer by the first few laser pulses, which increases the absorption at the irradiation wavelength (Küper and Stuke, 1989; Srinivasan, et al., 1990), e.g. the formation of double bonds in poly(methylmethacrylate) (PMMA). Incubation is normally only observed for polymers with low absorption coefficients at the irradiation wavelength.

Nevertheless, even taking into account these approaches, it is often the case that this dependence of the ablation rate cannot be described by a single set of parameters. In Fig. 1 the ablation rate versus irradiation fluence is schematically shown. Fig. 1 is a generic representation of polymer ablation rates indicating three fluence regions, which can be categorized as follows:

- Low fluence range, from which F_{th} can be defined. Incubation phenomena are also observed at these low fluences.
- Intermediate fluence range in which an increase of the slope of the ablation rate is often observed, which may be due to additional or more effective decomposition of the polymer by energy that has been gained from decomposing the polymer.
- High fluence range where the ablation rates of many polymers are similar, as the incident laser light is screened by ablation products and the plasma which are created during the ablation process (Lazare and Granier, 1989).

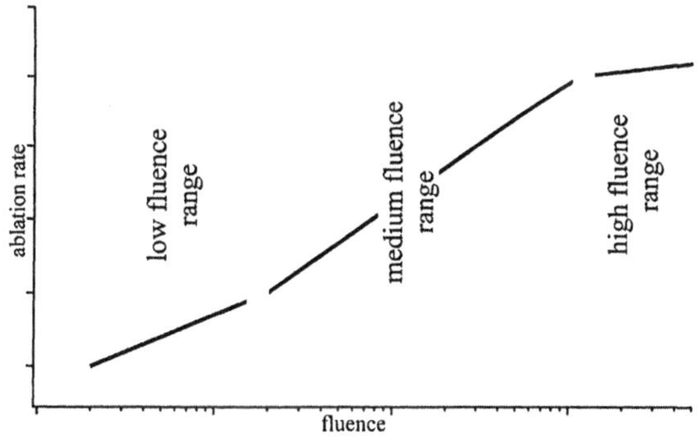

Figure 1. Schematic view of the three fluence regimes observed for most polymers.

It is worth mentioning in this context that "polyimide" (PI) is probably the most studied polymer in laser ablation and is also the material for which most ablation models are applied, but great care has to be taken for which type of polyimide the data has been obtained. Polyimide is not a single polymer, but describes a class of polymers, which contain at least one cyclic imide group per repetition unit. PI ranges from soluble polymers, to films and even photosensitive polymers which can all have very different properties (Lippert, 2005). Even products with the same name, such as Kapton™ are not one polymer, but there are also many different types of Kapton™ which are defined with additional letters, e.g. HN.

DOPED POLYMERS

In order to investigate the ablation mechanism on PMMA, various dopants have been added to the polymer ranging from polyaromatic compounds to compounds that contained photochemical active groups (Lippert, et al., 1997). Dopants based on the triazene group (-N=N-N<) have been tested, as they are photochemically well studied (Stasko, et al., 1993; Lippert, et al., 1994; Nuyken, et al., 1995a) and release a large amount of nitrogen during photochemical decomposition. The formation of gaseous products by photolabile dopants has been confirmed with a SEM analysis of the ablation craters of doped PMMA, which showed a pronounced swelling at low fluences. This swelling is caused by the decomposition of the dopants into gaseous products (Lippert and Dickinson, 2003). These authors have suggested that the released nitrogen and other gaseous ablation products act as carriers for larger ablation fragments.

At higher fluence an ablation rate of up to 80 μm can be achieved, but pronounced signs of surface melting are always visible (Lippert, et al., 1993), indicating the presence of a photothermal mechanism. One reason for this could be the limited amount of dopant (maximum of ≈10 %) that can be added to the polymer.

DESIGNED POLYMERS

In order to improve the quality of the ablation process, new polymers have been developed, where the triazene group was incorporated into the polymer main chain. One of the unique features of these polymers is the possibility to tune the absorption maximum to certain wavelengths by varying the "X" component in Fig. 2 (Nuyken, et al., 1995b). The absorption maximum of these triazene polymers (TP) can be tuned from 290 to 360 nm with a maximum linear absorption coefficient of up to almost 200000 cm^{-1}.

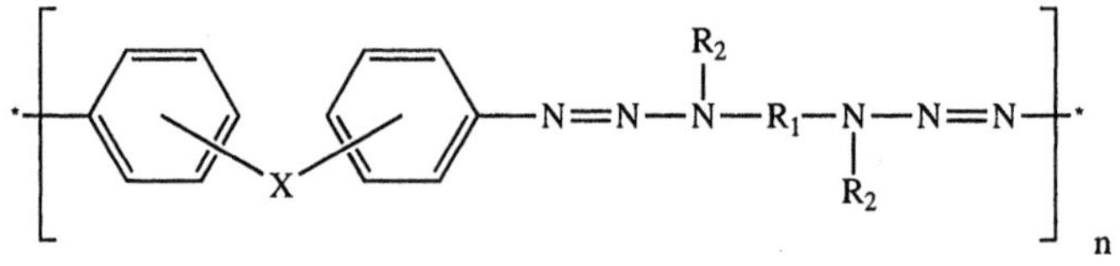

Figure 2. Chemical structure of the triazene polymers.

In Fig. 3, the chemical structure and the absorption spectra of a selected TP are shown. Two absorption maxima can clearly be distinguished in the absorption spectra, one around 200 nm which can be assigned to the aromatic system, and one around 330 nm, corresponding to the triazene unit. This allows excitation of different chromophores and the possibility to test their influence on the ablation behavior by using different irradiation wavelengths such as 193, 248, and 308 nm. A clear and well-defined F_{th} of 25 mJ cm^{-2} (± 5 mJ cm^{-2}) is observed for TP1 at an irradiation wavelength of 308 nm, while for irradiation with 248 nm a much broader range for F_{th} of 16 – 28 mJ cm^{-2} has been measured (Lippert, et al., 1996). The ablation rates measured for different wavelengths ranging from 193 to 351 nm also showed much higher ablation rates for the wavelengths that excite the triazene system, i.e. 266, 308, and also 351 nm (Lippert and Dickinson, 2003).

Another indication for different ablation mechanisms is the detection of carbonization for 248 nm irradiation, whereas the chemical composition remained unchanged after several pulses with 308 nm.

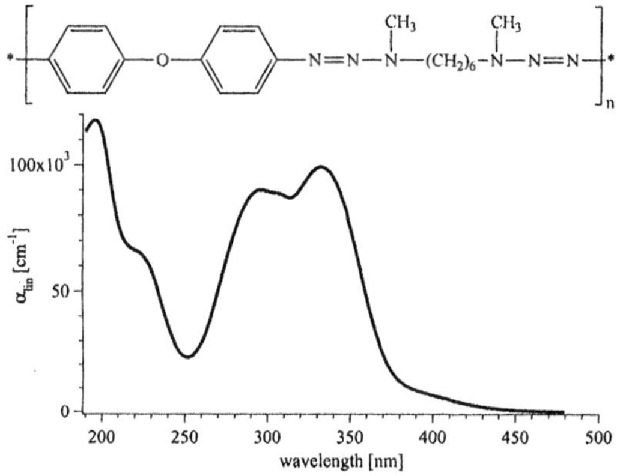

Figure 3. Chemical structure and linear absorption coefficient of TP1.

The ablation products have been studied by mass spectrometry to gain an insight into the ablation mechanism (Lippert, et al., 1999a; Lippert, et al., 1999b; Lippert, et al., 2001a). With the time-resolved mass spectrometry measurements at 248 and 308 nm irradiation wavelengths, all decomposition products for the decomposition of TP1 were identified (shown in Fig. 4 for irradiation with 308 nm), but for a thermal decomposition similar products were also observed (Nuyken, et al., 1995a). An important observation is the presence of three different species of nitrogen in the TOF (time-of-flight) signal, i.e. a very fast ground state neutral with up to 6 eV of kinetic energy, a slower neutral ground state species with a broad energy distribution, which is probably a thermal product, and possibly a metastable (excited) neutral N_2

species, that can only be created by an electronic excitation (Hauer, et al., 2002).

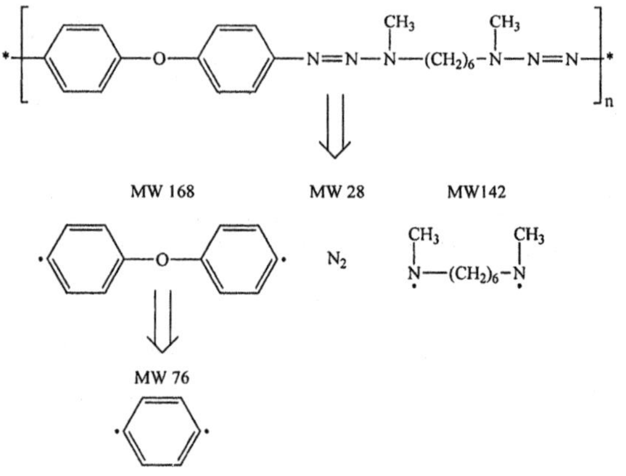

Figure 4. Decomposition pathway for TP1 measured by TOF-MS after irradiation with 308 nm.

The photochemical activity of the triazene group was also confirmed by irradiation with excimer lamps at low fluences, where linear photochemistry is expected (Lippert, et al., 2001a). At irradiation wavelengths of 308 nm and 222 nm, decomposition of the triazene chromophore was observed below F_{th}. In the case of the irradiation with 222 nm, a decomposition of the aromatic chrompohores has also been detected (Lippert, et al., 2002b) suggesting that decomposition of the aromatic part is related to carbonization. This indicates that the triazene group is clearly the most sensitive unit in the triazene polymer and that a photochemical part in the ablation mechanism is likely.

Another method to determine the present ablation mechanisms is ns-interferometry. It has been shown that thermal/photothermal ablation first results in pronounced swelling of the polymer surface, followed by etching (Furutani, et al., 1996; Furutani, et al., 1998). The etching takes place on time scales much longer than the pulse length of the excimer laser (up to the μs range).

However, for a photochemical process, as was observed for TP (see Fig. 5), etching starts and ends with the excitation pulse of the excimer laser (Lippert, et al., 2002b; Hauer, et al., 2003).

The triazene polymers were also tested for 157 nm irradiation and mass spectrometry measurements showed a higher fragmentation of the polymer than for 308 or 248 nm. At this shorter wavelength, even fragmentation of the aromatic groups was observed (Kuhnke, et al., 2003).

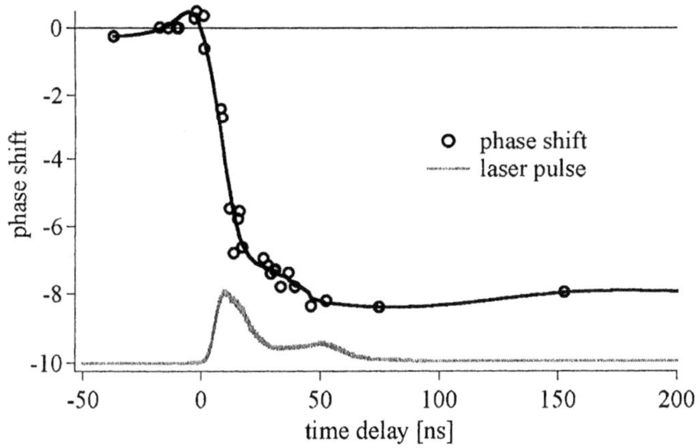

Figure 5. Interference measurement for TP1 during irradiation with 308 nm. The gray curve represents the laser pulse, while the black line corresponds to the phase shift, which is related to the ablation depth.

Irradiation with pulses in the pico and femtosecond range were performed in the near-IR at 800 nm. A lower ablation threshold fluence was found for the femtosecond pulse than for picosecond pulse indicating the presence of a thermal mechanism (Bonse, et al., 2005a). A clean removal of a thin TP film from a glass substrate was impossible with a 100 fs pulse. Therefore the structuring of TP with ultrashort pulses in the near-IR range is not an alternative to UV ablation (Bonse, et al., 2005b).

Compared to other designed polymers like polyesters or commercially available polymers such as polyimides, TP showed the highest ablation rate and lowest threshold fluence for selected irradiation wavelengths. Furthermore, structures obtained in TP (example in Fig. 6 (left)) are much sharper than in Kapton™ (see Fig. 6 (right)) and no redeposited carbon material is visible in the case of TP irradiated at 308 nm (Lippert, et al., 2002b). For comparison, Kapton™ was selected as a commercially available reference with a similar α_{lin} at 308 nm.

With ns-shadowgraphy measurements, it was shown that no solid products are produced for 308 nm irradiation of TP, which correlates well with the absence of redeposited material (Hauer, et al., 2003).

To summarize, the data obtained for the photochemical active TP strongly suggest that photochemical mechanisms play a significant role during laser ablation of polymers, but it is also clear that photothermal processes also take place. This is, for example, confirmed by the presence of a thermal N_2 product in the TOF curves. Photothermal processes will also always be present if the polymer decomposes exothermically during a photochemical decomposition and if the quantum yields of the photochemical reaction is not equal to one.

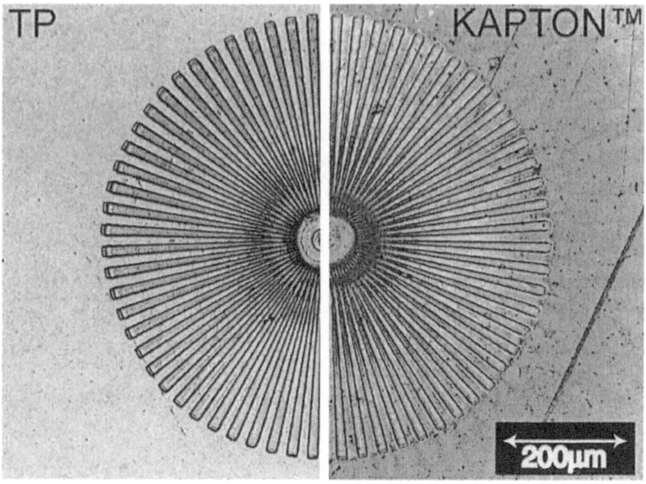

Figure 6. SEM of Siemens Stars in TP (left) and Kapton™ (right), both produced with five laser pulses at 308 nm.

Therefore, the ablation of polymers will always be a mixture of photochemical and photothermal reactions, where the ratio between the two is a function of the polymer and the irradiation wavelength.

Finally, from a material structuring point of view, it is worth noting that a photochemical mechanism leads to a more uniform decomposition of the polymer into small fragments. Furthermore, greater quantities of gaseous product are produced in the ablation plume and less material is redeposited. Therefore, the designed polymers of the TP group show clear advantages over commercially available polymers.

Selected novel applications

One of the novel applications for the designed polymers is the production of microoptical elements.

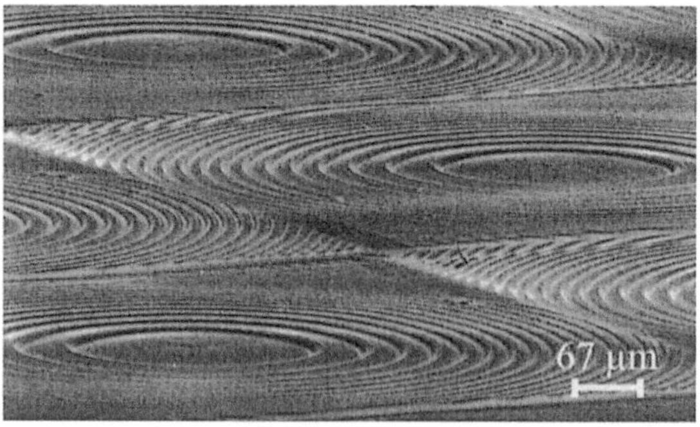

Figure 7. Fresnel lens array in photosensitive PI.

Complicated structures such as a Fresnel lens array (shown in Fig. 7) can be created with only a few pulses (Lippert, et al., 2001b), by using a specially designed diffractive gray tone phase mask that modulates the laser beam intensity (David, et al., 2001). The high sensitivity and etch rate of the TP or photosensitive PI allows the application of larger phase masks compared to other polymers such as polyimide. Alternatively, less pulses are necessary to fabricate an optical element with a given depth of the structure.

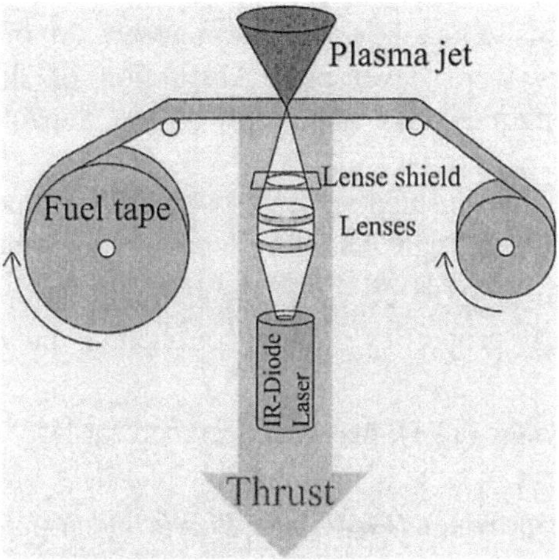

Figure 8. Scheme of the laser plasma thruster.

Another quite different application utilizes near-IR irradiation from diode lasers. The plasma created by laser ablation of the polymer acts as a micro thruster for small satellites (operating principle and setup are described in detail elsewhere (Lippert, et al., 2001b; Phipps, et al., 2002)).

The micro laser plasma thruster (μ-LPT) is a micropropulsion device designed for the steering and propelling of small satellites (1 to 10 kg) (scheme shown in Fig. 8). A laser is focused on to a polymer layer on a substrate to form a plasma. The thrust produced by this plasma is used to control the satellite motion (Phipps, et al., 2003; Phipps, et al., 2004a). Due to the specific demands (weight and power), the μ-LPT is driven by small powerful diode lasers, which emit in the near-IR (920-980 nm) with an available power of 1 to 15 W, and a pulse length from 100 μs to the millisecond range (Phipps, et al., 2004b). This pulse duration and wavelength require the utilization of materials for the fuel with low thermal conductivity, i.e. polymers (Lippert, et al., 2002a; Lippert, et al., 2003).

As most polymers are not absorbing in the near-IR range, it was necessary to add dopants, such as carbon or IR-dyes. Carbon nanopearls were selected as they showed the best results in thrust experiments, even though they tend to agglomerate in suspensions to form particles of 10 to 20 μm diameter (Richner, 2001). The IR-dye, on the other hand, is distributed on a molecular level. A different concentration for both dopants was selected to reach the same linear absorption coefficient at the irradiation wavelength of 1064 nm.

TP's also revealed good properties for this application, but the photochemical properties are only of minor importance, since exothermic decomposition, gas formation, and a well-defined decomposition temperature are more important for the LPT application. As exothermic decomposition seems to be a key element in generation of thrust, new higher energetic polymers were developed. Absorption of the polymer at the irradiation wavelength is less important, as the dopants are the primary absorbers.

To understand the influence of the specific properties of the fuel polymers, four different "high" and "low"energetic polymers were tested:

- Poly(vinyl chloride) (PVC) as a low-energetic reference polymer that showed the best properties among commercial polymers (Phipps and Luke, 2002),
- A triazene polymer (TP1), that was designed for laser ablation in the UV range, and
- Glycidyl azide polymer (GAP) and Poly(vinyl nitrate) (PVN) as high-energetic polymers (chemical structure in Fig. 9).

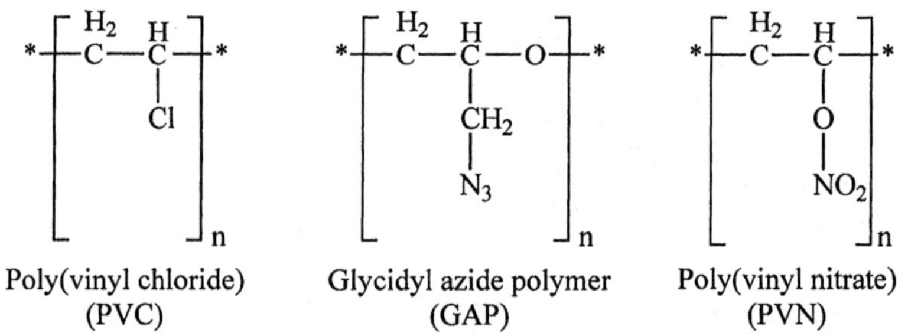

Figure 9. Chemical structure of PVC, GAP, and PVN.

GAP was originally developed as binder for solid propellant rockets (Frankel, et al., 1992; Köhler and Meyer, 1998) but it meets the demands for the LPT, such as easy handling, exothermic decomposition and well defined decomposition temperature. PVN is a thermoplastic polymer with a similar decomposition temperature but a much higher decomposition enthalpy than PVC, TP1, and GAP (see Table 1).

Table 1. Properties of polymers (Urech, et al., 2004)

Polymer	Decomposition temperature [°C]	Decomposition Enthalpy [J/g]
PVC	241, 288, 383	-418
TP1	227	-700
GAP	249	-2053
PVN	204	-3829

All polymers were doped with carbon (nanoparticles) and, in the case of GAP, also with an IR-dye (Epolight 2057), to achieve absorption in the near-IR (λ=1064 nm). In this section polymers will be referred to with the abbreviation of the polymer, and by indicating the dopant by "+ C" for carbon and "+ IR" for IR-dye.

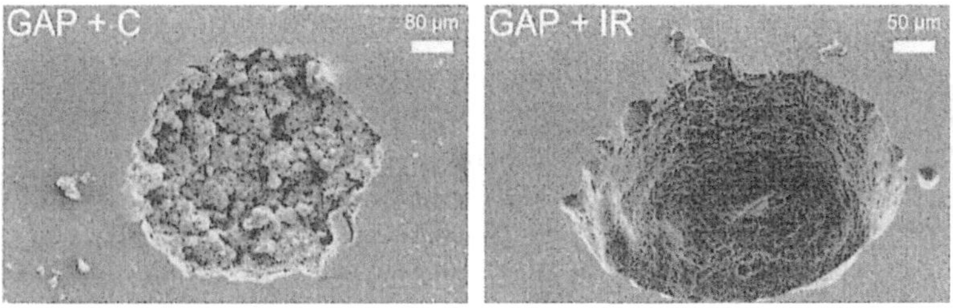

Figure 10. SEM images of the ablation spots of GAP + C (left) and GAP + IR (right).

The different dopants for GAP seem to have only a small influence on the ablation properties, such as ablation rate and threshold fluence (Urech, et al., 2004). The most pronounced differences are observed in the ejected fragments detected in the shadowgraphy measurements and the ablation crater appearance. An SEM investigation of the ablation crater confirmed these results, by showing an ablation crater with steep, smooth walls for GAP + IR, whereas the crater of GAP + C is quite rough, with deep holes and a very uneven bottom (see Fig. 10). Large fragments of solid and liquid ablation products are observed for GAP + C in the ns-shadowgraphy image (see Fig. 11 (left)), while almost no solid fragments are ejected by GAP + IR (see Fig. 11 (right)). It seems that the ablated material is transformed completely into gaseous products. This is a desired effect, as more energy is gained by decomposing more polymer.

The more homogeneous and non-agglomerating IR-dye leads to a more homogeneous decomposition, whereas the big carbon clusters produce local hotspots (Wen, et al., 1994). In the near surrounding of these clusters, the material is decomposed. This decomposition leads to the formation of gaseous products and the ejection of polymer fragments between the ablation spots. Energy gained by decomposing the polymer is lost by the ejection of solid or liquid polymer fragments.

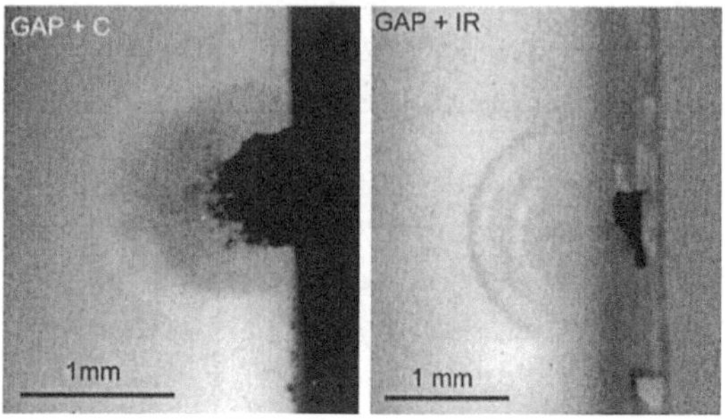

Figure 11. Shadowgraphy images of carbon (left) and IR-dye (right) doped GAP. The image was taken 1 μs after the laser pulse.

A pronounced difference has been observed in the shockwave and ejected products for the different polymers. In all cases, large polymer fragments are ejected. The largest fragments are observed when using PVC + C and PVN + C (see Fig. 12). The trajectory of these fragments is strongly directed towards the normal to the polymer surface, much more so than in the case of GAP + C and TP1 + C. The fragments also overtake the shockwave after ~500 ns in the case of PVN, and after 3 μs in the case of PVC. For GAP + C and TP + C, the product plume can be separated into two parts. The first part consists of small (gaseous) products, whereas the second part is formed by much larger polymer fragments. The particle plume in both cases expands almost hemispherical and is slower than the shockwave.

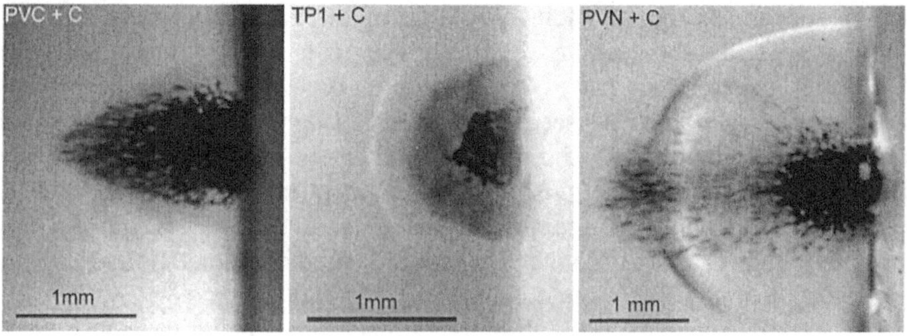

Figure 12. Ns-shadowgraphy images for carbon doped PVC, TP1, and PVN. The images were taken 1 μ s after the laser pulse.

The fastest shockwave was observed for PVN, followed by GAP, where only a small difference between the two dopants could be observed, and then TP and PVC (Hauer, et al., 2004; Urech, et al., 2004) (shown in Fig. 13). The slowest shockwave velocity was measured for PVC. This order correlates well with the decomposition enthalpy of the polymers, where the largest amount of chemically stored energy leads to the highest shockwave velocity.

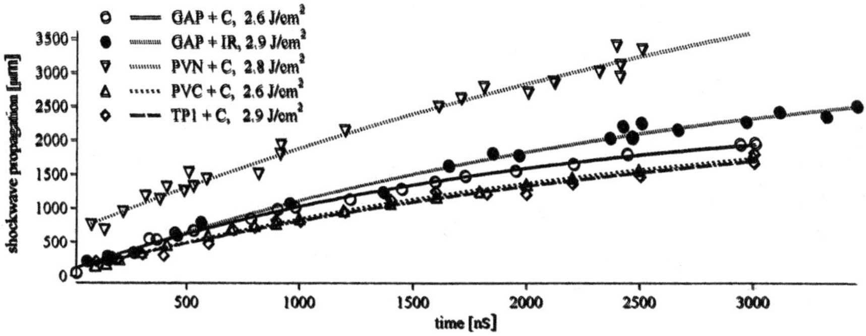

Figure 13. Shockwave front propagation versus time for GAP, PNV, TP1, and PVC doped with carbon, and for GAP doped with IR-dye

To analyze the performance of the polymers as fuel for the LPT, the target momentum was measured on a torsion balance as described in (Phipps and Luke, 2002). The target momentum was then used to calculate the momentum coupling coefficient C_m:

$$C_m = \frac{m\Delta v}{W} = \frac{F}{P} \ , \qquad (2)$$

where $m\ \Delta v$ is the target momentum produced during the ejection of laser ablated material. W is the incident laser pulse energy, F is the thrust, and P the incident power. The second important parameter for thrusters is the specific impulse I_{sp}, which is defined as:

$$I_{sp}g = v_E = C_m Q^* \ . \qquad (3)$$

Q^* is the specific ablation energy (incident power/mass ablation rate), v_E is the exhaust velocity, and g is the acceleration due to gravity.

As Eq. (3) demonstrates, I_{sp} and C_m are not independent. If, for example, a significant amount of incident energy is absorbed as heat in the target rather than producing material ejection, Q^* will be higher and C_m will be proportionally lower. This results in the same exhaust velocity and I_{sp} for both cases.

The thrust measurements revealed the highest I_{sp} for PVN + C (2320 s), whereas the other polymers showed much lower values (200 to 650 s) (dark grey in Fig. 14). The highest values for C_m were obtained for GAP + IR (2000 μN/W), followed by GAP + C, PVN + C, TP + C, and PVC + C (1170 to 120 μN/W) (light gray in Fig. 14). The high C_m value for the IR-doped polymer indicates a high transformation of polymer into the gaseous state, as also observed in the shadowgraphy and SEM measurements.

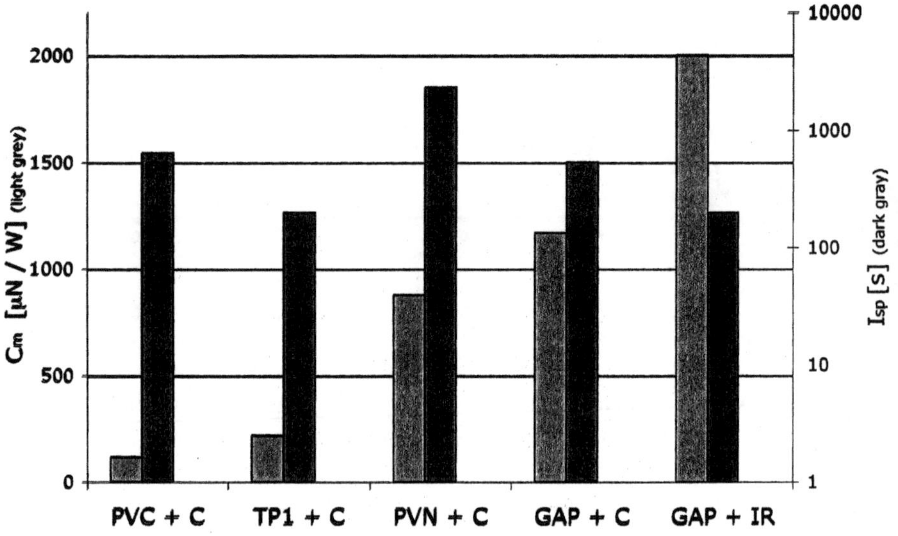

Figure 14. Coupling momentum coefficient (light grey) and the specific impulse (dark gray) for PVC, TP1, PVN, and GAP doped with carbon, and for GAP doped with IR-dye.

Surprisingly, the same order for the performance was observed for thrust and shadowgraphy measurements, even though the measurements were performed under totally different conditions, i.e. ambient conditions and low fluences for the shadowgraphy measurements and vacuum and high fluences for the thrust measurements. This allows evaluation of polymers under ambient conditions instead of vacuum.

For energetic polymers, a transformation of induced laser energy to measured thrust of more than 100% was observed. This indicates that energy gained from decomposing the polymer can be transferred into thrust, and that the ablation mechanism plays only a minor role.

These experiments demonstrate that the main demand on polymers as fuel for the μ-LPT is a high decomposition enthalpy. Furthermore, a high transition rate of the polymer into gaseous products is favorable, as more energy can be gained. As the influence of the dopant also needs to be considered, future developments may have to concentrate on polymer-dopant systems, not just on polymers.

ACKNOWLEDGMENTS

This work has also been partially supported by the Swiss National Science Foundation. The data on the LPT is based upon work supported by the European Office of Aerospace Research and Development, Air Force Office of Scientific Research, Air Force Laboratory, under contract FA8655-03-1-3058. The authors would also like to thank C. Phipps and J. Luke for the thrust data.

REFERENCES

Andrew, J. E., Dyer, P. E., Forster, D. and Key, P. H., 1983, Direct etching of polymeric materials using a XeCl laser, *Appl. Phys. Lett.* **43** (8): 717-719.

Aoki, H. (1998). U.S. Patent 5736999.

Arnold, N. and Bityurin, N., 1999, Model for laser-induced thermal degradation and ablation of polymers, *Appl. Phys. A-Mater. Sci. Process.* **68** (6): 615-625.

Bennett, L. S., Lippert, T., Furutani, H., Fukumura, H. and Masuhara, H., 1996, Laser induced microexplosions of a photosensitive polymer, *Appl. Phys. A-Mater. Sci. Process.* **63** (4): 327-332.

Bityurin, N., Luk'yanchuk, B. S., Hong, M. H. and Chong, T. C., 2003, Models for laser ablation of polymers, *Chem. Rev.* **103** (2): 519-552.

Bonse, J., Wiggins, S. M., Solis, J. and Lippert, T., 2005a, Phase change dynamics in a polymer thin film upon femtosecond and picosecond laser irradiation, *Appl. Surf. Sci.* **247** (1-4): 440-446.

Bonse, J., Wiggins, S. M., Solis, J., Lippert, T. and Sturm, H., 2005b, Femtosecond laser-induced decomposition in triazenepolymer thin films, *Appl. Surf. Sci.* **248** (1-4): 157-162.

Bäuerle, D., 2000, *Laser processing and chemistry*, Springer Verlag, Berlin

David, C., Wei, J., Lippert, T. and Wokaun, A., 2001, Diffractive grey-tone phase masks for laser ablation lithography, *Microelectron. Eng.* **57-8**: 453-460.

Frankel, M. B., Grant, L. R. and Flanagan, J. E., 1992, Historical development of glycidyl azide polymer, *J. Propul. Power* **8** (3): 560-563.

Furutani, H., Fukumura, H. and Masuhara, H., 1996, Photothermal transient expansion and contraction dynamics of polymer films by nanosecond interferometry, *J. Phys. Chem.* **100** (17): 6871-6875.

Furutani, H., Fukumura, H., Masuhara, H., Kambara, S., Kitaguchi, T., Tsukada, H. and Ozawa, T., 1998, Laser-induced decomposition and ablation dynamics studied by nanosecond interferometry. 2. A reactive nitrocellulose film, *J. Phys. Chem. B* **102** (18): 3395-3401.

Hauer, M., Dickinson, T., Langford, S., Lippert, T. and Wokaun, A., 2002, Influence of the irradiation wavelength on the ablation process of designed polymers, *Appl. Surf. Sci.* **197**: 791-795.

Hauer, M., Funk, D. J., Lippert, T. and Wokaun, A., 2003, Laser ablation of polymers studied by ns-interferometry and ns-shadowgraphy measurements, *Appl. Surf. Sci.* **208**: 107-112.

Hauer, M., Funk, D. J., Lippert, T. and Wokaun, A., 2004, Time resolved study of the laser ablation induced shockwave, *Thin Solid Films* **453-54**: 584-588.

Kawamura, Y., Toyoda, K. and Namba, S., 1982, Effective deep ultraviolet photoetching of poly(methyl methacrylate) by an excimer laser, *Appl. Phys. Lett.* **40** (5): 374-375.

Kuhnke, M., Cramer, L., Dickinson, J. T., Lippert, T. and Wokaun, A., 2003, TOF-MS study of photoreaktive polymers ablated by F_2 excimer laser (157 nm). *Poster at COLA* 2003.

Köhler, J. and Meyer, R., 1998, *Explosivstoffe*, Wiley-VCH, Weinheim

Küper, S. and Stuke, M., 1989, UV-excimer-laser ablation of polymethylmethacrylate at 248 nm - characterization of incubation sites with fourier-transform IR-spectroscopy and UV-spectroscopy, *Appl. Phys. A-Mater. Sci. Process.* **49** (2): 211-215.

Lazare, S. and Granier, V., 1989, Ultraviolet-laser photoablation of polymers - a review and recent results, *Laser Chem.* **10** (1): 25-40.

Lippert, T. (2004). Laser application of polymers. Polymers and light. Berlin, Springer-Verlag Berlin. 168: pp. 51-246.

Lippert, T., 2005, Interaction of photons with polymers: From surface modification to ablation, *Plasma Processes And Polymers* **2**: 525-546.

Lippert, T., Bennett, L. S., Nakamura, T., Niino, H. and Yabe, A., 1996, Single pulse threshold and transmission behaviour of a triazeno-polymer during pulsed UV-laser irradiation, *Appl. Surf. Sci.* **96-8**: 601-604.

Lippert, T., David, C., Dickinson, J. T., Hauer, M., Kogelschatz, U., Langford, S. C., Nuyken, O., Phipps, C., Robert, J. and Wokaun, A., 2001a, Structure property relations of photoreactive polymers designed for laser ablation, *J. Photochem. Photobiol. A-Chem.* **145** (3): 145-157.

Lippert, T., David, C., Hauer, M., Masubuchi, T., Masuhara, H., Nomura, K., Nuyken, O., Phipps, C., Robert, J., Tada, T., Tomita, K. and Wokaun, A., 2002a, Novel applications for laser ablation of photopolymers, *Appl. Surf. Sci.* **186** (1-4): 14-23.

Lippert, T., David, C., Hauer, M., Wokaun, A., Robert, J., Nuyken, O. and Phipps, C., 2001b, Polymers for UV and near-IR irradiation, *J. Photochem. Photobiol. A-Chem.* **145** (1-2): 87-92.

Lippert, T. and Dickinson, J. T., 2003, Chemical and spectroscopic aspects of polymer ablation: Special features and novel directions, *Chem. Rev.* **103** (2): 453-485.

Lippert, T., Dickinson, J. T., Hauer, M., Kopitkovas, G., Langford, S. C., Masuhara, H., Nuyken, O., Robert, J., Salmio, H., Tada, T., Tomita, K. and Wokaun, A., 2002b, Polymers designed for laser ablation-influence of photochemical properties, *Appl. Surf. Sci.* **197**: 746-756.

Lippert, T., Hauer, M., Phipps, C. R. and Wokaun, A., 2003, Fundamentals and applications of polymers designed for laser ablation, *Appl. Phys. A-Mater. Sci. Process.* **77** (2): 259-264.

Lippert, T., Langford, S. C., Wokaun, A., Savas, G. and Dickinson, J. T., 1999a, Analysis of neutral fragments from ultraviolet laser irradiation of a photolabile triazeno polymer, *J. Appl. Phys.* **86** (12): 7116-7122.

Lippert, T., Stebani, J., Nuyken, O., Stasko, A. and Wokaun, A., 1994, Photolysis of 1-aryl-3,3-dialkyltriazenes, *J. Photochem. Photobiol. A-Chem.* **78** (2): 139-148.

Lippert, T., Wokaun, A., Langford, S. C. and Dickinson, J. T., 1999b, Emission of neutral molecules during UV laser ablation of a photolabile triazeno polymer, *Appl. Phys. A-Mater. Sci. Process.* **69**: S655-S658.

Lippert, T., Wokaun, A., Stebani, J., Nuyken, O. and Ihlemann, J., 1993, Dopant-induced laser-ablation of PMMA at 308-nm - influence of the molecular-weight of PMMA and of the photochemical activity of added chromophores, *Angew. Makromol. Chem.* **213**: 127-155.

Lippert, T., Yabe, A. and Wokaun, A., 1997, Laser ablation of doped polymer systems, *Adv. Mater.* **9** (2): 105-119.

Nuyken, O., Stebani, J., Lippert, T., Wokaun, A. and Stasko, A., 1995a, Photolysis, thermolysis, and protolytic decomposition of a triazene polymer in solution, *Macromol. Chem. Phys.* **196** (3): 751-761.

Nuyken, O., Stebani, J., Lippert, T., Wokaun, A. and Stasko, A., 1995b, Synthesis and characterization of novel triazeno-group containing photopolymers, *Macromol. Chem. Phys.* **196** (3): 739-749.

Patel, R. S. and Wassick, T. A., 1997. *Proc. SPIE-Int. Soc. Opt. Eng.*

Phipps, C. and Luke, J., 2002, Diode laser-driven microthrusters: A new departure for micropropulsion, *AIAA Journal* **40** (2): 310-318.

Phipps, C., Luke, J. and Lippert, T., 2004a, Laser ablation of organic coatings as a basis for micropropulsion, *Thin Solid Films* **453-454**: 573-583.

Phipps, C., Luke, J., Lippert, T., Hauer, M. and Wokaun, A., 2004b, Micropropulsion using a laser ablation jet, *J. Propul. Power* **20** (6): 1000-1011.

Phipps, C. R., Luke, J. R., McDuff, G. G. and Lippert, T., 2002, Laser ablation powered mini-thruster, *Proc. SPIE* **4760**: 833-842.

Phipps, C. R., Luke, J. R., McDuff, G. G. and Lippert, T., 2003, Laser-driven micro-rocket, *Appl. Phys. A-Mater. Sci. Process.* **77** (2): 193-201.

Pique, A., Wu, P., Ringeisen, B. R., Bubb, D. M., Melinger, J. S., McGill, R. A. and Chrisey, D. B., 2002, Processing of functional polymers and organic thin films by the matrix-assisted pulsed laser evaporation (MAPLE) technique, *Appl. Surf. Sci.* **186** (1-4): 408-415.

Richner, R. P., 2001, Entwicklung neuartig gebundener kohlenstoffmaterialien für elektrische doppelschichtkondensatorelektroden, *DISS. ETH Nr. 14413*, ETH, Zürich

Srinivasan, R. and Braren, B., 1984, Ablative photodecomposition of polymer-films by pulsed far-ultraviolet (193 nm) laser-radiation - dependence of etch depth on experimental conditions, *J. Polym. Sci. Pol. Chem.* **22** (10): 2601-2609.

Srinivasan, R. and Braren, B., 1989, Ultraviolet-laser ablation of organic polymers, *Chem. Rev.* **89** (6): 1303-1316.

Srinivasan, R., Braren, B. and Casey, K. G., 1990, Nature of incubation pulses in the ultraviolet-laser ablation of polymethyl methacrylate, *J. Appl. Phys.* **68** (4): 1842-1847.

Srinivasan, R. and Mayne-Banton, V., 1982, Self-developing photoetching of poly(ethylene-terephthalate) films by far ultraviolet excimer laser-radiation, *Appl. Phys. Lett.* **41** (6): 576-578.

Stasko, A., Adamcik, V., Lippert, T., Wokaun, A., Dauth, J. and Nuyken, O., 1993, Photochemical decomposition of triazenes - (electron-paramagnetic-resonance study), *Makromolekulare Chemie-Macromolecular Chemistry And Physics* **194** (12): 3385-3391.

Suzuki, K., Matsuda, M., Ogino, T., Hayashi, N., Terabayashi, T. and Amemiya, K., 1997. *Proc. SPIE.*

Urech, L., Hauer, M., Lippert, T., Phipps, C. R., Schmid, E., Wokaun, A. and Wysong, I., 2004, Designed polymers for laser-based microthrusters - correlation of thrust with material, plasma, and shockwave properties, *Proc. SPIE* **5448**: 52-64.

Wen, X. N., Hare, D. E. and Dlott, D. D., 1994, Laser polymer ablation threshold lowered by nanometer hot-spots, *Appl. Phys. Lett.* **64** (2): 184-186.

Chapter 12

FABRICATION OF WAVEGUIDES BY LASER DEPOSITION

Optical waveguide growth on Si with NGA-PLD method

Takeshi Okato and Minoru Obara*

*Department of Electronics and Electrical Engineering, Keio University 3-14-1, Hiyoshi, Kohoku-ku, Yokohama-shi, 223-8522 Japan, *E-mail: obara@obara.elec.keio.ac.jp*

1. INTRODUCTION

Rapidly growing photonic network technologies have placed demands on the development of a variety of functional optical waveguides, such as large bandwidth amplifiers, optical switches, wavelength converters, and narrow line-width lasers, based on thin-film or diffusion technologies. The diffusion techniques, including ion-implantation, thermal diffusion, and ion-exchange, are able to cause refractive index modification on the bulk surface, so that the optical waveguides are obtained (see for example, Field, et al., 1992; Suche, et al., 1999; Florea and Winick, 1999). In this these methods, the presented optical waveguide losses are very low (less than 1 dB cm^{-1}). However, the waveguiding material is inevitably limited by the substrate used, and the monolithic integration with other materials is difficult.

Thin-film optical waveguiding structure has great advantages over the diffusion processes in terms of flexibility for both waveguide and substrate materials. Thin-film optical waveguides can be grown by various methods like liquid phase epitaxy (LPE), chemical vapor deposition (CVD), molecular beam epitaxy (MBE) (see for example, Gawith, et al., 1999; Guldberg-Kjaer, et al., 1999; Daran, et al., 1999), and pulsed-laser deposition (PLD). Table 1 shows a summary of recent optical waveguiding films grown by PLD method. More overview of the optical waveguiding films can be found elsewhere (Jelinek, 2003). As can be seen in Table 1, PLD of optical waveguides has been applied for wide-ranging device materials. But, note that the optical waveguide on Si has not yet been extensively studied.

Table 1. Overview of optical waveguiding films grown by PLD (*published in 2004*).

Material/ film	substrate	deposition conditions				characterization		author
		laser	T_{sub} deg.	E J cm^{-2}	$P(O_2)$ mbar	loss dB cm^{-1}	other	
Ti:sapphire	sapphire	KrF	975	4	$3x10^{-4}$	1.6	lasing	Grivas
Nd:GGG	YAG	KrF	650	2	0.02	0.1	lasing	Maysmith
$LiNbO_3$	sapphire	KrF	400	1.2	0.13	3.1	SHG	Nakata
$PbTiO_3$	MgO	Nd:YAG	700	0.7	0.3	< 1	*n*	Roemer
PZT	BMT/Si	KrF	350	2-3	0.18	2.86	*n*	Chu
PLZT	ITO/glass	KrF	550	1.5	0.013	-	E/O	Gaidi
$Eu:Y_2O_3$	SiO_2	KrF	730	0.5	10^{-4}	~ 1	PL	Lancok
$Er,Yb:Y_2O_3$	SiO_2	XeCl	500	2.0	0.1	-	CL	Dikovska
Nd:KGW	CeO_2/Si	See in text.						

[GGG: $Gd_3Ga_5O_{12}$, YAG: $Y_3Al_5O_{12}$, PLZT: $(Pb,La,Zr,Ti)O_3$, PZT: $Pb(Zr,Ti)O_3$, BMT: $Ba(Mg,Ta)O_3$, KGW: $KGd(WO_4)_2$, SHG: second harmonic generation, E/O: Electro-optic property, PL: photoluminescence, CL: cathode luminescence, *n*: refractive index]

In PLD method, thin-films are typically grown inside a stainless steel vacuum chamber. A target placed inside the chamber is irradiated by high peak intensity laser pulses. After laser irradiation, the ablation plume consisting of ionic atoms/molecules is generated. Later, thin-film is deposited on the substrate placed at the opposite side. Among the well known film growth methods, PLD is known as the most rapid and efficient technique for fabrication of multicomponent thin-films and hetero structures with an atomic control of the composition and thickness. However, some of the disadvantages of this method, such as nonexact stoichiometric transfer of the light elements from the target to the films and micrometer-sized droplets/clusters on the surfaces, are observed. Here, we will demonstrate three important issues for PLD of optical waveguides by focusing on the deposition of Nd:KGW [Nd:$KGd(WO_4)_2$] waveguides on Si substrates. First, Si does not allow for a waveguide structure on it by direct growth because of its high refractive index. To overcome this problem, we introduced a buffer layer between the waveguide and substrate. Second, KGW contains a light alkali metal, potassium (K). We studied the compensation of the non-stoichiometric transfer effect. Third, we demonstrated a method to reduce the clusters on the film surface through a nozzle-gas-assist. In Section 2 of this Chapter, we review recent researches and developments of the waveguide lasers grown by PLD. In Section 3, we present the growth of the highly crystallized Nd:KGW waveguides on Si substrate with CeO_2 buffer layer using conventional PLD (C-PLD). It is followed by the demonstration of a reduction of the surface roughness by use of a newly-developed nozzle-gas-assisted PLD (NGA-PLD) in Section 4. Section 5 is the summary.

2. WAVEGUIDE LASERS FABRICATED BY PLD

One of the attractive applications of optical waveguide is waveguide laser, because not only will the optical confinement lead to a lower laser oscillation threshold and higher gain as an amplifier, but it also has many important applications, such as integrated optics and optoelectronics for optical communications, human soft tissue spectroscopy/treatment, environmental sensing, spectroscopy, etc. Ezaki, et al., for the first time, reported the fabrication of Nd:YAG films on various substrates using PLD (Ezaki, et al., 1995). Subsequently, many papers have focused on the PLD of (active layer)/(dielectric substrate) structure waveguide lasers, e.g., Nd:GGG/YAG (Anderson, et al., 1997), Nd,Cr:GGG/YAG (Fukaya, et al., 2001), Yb:YAG/YAG (Shimoda, et al., 2001), Ti:sapphire/sapphire (Anderson, et al., 1997), and so forth. Recently, the deposition of Nd:KGW films on sapphire, YAG, YAP, and MgO substrates has been reported (Atanasov, et al., 2000; Jelinek, et al., 2002).

Rare earth doped GGG on YAG structure has a good lattice matching with a numerical aperture of 0.75, which is suitable for low-loss and high optical confinement waveguides. High optical confinement is useful for low-gain or quasi-three-level transition Yb lasers. Nd and Cr codoped film is expected as a self-Q-switched laser, which is applicable to medical processing, remote sensing, and optical circuit integration with SHG crystal waveguides. Among the commercially available Nd ion doped laser crystals, KGW demonstrates the highest stimulated emission performance (Mochalov, 1995; Findeisen, et al., 1999). Furthermore, it has large nonlinear optical susceptibilities $\chi^{(3)}$, and the first Stokes emission at 1.53 μm wavelength is the result of the fundamental $^4F_{3/2} \rightarrow {}^4I_{13/2}$ channel stimulated emission field. These points show that Nd:KGW optical waveguide has huge possibilities for industrial applications. However, the difficulty of preserving the chemical composition between the ablation target and thin-film makes it difficult to demonstrate Nd:KGW waveguides by PLD method to date.

Very recently, focuses are not only on the crystal growth, but also other new processing technologies. Grivas, et al. have reported Ti:sapphire single-mode rib waveguides which have yielded strong optical confinement, together with sufficient levels of fluorescence output. This makes the Ti:sapphire rib waveguides a very interesting candidate as a fluorescence source for optical coherence tomography (Grivas, et al., 2004). May-Smith, et al. reported the epitaxial single-crystal Nd:GGG films on YAG substrates with thicknesses up to 135 μm through a multiple deposition process, resulting in a low propagation loss of 0.1 dB cm^{-1} (May-Smith, et al., 2004). At the same time, technologies for optical integration are emerging. For example, surface relief type gratings for input laser coupling were fabricated by femtosecond laser interferometric processing for Nd:GGG thin-film grown by PLD (Oi, et al., 2003).

3. Nd:KGW WAVEGUIDE FABRICATION ON SILICON SUBSTRATE

In Section 3, we demonstrate the Nd:KGW optical waveguide on Si substrate. Because Si devices and their applications are technically matured and widely used in the industries, the crystal growth on Si can also be applicable to magnetic, ferroelectric, semiconducting, and superconducting materials for device and conductor applications, not only to monolithic optical waveguide integration. We first show a scheme to realize an optical waveguide on Si substrate. Then, we show the compensation of the non-stoichiometric transfer effect during PLD to obtain Nd:KGW films.

3.1 Waveguide structure design

The physical and optical properties of Nd:KGW, CeO_2, and Si are summarized in Table 2 (Mochalov, 1995; Hayashi, et al., 2000; Hartridge, et al., 1998). As can be seen in Table 2, Si particularly has a high refractive index which does not allow for an optical waveguide structure on it by direct growth. We introduced a CeO_2 buffer layer between Nd:KGW optical waveguide and Si substrate, so that the waveguiding mode was allowed within the waveguide layer. However, we should note that the propagation light is absorbed by the Si substrate, unless the buffer layer has a sufficient thickness. If the evanescent waves, which decay exponentially with distance from the interface, reach the under layer, they are not confined inside the waveguide region anymore. This is an evanescent wave coupling alike. On the other hand, a too thick layer preparation is a time consuming work. In this study, beam propagation method was used to simulate the waveguide loss caused by evanescent waves. Calculation was performed in 2D based on the unconditionally stable finite difference method. The TE polarized waveguide mode was propagated under the Crank-Nicolson scheme without Padé approximant. The boundary was a transparent boundary condition.

Fig. 1 shows the calculated waveguide loss as a function of CeO_2 buffer layer thickness, where the light wavelength is 1.07 μm, refractive index and thickness of each layer is $n_{KGW} = 2.0$ (1.0 μm), $n_{CeO2} = 1.7$, and $n_{Si} = 3.5$ (50 μm), respectively. The Si thickness has enough margin to reduce the effect of reflected light from the bottom of the substrate. This result suggests that the waveguide on Si substrate is realized with the sub-μm-thick buffer layer.

Table 2. Comparative properties of KGW, CeO_2, and Si.

formula	$KGd(WO_4)_2$	CeO_2	Si
lattice	monoclinic	cubic	cubic
cell	$a = 8.10$, $b = 10.35$, $c = 7.55$; $\beta = 94.43°$	$a = 5.43$	$a = 5.41$
thermal expansion for 100 °C [/K^{-1} x10^6]	4.0[100],3.6[010], 8.5[001]	~ 7	~ 3.2
refractive indices for 1.07 μm	$n_p = 1.978$, $n_m = 2.014$, $n_g = 2.049$	1.68	3.52

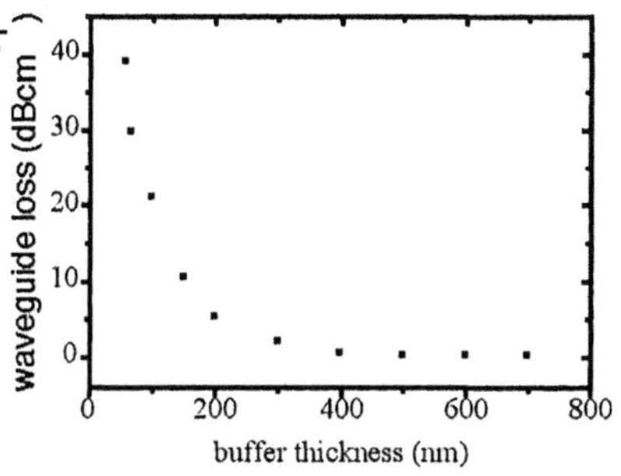

Figure 1. Calculated waveguide losses as a function of CeO_2 layer thickness

3.2 CeO_2 buffer layer preparation

Epitaxial growth during the KrF excimer PLD is very sensitive to the deposition conditions such as substrate temperature T_{sub}, oxygen pressure $P(O_2)$, laser fluence E, and target-substrate distance $D_{T\text{-}S}$ (Wu, et al., 1991).

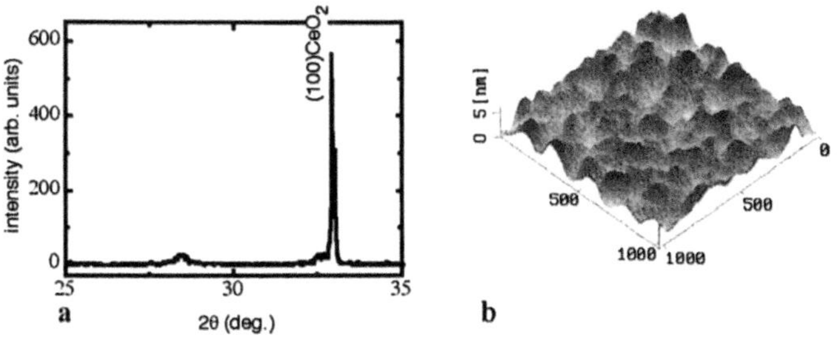

Figure 2. XRD scan – (a) AFM image – (b) of (100) CeO_2 thin film deposited on (100) Si substrate: $E = 4$ J cm^{-2}; $P(O_2) = 0.012$ mbar; $T_{sub} = 500$ °C, and $D_{T\text{-}S} = 50$ mm.

Also, it strongly depends on the quality of the amorphous SiO_2 layer formed on the (100)Si substrate. Beyond the exact optimal growth conditions, a preferential (111)CeO_2 is formed.

X-ray diffraction (XRD) scan and atomic force microscopy (AFM) image of as-grown (100)CeO_2 buffer layer is presented in Fig. 2 (a) and (b). The thickness of the CeO_2 layer is more than 200 nm, so as to prevent the light leaking out of the waveguide layer. In addition, a small amount of Gd is added to the target to prevent cracking (Atanasov, et al., 2004). The typical value of $\Delta\omega = 0.20°$ at FWHM for (100)CeO_2 suggests that the film has good crystallinity. The RMS surface roughness measured is less than 1 nm.

3.3 Compensation of non-stoichiometric transfer effect

After the laser ablation of a multicomponent target, collisions among ablated particles are significant, and the heavy atoms or ionic molecules accelerate or scatter the light atoms. Subsequently, the lighter atoms which are subject to scatterings forward off-axis direction, while the heavier atoms tend to reach the substrate along on-axis direction. If there is a significant difference in the atomic weight, this effect produces a non-stoichiometric film from a stoichiometric ablation target. Such an unfavorable process is one of the problems of PLD, which is avoidable by adjusting the target composition and ambient pressure as described below.

In the case of KGW [$KGd(WO_4)_2$], the lightest atomic weight is 39.098 of potassium (K), and the heaviest is 183.84 of tungsten (W). Therefore, it is expected that the preferential K-scattering in the ablation plume causes K-deficiencies in the films, if a single crystal KGW target is used for ablation. The K-deficiency results in the unfavorable Gd_2WO_6 crystalline phase growth, exclusively. Better crystallinity and lower optical loss waveguides can be achieved by use of K-rich ceramic targets, which are precisely described elsewhere (Okato, et al., 2003a; Okato, et al., 2003b).

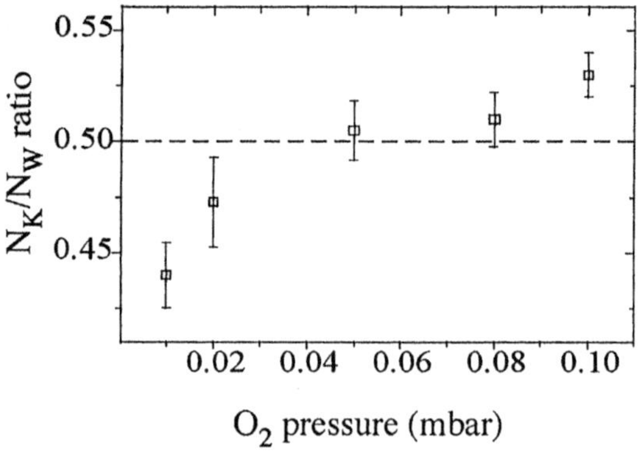

Figure 3. Plots of the N_K/N_W ratio of the as-grown films as a function of $P(O_2)$: $E = 1.2$ J cm^{-2}, $P(O_2) = 0.005$ - 0.1 mbar, $T_{sub} = 700$ °C, and $D_{T\text{-}S} = 40$ mm. The dashed line corresponds to the stoichiometric ratio.

Here, we show the evaluation of the K content in the films as a function of $P(O_2)$ in Fig. 3, where a K-rich target ($N_K/N_W = 1.16$) has been used. The films were deposited on the $(100)CeO_2/(100)Si$ substrates. As can be seen in Fig. 3, the K content is very little at a low pressure regime. However, it is also observed that the ratio increases as the $P(O_2)$ increases, and it approaches the stoichiometric value ($N_K/N_W = 0.5$) between 0.05 and 0.08 mbar.

The increment of N_K/N_W may be explained by two physics. First, the K atoms are prevented from going outside of the plume due to the high ambient pressure. Second, the increased plasma density inside the ablation plume promotes chemical reactions between respective species. The increased concentration of ionic complexes or molecules between K, Gd, W, and O may reduce the difference between heavier and lighter elements, which equalize the scattering yields.

3.4 Optical properties of Nd:KGW waveguides

Photoluminescence (PL) strongly depends on the crystallinity of the films, so that the film quality can be tested by the PL measurement. At the optimized growth conditions, Nd:KGW films preferentially grow (110)-orientation on CeO_2/Si substrates. The PL spectrum of the as-grown film and Nd:KGW single crystal is shown in Fig. 4. The intensity of the emission peaks of the films is an order of magnitude lower than those of the single crystal; however, as the film is only about 0.82 μm thick, its lower emission intensity is due to the excited active volume. The emission spectra of film have similar features with the exception of difference in the fluorescence intensity for each Nd transition.

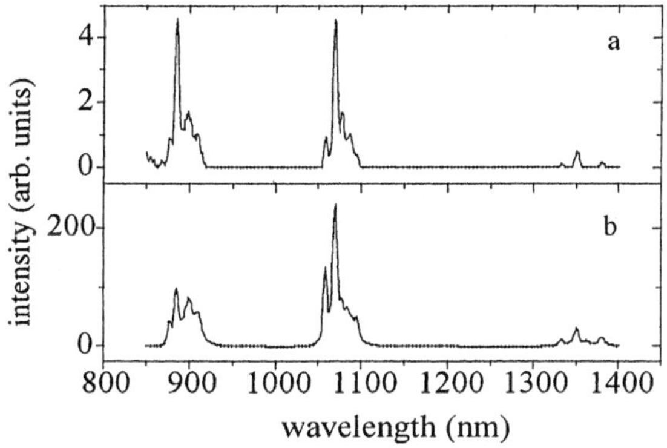

Figure 4. Photoluminescence spectra of Nd:KGW film and single crystal: (a) as-grown at $P(O_2) = 0.08$ mbar, $T_{sub} = 700$ °C; (b) Nd:KGW single crystal. The PL measurements were performed at room temperature using an 810 nm laser diode for pumping.

This is caused by large lattice mismatches between KGW film and CeO_2 substrate, which induce stresses and distortions into the KGW crystal and subsequently to the incorporated Nd. The optical waveguide loss was measured by recording the attenuation of the scattered light along with the propagation path. A rutile prism was used to couple the *s*-polarized He-Ne laser into the film. The incident angle of the laser light can be determined with the following refractive indices of CeO_2 and KGW films at 633 nm, being 1.68 and 1.9, respectively. The incident angle was determined to be $\theta = 33.8^{\circ}$ using the following equations:

$$\theta_{in} = \sin^{-1}(n_{eff} / n_p) \text{ and } n_o \sin(\theta - 45°) = n_p \sin(\theta_{in} - 45°),$$

where θ and θ_{in} are the incident angles of the prism and of the film, n_{eff} is the film effective refractive index determined by mode dispersion curve, and n_p and n_o are the refractive indices of the prism and air, respectively. The recorded attenuation of the scattered light is shown in Fig. 5 (a). The optical waveguide loss is evaluated to be about 3.53 dB cm^{-1} at 633 nm.

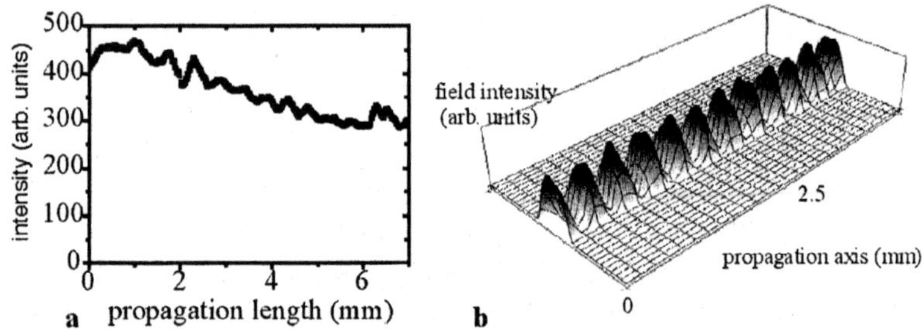

Figure 5. (a) Experimental observation of He-Ne laser propagation properties of Nd:KGW film (0.82 μm-thick) grown on CeO_2/Si substrate. (b) Simulated spatial TE field distribution of KGW/CeO_2/Si structure (λ = 633 nm), where the thicknesses of waveguide and buffer are 0.82 μm and 0.5 μm, respectively.

Fig. 5 (b) shows the simulated 2D TE mode intensity profile along the propagation path. The observed waveguide propagation property is in good agreement with the simulated one, i.e. peaks and troughs spacing is several hundred μm in both cases, giving the positive evidence of waveguide structure realization. The relatively high waveguide loss obtained from the film is due to the large mismatch between the film and substrate. In addition, we have high density of particles and droplets on the film surface, which are discussed in Section 4.

4. IMPROVEMENT OF SURFACE SMOOTHNESS

The background pressure during conventional PLD (C-PLD) is adjusted to oxidize growing films properly. However, high background pressure sometimes deteriorates the film surface morphology. Conversely, under extremely low background pressures, the ablated particles are deposited on a substrate as the state near single atoms, which enables MBE-like epitaxial growth at the expense of film oxidation. In Section 4, a nozzle-gas-assisted PLD (NGA-PLD) is demonstrated to achieve low-loss waveguide using only a simple oxygen gas nozzle, which enables effective film oxidation under lower ambient pressures.

4.1 Nozzle-gas-assisted PLD system

Despite the fact that PLD has been known as a powerful tool for the film growth of a variety of materials, especially for the multicomponent oxides, the films usually have a high optical waveguide loss. To date, great efforts have been devoted to produce high quality crystalline thin-films by PLD. For example, *in-situ* pulsed-laser annealing (Ito, et al., 1997), off-axis PLD (Strikovsky, et al., 1993), plasma-assisted PLD (Giacomol and Pascale, 2004), and so forth have been reported. *In-situ* pulsed-laser annealing is a scheme for film crystallization resulting from the thermal annealing effect due to energy transformation from photons to heat. The off-axis PLD technique successfully prevents droplets from depositing on the substrate, because only light ions or atoms can reach the substrate placed at the off-axis position, while heavy particles pass through without changing their direction. Plasma-assisted PLD is a technique which enhances the chemical reactivity between materials and the RF plasma gas source. The oxygen plasma gas source allows effective film oxidation, resulting in a higher crystallinity and better surface morphology.

Here, a nozzle-gas-assisted PLD (NGA-PLD) is demonstrated to meet the specific requirements of efficient film oxidization at low $P(O_2)$ using only a simple oxygen gas nozzle. A gas nozzle is placed near the substrate surface to accelerate film oxidation and to avoid the use of unnecessary high pressure. Oxygen gas is introduced into the process chamber via the gas nozzle, incident at 45°, placed near the substrate surface, as shown in Fig. 6. The nozzle diameter ϕ is 2 mm. By changing the nozzle-to-substrate distance $D_{N\text{-}S}$ and mass flow rate, we have optimized the NGA-PLD system.

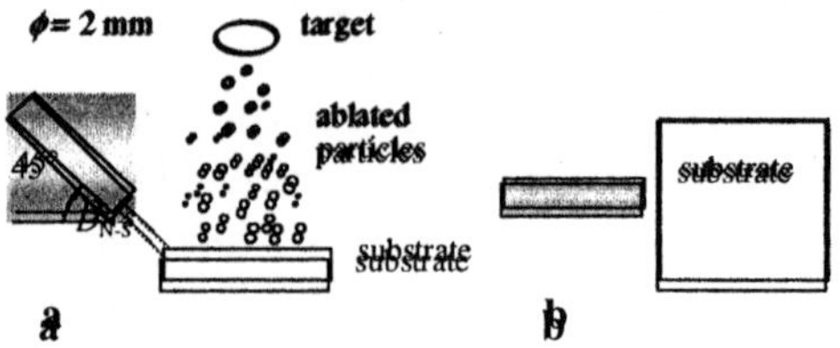

Figure 6. Schematic illustration of the NGA-PLD system for pumping O_2 gas onto the film surface: (a) side view; (b) top view.

4.1.1 DSMC simulation

To design a NGA-PLD system, it is necessary to consider where to set the nozzle and how to pump the gas. Before placing a gas nozzle inside the chamber, we have estimated the gas density profile of the ejected gas using the Direct Simulation Monte Carlo (DSMC) method (Usami and Teshima, 1999). Here, supersonic jets through a thin orifice are assumed. The molecules are a VHS (variable hard sphere) for a monatomic gas of Ar; however, we believe the results are still effective, because calculated results are not strongly influenced unless one assumes a mixed gas. Plus, the atomic weight of O_2 and Ar are close. Fig. 7 (a) shows the simulated supersonic expansion of room temperature Ar gas from an orifice, where the original point corresponds to the center of the orifice; z-axis and contour indicate the pressure ratio of upstream and downstream; and the represented contour interval is 0.005. The calculation is performed under the following conditions: stagnation pressure at the orifice P_1 = 1 mbar; the background pressure P_0 = 0.01 mbar; the orifice diameter ϕ = 2 mm; and the Knudsen number K_n = 1/30. K_n is obtained by dividing the mean free path by the orifice diameter. *Kn* is calculated to be ~ 1/30 by taking account of the mean free path of the 1 mbar Ar gas at room temperature. From Fig. 7 (a), we can estimate that the gas density in the upstream drops suddenly as the jet expands and a sufficient and uniform pressure distribution is achieved when the nozzle (ϕ = 2 mm) is set 5 mm distant from the substrate surface, provided that the background pressure is maintained at 0.01 mbar.

4.1.2 Effect of nozzle-gas on the film oxidation

Shown in Fig. 7 (b) is the dependence of applied conditions of the nozzle-to-substrate distance D_{N-S} and the mass flow rate. The middle of the left edge in the substrates corresponds to D_{N-S} = 0 mm. Thin-films grown under different D_{N-S} of 0 and 5 mm are shown in Fig. 7 (b) i) and ii) at the fixed mass flow rate of 200 sccm. As can be clearly seen in Fig. 7 (b) i), the film shows interference pattern and color non-uniformity, in the case where the nozzle is very close to the substrate. The interference pattern is caused by the gradient of the film thickness. As is shown in Fig. 7 (a), the gas density in the upstream is extremely high. Therefore, deposition rate around the orifice may be distorted by the ejected species. The non-uniformity in color is likely due to the rearrangement of the ejected gas pressure distribution affected by the scatterings against the substrate surface and the ablation plume created by successive laser pulses, which are not assumed in the simulation. By taking account of them, the $D_{N\text{-}S}$ applied in this study is determined to be 5 mm. The thin-films grown under different mass flow rates of 0, 50, and 200 sccm at the fixed nozzle separation of $D_{N\text{-}S}$ = 5 mm are shown in Fig. 7 (b) iii), iv), and ii). Here, 0 sccm is equivalent to C-PLD system. The thin-film is colored without nozzle-gas-assist, as shown in Fig. 7 (b) iii). By increasing the mass flow rate, the film colors changed from dark brown to transparent. The colored film is due to the oxygen deficiency, because the transparent film is synthesized with increasing the O_2 mass flow rate. The film coloring caused by oxygen deficiency was also noticed by Belloto, et al., (Bellotto, et al., 1993). As can clearly be seen, its oxygen deficiency fully disappeared in Fig. 7 (b) ii), i.e. mass flow rate is 200 sccm, as compared to the thin-films deposited with C-PLD technique. The ablated materials diffuse with repetitive collisions and make an elliptical cloud. It has the lowest collision probability onto the vertical direction to the laser spot, where the reactions with ambient gas are most unlikely to occur. Therefore, it is reasonable that the film has the color gradation from the center to the edge, as shown in Fig. 7 (b) iv).

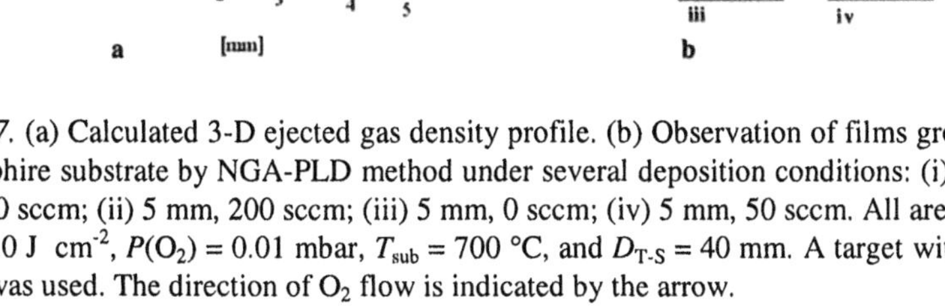

Figure 7. (a) Calculated 3-D ejected gas density profile. (b) Observation of films grown on r-cut sapphire substrate by NGA-PLD method under several deposition conditions: (i) $D_{N\text{-}S}$ = 0 mm, 200 sccm; (ii) 5 mm, 200 sccm; (iii) 5 mm, 0 sccm; (iv) 5 mm, 50 sccm. All are grown at E = ~ 1.0 J cm^{-2}, $P(O_2)$ = 0.01 mbar, T_{sub} = 700 °C, and $D_{T\text{-}S}$ = 40 mm. A target with N_K/N_W = 1.20 was used. The direction of O_2 flow is indicated by the arrow.

4.2 Characterization of the film properties

At the optimized experimental conditions: $P(O_2)$ = 0.01; T_{sub} = 700 °C; mbar; $D_{N\text{-}S}$ = 5 mm; and mass flow rate of 200 sccm, Nd:KGW films on r-cut sapphire and CeO_2/Si substrates are obtainable. By using this technique, the optical waveguide loss was reduced to 2.70 dB cm^{-1}. We note that if the film is deposited at lower oxygen pressures and the oxygen deficient phase is predominant, the thin-film is not crystallized (Collongues, 1993). Indeed, the colored film shown in Fig. 7 (b) iii) does not indicate any XRD peaks.

The film surface morphology of the Nd:KGW film is compared to the one made by the C-PLD method. The C-PLD is performed using a N_K/N_W = 0.94 target at E = ~ 1.0 J / cm^2, $P(O_2)$ = 0.05 mbar, and T_{sub} = 700 °C. The KGW diffraction peaks are confirmed by the XRD. Fig. 8 shows the SEM images of the thin-films grown by the NGA-PLD and C-PLD. As can be seen in Fig. 8, the surface morphology is considerably improved by use of the NGA-PLD method. The film surface fabricated by the C-PLD method consists of many grains. The mechanism for the grain formation is described as follows. After initial free expansion from the target surface, the mean free path of the ablated particles is reduced in the presence of a buffer gas. More specifically, at higher ambient pressures, the more collisions and scatterings occur. Then the particles lose energy to the level adequate for forming ionic complexes or molecules (Martin, et al., 1997). If these clusters reach the substrate surface, small grains start to grow as they become the nucleus. On the other hand, most of the ablated particles can reach the substrate in the state near the single atoms if the ambient pressure is extremely low. However, the improvement of the film morphology is too drastic to explain only by the reduction of mean free path. In the pressure range applied in this study (between 0.01 and 0.05 mbar), these two phenomena may occur simultaneously. Further explanation may be made by taking into consideration the dynamics of laser ablation in the diluted gas.

In the presence of a buffer gas and after the initial free expansion region, it is suggested that the plasma expansion occurs in the two consecutive stages: (i) Desorption and initial expansion when the ablated particle collisions among themselves are appreciable and the influence of the ambient gas can be neglected; and (ii) The expansion of the gas cloud in the ambient gas when the collisions between the ablated particles and ambient gas are significant. Dyer, et al., observed the oxide formations in the boundary of these two stages (Dyer, et al., 1990). They estimated the plume length of the first stage L_p within the model of adiabatic expansion as:

$$L_p = A\,[(\gamma - 1)\,E]^{1/3\gamma}\,P^{-1/3}\,V^{(\gamma-1)3\gamma}, \tag{1}$$

where A is a geometrical factor related to the shape of the laser spot at the target surface; γ is the ratio of specific heats (C_P/C_v); E is the laser energy per pulse; P is the gas pressure; and V is the initial volume of the plasma ($V = v_0 \times \tau \times$ spot size, v_0 being the initial species velocity, and τ the laser pulse duration). Taking into account the experimental parameters used in this work, $A = 1.6$, $\tau = 27$ ns, and a spot size of approximately 3 mm^2, together with the expansion velocities of the ejected species measured by Gonzalo, et al., ($v_0 = 0.8 \times 10^4$ m/s) (Gonzalo, et al., 1997) and a value of $\gamma = 1.3$, the above formula leads to plume length L_p. Here, the L_p indicates the distance that the ablated particles can fly before oxide forms. From this formula, L_p becomes ~ 40 mm at $P(O_2)$ = 0.01 mbar, corresponding to the target-to-substrate distance $D_{T\text{-}S}$, while it is much shorter than $D_{T\text{-}S}$ at $P(O_2)$ = 0.05 mbar. In other words, the ablated particles reach the substrate after the formation of many oxide clusters in the latter case. It is then concluded that the derivation of the many grains on the film surface is due to the oxide clusters generated in the ablation plume and their growth on the surface.

a b

Figure 8. SEM images of Nd:KGW films grown on sapphire substrates: (a) deposited by NGA-PLD at $P(O_2)$ = 0.01 mbar; (b) deposited by C-PLD at $P(O_2)$ = 0.05 mbar.

5. SUMMARY

We presented the growth of Nd:KGW waveguides on Si substrates. The waveguide structure was realized by the CeO_2 buffer layer, and the as-grown Nd:KGW films were optically active. We also demonstrated the nozzle-gas-assisted PLD (NGA-PLD) method for efficient film oxidation. These results are applicable to the general problems of PLD, and are of great interest.

6. REFERENCES

Anderson, A. A., Bonner, C. L., Shepherd, D. P., et al., 1997, Low loss (0.5 dB/cm) $Nd:Gd_3Ga_5O_{12}$ waveguide layers grown by pulsed laser deposition, *Opt.Commun.* **144**: 183-86.

Anderson, A. A., Eason, R. W., Hickey, L. M., et al., 1997, Ti:sapphire planar waveguide laser grown by pulsed laser deposition, *Opt. Lett.* **22**: 1556-58.

Atanasov, P. A., Okato, T., Tomov, R. I., Obara, M., 2004, (110)Nd:KGW waveguide films grown on CeO_2/Si substrates by pulsed laser deposition, *Thin Solid Film.* **453/454**: 150-53.

Atanasov, P. A., Tomov, R. I., Perriere, J., et al., 2000, Growth of Nd:potassium gadolinium tungstate thin-film waveguides by pulsed laser deposition, *Appl. Phys. Lett.* **76**: 2490-92.

Bellotto, M., Caridi, A., Cereda, E., et al., 1993, Influence of the oxygen stoichiometry on the structural and optical properties of reactively evaporated ZrO_x films, *Appl. Phys. Lett.* **63**: 2056-58.

Chu, Y.-H., Liang, C.-W., Lin, S.-J., et al., 2004, Low-temperature deposition of $Pb(Zr,Ti)O_3$ thin-films on Si substrates using $Ba(Mg_{1/3}Ta_{2/3})O_3$ as buffer layer, *Jpn. J. Appl. Phys.* **43**: 5409-13.

Collongues, R., 1993, Nonstoichiometry in ternary oxides and materials science, *Jpn J. Appl. Phys. Suppl.* **32**: 442-47.

Daran, E., Shepherd, D. P., Bhutta, T., Serrano, C. E., 1999, Laser operation of $Nd:LaF_3$ thin-film grown by molecular beam epitaxy, *El. Lett.* **35**: 398-400.

Dikovska, A. Og., Okato, T., Atanasov, P. A., Obara, M., 2004, Cathodoluminscent properties of pulsed laser deposited Er,Yb co-doped Y_2O_3 thin-films, *J. Phys. D: Appl. Phys.* **37**: L41-44.

Dyer, P.E., Issa, A., Key, P. H., 1990, An investigation of laser ablation and deposition of Y-Ba-Cu-O in an oxygen environment, *Appl. Surf. Sci.* **46**, 89-95.

Ezaki, M., Kumagai, H., Kobayashi, K., et al., 1995, Crystal growth of Nd:YAG laser films on various substrates by pulsed laser deposition, *Jpn. J. Appl. Phys.* **34**: 6838-41.

Field, S. J., Hanna, D. C., Large, A. C., et al., 1992, Ion-implanted Nd:GGG channel waveguide laser, *Opt. Lett.* **17**: 52-54.

Findeisen, J., Eichler, H. J., Kaminski, A. A., 1999, Efficient Picosecond $PbWO_4$ And Two-wavelength $KGd(WO_4)_2$ Raman Lasers In The IR And Visible, *IEEE J. Quant. Electron.* **35**: 173-78.

Florea, C., Winick, K. A., 1999, Ytterbium-doped glass waveguide laser fabricated by ion exchange, *J. Lightwave Tech.* **17**: 1593-1601.

Fukaya, S., Hasegawa, T., Ishida, Y., et al., 2001, Fabrication of Nd^{3+}, Cr^{4-} co-doped $Gd_3Ga_5O_{12}$ thin-film waveguide by two-target pulsed laser deposition, *Appl. Sur. Sci.* **177**: 147-51.

Gaidi, M., Amassian, A., Chaker, M., et al., 2004, Pulsed laser deposition of PLZT films: structural and optical characterization, *Appl. Surf. Sci.* **226**: 347-54.

Gawith, C. B. E., Bhutta, T., Shepherd, D. P., et al., 1999, Buried laser waveguides in neodymium-doped BK-7 by K^+-Na^+ ion-exchange across a direct-bonded interface, *Appl. Phys. Lett.* **75**: 3757-59.

Giacomol, A. De, Pascale, O. De, 2004, The effect of oxygen rf discharge on pulsed laser deposition of oxide films, *Appl. Phys. A* **79**: 1405-07.

Gonzalo, J., Afonso, C. N., Ballesteros, J. M., et al., 1997, Li deficiencies in $LiNbO_3$ films prepared by pulsed laser deposition in a buffer gas, *J. Appl. Phys.* **82**: 3129-33.

Grivas, C., May-Smith, T. C., Shepherd, D. P., et al., 2004, Broadband single-transverse-mode fluorescence sources based on ribs fabricated in pulsed laser deposited Ti: sapphire waveguides, *Appl. Phys. A* **79**: 1195-98.

May-Smith, T. C., Grivas C, Shepherd, D. P., et al., 2004, Thick film growth of high optical quality low loss (0.1 dB cm^{-1}) $Nd:Gd_3Ga_5O_{12}$ on $Y_3Al_5O_{12}$ by pulsed laser deposition, *Appl. Surf. Sci.* **223**: 361-71.

Guldberg-Kjaer, S., Hubner, J., Kristensen, M., et al., 1999, Planar waveguide laser in Er/Al-doped germanosilicate, *El. Lett.* **35**: 302-03.

Hartridge, A., Krishna, M. G., Bhattachrya, A. K., 1998, Structure and optical properties of nanocrystalline yttria doped ceria thin-films, *Inter. J. Mod. Phys. B* **12**: 1573-83.

Hayashi, H., Kanoh, M., Quan, C. J., et al., 2000, Thermal expansion of Gd-doped ceria and reduced ceria, *Solid State Ionics* **132**: 227-33.

Ito, A., Machida, A., Obara, M., 1997, Cobalt doping in $BaTiO_3$ thin-films by two-target pulsed KrF laser ablation with in situ laser annealing, *Appl. Phys. Lett.* **70**: 3338-40.

Jelinek, M., 2003, Progress in optical waveguiding thin-films, *Czech. J. Phys.* **53**: 365-77.

Jelínek, M., Lančok, J., Pavelka, M., et al., 2002, Optical and waveguiding properties of Nd:KGW films grown by pulsed laser deposition, *Appl. Phys. A* **74**: 481-85.

Lancok, J., Garapon, C., Martinet, C., et al., 2004, Influence of the PLD parameters on the crystalline phases and fluorescence of $Eu:Y_2O_3$ planar waveguides, *Appl. Phys. A* **79**: 1263-65.

Martin, M. J., Alfonso, J. F., Mendiola, J., Zaldo, C., 1997, Pulsed laser deposition of $KNbO_3$ thin-films, *J. Mater. Res.* **12**: 2699-706.

Mochalov, I. V., 1995, Laser and nonlinear properties of the potassium gadolinium tungstate laser crystal $KGd(WO_4)_2:Nd^{3+}$-(KGW:Nd), *Opt. Eng.* **36**: 1660-69.

Nakata, Y., Gunji, S., Okada, T., Maeda, M., 2004, Fabrication of $LiNbO_3$ thin-films by pulsed laser deposition and investigation of nonlinear properties, *Appl. Phys. A* **79**: 1279-82.

Oi, K., Obara, M., Sumiyoshi, T., 2003, Femtosecond laser interferometric processing of Nd:GGG planar optical waveguide, *Proc. of SPIE* **4977**: 381-85.

Okato, T., Atanasov, P. A., Obara, M., 2003a, Pulsed-laser deposition of a $Nd:KGd(WO_4)_2$ waveguide in Ar andO_2 environments, *Appl. Phys. A* **77**: 395-98.

Okato, T., Atanasov, P. A., Tomov, R. I., Obara, M., 2003b, Fabrication of Nd:KGW film on a Si substrate with a CeO_2 buffer layer, *Appl. Phys. A* **77**: 775-78.

Roemer, A. Millon, E., Vincent, B., et al., 2004, Epitaxial $PbTiO_3$ thin-films grown on (100) MgO by pulsed-laser deposition for optical waveguiding properties, *J. Appl. Phys.* **95**: 3041-47.

Shimoda, T., Ishida, Y., Adachi, K., Obara, M., 2001, Fabrication of highly ytterbium (Yb^{3+})-doped YAG thin-film by pulsed laser deposition, Opt.Commu. **194**: 175-79.

Strikovsky, M. D., Klyuenkov, E. B., Gaponov, S. V., et al., 1993, Crossed fluxes technique for pulsed laser deposition of smooth $YBa_2Cu_3O_{7-x}$ films and multilayers, *Appl. Phys. Lett.* **63**: 1146-48.

Suche, H., Oesselke, T., Pandavenes, J., et al., 1998, Efficient Q-switched $Ti:Er:LiNbO_3$ waveguide laser, *El. Lett.* **34**: 1228-30.

Usami, M., Teshima, K., 1999, Molecular simulation of rarefied supersonic free jets by DSMC method, *JSME Inter. J. Series B Fluids and Thermal Eng.* **42**: 369-76.

Wu, X. D., Dye, R. C., Muenchausen, R. E., et al., 1991, Epitaxial CeO_2 films as buffer layers for high-temperature superconducting thin-films, *Appl. Phys. Lett.* **58**: 2165-67.

Chapter 13

PULSED LASER DEPOSITION FOR FUNCTIONAL OPTICAL FILMS

C.N. Afonso, J. Gonzalo, R. Serna, J. Solís
Instituto de Optica, CSIC, Serrano 121, 28006 Madrid, Spain

1. MAIN FEATURES OF PULSED LASER DEPOSITION

The widespread use of laser ablation for thin-film deposition, referred to from now on as pulsed laser deposition (PLD), has experienced an enormous growth in the 90's. The reason for this growth is twofold. On the one hand, there have been many works reporting fundamental studies that have contributed to understanding its physical basis and optimising the deposition process [Chrisey and Hubler, 1994; Miller and Haglund, 1998]. On the other hand, films of materials for which more standard techniques have shown limited success have successfully been produced by PLD and some optical devices based on heterostructures produced by PLD have been proposed in the literature [Yilmaz, et al., 1994; Wu, et al., 1998; Koinuma, et al., 2000]. Today, commercial systems are available and several device prototypes using layers fabricated by PLD have been demonstrated.

From the conceptual point of view, the PLD technique is very simple and only requires a pulsed laser, preferentially delivering nanosecond pulses or shorter in the UV, a target with no electrical, geometrical or thermal constraints, and a substrate. The basic setup of the conventional configuration has been sketched in Fig. 1a. The laser beam is focused to the target typically at an angle of incidence of 45° with respect to its perpendicular direction. The ablation of the target produces a visible plasma (commonly known as "plume") that expands from the target. In order to avoid its fast deterioration and keep its surface under approximately constant conditions, the target is typically rotated. The substrate is located in front of the target at a distance in the range 30-100 mm. Both the substrate and the target are usually placed inside a chamber that can be evacuated. One of the advantages of PLD is that the process can be performed at any pressure ranging from ultra high vacuum to high pressures (typically up to 10 mbar) and that any kind of atmosphere either inert or reactive can be used.

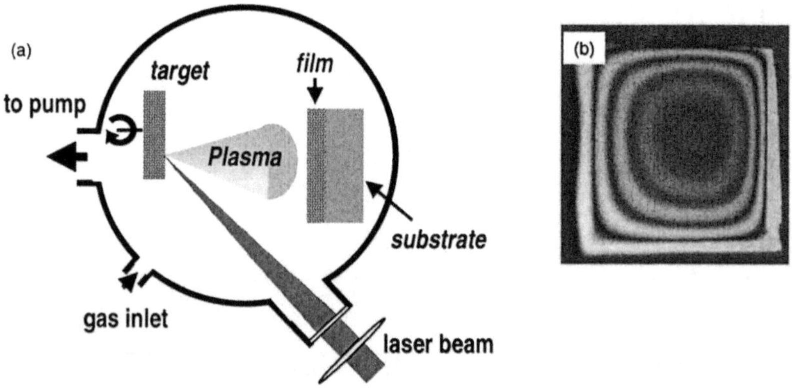

Figure 1. (a) Schematics of a standard PLD experimental setup. (b) Optical photograph of a ~ 1 cm^2 $LiNbO_3$ film deposited by PLD on Si. The fringes are associated to the material thickness distribution.

The nature of the target in itself can have an important impact on the final quality of the films. One of the main drawbacks of the PLD technique addressed in the literature is the generation of particulates of sizes up to a few μm that are deposited on the growing film, leading to rough and thus low quality films [Chen, 1994; Willmott and Huber, 2000]. One way to overcome this problem is through the use of very dense or single crystalline materials as targets, together with selection of appropriate experimental conditions.

The plasma is forward directed over a relatively small angular distribution and thus, film thickness is not homogeneous as shown in Fig.1b, where a photograph of a transparent oxide film grown on a Si substrate is shown. A series of dark and bright regions are observed that relate to interference fringes produced as a consequence of the decrease of thickness from the center to the edges. Only a reduced central region of the film has homogeneous thickness. To increase its size, one possible alternative would be to expand the beam at the target site. However, this solution would require pulse energies not available in most of the commonly used lasers. Solutions widely used to overcome this problem are either to rotate the substrate or to use complex deposition geometries. Homogenous films up to

3 inches in diameter have been reported [Prins, et al., 1996; Pignolet, et al., 1998].

PLD has several characteristics of interest for the production of functional optical thin-films. Table 1 shows average values for the kinetic energy of the species and the deposition rate in PLD, and in two standard physical vapour deposition processes: thermal evaporation and sputtering. The species involved in PLD are the ones having the highest kinetic energy, the values being typically one order of magnitude higher than in the case of sputtering. This feature is essential for achieving films with very high density and good adherence to the substrate, both features being required for high performance coatings [Voevodin, et al., 1997; Lackner, et al., 2004]. Table 1 also shows that the time averaged deposition rate in PLD is small and close to the smaller values that are achieved in thermal evaporation. If the amount of material deposited per pulse (low energy density regime) is low enough ($\leq 10^{13}$ at. cm^{-2}), an excellent control of the growth process at submonolayer level can be achieved [Blank, et al., 2000; Koinuma and Takeuchi, 2004]. However, PLD is in itself a pulsed process since most deposition takes place over several tens of microseconds after each nanosecond laser pulse. This leads to the "instantaneous" high deposition rate per pulse included in brackets in Table 1. This is a conservative minimum value correlated to the rate of energy transfer from the gas phase to the substrate that eventually determines the quenching rate at the substrate. This high deposition rate or arrival flux of species to the substrate is responsible for favouring the production of metastable phases or materials [Krebs, et al., 1996; Gonzalo, et al., 2003].

Table 1. Kinetic energy of species and typical average deposition rates for PLD and two standard film deposition techniques. The "instantaneous" or per pulse deposition rate is also included within brackets for the case of PLD.

Deposition technique	**Kinetic energy (eV)**	**Deposition rate ($nm.s^{-1}$)**
Pulsed laser deposition	10-100	0.1// (>10^4)
Thermal evaporation	< 1	0.01-0.1
Sputtering	5-10	0.1-1.0

Another advantage of PLD is its flexibility to use several targets in a single film growth process. The laser beam can easily be focused alternatively to two or more targets, leading to a sequential ablation process. The firing of the laser and the movement of the targets can then be controlled in order to define a tailored film profile in depth. This configuration referred to from now on as alternate PLD, is ideal for producing complex structures such as multilayers [Koinuma and Takeuchi, 2004] or nanostructured [Afonso, et al., 2006] films. An example of the latter is shown in Fig. 2 that includes both a Rutherford Backscattering spectrum and a cross-section transmission electron microscopy (TEM) image of a nanocomposite film containing 10 layers of Cu nanoparticles (NPs) embedded in amorphous Al_2O_3 [Serna, et al., 2000]. The films have been produced by alternate PLD of the metal and the matrix targets. Whereas the former evidences the presence of Cu rich layers, the latter shows that the layers are indeed formed by NPs that are of approximately 3 nm in diameter and are separated by amorphous Al_2O_3. The deposition sequence was designed in order to have the NPs layers equally spaced with a separation of 20 nm. Fig. 2 evidences the excellent control that PLD offers for organising nanostructures in depth with nanometer resolution.

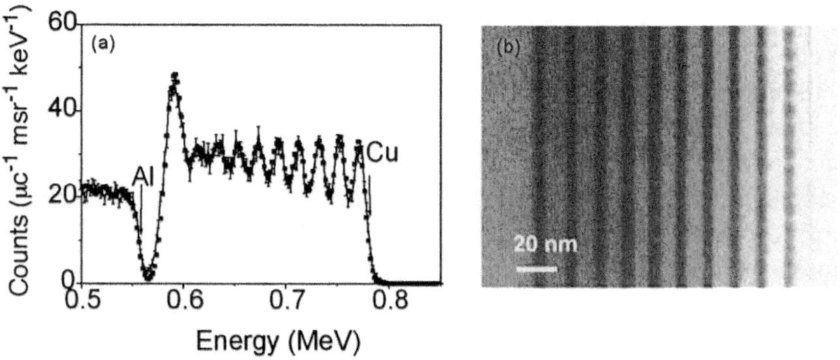

Figure 2. (a) High energy region of the Rutherford Backscattering spectrum (Cu profile) of a nanocomposite film containing 10 layers of Cu NPs. (b) Cross-section TEM image corresponding to the same film in (a), where dark and bright areas correspond respectively to the Cu NPs and the amorphous Al_2O_3 host. [Adapted from Serna, et al., 2000].

2. CRITICAL DEPOSITION PARAMETERS FOR THE PRODUCTION OF OXIDES

PLD of complex oxides usually requires the presence of a gas background in order to reproduce the oxygen stoichiometry of the target in the films. Although the use of gases such as NO_2, N_2O, or Ar has been reported to lead to good composition and crystallinity, O_2 is the most widely used gas [Gupta, 1994; Blank, et al., 2000]. The presence of a gas background and its possible reactive character influences the kinetic energy, the spatial distribution and even the nature of the species present in the plasma. It thus determines not only the plasma expansion dynamics but also the morphology, composition, and structure of the films [Geohegan, 1994; Saenger, 1994; Gonzalo, et al., 1996]. Related experimental parameters that also have a relevant role in the production of oxides by PLD are the laser energy density that modifies the kinetic energy and density of the ejected species, and thus influences the expansion dynamics, as well as the target-substrate distance since the plasma features drastically change with distance in the presence of a gas background [Xu, et al., 1994; Gonzalo, et al., 1996; Gottmann and Kreutz, 2000].

Shortly after the laser pulse, the visible plasma is limited to a region close to the target. Subsequently, the plasma expands increasing its size and decreasing its density while the expansion velocity remains constant. The basic features of the expansion in vacuum have successfully been described by theoretical models that generally consider that the plasma expands adiabatically [Singh and Narayan, 1990; Anisimov, et al., 1993]. Typical plasma expansion velocities are 10^5-10^6 cm s^{-1} that lead to the characteristic kinetic energies shown in Table 1.

The presence of a gas background has a strong influence on the plasma expansion dynamics for gas pressures higher than 10^{-2} mbar, the main difference being its confinement. If the gas pressure is high enough, the plasma remains confined to a region close to the target surface. Common models used in the literature to describe the plasma expansion in the presence of gases have been the *"drag force"* and the *"shockwave"* models for pressures respectively below and above a few mbar [Dyer, et al., 1990; Geohegan, 1994].

Qualitatively, the plasma expansion in a gas background can be described in three consecutive steps. Initially, the density of the plasma is much higher than that of the gas leading to initial velocities similar to the ones when expanding in vacuum. As the plasma expands, the mass of gas displaced increases and the plasma slows down its velocity. Once the latter have lost most of its kinetic energy by collisions with the gas, the species moves by diffusion into the background gas. The distance from the target at which this happens (L), referred to in the literature as *plasma length*, *plasma range,* or *stopping distance*, among other terms [Dyer, et al., 1990; Geohegan, 1994; Strikovski and Miller, 1998; Amoruso, et al., 2003], is an important parameter when growing oxide films. It has experimentally been observed that the composition and structure of the films depend on the target-substrate distance, the best quality films being produced when placing the substrate at distances $D \approx L$ from the target. This is illustrated in Fig. 3, which shows the XRD spectra corresponding to two $LiNbO_3$ films grown at target-substrate distances either $D > L$ or $D \approx L$ [Chaos, et al., 2001]. Whereas the former film exhibits several diffraction peaks indicating it is polycrystalline, the latter shows only one peak evidencing its textured growth. There have been different approaches to estimating the *plasma length* by considering it is the distance from the target at which the rate of thermal decomposition of the ensemble of species dominates over its forward movement [Strikovski and Miller, 1998], the plasma pressure equals the gas pressure [Dyer, et al., 1990; Geohegan, 1994] or the expansion velocity reaches a certain optimum value [Kim and Kwok, 1992]. All models lead to lengths that are in the 100 mm to 10 mm range for the gas pressures typically used in PLD (10^{-2} to 10 mbar).

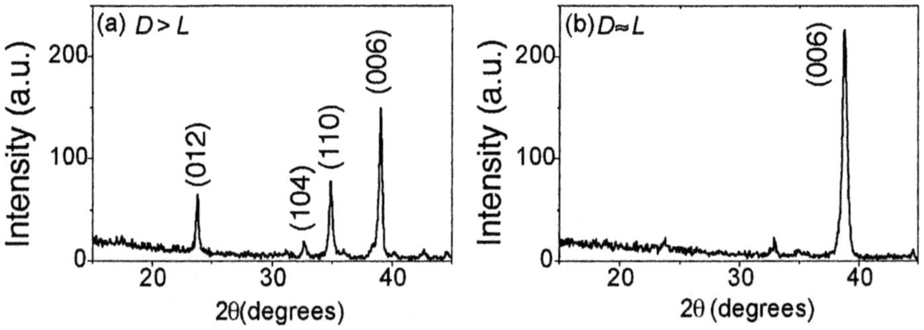

Figure 3. X-ray diffraction patterns of $LiNbO_3$ films deposited on Si at 650° C and target-substrate distances (D) of (a) 35 mm > L and (b) 23 mm ≈ L. The plasma length (L) has been estimated to be L = 27±5 mm according to [Dyer, et al., 1990]. The rest of the deposition parameters (Pressure = 1 mbar of O_2, Laser energy density = 1.2 J cm^2 and cooling down gas environment = 100 mbar of O_2) were kept constant. All the peaks correspond to $LiNbO_3$ (Adapted from Chaos, et al., 2001).

An additional issue to be considered when producing films in a gas background is the modification of the angular distribution of ejected species; it broadens and there is a preferential dispersion of the lighter elements with respect to the heavier ones. This effect becomes significantly important for gas pressures > 10^{-2} mbar at the usual laser energy density values (a few J cm^{-2}) [Lichtenwalner, et al., 1993; Gonzalo, et al., 1996]. This effect has been reported for the case of complex oxides composed of cations having very different masses as it is the case of high Tc superconductors or heavy metal oxide glasses [Saenger, 1994; Gonzalo, et al., 1996; Sanz, et al., 2004]. Under these conditions, films produced using the standard PLD configuration, shown in Fig. 1a tends to be enriched in the heavier elements. However, this situation can be reversed since the opposite effect (plasma narrowing) has been reported to occur when using laser energy densities well above the ablation threshold or very high pressures [Saenger, 1994; Lichtenwalner, et al., 1993].

Finally, if the gas background is reactive, chemical reactions between the ejected species and the gas species, or surface oxidation, may enhance the incorporation of oxygen to the films. There is broad experimental evidence for the former processes when they are energetically favorable (exothermic) such as the formation of YO or BaO when ablating $Y_1Ba_2Cu_3O_{7-x}$ in O_2. However, evidences have been scarce and restricted to a limited range of gas pressures when these reactions are endothermic [Gupta, 1994].

The laser energy density (*E*) has a strong influence on the initial density and kinetic energy of the ejected species as well as on the ionization fraction of the plasma. Its effect can be related to that of the gas pressure (*P*) and the target-substrate distance (*D*) in the frame of the "*shockwave*" model if it is assumed that the optimum growth conditions can be expressed in terms of an optimum arrival speed of the species to the substrate. Then, it is possible to correlate these parameters in the form [Kim and Kwok, 1992; Xu, et al., 1994]:

$$\frac{E^{\frac{1}{5n}}}{P^{\frac{1}{5n}} D^{\frac{1-n}{n}}} = C \quad , \tag{1}$$

where C is a constant and n depends on the symmetry of the shockwave front. According to this expression, an increase of the gas pressure requires an increase of either the laser energy density or a reduction of the target-substrate distance in order to achieve the optimum growth conditions. A similar reasoning can be made when modifying the laser energy density. This qualitative reasoning has been proven experimentally in different materials such as $LiNbO_3$ or high Tc superconductors, among others [Kim and Kwok, 1992; Xu, et al., 1994; Chaos, et al., 2001].

The impact that the kinetic energy of the ejected species has on the nucleation and growth of oxide films for optical applications depends on its actual values [Kools, et al., 1994; Willmott and Huber, 2000]. For kinetic energies lower than the displacement threshold energy or the bond strength ($\leq$ 30-50 eV), the arrival of the species has no significant impact on the surface quality of the growing film. However, these values are high enough to enhance the surface mobility and reactivity at the substrate, features that are essential to promote the growth of films with preferential crystalline orientation or with improved optical density [Afonso, 1995; Gottmann and Kreutz, 2000]. For kinetic energies above these values, phenomena such as implantation or surface sputtering may appear. Implantation phenomena in PLD have unambiguously been evidenced when depositing high mass species such as Bi [Barnes, et al., 2003] or Au [Gonzalo, et al., 2005] on Al_2O_3 substrates. Resputtering of a fraction of the material already deposited induced by the bombardment of highly energetic incoming species has also been reported in the literature for the case of metals [Fähler, et al., 1999] and complex oxides [Frantti and Lantto, 1994; Hau, et al., 1995]. Both processes can have a high impact on the quality of deposited films since they can lead to damage of the film surface or its structure, or they can even alter the composition of the film due to the different sputtering yield of the elements forming the films.

3. PULSED LASER DEPOSITION OF OXIDES OF INTEREST FOR OPTICAL APPLICATIONS

Pulsed laser deposition has proven to be an excellent method for producing high quality oxide films. This is particularly true for the case of crystalline multicomponent oxides for optical applications such as $LiNbO_3$, PZT, $BaTiO_3$, and more recently to ZnO and related materials. Since these types of materials are the ones to which PLD has first been applied, there are a number of comprehensive reviews describing the key aspects and features [Chrisey and Hubler, 1994; Afonso, 1995; Willmott and Hubler, 2000; Ohtomo and Tsukazaki, 2005]. In the limited space of this chapter, we will thus focus on other materials for which PLD is emerging as a promising production route, namely glassy films and films structured in the nanometer scale. These films can be produced at room temperature and in a single step process, features that are not easily achievable by direct competing technologies. Although this selection has been made on the basis of their interest for optical applications, some of the materials or approaches referred to also have a high potential for other fields of applications.

Glasses are very attractive for photonics applications such as ultralow-loss optical waveguides and efficient gain or nonlinear optical devices. In spite of the success of PLD in producing complex oxide films, it has seldom been applied to the production of glassy oxide films. The reasons for this limited use of PLD are most likely related to difficulties in producing films with good surface quality and to the fact that more conventional deposition procedures such as sol-gel, plasma or laser enhanced CVD, or sputtering lead, in general, to satisfactory results [La Serra, et al., 1990; Mazzoldi and Righini, 1995; Ay, et al., 2003]. However, the inherent capabilities of PLD for producing conditions far from thermodynamical equilibrium, the ability to reproduce the target stoichiometry, and finally, the possibility of designing the films at the nanoscale while having complex oxides as host, are additional features that make PLD an attractive technique for the synthesis of oxide glasses.

Nanostructured materials are to be understood as materials artificially designed in the nanometer scale to achieve certain physical properties. In this section, we will focus on the modification of the optical properties of dielectric (oxide) thin-films by embedding either metal NPs or rare-earth (RE) ions, an approach that has sometimes been referred to as *"optical"* doping. Metal NPs embedded in dielectric hosts are interesting for several applications ranging from passive optical elements such as polarizers and filters, to active functions based on their nonlinear optical properties such as ultrafast all-optical switches or optical power limiters [Kreibig and Vollmer, 1995; Hirao, et al., 2001]. The interest on RE doping of dielectrics has been developed as a result of its success for the production of lasers (i.e. Nd:YAG) or amplifiers (i.e. the Er-doped fiber amplifier in long distance fiber communications). Integrated gain devices are foreseen as key components for the development of integrated photonic circuits [Miniscalco, 1993; Polman, 1997; Hirao, et al., 2001].

3.1 Glassy films

After the pioneering work of Vogel and coworkers on PLD of gallate and titanium-niobium silicate and borosilicate glasses, [Vogel, et al., 1989] the activity on PLD for producing glassy films has been limited to a few simple glasses such as silica [Ford, et al., 2003] and lead oxide [Luo, et al., 2004], or complex oxide glasses such as phosphate [Afonso, et al., 1996], lead-germanate [Mailis, et al., 1999], lead-niobium germanate [Gonzalo, et al., 2003], tellurite [Martino, et al., 2003] and bioactive [Serra, et al., 2004] glasses.

Surface roughness is one of the main concerns when preparing complex oxide glassy films by PLD. The appearance of particulates with sizes in the range of 0.1 μm to a few μm has often been observed when producing glassy films and is related to the laser parameters. For instance, the use of a wavelength of 193 nm has been found to lead to smooth transparent phosphate glasses, while the same glass was hardly produced by using 248 nm [Mailis, et al., 1999]. Moreover, even when the appropriate wavelength is used, the laser energy density must be carefully selected in order to achieve smooth films. Energy density values that are either very high or close to the ablation threshold lead to rough surfaces, mainly due to the generation of particulates in the ablation process. At high energy densities, subsurface boiling or recoil ejection induce the ejection of molten particulates that eventually reach the substrate. At low energy densities, the increase of the ablation time required to deposit films leads to the degradation of the target surface and favors the ejection of solid particulates [Chen, 1994; Willmott and Huber, 2000]. These results suggest that intermediate energy densities have to be used in order to achieve smooth films [Serra, et al., 2004].

Another important issue is the transparency of the films in the wavelength range of interest for optical applications (visible and near infrared). Oxygen is the main constituent of glasses and can enter in the glass network either as a bridging or non-bridging ion [Lines, 1991; Clement, et al., 1995]. Oxygen deficiency modifies the glass structure and can even prevent its formation if it is too high, this generally leading to absorbing films [Mailis, et al., 1999; Sanz, et al., 2004]. Fig. 4a shows the evolution of the relative oxygen content in glassy films produced by PLD in O_2 gas from a lead niobium germanate (LNG) glass [Sanz, et al., 2004]. At low pressures, the films are clearly oxygen deficient, which is a typical result when producing films of complex oxides having cations much heavier than oxygen [Gupta, 1994; Chaos, et al., 2001]. The oxygen content remains low for pressures up to 10^{-3} mbar and it increases significantly for pressures above 10^{-2} mbar to become close to that of the target for pressures in the range 0.5-1.0 x 10^{-1} mbar. For these pressures, at least 25% of the oxygen in the film comes from the gas environment as demonstrated by comparing the oxygen content of films grown in O_2 and Ar under the same conditions.

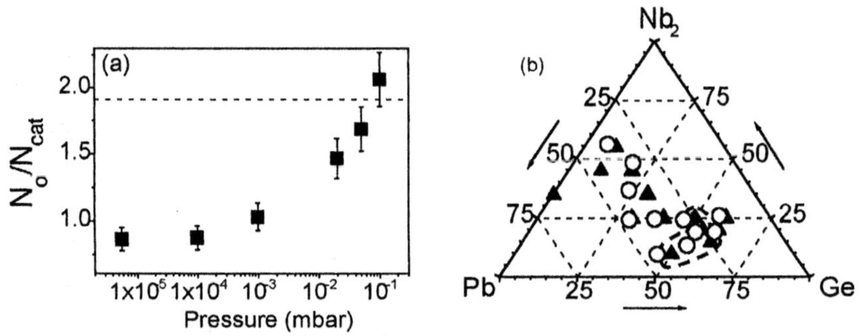

Figure 4. (a) Relative oxygen content (N_O/N_{cat}) of LNG glassy films as a function of the oxygen pressure, where N_O and N_{Cat} are respectively the number of oxygen atoms and cations. The dashed line corresponds to the relative oxygen content of the target. (b) Relative cation composition of LNG bulk targets (▲) and the corresponding films (O), produced by PLD in an O_2 pressure of 5 x10^{-2} mbar (Adapted from Gonzalo, et al., 2003; and Sanz, et al., 2004).

Although the use of an O_2 background gas is necessary for the production of glassy oxide films by PLD, it is important to bear in mind that it can modify the relative cation content. The presence of a gas in the optimum pressure range leads to an enrichment of the heavier elements with respect to the lighter ones as described in Section 2. This effect is illustrated in Fig. 4b, where the composition of both LNG targets and films produced at 5x10^{-2} mbar of O_2 is shown [Gonzalo, et al., 2003]. A general slight deficiency of Ge (lightest element) and excess of Pb (heaviest element) is observed in the films, while the Nb content remains approximately constant.

Fig. 4b also shows that only targets having compositions in the region within the dashed line are vitreous and thus transparent. Instead, all films produced are vitreous and transparent. This result further illustrates the interest of PLD for the synthesis of complex multicomponent oxide glasses as it enables the production of transparent films in an extended composition range when compared to the bulk material. This result is related to the high instantaneous deposition and quenching rates achievable by PLD as shown in Table 1 and described in Section 1, that lead to synthesis conditions far from thermodynamical equilibrium.

The refractive index of a glass is generally influenced by the bond polarizabilities of its constituents. In the case of conventional glasses, it has been observed that both the linear (*n*) and nonlinear (n_2) refractive indices increase with the concentration of non-bridging oxygen bonds that are induced by the addition of network modifiers to the glass composition [Vogel, et al., 1991]. However, this is not the case for glasses with a significant fraction of heavy metal cations with ns^2 electron pairs (Pb^{2+}, Bi^{3+}, Te^{4+}) or d^0 ions (Ti^{4+} and Nb^{5+}, among others) [Lines, 1991; Santran, et al., 2004]. In this case, the concentration of heavy metal cations, and in particular their high polarizability, is the factor determining the high values of the refractive index. This behaviour is illustrated in Fig. 5, which shows the evolution of *n* and the energy gap E_g in LNG glassy films with the [Pb]+[Nb] content [Gonzalo, et al., 2003]. It is seen that *n* increases linearly up to a maximum value that is clearly higher than that achievable for bulk transparent glasses. Instead, E_g shows a more complex behavior that is related to the different role played by Pb and Nb in the glass network.

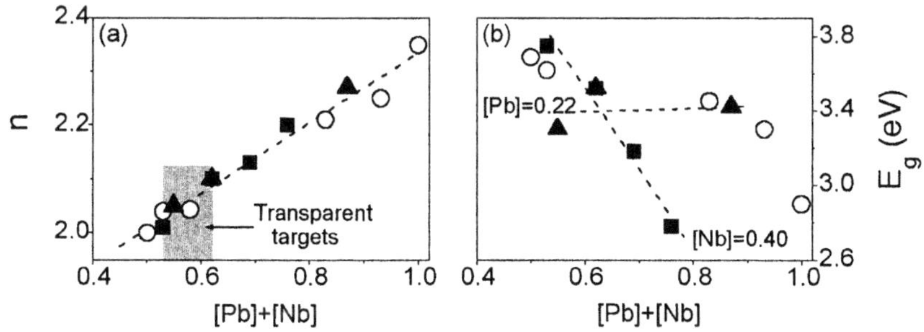

Figure 5. (a) Real part of the refractive index (n) at $\lambda = 630$ nm and (b) Optical energy gap (Eg) as a function of lead and niobium fraction ([Pb]+[Nb]) in LNG glassy films produced by PLD in an O_2 pressure of 5×10^{-2} mbar. Results corresponding to series of films with constant lead (▲) or niobium (■) contents are highlighted. The dashed lines are linear fits of the dependence of (a) n and (b) E_g on ([Pb]+[Nb]), for the series of films with constant lead or niobium contents. The composition range in which transparent bulk targets are produced is indicated in (a) as a grey rectangle (Adapted from Gonzalo, et al., 2003).

Semi-empirical models used for the development of new glass compositions of interest for nonlinear applications [Vogel, et al., 1991; Lines, 1991; Santran, et al., 2004] predict an increase of the nonlinear refractive index n_2 with the increase of n and the decrease of E_g.

This prediction has recently been confirmed for glassy films of the LNG system with large heavy metal content. Values of n_2, more than 10^3 times higher than that of SiO_2, have recently been measured [Gonzalo, et al., submitted] at $\lambda = 800$ nm.

3.2 Nanostructured films

Nanostructuring is a powerful tool to artificially design materials and is becoming a common approach in current nanotechnology. This section illustrates the use of PLD for this purpose through examples of materials *"optically"* doped either with metal NPs or RE ions. In the former case, the dimensions, shape, and distribution of the NPs are essential to achieve the desired optical response both in wavelength and intensity; the control of the RE ion concentration and distribution is essential for achieving an optimized photoluminescence intensity and lifetime in the latter one. The optimization of these features usually starts from a common approach that is based on the control of laser parameters and the deposition sequence. This is well documented for the case of films containing metal NPs through images such as that shown in Fig. 2, which illustrates the nanostructuring process and the good degree of control that can be achieved.

Most of the examples that are discussed in this section relate to films in which the host or dielectric is amorphous aluminium oxide (Al_2O_3). This host has first been chosen because high-quality low-loss waveguides of this material on silica-on-silicon substrates have been produced [van den Hoven, et al., 1993]. The high contrast between its refractive index and that of the SiO_2 cladding layer (~0.2) leads to a high confinement of the guided light, making the design of small device structures possible. An additional aspect that makes this host attractive is the fact that, unlike most oxides, the production by PLD of this type of films can be performed in vacuum, thus making the process simpler. However, the approach described in this Section can easily be extended to other dopants or hosts such as the glassy films described in the previous section or the perovskite-type oxides successfully produced by PLD. To illustrate this potential, a comparison of the results achieved in doping with Er amorphous Al_2O_3 and epitaxial $LiNbO_3$ films will finally be discussed.

3.2.1 Nanocomposite films

The control of the features of NPs embedded in a solid host has been a challenge for long time and, in fact, the lack of production methods with the required degree of control has been one of the major drawbacks for the development of technological applications based on this type of materials. A strong experimental effort has been made in recent years to develop methods with the required level of control, the most widely used ones being ion implantation and thin-film technologies, in some cases combined with lithography techniques. While the former produces or segregates the NPs into the bulk or previously deposited host, the latter are based on the production of both the host and the NPs using the same technique.

Two main approaches have been reported in the literature for producing NPs by PLD, the main difference being the place where the NPs nucleate: in the gas phase by ablating the material at high pressures (10's of mbar) of an inert gas, typically Ar [Yoshida, et al., 1996; Geohegan, et al., 1998; Marine, et al., 2000]; or at the substrate surface by taking advantage of the Vollmer-Weber island growth. Whereas the former approach has mainly been used to produce semiconductor NPs that are addressed in a different Chapter of this book, the latter has been the preferred route for producing metal ones due to the low adsorption energy of metals on most oxides and insulating substrates that favor the island growth [Henry, 1998]. In this section, we will summarize the features and possibilities of the latter approach. For a more detailed description, we refer the reader to a recent comprehensive review on the production of metal nanocomposite films by PLD [Afonso, et al., 2006].

Fig. 6 shows transmission electron microscopy (TEM) images of Ag and Cu metal NPs (dark contrast regions) produced by PLD. The NPs have a round shape for low metal contents (Fig. 6a) and become larger and irregular as the metal content increases due to coalescence and coarsening (Fig. 6b).

Similar shape and dimension evolution has been reported for NPs produced by PLD of other metals like Au, Bi, and Fe [Serna, et al., 1998a; Serna, et al., 1999; Dempsey, et al., 2001; Barnes, et al., 2002; Gonzalo, et al., 2005]. However, noble metals (Au and Ag) tend to form larger NPs than other metals since it is well-known that the higher the interface energy, the lower the nucleation probability. This effect can be appreciated by comparing the dimensions and number density of Ag and Cu NPs in Figs. 6b and 6c in specimens containing the same overall metal content.

The most widely studied optical feature of metal-dielectric nanocomposite materials is the appearance of an enhanced absorption band at the so-called surface plasmon resonance (SPR) due to *classical confinement* effects. This resonance appears whenever the condition $\varepsilon_m + 2\,\varepsilon_d$

= *0* is satisfied, where ε_m and ε_d are the real parts of the dielectric constant of the metal and the dielectric host, respectively.

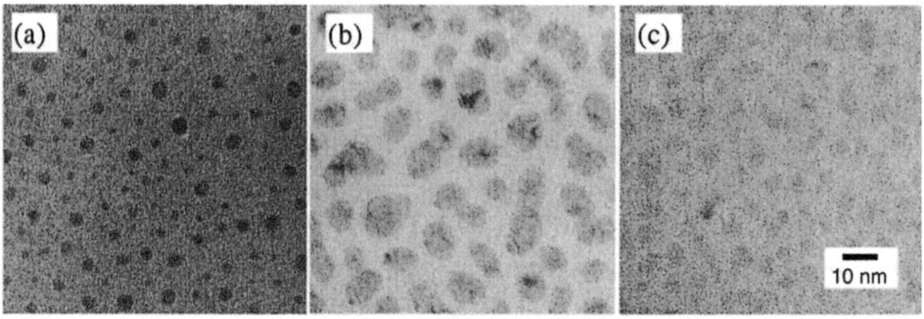

Figure 6. Plan view TEM images of nanocomposite films containing Ag or Cu NPs embedded in amorphous Al_2O_3 with increasing Ag areal density per layer (a) 3.5 x10^{15} at/cm^2, (b) 12 x 10^{15} at/cm^2, and (c) a Cu areal density of 12 x 10^{15} at/ cm^2.

Fig. 7a shows the real (*n*) and imaginary (*k*) parts of the refractive index of a nanocomposite film containing Ag NPs as a function of wavelength, together with the *n* values of a reference host film with no NPs. The corresponding curve for *k* in the latter case is not included since its absorption is negligible. For the nanocomposite film, whereas *k* shows a broad band peaking around the SPR wavelength (~400 nm), *n* instead shows anomalous dispersion behaviour in the same spectral region. It becomes clear that neither the real part nor the imaginary part of the refractive index of nanocomposite materials can be approximated by that of the host [Serna, et al., 2001a]. Nevertheless, as the real part of the refractive index has seldom been measured in nanocomposite systems, this approximation has been a common practice especially for determining the nonlinear optical properties of these materials, something that can lead to serious inaccuracies [del Coso and Solis, 2004a].

The control of the overall light transmission of metal-dielectric nanocomposite films in the neighbourhood of the SPR wavelength is an important issue, since an excessive optical density can prevent their use in some applications. A possible way to tailor its value to the application envisaged is the design of the in-depth distribution of the NPs [Suarez-Garcia, et al., 2003]. This approach has successfully been proved for the case of Cu NPs embedded in amorphous Al_2O_3. It uses two photonic units formed by three NPs layers spaced 7 nm, as building blocks for the nanocomposite film. This is illustrated in Fig. 7b, which shows a reduction of the effective extinction larger than 30 % in the neighborhood of the SPR for a film having two photonic units separated $L_2 = 154$ nm with respect to the nanocomposite film, having the Cu NPs layers homogeneously distributed in-depth.

The local field enhancement effect associated to the presence of metal NPs in a dielectric matrix is mirrored by the presence of strong third-order optical nonlinearities [Vogel, et al., 1991]. In the particular case of the nanocomposite films produced by PLD and containing Cu NPs, a nonlinear refractive index n_2 as high as 10^{-8} cm^2 W^{-1} has been measured at wavelengths close to the SPR with a buildup time of a few picoseconds [del Coso, et al., 2004b].

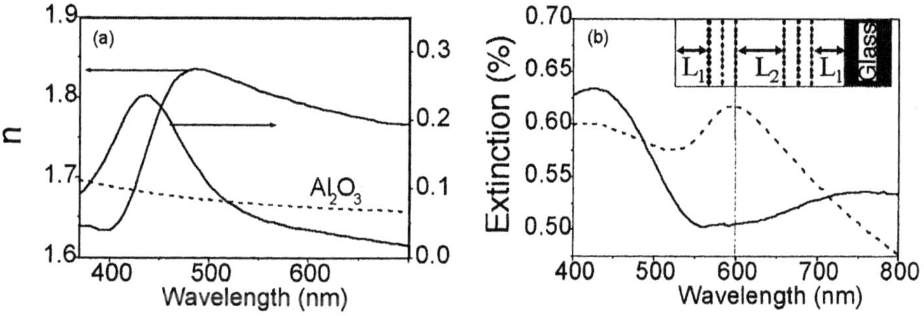

Figure 7. (a) Real (n) and imaginary (k) parts of the refractive index of a nanocomposite film containing Ag NPs embedded in amorphous Al_2O_3. The n value for a host reference film is included for comparison. (b) Extinction of films having six equally spaced Cu NPs layers (- - -) at the center of the film, and with NPs layers distributed forming a photonic structure described by the scheme shown in the inset (—) with L_1=77 nm and L_2=2L_1 (Adapted from de Sande, et al., 2002; and Suarez-Garcia, et al., 2003).

3.2.2 Rare-earth doped films

There have been a relatively large number of reports on the preparation of RE ions doped thin-films by PLD [Serna, et al., 1998b; Lanzerstorfer, et al., 1998; Komuro, et al., 2000; Bär, et al., 2003; Lancok, et al., 2004]. In these works, a single RE-doped target was used, and thus varying the dopant concentration required the use of different targets. This is the normal approach followed when using sputtering [Yan, et al., 1997; Lazarouk, et al., 1998], and no straightforward means to control the RE concentration or distribution is feasible. Alternate PLD offers the possibility to control the features of the host and the dopant independently. This section will illustrate the impact of these features on the photoluminescence (PL) performance of the films. Most of the examples will be given for Er as RE dopant and amorphous Al_2O_3 as host. To further illustrate the importance of nanostructuring and its potential for other materials, examples of co-doping with another RE (Yb) or doping with Er another host ($LiNbO_3$) will finally be given.

The characteristic spectrum of the Er doped films shows an emission band with a maximum at around 1.53 μm, corresponding to the transition between the $^4I_{13/2}$ and $^4I_{15/2}$ states of Er^{3+} that makes Er an attractive dopant for amplifying optical signals in fibers [Polman, 1997]. One of the key issues in developing materials suitable for gain devices is to achieve large RE concentrations (>10^{20} ions cm^{-3}) [Polman, 1997]. In PLD, the concentration can be varied using two approaches. The most straightforward one is by changing the number of pulses on the RE target, which is the same method used for varying the metal content per layer in nanocomposite films (Figs. 6a and b). The other approach is by varying the laser energy density on the RE target. Both approaches lead indeed to an increase of the PL intensity as the Er concentration is increased, consistently with an effective increase of the number density of active Er^{+3} ions [Serna, et al., 2001b]. The PL lifetime follows instead a different behavior depending on the approach used to increase the RE concentration as seen in Fig. 8a; it decreases when increasing the number of pulses, while it increases up to a saturation value when increasing the laser energy density. The decrease of the intensity in the former case can easily be understood in terms of concentration quenching associated to the in-plane concentration increase. Note that for the film having the largest areal density, the average in-plane distance is about 0.8 nm and thus, concentration quenching of the PL may occur most likely by excitation energy migration [Miniscalco, 1993]. When varying the laser energy density, in spite that the in-plane average nominal distance is similar (1.0 nm), the lifetime increases and reaches much higher values. The most likely explanation for this different behavior relates to the importance of low-depth implantation when increasing the laser energy density as described in Section 2. Taking into account the high mass of Er, the estimated kinetic energy of the species arriving at the substrate can be as large as 175 eV. They can thus reach a depth of 1 nm similarly to what has been observed when producing NPs of high mass metals such as Bi [Barnes, et al., 2003] or Au [Gonzalo, et al., 2005]. Er^{3+} are thus no longer exclusively deposited on the surface, this shallow implantation leading to a higher average ion separation that prevents clustering and coupling of Er^{3+} ions [Serna, et al., 2001b].

The importance of nanostructuring the films is further illustrated in Fig. 8b where the PL lifetime is plotted as a function of Er^{3+}-Er^{3+} layer in-depth separation. It is seen that the PL lifetime increases up to a saturation value that is reached for a separation close to 6 nm thus evidencing that PL quenching can in part be controlled by choosing an adequate Er^{3+}-Er^{3+} separation in the perpendicular direction. Fig. 8b also includes the result of a similar study performed using $LiNbO_3$ as host where it is seen that the PL lifetime follows the same trend as a function of the ion-ion in-depth separation [Gonzalo, et al., 2002], thus supporting that the PL enhancement when Er^{3+}-Er^{3+} separation is increased is independent on the host.

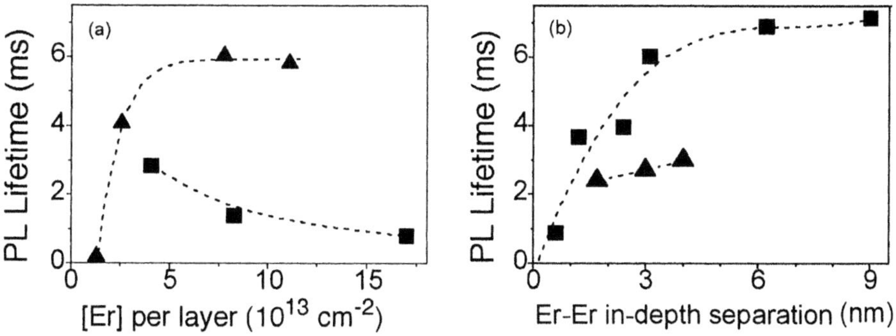

Figure 8. Photoluminescence lifetime at 1.53 μm as a function of: (a) Er areal density per layer for amorphous Al_2O_3 films doped with Er and prepared varying (■) the number of pulses or (▲) the laser energy density on the Er target. (b) Er^{3+}-Er^{3+} layers in-depth separation for the Er doped (■) amorphous Al_2O_3 and (▲) $LiNbO_3$ crystalline films (Adapted from Serna, et al., 2001a and Gonzalo, et al., 2003).

The last example to illustrate the high potential of nanostructuring is the co-doping of a film with two RE ions. It is well-known that Er ions have a low absorption cross-section in the 980 nm range, where integrated semiconductor diode lasers operate. Yb^{3+} has instead one order of magnitude higher absorption cross-section [Strohhöfer and Polman., 2001] and is known to resonantly transfer its energy to Er^{3+} [Miniscalco, 1993; Polman, 1997]. The main concern in co-doping is to determine the Yb to Er (Yb:Er) concentration ratio and/or Yb to Er distance that leads to optimum energy transfer [Lester, et al., 1995; Federighi and Dipasquale, 1995; Kozanecki, et al., 1999; da Vila, et al., 2003]. The described capabilities of PLD for nanostructuring the RE in-depth distribution thus offer a unique opportunity for determining the mechanisms for energy transfer through the independent study of the role of Yb^{3+}-Er^{3+} separation and Yb:Er concentration ratio. Fig. 9 shows both a sketch of the structure of films having Er-doped "layers" separated 6 nm, and Yb-doped "layers" separated a distance S from Er ones, and the PL intensity of such co-doped films as a function of S. It is seen that the PL intensity increases as S increases, irrespective of the concentration ratio. The improvement when increasing S from 0 nm to 1 nm relates to the decrease of clustering when separating the two types of dopants. The improvement for S between 1 nm to 3 nm highlights instead that energy transfer from Yb to Er up to a distance of 3 nm can still be an efficient process, in spite of the general idea that the main energy transfer mechanism is dipole-dipole interaction that has a S^{-6} dependence [da Vila, et al., 2003]. A possible mechanism to explain this improvement is the existence of radiative energy transfer. The pairing of Er and Yb can be further improved by distributing the Yb in two "layers", each at either side of the Er "layer". The PL intensity measured in this configuration is indeed more than 2 times higher than that of the films having the same amount of Yb in a single layer [Suarez-Garcia, et al., 2004].

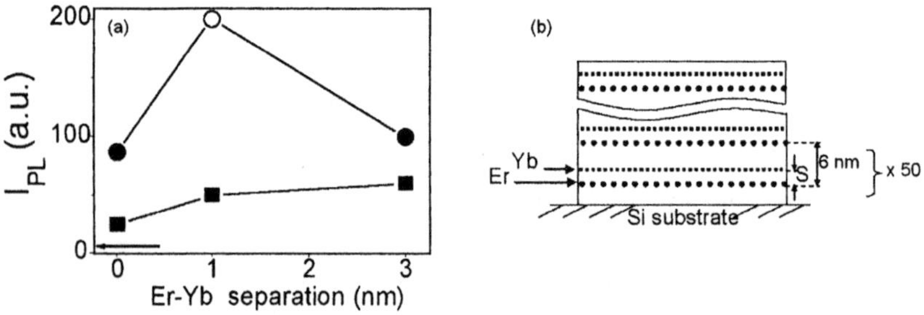

Figure 9. Photoluminescence intensity values at 1.53 μm as a function of Er^{3+}-Yb^{3+} in-depth separation (S) for amorphous Al_2O_3 films co-doped with Er and Yb, and having an Yb to Er concentration ratio of (■) 1.8 and (●, ○) 3.6. The layers of Er and Yb are distributed following the sketch in (b) except for the case of (○) in which the Yb ions have been distributed in two layers each located at either side of the Er one. Reused with permission from Suarez-Garcia, et al., *Applied Physics Letters*, **84**, 2151 (2004). Copyright 2004, American Institute of Physics.

4. SUMMARY

Pulsed laser deposition is a versatile technique with high capabilities to produce complex oxide films. Several parameters must be controlled and optimised in order to achieve the desired film quality or structure, since there is a narrow range of laser energy density values, gas pressures and target-substrate distances that lead to optimum deposition conditions. The goal to reach is the control of the kinetic energy of the species arriving to the substrate to avoid undesired processes (such as surface damage or sputtering), while keeping a value high enough to enhance beneficial processes at the substrate (such as film density enhancement, surface mobility and reactivity and, in some cases, even implantation).

Additionally, PLD has a high potential to produce complex glassy films for integrated optical applications with improved optical performances. The main advantages of PLD when compared to other deposition methods are the capability of producing glassy films in an extended compositional range with respect to bulk materials and the possibility of avoiding oxygen deficiencies in the glass network by using an oxygen pressure during deposition. Finally, PLD has a high potential to produce complex nanostructured films formed by *layers* with thicknesses ranging from submonolayers (10^{13} atoms cm^{-2} level) that can be associated to atomic doping to clusters and nanoparticles (up to 10^{16} atoms cm^{-2}). The main advantages of PLD in these cases when compared to other deposition methods are the capability of producing complex oxide hosts and the possibility of structuring the distribution of the dopant in the nanometer scale in a single step process.

ACKNOWLEDGMENTS

This work was partially sponsored by CICYT (Spain) under TIC1999-0866, MAT2000-1135-C02-02, TIC2002-03235 and MAT2003-01490 Projects and EU under BRPR-CT98-0616 Project. We wish to thank past and present members of the Laser Processing Group at Optics Institute who have contributed with their work to the progress and application of PLD. The enthusiastic collaboration and fruitful discussions with many other researchers at several foreign institutions is similarly recognised and appreciated, among which the ones with A. Petford-Long Group from University of Oxford (UK), P. D. Townsend Group from University of Sussex, and J. Perriere and I. Vickridge from University of Paris Universite de Paris 6 et 7 (France) and R. F. Haglund Jr. from Vanderbilt University (USA) have been particularly long lasting.

REFERENCES

Afonso, C.N., 1995, Pulsed laser deposition of films for optical applications, in: *Insulating Materials for Optoelectronics: New Developments*, F. Agulló-López, ed., World Scientific, Singapore, pp. 1-28.

Afonso, C.N., Ballesteros, J.M., Gonzalo, J., Righini, G.C., and Pelli, S., 1996, Rare-earth doped glass waveguides prepared by pulsed laser deposition, *Appl. Surf. Sci.* **96-98**: 760-763

Afonso, C. N., Gonzalo J., Serna R., and Solís J., in press, Metal-dielectric nanocomposites produced by pulsed laser deposition: a route for new functional materials, in: *Recent Advances on Laser Processing*, J. Perrière, E. Millon and E. Fogarassy, eds., Elsevier.

Amoruso, S., Bruzzese, R., Spinelli, N., Velotta, R., Vitiello, M., and Wang, X., 2003, Dynamics of laser-ablated MgB_2 plasma expanding in argon probed by optical emission spectroscopy, *Phys. Rev. B* **67**: Art. 224503

Anisimov, S.I., Bäuerle, D., and Lukyanchuk, B.S., 1993, Gas-dynamics and film profiles in pulsed-laser deposition of materials, *Phys. Rev. B* **48**: 12076-12081

Ay, F., Aydinli A., and Agan, S., 2003, Low-loss as-grown germanosilicate layers for optical waveguides, *Appl. Phys. Lett.* **83**: 4743-4745.

Bär, S., Huber, G., Gonzalo, J., Perea, A., Climent, A., and Paszti, F., 2003, Europium-doped sesquioxide thin-films grown on sapphire by PLD, *Mat. Sci. Eng. B* **105**: 30-33.

Barnes, J.-P., Petford-Long, A.K., Doole, R.C., Serna, R., Gonzalo, J., Suarez-Garcia, A., Afonso, C.N., and Hole, D., 2002, Structural studies of Ag nanocrystals embedded in amorphous Al_2O_3 grown by pulsed laser deposition, *Nanotechnology* **13**: 465-470.

Barnes, J.-P., Petford-Long, A.K., Suarez-Garcia, A., and Serna, R., 2003, Evidence for shallow implantation during the growth of bismuth nanocrystals by pulsed lased deposition, *J. Appl. Phys.* **93**: 6396-6398.

Blank, D.H.A., Koster, G., Rijnders, G.A.J.H.M., van Setten, E., Slycke, P. and Rogalla, H., 2000, Epitaxial growth of oxides with pulsed laser interval deposition, *J. Cryst. Growth* **211**, 98-105.

Chaos, J.A., Gonzalo, J., Afonso, C.N., J. Perrière and García-González, M.T., 2001, Growth of stoichiometric and textured $LiNbO_3$ films on Si by pulsed laser deposition, *Appl. Phys. A* **72**: 705-710.

Chen, Li-Chyong, 1994, Particulates generated by pulsed laser ablation, in: *Pulsed Laser Deposition of Thin-films*, D. B. Chrisey and G. K. Hubler, eds., John Wiley & Sons, New York, pp. 167-198.

Chrisey D.B. and Hubler G.K., eds, 1994, *Pulsed Laser Deposition of Thin-films*, John Wiley and Sons Inc., New York.

Clement, M.K.Th., Hayden, J.S., Hayden, Y.T., Hoffmann H.-J., Lentes F.-T., and Neuroth, N., 1995, Optical properties, in: *The Properties of Optical Glass*, H. Bach and N. Neuroth, eds., Springer-Verlag, Berlin, pp. 59-165.

del Coso, R. and Solis, J., 2004a, Relation between nonlinear refractive index and third-order susceptibility in absorbing media, *J. Opt., Soc. Am. B* **21**: 640-644.

del Coso, R., Requejo-Isidro, J, Solis, J., Gonzalo, J. and Afonso, C.N., 2004b, Third order nonlinear optical susceptilibity of Cu:Al2O3 nanocomposites: from spherical nanoparticles to the percolation threshold, *J. Appl. Phys.* **95**, 2755-62.

Dempsey, N.M., Ranno, L., Givord, D., Gonzalo, J., Serna, R., Fei, G.T., Petford-Long, A.K., Doole, R.C., and Hole, D.E., 2001, Magnetic behavior of $Fe:Al_2O_3$ nanocomposite films produced by pulsed laser deposition, *J. Appl. Phys.* **90**: 6268-6274.

Dyer, P.E., Issa, A., and Key, P.H., 1990, Dynamics of excimer laser ablation of superconductors in an oxygen environment, *Appl. Phys. Lett.* **57**: 186-188.

Fähler S., Sturm, K., and Krebs, H-U., 1999, Resputtering during the growth of pulsed-laser-deposited metallic films in vacuum and in an ambient gas. *Appl. Phys. Lett.* **75**: 3766-3768.

Federighi, M. and Dipasquale, F., 1995, The effect of pair-induced energy-transfer on the performance of silica wave-guide amplifiers with high Er^{3+}/Yb^{3+} concentrations, *IEE Photonic. Tech. L.* **7**: 303-305.

Ford, A.C., Tepper, T., and Ross, C.A., 2003, Reactive pulsed laser deposition of silica and doped silica thin-films, *Thin Solid Films* **437**: 211-216

Frantti, J., and Lantto V., 1994, Characterization of $Pb_{0.97}Nd_{0.02}(Zr_{0.55}Ti0._{45})O_3$ thin-films prepared by pulsed laser deposition, *J. Appl. Phys.* **76**: 2139-2142

Geohegan, D.B., 1994, Diagnostics and characteristics of pulsed laser deposition of laser plasmas, in: *Pulsed Laser Deposition of Thin-films*, D. B. Chrisey and G. K. Hubler, eds., John Wiley & Sons, New York, pp. 115-165.

Geohegan, D.B., Puretzky, A.A., Duscher, G., and Pennycook, S.J., 1998, Time-resolved imaging of gas phase nanoparticle synthesis by laser ablation, *Appl. Phys. Lett.* **72**: 2987-2989.

Gonzalo, J., Afonso, C.N., and Perrière, J., 1996, Influence of the laser energy density on the plasma expansion dynamics and film stoichiometry during laser ablation of BiSrCaCuO, *J. Appl. Phys.* **79**: 8042-8046.

Gonzalo J., Chaos J.A., Suarez-Garcia A, Afonso C.N., and Pruneri V., 2002, Enhanced second-order nonlinear optical response of LiNbO3 films upon Er doping, *Appl. Phys. Lett.* **81**: 2532-2534.

Gonzalo, J., Sanz, O., Perea, A., Fernández-Navarro, J.M., Afonso, C.N., and García López, J., 2003, High refractive index and transparent heavy metal oxide glassy thin-films, *Appl. Phys. A* **76**: 943-946.

Gonzalo, J., Perea, A., Babonneau, D., Afonso, C.N., Beer, N., Barnes, J.P., Petford-Long, A.K., Hole, D.E., and Townsend, P.D., 2005, Competing processes during the production of metal nanoparticles by pulsed laser deposition, *Phys. Rev. B* **71**: Art. 125420.

Gonzalo, J., Fernández, H., Sanz, O., Solís J, and Fernández-Navarro, J.M., submitted, Nonlinear optical properties of lead niobium germanate glass films, *Appl. Phys. Lett.*

Gottmann, J., and Kreutz, E.W:, 2000, Controlling crystal quality and orientation of pulsed-laser-deposited $BaTiO_3$ thin-films by the kinetic energy of the film forming particles, *Appl. Phys. A* **70**: 275-281.

Gupta, A., 1994, Novel pulsed laser deposition approaches, in: *Pulsed Laser Deposition of Thin-films*, D. B. Chrisey and G. K. Hubler, eds., John Wiley & Sons, New York, pp. 265-291.

Hau, S.K., Wong, K.H., Chan P.W., and Choy, C.L., 1995, Intrinsic resputtering in pulsed-laser deposition of lead-zirconate-titanate thin-films, *Appl. Phys.Lett.* **66**: 245-247.

Henry, C.R., 1998, Surface studies of supported model catalysts, *Surf. Sci. Reports* **31**, 231-326.

Hirao, K., Mitsuyu, T., Si, J., and Qiu J., 2001, *Active glass for photonic devices*, Springer, Berlin

Kim, H.S., and Kwok, H.S., 1992, Correlation between target-substrate distance and oxygen pressure in pulsed laser deposition of $YBa_2Cu_3O_7$, *Appl. Phys. Lett.* **61**: 2234-2236.

Koinuma, H., Aiyer, H., and Matsumoto, Y., 2000, Combinatorial solid state materials science and technology, *Sci. and Tech. of Advanced Materials* **1**, 1-10.

Koinuma, H. and Takeuchi, I., 2004, Combinatorial solid-state chemistry of inorganic materials, *Nature Materials* **3**: 429-438.

Komuro, S., Katsumata, T., Morikawa, T., Zhao, X., Isshiki, H., and Aoyagi, Y., 2000, 1.54 μm emission dynamics of erbium-doped zinc-oxide thin-films, *Appl. Phys. Lett.* **76**: 3935-3937.

Kools, J.C.S. 1994, Pulsed laser deposition of metals, in: *Pulsed Laser Deposition of Thin-films*, D. B. Chrisey and G. K. Hubler, eds., John Wiley & Sons, New York, pp. 455-471.

Kozanecki, A., Przybylinska, H., Jantsch, W., and Palmetshofer, L., 1999, Room-temperature photoluminescence excitation spectroscopy of Er3+ ions in Er- and (Er plus Yb)-doped SiO_2 films, *Appl. Phys. Lett.* **75**: 2041-2043.

Krebs, H-U, Bremert, O., Luo, Y., Fähler, S., and Störmer M., 1996, Structure of laser-deposited metallic alloys and multilayers, *Thin Solid Films* **275**, 18-21.

Kreibig, U., and Vollmer, M., 1995, *Optical properties of metal clusters*, Springer, Berlin.

Lackner J.M., Wakdhauser, W., Ebner, R., Major, B., and Schoberl, T., 2004, Pulsed laser deposition of titanium oxide coatings at room temperature-structural, mechanical and tribological properties, *Surf. Coat. & Tech.* **180-81**, 585-90.

Lancok, J., Garapon, C., Martinet, C., Mugnier, J., and Brenier, R., 2004, Influence of the PLD parameters on the crystalline phases and fluorescence of Eu:Y_2O_3 planar waveguides, *Appl. Phys. A* **79**: 1263-1265.

Lanzerstorfer, S., Pedarnig, J. D., Gunasekaran, R. A., Bauerle, D., and Jantsch, W., 1998, 1.54 μm emission of pulsed-laser deposited Er-doped films on Si, *J. Lumin.* **80**: 353-356.

Lazarouk, S. K., Mudryi, A. V., and Borisenko, V. E., 1998, Room-temperature formation of erbium-related luminescent centers in anodic alumina, *Appl. Phys. Lett.* **73**: 2272-2274.

Lester, C., Bjarklev, A., Rasmussen, T., and Dinesen, P. G., 1995, Modeling of Yb^{3+}-sensitized Er^{3+}-doped silica wave-guide amplifiers, *J. Lightwave Technol.* **13**: 740-743.

Lichtenwalner, D.J., Auciello, O., Dat, R., and Kingon A.I., 1993, Investigation of the ablated flux characteristics during pulsed laser ablation deposition of multicomponent oxides, *J. Appl. Phys.* **74**: 7497-7505.

Lines, M.E., 1991, Oxide glasses for fast photonic switching: A comparative study, *J. Appl. Phys.* **69**: 6876-6884.

Luo, Y., Biswas, A., Frauenglass, A., and Brueck S.R.J., 2004, Large second-harmonic signal in thermally poled lead glass-silica waveguides, *Appl. Phys. Lett.* **84**: 4935-4937.

Mailis S., Riziotis Ch., Wang, J., Taylor E., Anderson A.A., Berrington S.J., Rutt H.N., Eason R.W., Vainos, N.A., and Grivas Ch., 1999, Growth and characterization of pulsed laser deposited lead germanate glass optical waveguides, *Opt. Mat.* **12**:27-33.

Marine, W., Patrone, L., Luk'yanchuk, B., and Sentis, M., 2000, Strategy of nanocluster and nanostructure synthesis by conventional pulsed laser ablation, *Appl. Surf. Sci.* **154**, 345-352.

Martino, M., Caricato A.P., Fernández, M., Leggieri, G., Jha A., Ferrari, M., and Mattarelli M., 2003, Pulsed laser deposition of active waveguides, *Thin Solid Films* **433**: 39-44

Mazzoldi, P., and Righini, G.C., 1995. Glasses for Optoelectronic Devices, in: *Insulating Materials for Optoelectronics: New Developments*, F. Agulló-López, ed., World Scientific, Singapore, pp. 367-392.

Miller, J.C. and Haglund, R.F., eds., 1998, *Laser ablation and desorption,* Academic Press, San Diego.

Miniscalco, W.J., 1993 Optical and electronic properties of rare-earth ions in glasses, in: *Rare-Earth doped fiber lasers and amplifiers*, ed. by M.J.F. Digonnet, Dekker, New York.

Ohtomo, A. and Tsukazaki, A., 2005, Pulsed laser deposition of thin-films and superlattices based on ZnO, *Sem. Sci. and Tech.* **20**, S1-S12.

Pignolet, A., Curran, C., Alexe, M., Senz, S., Hesse, D., and Gosele, U., 1998, Epitaxial and large area PLD ferroelectric thin-film heterostructures on silicon substrates, *Integr. Ferroelectr.* **21**: 485-498.

Polman, A., 1997, Erbium implanted thin-film photonic materials, *J. Appl. Phys.* **1997**, 1-39.

Prins, M.W. J., GrosseHolz, K.O., Muller, G., Cillessen, J.F.M., Giesbers, J.B., Weening, R.P., and Wolf, R.M., 1996, A ferroelectric transparent thin-film transistor, *Appl. Phys. Lett.* **68**: 3650-3652.

de Sande, J.C.G., Serna, R., Gonzalo, J., Afonso, C.N., Hole, D.E. and Naudon, A., 2002, Refractive index of Ag nanocrystals composite films in the neighborhood of the surface plasmon resonance, *J. Appl. Phys.* **91**, 1536-41.

Saenger, K.L. 1994, Angular distribution of ablated material, in: *Pulsed Laser Deposition of Thin-films*, D. B. Chrisey and G. K. Hubler, eds., John Wiley & Sons, New York, pp. 199-227.

Santran, S., Canioni, L., Sarger, L., Cardinal, Th., and Fargin, E., 2004, Precise and absolute measurements of the complex third-order optical susceptibility, *J. Opt. Soc. Am. B* **21**: 2180-2189

Sanz, O., Gonzalo, J., Perea, A., Fernández-Navarro, J.M., Afonso, C.N., and García López, J., 2004, Wide transparency range and high refractive index and lead-niobium-germanate glass thin-films, *Appl. Phys. A* **79**: 1907-1911

Serna, R., de Sande, J.C.G., Ballesteros, J.M., and Afonso, C.N., 1998a, Spectroscopic ellipsometry of composite thin-films with embedded Bi nanocrystals, *J. Appl. Phys.* **84**: 4509-4516.

Serna, R., Ballesteros, J. M., de Castro, M. J., Solis, J., and Afonso, C. N., 1998b, Optically active Er-Yb doped glass films prepared by pulsed laser deposition, *J. Appl. Phys.* **84**: 2352-2354.

Serna, R., Afonso, C.N., Ballesteros, J.M., Naudon, A., Babonneau, D., and Petford-Long, A.K., 1999, Size, shape anisotropy, and distribution of Cu nanocrystals prepared by pulsed laser deposition, *Appl. Surf. Sci.* **139**: 1-5.

Serna, R., Afonso, C.N., Ricolleau, C., Wang, Y., Zheng, Y., Gandais, M., and Vickridge, I., 2000, Artificially nanostructured Cu:Al_2O_3 films produced by pulsed laser deposition, *Appl. Phys. A* **71**, 583-586.

Serna, R., Gonzalo, J., Afonso, C.N., and de Sande, J.C.G., 2001a, Anomalous dispersion in nanocomposite films at the surface plasmon resonance, *Appl. Phys. B* **73**: 339-343.

Serna, R., de Castro, M.J., Chaos, J.A., Suarez-Garcia, A., Afonso, C.N., Fernandez, M., and Vickridge, I., 2001b, Photoluminescence performance of pulsed-laser deposited Al_2O_3 thin-films with large erbium concentrations, *J. Appl. Phys.* **90**: 5120-5125.

la Serra, E.R., Charbouillot, Y., Baudry, P., and Aegerter, M.A., 1990, Preparation and characterization of thin-films of TiO_2-PbO and TiO_2-Bi_2O_3 compositions, *J. of Non Cryst. Solids* **121**: 323-328.

Serra, J., Liste, S., González P., Serra, C., Borrajo, J.P., Chiussi, S., León, B., and Pérez-Amor, M., 2004, The role of the temperature and laser fluence on the properties of PLD bioactive glass films, *Appl. Phys. A* **79**: 983-986.

Singh R.K., and Narayan, J., 1990, Pulsed-laser evaporation technique for deposition of thin-films: Physics and theoretical model, *Phys. Rev. B* **41**: 8843-8859.

Strikovski, M. and Miller Jr., J.H., 1998, Pulsed laser deposition of oxides:Why the optimum rate is about 1Å per pulse, *Appl. Phys. Lett.* **73**: 1733-1735.

Strohhofer, C. and Polman, A., 2001, Relationship between gain and Yb^{3+} concentration in Er^{3+}-Yb^{3+} doped waveguide amplifiers, *J. Appl. Phys.* **90**: 4314-4320.

Suárez-García, A., del Coso, R., Serna, R., Solís, J. and Afonso C.N. 2003, Controlling the transmission at the surface plasmon resonance of nanocomposite films using photonic structures, Appl. Phys. Lett. **83**, 1842-4.

Suarez-Garcia, A., Serna, R., de Castro, M.J., Afonso, C.N. and Vickridge, I., 2004, Nanostructuring the Er-Yb distribution to improve the photoluminescence response of thin-films, *Appl. Phys. Lett.* **84**: 2151-2153.

Van den Hoven, G.N., Snoeks, E., Polman, A., Vanuffelen, J.W.M., Oei, Y.S., and Smit, M.K., 1993, Photoluminescence Characterization of Er-Implanted Al_2O_3 Films, *Appl. Phys. Lett.* **62**, 3065-3067.

da Vila, L. D., Gomes, L., Tarelho, L. V. G., Ribeiro, S. J. L. and Messadeq, Y., 2003, Mechanism of the Yb-Er energy transfer in fluorozirconate glass, *J. Appl. Phys.* **93**: 3873-3880.

Voevodin, A.A., Donley, M.S., and Zabinski, J.S., 1997, Pulsed laser deposition of diamond-like carbon wear protective coatings: a review, *Surf. & Coat. Tech.* **92**, 42-49.

Vogel, E.M., Chase, E.W., Jackel, J.L. and Wilkens, B.J., 1989, Fabrication of thin-film nonlinear optical glasses using pulsed excimer laser deposition, *Appl. Opt.* **28**: 649-650.

Vogel, E.M., Weber, M.J. and Krol D.M., 1991, Nonlinear optical phenomena in glass, Phys. *Chem. Glasses* **32**: 231-254.

Willmott, P.R. and Huber, J.R., 2000, Pulsed laser vaporization and deposition, *Rev. Mod. Phys.* **72**: 315-328.

Wu, N. J., Chen, Y. S., Fan, J. Y. and Ignatiev, A., 1998, Infrared photoresponse of (Mn,Sb) doped Pb(Zr,Ti)O_3/$YBa_2Cu_3O_7$ heterostructure detectors, *J. Appl. Phys.* **83:** 4980-4984.

Yan, Y.C., Faber, A.J., deWaal, H., Kik, P.G. and Polman, A., 1997, Erbium-doped phosphate glass waveguide on silicon with 4.1 dB/cm gain at 1.535 μm, *Appl. Phys. Lett.* **71**: 2922-2924.

Yilmaz, S., Gerhardmulthaupt, R., Bonner, W.A., Hwang, D.M., Inam, A., Martinez, J.A., Ravi, T.S., Sands, T., Wilkens, B.J., Wu, X. D. and Venkatesan, T., 1994, Electrooptic potassium-tantalate-niobate films prepared by pulsed-laser deposition from segmented pellets, *J. Mat. Res.* **9**: 1272-1279.

Yoshida, T., Takeyama, S., Yamada, Y. and Mutoh, K., 1996, Nanometer-sized silicon crystallites prepared by excimer laser ablation in constant pressure inert gas, *Appl. Phys. Lett.* **68**: 1772-1774.

Xu, S.F., Tian, Y.J., Lü, H.B., Cui, D.F., Chen, Z.H., Li, L., and Yang G.Z., 1994, The effect of laser energy density and target-substrate distance on the quality of of $YBa_2Cu_3O_{7-x}$ thin-films, *Supercond. Sci. Technol.* **7**: 435-437.

Chapter 14

LASER FORWARD TRANSFER OF ELECTRONIC AND POWER GENERATING MATERIALS

Alberto Piqué[1], Heungsoo Kim[1] and Craig B. Arnold[2]
[1]U.S. Naval Research Laboratory, Washington DC; [2]Dept. of Mechanical and Aerospace Engineering, Princeton University, Princeton, NJ.

1. INTRODUCTION

The application of laser forward transfer techniques has steadily grown since the first reports of patterned copper deposition by Bohandy, et al. 20 years ago (Bohandy, 1986). These general techniques employ a pulsed laser to locally transfer material from a source film onto a substrate in close proximity or in contact with the film. The source is typically a coated laser-transparent substrate, referred to as the target, donor, or ribbon. Laser pulses propagate through the transparent ribbon until they are absorbed by the film. Above an incident laser energy threshold, material is ejected from the film and transferred toward the acceptor, waiting or receiving substrate. These laser forward transfer techniques, known as laser direct-write (LDW), belong to a class of processes capable of generating a high-resolution pattern without the need for lithogaphic processes afterwards. Other examples of laser direct-write techniques include pyrolitic or photolytic decomposition of gas- or liquid-phase precursors, also known as laser CVD or laser-assisted deposition (Osgood, 1985; Herman, 1989). However, these non-forward transfer techniques will not be discussed in this chapter because they are not ablative in nature.

Translation of the ribbon and receiving substrate, and/or scanning and modulating the laser beam enables complex pattern formation. Commercially available, computer-controlled translation stages or galvonometric scanning mirrors enable high-resolution patterns from the individually written 3D pixels or voxels resulting from the laser forward transfer process. Fig. 1 shows a schematic illustrating the basic elements required for the LDW apparatus.

One may consider the LDW technique to be analogous to ink-jet deposition of functional materials without the constraints of a nozzle, and with the added benefits of laser processing, such as material modifications and micromachining. This added versatility in comparison to other printing methods enables LDW to find opportunities in fields ranging from metals and power generation materials to biological and soft condensed matter.

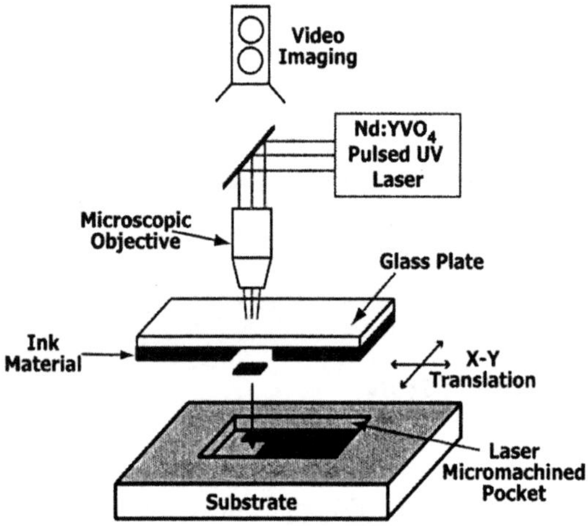

Figure 1. Schematic illustration of the setup used for laser direct-write.

This chapter will provide a brief overview of the laser forward transfer and LDW literature, with a particular emphasis towards its application in electronic materials, followed by a more detailed review of LDW for the fabrication of micropower systems such as ultracapacitors, microbatteries and dye-sensitized solar cells. Finally, the chapter concludes with a discussion of the role that the LDW technique may play in the future of microelectronic manufacturing and other applications.

2. BACKGROUND

In order to discuss the recent developments in LDW of electronic and power generating materials for microelectronic applications, it is necessary to begin with an overview of the original work performed in laser forward transfer of the late 1980's.

2.1 Laser Induced Forward Transfer

The laser forward transfer process was first used in depositing copper metal patterns inside a vacuum chamber (Bohandy, 1986). Excimer laser pulses (λ = 193 nm, 15 ns) were focused with a cylindrical lens to a 25 mm long by 50 μm wide line on a source substrate containing a thin copper film. The Cu was transferred to Si and fused silica substrates, where further examination revealed resistivities ranging between 3 to 50 times the value for bulk copper with adhesion behavior that passed the "tape test". Bohandy's group coined the term laser-induced forward transfer (LIFT) to denote the process and developed a model to describe how it worked. According to this model; (1) the laser pulse heats the interface of the film at the source substrate; (2) a resulting melt front propagates through the film until it reaches the free surface; (3) at about this time, the material at the interface is superheated beyond its boiling point until, (4) the resulting vapor induced pressure at the interface propels the molten film forward towards the acceptor substrate (Adrian, 1987). The same group then demonstrated that this process can be carried out in air, i.e. under atmospheric conditions, without the need for a vacuum (Bohandy, 1988).

The LIFT technique is simple and can be used with a wide variety of film materials, mainly metals such as Cu and Ag (Bohandy, 1988), Al (Schultze, 1991), W (Tóth, 1993; Kántor, 1994) and Cr (Zergioti, 1998a). Reports of LIFT for other materials such as Al_2O_3 (Greer, 1988), In_2O_3 (Zergioti, 1998a), and even high temperature superconductors (Fogarassy, 1989) are worth mentioning, although the quality of the transferred ceramics was not as good as those deposited by traditional film growth techniques. The thickness of the transferred film on the acceptor substrate can be adjusted by repetitive transfers from the ribbon. In a similar way, multilayer structures can be deposited. In general, to achieve uniform transfers with good morphology and spatial resolution on the acceptor substrate; (1) the thickness of the film on the ribbon should not exceed 100 nm; (2) the ribbon film should strongly absorb the laser wavelength in use; (3) the ribbon should be in contact or very close (~ 10's of micrometers) to the acceptor substrate; and (4) the laser fluence should barely exceed the threshold fluence for removal of the film from the ribbon. This last parameter is very important in order to control the quality of the films, since too high fluences result in excessive explosive vaporization of the film affecting the morphology and resolution of the transfers, while too low fluences result in incomplete transfers.

2.2 Development of the Laser Direct-write Process

Despite its successful application to the deposition of thin metal layers, the actual uses of the LIFT process are limited due to several shortcomings. In LIFT, metal films are required to be deposited on the ribbon by conventional vapor deposition techniques that require vacuum deposition and other expensive processes. Since these metal films tend to be very thin (a few hundred nanometers), the individual layers deposited by LIFT are similarly thin, thus limiting its application to lithographic mask repair and other niche areas. During LIFT, the melting and solidification of the transferred material results in the formation of interfaces between adjacent voxels, which can have deleterious effects in the electrical transport properties of the patterned structure being fabricated. Furthermore, the melting of the transferred material becomes a serious issue when LIFT is performed under atmospheric conditions, because most metals are easily oxidized when melted in air. Moreover, the rapid quenching of the metal voxels can result in high intrinsic stresses between the transferred metal and the substrate, ultimately leading to poor adhesion and delamination of the transferred layers. Finally, LIFT is not suited for the transfer of ceramics and other inorganic phases given the irreversible phase changes and decomposition that tend to be exhibited by these materials upon melting and solidification.

The above limitations explain why LIFT is not suitable for the laser transfer of complex, sensitive materials such as those found in electronic and power generation devices. In fact, this realization is underscored by the decline in the number of publications reporting on LIFT of metals and other inorganic materials that followed in the late nineties. Despite these shortcomings, the capability for laser transferring materials in patterned form is extremely important for many commercial, aerospace and military applications.

The first variation of the LIFT process to attempt to overcome these limitations employed multilayered films (Tolbert, 1993a). The multilayers are formed by depositing a thin laser-absorbing layer (usually a metal) on the transparent support, followed by a layer of the material to be transferred. During transfer, the laser pulse interacts with the absorbing layer, referred to as the dynamic release layer, causing it to vaporize in a similar manner as in LIFT. This confined layer forces the material in the second layer to be removed from the film and transferred to the acceptor substrate. The advantage of this approach is that it allows the transfer of materials for which the previously described LIFT process is not effective. For instance, this can include materials with weak absorption of the laser radiation, or materials that can be damaged by their interaction with the laser pulse. This variation on the LIFT process was originally intended for high speed laser color printing applications, and was called laser ablation transfer (Tolbert, 1993a). More recently this approach has been used for the laser direct-write of phosphor powders (Fitz-Gerald, 2000) and extremely laser sensitive materials such as DNA biomolecules (Fernández-Pradas, 2004; Serra, 2004).

The work by Tolbert using the dynamic release layer approach to LIFT, showed for the first time that the pulse duration has an effect on the transfer process. Their experiments indicated that picosecond laser pulses are more efficient in ablating the multilayered films than nanosecond pulses, with transfer laser fluence thresholds an order of magnitude smaller (Tolbert, 1993b). Since then, various reports on the use of femtosecond lasers for laser forward transfer have validated Tolbert's results. Most of the applications of femtosecond LIFT have been related to the laser transfer of metals (Zergioti, 1998b; Papakonstantinou, 1999; Bähnisch, 2000). However, more recently, femtosecond LIFT has successfully been applied to the transfer of biological materials (Karaiskou, 2003; Zergioti, 2005).

A second variation on the LIFT process employs an organic precursor compound (palladium acetate) rather than a pure metal for transfer (Esrom, 1995). This is significant as it demonstrates the laser forward transfer process can take advantage of the photoinduced decomposition of a precursor at the ribbon/coating interface to propel the reminder of the film forward and create a metal pattern. By irradiating the transferred mixture with additional laser pulses, a very thin (< 100 nm) but grainy Pd coating is produced, on which a thicker metal layer can be obtained by electroless plating. In their paper, the authors consider the laser transfer and subsequent electroless metal plating as a two-step process enabling the selective deposition of metal patterns. However, their key contribution shows that the LIFT process can be used to deposit a precursor of the desired material, which once transferred can be further decomposed or processed into its final form.

More recently, the LIFT process has been adapted for the transfer of various materials in powder form mixed with an organic binder. Such mixtures or matrices are then applied as a uniform coating on the transparent donor substrate to form a ribbon. The advantage of this approach resides in the fact that the transfer of thicker films from the ribbon can be achieved at lower laser fluences than with LIFT of homogenous films. Furthermore, the composite film is transferred in its solid phase, thus avoiding the melting and re-condensation steps that take place in traditional LIFT. This modified LIFT approach has been used for the LDW of diamond nanopowders (Pimenov, 1995), ferroelectric ($BaTiO_3$ and $SrTiO_3$) and ferrite ($Y_3Fe_5O_{12}$) micron sized powders (Piqué, 1999, 1999b), and carbon composite polymers for gas sensing elements (Piqué 2000a). A limitation of this approach for these applications is that the adhesion of the transferred layers can be poor and the organic binder remains after the transfer, thus high temperature calcination steps are necessary to achieve dense layers. However, in many applications, dense layers are not required and the particular properties of the deposited films are ideal.

As a further generalization, the idea of using an organic binder in the process can be applied to the direct transfer of rheological systems (Piqué, 2000b; Auyeung 2000). The realization that pastes, inks, suspensions, and sol-gel solutions can be deposited in patterns without physical or chemical modification during their transfer from the ribbon to the acceptor substrate, opened the door to a number of applications with a wider range of materials and formulations. Originally, this "wet" laser forward transfer technique was named matrix assisted pulsed laser evaporation direct-write (MAPLE-DW), in reference to the transfer of a loosely defined matrix comprising of powders or particles dispersed in a liquid. However, since the transfer process does not necessarily involve the evaporation of a matrix, this terminology has been discarded in favor of the less confusing laser direct-write or LDW.

3. LASER DIRECT-WRITE OF ELECTRONIC MATERIALS

The use of laser direct write for the deposition of high quality electronic materials requires the generation of structures comprising of multiple voxels, adjacent or on top of each other, that readily merge to form a single, continuous pattern. Electrical interconnects provide a perfect example of this requirement as heterogeneous interfaces between voxels can degrade the overall conductivity. By enabling the transferred material to remain fluid, adjacent voxels on the receiving substrate will merge into one continuous segment. Fig. 2 shows a simple schematic illustrating the basic steps on the laser direct-write of such rheological systems.

In reality, the process represents an important difference from prior LIFT technology as functional materials are deposited without direct vaporization, which could affect their desirable physical or chemical properties such as electrical conductivity, dielectric properties or electrochemical activity. As shown schematically in Fig. 2, for absorbing materials, a small region of the ink interacts with a low fluence (< 100 mJ/cm^2) laser pulse causing a small amount of the ink to evaporate. As the resulting vapor expands, it generates shear forces that result in the ejection of a droplet from the film which is unaffected by the incident laser. The droplet is then propelled in jet-like fashion towards the receiving substrate, where it is deposited with its original rheological properties intact. The claim that most if not all of the transferred fluid does not interact with the laser pulse has been substantiated by the results obtained with transfers of extremely laser sensitive systems, such as buffer solutions containing biomaterials, proteins and living cells (Wu, 2001), or electrochemically sensitive materials discussed below (Arnold, 2004a).

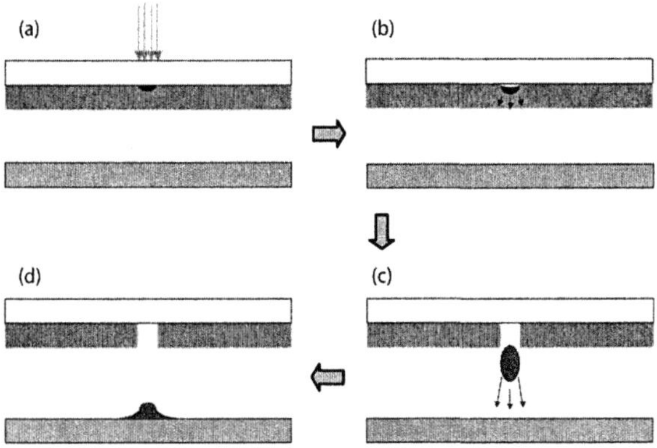

Figure 2. Schematic representation of the steps involved in the forward transfer of rheological systems during the LDW process. (a) The laser pulse is absorbed in the ink layer adjacent to the transparent glass plate. (b) The absorbed energy causes vaporization to occur and (c) propel a droplet of ink forward from the film. (d) Finally, the droplet splashes down on the substrate.

Given the diverse nature and large number of parameters affecting the LDW process, a simple model as the one described in the previous paragraph cannot be expected to completely explain its behavior. For instance, it is known that laser parameters such as fluence, pulse duration and wavelength play an important role in the LDW process. Additionally, ink parameters such as the composition of the ink in the ribbon, its thickness, viscosity, solids content, and solids particle size greatly affect the ability to transfer a particular fluid and the resulting morphology of the transferred voxels.

Parameters such as the distance between the ribbon and the receiving substrate, substrate material, substrate temperature, and the surface chemistry and morphology of the substrate need to be considered as well. This wide parameter space for LDW leads to greater versatility in material choice than available with other direct-write approaches. Relevant insight for the influence of incident laser parameters has been gained by analyzing the images obtained through time-resolved optical microscopy of the LDW transfers.

These studies underscore the differences between traditional LIFT and LDW of rheological fluids, and demonstrate the fluid nature of the transfers responsible for the coalescence of the individual voxels into continuous, pinhole-free layers on the surface of the receiving substrate (Young, 2001a, 2002). Dlott and coworkers studied the effect of various laser fluences on the transfer of a $BaTiO_3$ ink, comprised of nanopowders (150 nm dia.) suspended in α-terpineol. Analysis of the time-resolved microscopy images of the transfers confirmed their strong dependence on laser fluence, and showed that jetting-like transfer behavior (schematically shown in Fig. 2) can be obtained at fluences just above the transfer threshold for a particular ink formulation. Fig. 3 shows time-resolved microscopy images from two views, front and transverse, obtained from these studies.

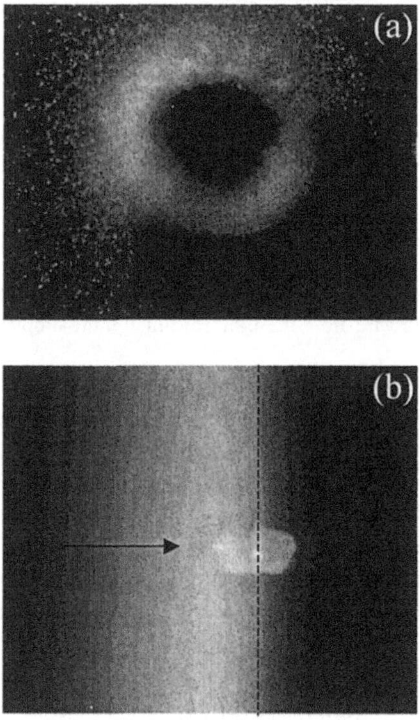

Figure 3. Time-resolved microscopy images taken 100 ns after the laser pulse, showing the laser transferred fluid from (a) front and (b) transverse views. In (b), the arrow indicates the laser direction and the dashed line indicates the ribbon edge.

The capability offered by the LDW technique to conformally transfer viscous fluids, pastes or inks has been used with great success for the fabrication of metal interconnects, vias and antenna structures (Piqué, 2003, 2005). In fact, the LDW process has been used to deposit metallic screen printable inks over complex 3-D surfaces, which has always been extremely difficult if not impossible using traditional lithographic processes. Typically, a commercially available screen printable silver paste is used for the ink. The laser spot size is adjusted depending on the required line-width of the metal lines. Once the transfers are completed, the acceptor substrate (usually printed circuit board) is baked at 100 – 150 °C to obtain the final metallic silver patterns. The electrical resistivity of these patterns ranges between 3 to 50 times higher than that of bulk silver depending on the silver ink used and the baking temperature. The adhesion and mechanical properties of the LDW patterns are very good, as indicated by tape and flexing tests. Overall, the ability to deposit conformal metal patterns on substrates at low temperatures allows for the fabrication of novel types of electronic designs such as embedded circuits. This important application of LDW technologies will be discussed at the end of this chapter.

Similarly, LDW of thick film polymer or ceramic pastes has been used to fabricate passive electronic components such as resistors (Modi, 2001) and interdigitated capacitors (Young, 2001b). The use of LDW to fabricate simple electronic circuits comprising of several passive components and their interconnects has been demonstrated as in the case of a simple chemoselective gas sensor circuit (Piqué, 2002) and RF filter test structures (Zhang, 2003).

4. LASER DIRECT-WRITE OF ELECTRO-CHEMICAL MICROPOWER SOURCES

The recent advances in integration and processing techniques have enabled microelectronic and microelectromechanical devices to shrink in size dramatically. While these devices have been used as components in larger sensor, actuator, and control systems, the lack of commensurately sized power sources has limited their ability to function autonomously. One solution to this challenge is to integrate the micropower sources directly into or adjacent to the microdevice thus forming a self-sufficient system. The resulting savings in volume and weight would allow the development of truly miniaturized autonomous systems, such as those needed for distributed wireless sensor networks. In these types of applications, the necessary size scale is on the order of microns to millimeters. Such a vision might become a reality with the use of LDW techniques.

For the purposes of this chapter, we may consider energy related electrochemical systems, such as ultracapacitors, batteries, or solar cells, to be composed of three main internal components: the negative electrode or anode, positive electrode or cathode, and electrolyte/separator. Each of these materials typically has a large degree of structural complexity, such as nanocomposites, solid-state polymers, liquids, or mesoporous mixtures of electrochemically active materials. Micropower sources employ similar types of materials, but the unique challenge is to maintain their electrochemical activity and structural integrity in a confined space subject to the limitations, such as temperature and pressure, imposed by the microdevices, their substrates, and packaging.

To meet the power demands of a given microdevice, combinations of different types of power sources are needed (Koeneman, 1997). For instance, an application may require constant low power, supplied by a lithium or alkaline microbattery, or an occasional short burst of energy, supplied by an ultracapacitor, combined with long intervals in "stand-by" mode, during which recharging takes place by harvesting energy from the environment supplied by a solar cell. The LDW techniques described in this chapter are ideally suited to prototype, optimize, and fabricate these electrochemical components. Advanced laser techniques can even embed the electrochemical components directly within a substrate, further reducing the packaged size of an entire microdevice, while allowing their geometry to be adapted to fit virtually any form factor.

In constructing electrochemical cells for energy storage and power generation, it is instructive to consider the desired operating geometries. The two main approaches include placing the anode and cathode adjacent to each other in the same plane (planar), or layering the anode and cathode on top of one another (stacked). There are particular advantages and disadvantages to each of these different geometries. In the case of stacked geometries, one can obtain higher area densities and lower resistances owing to the relatively thin separator layer, but this layer must be structurally stable enough to support the anode/cathode/current collectors. This feature rules out highly conductive gel and wet electrolytes. For planar geometries, these are easier to process in that they can handle softer and rougher materials, but they have higher internal resistances due to the greater distance between the electrodes, and therefore tend to have lower performance at high currents. From a practical standpoint, two-dimensional planar structures are easier to fabricate using LDW, although stacked configurations have also been demonstrated and will be discussed in Section 6.

5. LASER DIRECT-WRITE OF ULTRA-CAPACITORS

The first system explored for energy storage and power generation is the sulfuric acid/hydrous ruthenium oxide ultracapacitor (Arnold, 2004a). This electrochemical system is conceptually the easiest to understand, as it is a planar structure that is symmetric, namely, the anode and cathode are made from the same materials. Furthermore, hydrous ruthenium oxide exhibits one of the largest specific capacitance of any material (Zheng, 1995), which will be exploited in these small-scale systems. In this section, we will introduce the concept of an ultracapacitor, followed by a description of how LDW enables new advances in these systems. Finally, advanced laser techniques that further streamline the fabrication process will be explored.

5.1 What are Ultracapacitors?

Ultracapacitors, also known as supercapacitors or pseudocapacitors, are a class of electrochemical capacitor that exhibits large specific capacity (Conway, 1999). These systems display electronic properties similar to both batteries and capacitors. Like a capacitor, an ultracapacitor has the ability to very rapidly discharge its energy leading to a high power density. However, like a battery, it has the ability to store a large amount of energy in the charge state of the active materials. In the most basic manner, we can think of an ultracapacitor as a battery with a high discharge rate. These devices are typically used for load leveling and applications where a short burst of power is needed.

In its most straightforward form, an ultracapacitor can be constructed of two identical electrodes separated by an appropriate electrolyte. The fundamental mechanisms for charge storage in these devices come from two possible sources. The first effect is similar to a typical capacitor in which the electrostatic double-layer at the interface between the electrode and electrolyte stores charge. A high surface area leads to the large amounts of charge storage in the system. These effects are common in carbon based ultracapacitor systems. The second effect stores charge through a Faradaic transfer at the surface of the electrode material. This is similar to the oxidation/reduction reactions that occur in a regular battery, but once again, the high surface area enables a large amount of charge storage and rapid charge transfer through this mechanism. These effects become relevant in metal oxide ultracapacitor systems. In either case, the discharge properties of the device resemble that of a capacitor. Namely, at constant current, the voltage across the device will decrease linearly with time.

Hydrous ruthenium oxide is an ideal electrode material for a micro-ultracapacitor due to its high specific capacitance (capacitance per unit mass). The rapid insertion and release of protons and electrons through the material enables a large pseudocapacitance effect, which, in combination with a high specific surface area, leads to the large amounts of charge stored in the material (Trasatti, 1994; Sarangapani, 1996). This effect is enhanced by the presence of structural water in the lattice that provides nanostructured percolation pathways for proton conduction into the bulk of the material (Dmowski, 2002).

In the context of a small-scale ultracapacitor system, the main challenge is to produce electrodes that preserve the desirable electrochemical properties of the active material while maintaining the structure necessary for high surface area, subject to the process temperature/pressure limitations. These stringent requirements have made the deposition of hydrous ruthenium oxide films incompatible with standard vacuum techniques such as physical or chemical vapor deposition for thin-film growth. Other thin-film techniques such as sol-gel techniques (Fang, 2001) or electrostatic spray deposition (Kim, 2001) are compatible with the material constraints, but require multiple processing steps, including additional lithography to produce the two-dimensional structures required for making microultracapacitors. LDW provides a solution to these challenges.

5.2 Making Ultracapacitors with LDW

As with all LDW deposition, the process starts with the preparation of an ink composed of an active material and a suitable transfer liquid. Hydrous ruthenium oxide is pretreated at 150 °C for 18 hours to give it an optimal water content (McKeown, 1999; Zheng 2002). The appropriate transfer liquid in this case is the electrolyte itself, 5M sulfuric acid. The ability to premix the electrode material with electrolyte gives LDW a distinct advantage over other direct-write methods such as ink jet or LIFT as it enables better control over impurities in the electrodes and has been shown to significantly improve the discharge behavior of these cells (Arnold, 2002).

The process by which planar microultracapacitors are fabricated is straightforward. A 1 cm^2 gold-coated glass substrate is laser micromachined into four electrically isolated regions. Two pads of electrode ink are LDW deposited on the substrate, and finally, laser micromachining is used to isolate the electrodes on the prepatterned substrate. In general, any substrate and electrode geometry is possible, provided two electronically isolated current collectors are formed.

In reality, the details of the deposition and laser micromachining are important to the finished device properties. For instance, the amount of electrolyte in the ink will affect the ability to micromachine electrodes from the initial deposited pad. In these experiments, the deposited material is baked on a hot plate at 100 °C for 10-15 minutes to drive off excess liquid.

This enables improved resolution in laser micromachining and subsequent improvement in device performance.

Figure 4. SEM image of a hydrous ruthenium oxide film deposited by LDW. Note the mesoporous structure and interconnected ruthenium oxide backbone. (a) High magnification, top-view of electrode. (b) Angle view of laser micromachined groove between electrodes.

Fig. 4 shows the microstructure of a LDW hydrous ruthenium oxide film. The material is mesoporous exhibiting a random network of connected grains. This structure is ideally suited for the penetration of liquid or gel electrolyte which enables a large surface area in contact with the electrolyte for charge transfer. This in turn leads to better utilization of the electrode material and improved transport throughout the electrode. In Fig. 4b, a view of the laser micromachined edge shows no damage to the structure at the interface between the two electrodes. LDW is unique in comparison to other deposition techniques in its ability to create these electrochemically desirable structures on the mesoscale. Once the planar structure is fabricated, electrochemical evaluation is performed to determine the performance characteristics of the ultracapacitor cell. An encapsulating layer of Nafion® is placed on top of the structure and a droplet of sulfuric acid is added to assure a fully saturated membrane.

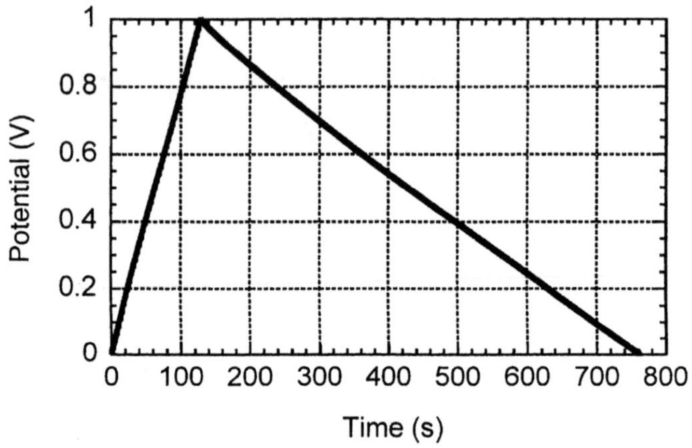

Figure 5. Plot showing the charging/discharging behavior of a LDW hydrous ruthenium oxide ultracapacitor. Cell is charged at 50 μA and discharged at 10 μA. Straight lines are indicative of ideal capacitor behavior. Total electrode mass is ~100 μg with a footprint of 2 mm^2.

Fig. 5 shows a typical charging and discharging curve for this system. The cell is charged at 50 μA until the voltage across the cell reaches 1 V. A constant current is applied for discharging the cell until the voltage returns to zero. The linear charging and discharging behavior at constant current is expected behavior for a capacitor. However, in this case the extended period of time over which the cell is discharged corresponds to the large amount of energy stored in the system. For the particular example shown in the figure, the discharge current is 10 μA, and the resulting capacitance is 6.5 mF or 3100 F/m^2. Additional studies (Arnold, 2003) have shown that these cells can be discharged at currents above 50 mA and can be connected in series and parallel combinations to yield the proper additive values.

5.3 Advanced Laser Techniques for Micro-ultracapacitors: Self-filling Ultracapacitors

One of the key aspects of LDW for microultracapacitors is that the entire process is performed using a single tool. After LDW is used to deposit material, this material is baked prior to micromachining with the same laser. It is then necessary to add an encapsulant and excess electrolyte to perform the electrochemical tests. Although the main tool remains the same, these two processes require the ultracapacitor cell to be removed from the experimental apparatus costing additional time and effort. Furthermore, excess electrolyte makes packaging and handling more difficult.

By employing advanced laser techniques, the process can be dramatically simplified and optimized. Rather than bake the material prior to laser micromachining, it is possible to isolate the two electrodes without the need to remove electrolyte from the ink by employing two separate lasers. An IR laser can be used to drive electrolyte away from the location of interest. This is immediately followed by the UV laser for removing a stripe of electrode material (Arnold, 2004c).

The most interesting aspect of this dual laser approach is that after the two electrodes have been separated, the dividing groove refills with electrolyte. Although the fundamental mechanism for this is not well understood, the end result is a working ultracapacitor cell without excess electrolyte, and without the need to remove the substrate from the LDW apparatus.

LDW has been successfully used to fabricate small-scale ultracapacitor cells. Although the process has not been completely optimized, existing cells have been shown to have outstanding performance for energy and power density. Dual laser techniques enable a level of process optimization toward the ultimate goal of a commercialized product.

6. LASER DIRECT-WRITE OF MICROBATTERIES

The second type of electrochemical energy storage and power generation systems to be considered are batteries. In batteries, the stored chemical energy is converted to electrical energy through oxidation and reduction reactions at the electrodes. Electrons generated by oxidation of the anode material generate work in the electrical circuit before recombining to reduce the cathode material. The separator/electrolyte enables the flow of ions between the anode and cathode to complete the circuit. The myriad of chemical possibilities are discussed and reviewed in readily available reference books (Vincent, 1997; Linden, 2001).

In the case of small-scale microbatteries, the main challenge is to create mesoscale patterns of these active materials while maintaining the necessary physical and chemical properties of a functioning cell. There has been significant effort to develop small microbatteries for various primary and secondary chemistries. Examples include secondary solid-state lithium-ion cells in a stacked configuration (Bates, 1995, 2000, Arnold 2004a) and primary alkaline microbatteries in a planar configuration where the anode and cathode are located adjacent to each other on the substrate (Humble, 2001; LaFollette, 2001; Arnold 2004c).

The next two sections describe the fabrication of single use (primary) and rechargeable (secondary) microbatteries using LDW processes. The primary microbatteries consist of zinc/silver-oxide alkaline cells. The secondary microbatteries are comprised of lithium and lithium-ion cells. Both types of microbatteries exhibit the appropriate open-circuit voltages characteristic of their chemistries. The primary cells are able to operate at relatively high discharge currents, while the secondary cells maintain high efficiencies and show limited charge fading through numerous charge/discharge cycles.

6.1 LDW of Primary Microbatteries

Alkaline batteries are one of the most ubiquitous battery chemistries in use today. Their low cost, relative environmental safety, and ability to handle high discharge rate applications make them readily available for consumer applications. In comparison to other alkaline chemistries, the zinc/silver-oxide (Zn-Ag_2O) system has found extensive use in applications which require maintaining a constant voltage throughout the battery discharge. Primary Zn-Ag_2O alkaline microbatteries have been demonstrated using LDW techniques by laser printing zinc and monovalent silver oxide (Ag_2O) powders in both planar and stacked geometries on a variety of substrates (Arnold 2004a, 2004c; Pique 2004b), as shown in Fig. 6.

The planar alkaline microbattery cells are constructed by first preparing the substrates so that two electrically isolated current collector pads are obtained. The cathode ink (consisting of Ag_2O powders plus 5 wt% graphite to improve conductivity, and a polymer binder mixed with an organic solvent) is first deposited followed by baking at 150 °C for 5-10 minutes to remove the solvent and to bind the cathode to the substrate. Then the Zn ink (formed by mixing zinc powders and a polymer binder with an organic solvent) is deposited to form the anode and subsequently baked at 150 °C for

5-10 minutes. After deposition and processing of the cathode and anode, the separation gap between them is UV laser micromachined to remove any additional material that could cause shorting of the electrodes. The typical weight of all transferred material is 250 μg, while the thickness of the anode and cathode layers is about 10-20 μm thick. In order to activate the planar cells, a drop of 25 wt.% KOH solution is placed on the electrodes.

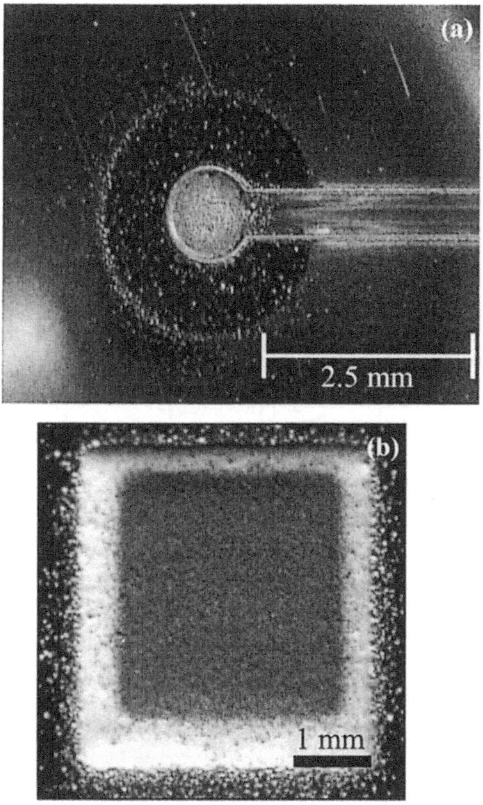

Figure 6. Optical micrographs showing (a) planar and (b) stacked alkaline microbatteries made by LDW. In (b), each of the layers required for the stacked structure are shown after being laser printed.

The fabrication of stacked alkaline microbatteries using LDW is performed in similar fashion in terms of the electrode materials and processing. However, in this case, a separator layer, consisting of barium titanante 30 nm dia. nanoparticles and a cellulose binder, is laser printed on top of the cathode. Alternatively, a commercially available separator material can be laser micromachined to the desired shape and size and manually placed on top of the cathode. To complete the stack, the Zn ink is laser printed on top of the separator. To avoid shorting between the cathode and anode layers, the separator layer is made larger in area than either of the electrode layers.

Electrochemical evaluation of cell voltage and discharge behavior of the alkaline microbatteries demonstrated similar behavior to that of larger cells (see Fig. 7). The open-circuit potential for both planar and stacked microbatteries was 1.55 V, while their discharge at constant currents showed the characteristic flat response of the Zn-Ag_2O alkaline system. High capacity cells are obtained in this manner. For instance, planar cells have been fabricated with capacities greater than 450 μAhr/cm^2. The relatively high discharge rate behavior is demonstrated in the stacked geometries by applying 1 mA for more than 12 minutes of discharge, corresponding to a power density of 1.9 mW/cm^2.

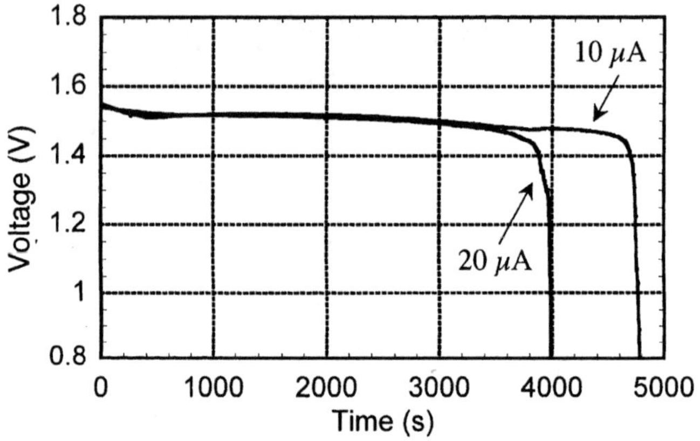

Figure 7. Discharge plot from a planar alkaline microbattery at currents of 10 and 20 μAmps.

An important challenge remains in how to package these cells for commercial use. Since a liquid consisting of a highly corrosive alkaline solution is used as the electrolyte of the Zn-Ag_2O microbatteries, sealing of these tiny structures is not easily achieved without compromising their size and wasting significant mass and footprint on the packaging. A better alternative would be to fabricate "dry" microbatteries, where issues such as leakage and contamination of the electrolyte are not as critical, or employ embedding strategies as discussed later in the chapter.

6.2 LDW of Secondary Microbatteries

Lithium microbatteries represent a relatively new battery chemistry that offers significantly higher energy densities than other rechargeable battery systems owing to the small size of Li and its high reduction potential. Secondary lithium ion intercalation microbatteries have been fabricated by laser printing $LiCoO_2$ powders for the cathode and graphite powders for the anode in stacked geometries.

Two types of separators have been used to make the lithium cells. In one case, porous polymer membranes were laser cut and placed between the cathode and anode layers (Wartena, 2004), while on the other an ionically conductive polymer nanocomposite membrane was laser printed on top of the cathode layer (Sutto, 2006).

The $LiCoO_2$ cathodes and carbon anodes were deposited by LDW onto aluminum and copper foils (15 mm x 15 mm x 40 μm thick), respectively. Both the cathode and anode inks were prepared by mixing the powders with a polymer binder, poly(vinylidene fluoride-hexafluoropropylene) (PVDF-HFP), and an organic solvent, (dibasic ester). In this particular case, the area of the transferred electrodes is 10 mm x 10 mm. Transferred films are dried in a vacuum oven at 120 °C for 2 days prior to placing them inside a dry glove box (oxygen less than 1 ppm), where they are assembled into cells using one of the two types of separator membranes. An optical micrograph of one of these solid-state Li-ion microbatteries is shown in Fig. 8(a).

Fig. 8 (b) shows the charge/discharge cycling performance (C/18 rate) of a packaged Li-ion microbattery, demonstrating the expected behavior for the $LiCoO_2/Li$ system. This microbattery had a charge/discharge efficiency of ~98% and a capacity per unit area of 205 $\mu Ahr/cm^2$, which compares favorably to the value of 163 $\mu Ahr/cm^2$ achieved for sputter deposited Li microbatteries reported by Bates, et al. (Bates, 2000).

6.3 Advanced Laser Techniques for Microbatteries: Developing Embedded Micropower Sources

One of the important issues for microbatteries and other small-scale electrochemical systems is the development of an appropriate packaging scheme that protects the device from environmental degradation and preserves the small size. In order to address this issue, one can utilize more advanced surface restructuring and processing that is possible with the LDW process.

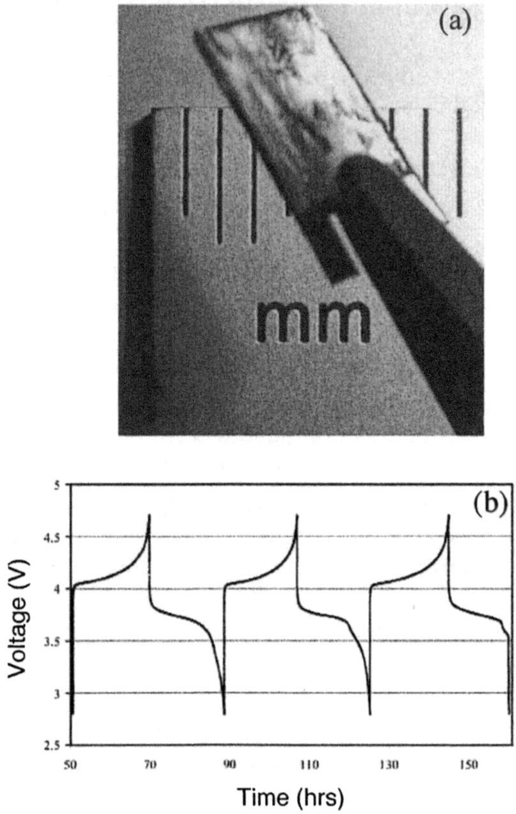

Figure 8. (a) Optical micrograph of a packaged Li-ion microbattery made by LDW. (b) Charge/discharge profile from one of these microbatteries (C/18 rate, cycles 3-5).

In many cases where microbatteries are required, one desires to place the cell directly on the substrate of the microdevice to be powered. In these cases, one can take advantage of the existing substrate as a packaging material by placing the battery *within* the substrate rather than on top of the substrate. For instance, in the case of fiberglass or polyimide circuit boards, one can laser micromachine pockets of desired dimensions and deposit each of the electrode layers inside as shown in Fig. 9(a). The problem is then reduced to sealing the top of the battery with a metallic current collector to produce a fully embedded solution. With further refinements, one can create a separator layer from the substrate itself by laser micromachining both top and bottom to leave a thin layer of polymer or fiberglass before refilling it from both sides. This is illustrated schematically in Fig. 9(b). In the case of solid-state Li-ion microbatteries, it is also possible to LDW all the layers inside a well within the substrate as shown in Fig. 9(c). The ability to completely transfer and assemble electrochemically and thermally stable micropower sources into bulk materials can be extended to more advanced applications in which the power source is directly incorporated and sealed within the substrates used by the microdevice (Arnold, 2004c).

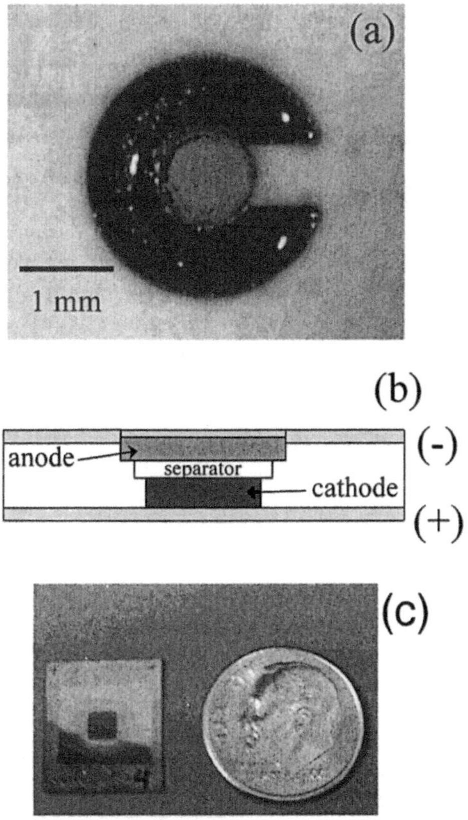

Figure 9. (a) Optical micrograph showing the planar view of an embedded alkaline microbattery made by LDW. (b) Schematic showing an embedded stacked Li-ion microbattery in cross section view. (c) Optical micrograph of a solid-state Li-ion microbattery embedded in a polyimide substrate by LDW.

In summary, LDW processing is an enabling technology for the fabrication of primary and rechargeable microbatteries as the prior work demonstrates. Notably, the LDW is not chemistry specific, allowing different types of materials to be deposited using the same tool by simply changing the source ribbon. More advanced techniques such as deposition of a liquid phase polymer nanocomposite that dries into a solid, ionically conductive micron thick solid separator, or laser surface modifications for embedded designs makes possible other types of microelectrochemical cells that can be incorporated into a wide variety of microdevices.

7. LASER DIRECT-WRITE OF DYE-SENSITIZED SOLAR CELLS

The dye-sensitized solar cell (DSSC) is a photosensitive electrochemical cell with a more complicated electrode configuration in comparison to the previous two systems. The anode in this case comprises of light-absorbing dye molecules attached to the surface of nanocrystalline-TiO_2 (nc-TiO_2) particles. The electrons generated by oxidation of the dye molecules are injected into the conduction band of the wide band-gap TiO_2 and transported to the external circuit through a transparent conducting oxide layer. On the cathode side, a metal catalyst enables the direct reduction of the electrolyte itself (I^-/I_3^-), which subsequently reduces the oxidized dye molecules to their initial state. The schematic structure of a dye-sensitized solar cell is shown in Fig. 10.

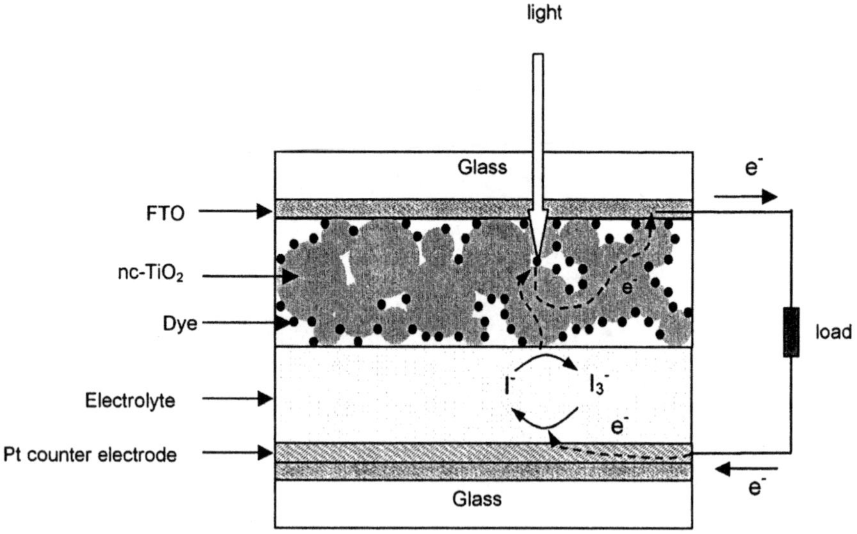

Figure 10. Schematic showing the basic structure and operation of a typical dye-sensitized solar cell.

Over the last decade, DSSCs based on nc-TiO_2 films have been extensively studied as possible alternatives to silicon-based solar cells due to their high power conversion efficiencies (~ 10%) and expected low materials and manufacturing costs (O'Regan, 1991; Nazeeruddin, 1993; Barbe, 1997; Nakada, 2002). Mesoporous nc-TiO_2 layers with high surface area are essential for achieving high efficiency in the DSSC system because a large amount of dye can be adsorbed on the surface of the nc-TiO_2 particles resulting in an increased solar light absorption and an increased reacting interface per unit area. As described in earlier sections, the LDW process is ideally suited for generating these microstructures with arbitrary patterns on a variety of substrates. In this section, we will describe the use of LDW techniques for the fabrication of mesoporous nc-TiO_2 electrodes for dye-sensitized solar cells (the remaining components are fabricated using traditional techniques). In addition, advanced laser processing techniques for *in situ* annealing and sintering of the films without damaging the underlying substrates are discussed with implications for active nc-TiO_2 layers fabricated on flexible plastic substrates.

7.1 LDW of the nc-TiO_2 Layer

The general structure of a LDW dye-sensitized solar cell is similar to the stacked geometry described in the previous section on microbatteries. However, in this case, we use traditional vapor deposition to create the transparent conducting oxide layers and Pt cathode. The inks used are made from water based colloidal suspensions of nc-TiO_2 powders mixed with organic additives. Mesoporous, nc-TiO_2 films are deposited onto fluorine-doped tin oxide (FTO)-coated glass substrates by LDW of these inks (Kim, 2004, 2005). The transferred films are dried in air and sintered in the oven at 450°C for 30 minutes. Once the films are sintered, the electrode is soaked in the dye solution in order to coat the TiO_2 surface. Finally, the completed cells are assembled and sealed using a Surlyn gasket to separate the anode and cathode layer and provide a reservoir for the I^-/I_3^- electrolyte.

The thickness and mass of the nc-TiO_2 layers can be easily controlled by varying the number of LDW passes over the substrate. By looking at the cross-sectional SEM of Fig. 11(a), it is clear that no interfacial gaps are formed between the nc-TiO_2 layers because the transferred material remains in the form of a viscous fluid after the LDW transfer. As with the previous cases, the transferred materials maintain a homogeneously distributed network of particles with a high degree of porosity, consistent with a high surface area structure. The 3-dimensional network of interconnected TiO_2 nanoparticles enables good electron conduction while the high surface area structure can maximize the amount of dye adsorbed on the surface of the

TiO_2 particles. Both properties are essential for the fabrication of efficient dye-sensitized solar cells.

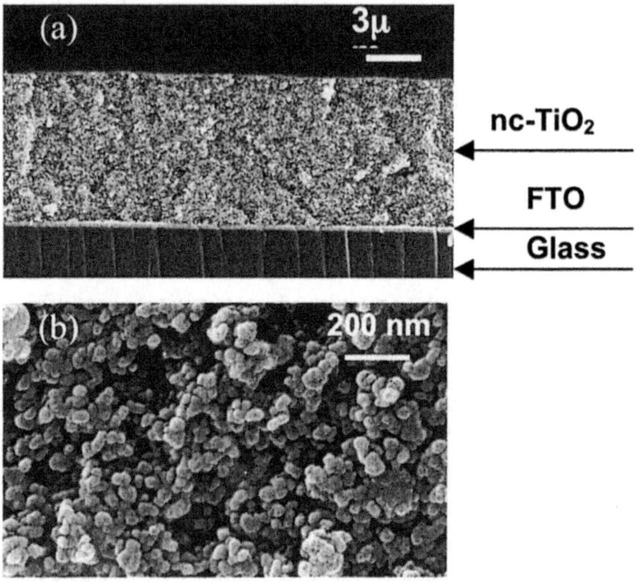

Figure 11. SEM micrographs showing (a) the cross section and (b) the surface of a 12-μm thick nc-TiO_2 layer transferred by LDW on FTO coated glass. The film was sintered at 450 °C for 30 min.

Fig. 12 shows the current density (*J*) versus voltage (*V*) characteristics for several dye sensitized solar cells containing LDW nc-TiO_2 layers of varying thickness. The initial increase in J_{sc} with the nc-TiO_2 film thickness can be related to the increased surface area of the TiO_2 films and concomitantly the increased amount of adsorbed dye molecules. Although the thicker films allow for more adsorbed dye molecules and would be expected to produce higher photocurrents, in practice it is found that the very thick nc-TiO_2 films also contain a large number of defect/recombination sites resulting in an overall decrease in the cell operational characteristics. Because of this, the thickness of the nc-TiO_2 films must be optimized for improving both J_{sc} and V_{oc} parameters simultaneously. In the current work, cells made with an optimum thickness of 15 μm exhibited a J_{sc} of 10.1 mA/cm^2, V_{oc} of 0.64 V, a fill factor (*ff*) of 0.65, and a light power conversion efficiency (η) of ~ 4.3 %. The conversion efficiency is comparable to those reported earlier for analogous cells fabricated with commercial TiO_2 powders (P25) using traditional techniques (Okuya, 2002; Nakada, 2002; Hinsch, 2001).

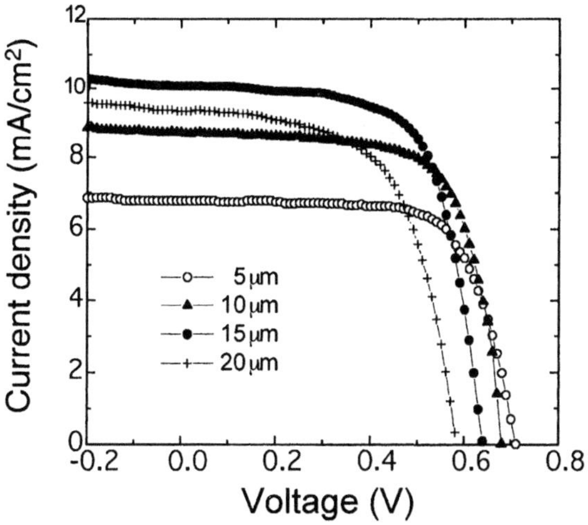

Figure 12. Current density versus voltage (*J-V*) characteristics of dye-sensitized solar cells fabricated by laser based techniques with different thicknesses of active TiO_2 layers under the light intensity of 100 mW/cm^2 (AM 1.5 simulated solar illumination). The active cell area of all samples was 0.25 cm^2.

7.2 Advanced Laser Techniques for DSSCs: *in situ* Laser Sintering of the nc-TiO_2 Layer

To date, the use of organic additives, typically a poly(ethylene glycol)-based surfactant, is necessary for the formation of a viscous colloidal suspension that is essential in building crack-free thick films with increased porosity after sintering when using LDW or traditional fabrication approaches. In order to maximize the device efficiency, the bladed or printed nc-TiO_2 electrodes must be sintered at relatively high temperatures (~ 450°C) to remove the organic additives and to form an electrically connected nanostructure network within the mesoporous electrode. Due to the high sintering temperature required, most of the research and development (R&D) work on DSSCs has been limited to substrates that can withstand high temperatures such as glass and stainless steel. Considering manufacturing cost, flexibility and safety issues, the use of plastic substrates can offer many advantages over glass substrates because the DSSC devices based on plastic substrates would be cheaper, lighter, and thinner than the DSSC devices on glass substrates. For this reason, the traditional high temperature sintering should be replaced by a low temperature sintering process in the fabrication steps of nc-TiO_2 electrodes.

Using additional laser techniques within the LDW apparatus, it is possible to remove the high temperature sintering step from the overall process (Kim, 2005). Deposited films are first heat treated at 100 °C to remove excess solvent from the paste. This low temperature is not sufficient to remove organic additives from the deposited film (see Fig. 13(a)), but it is sufficient to form clusters approximately 100 - 200 nm in diameter due to the drying of the organic material. Such agglomerates are not well suited to form the electronically connected nanostructured network required for the proper operation of DSSCs. Typical laser sintering at IR wavelengths is insufficient for TiO_2 due to its large band gap (3.0-3.2 eV) and low absorption in the IR region.

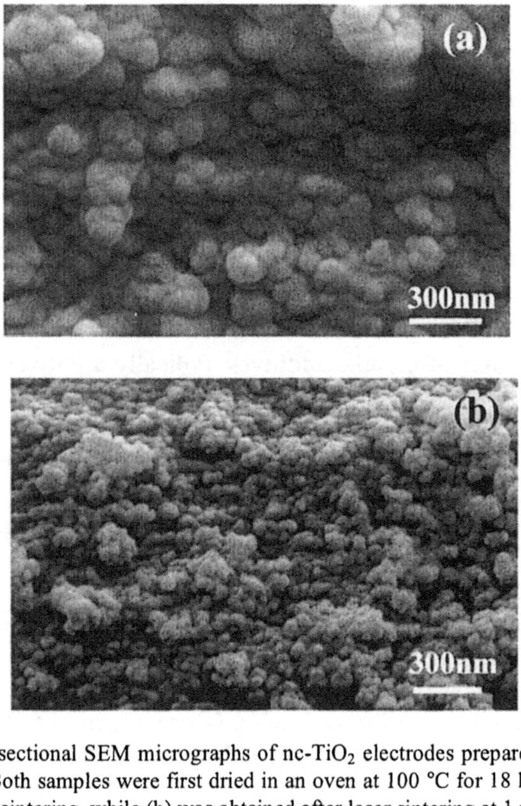

Figure 13. Cross sectional SEM micrographs of nc-TiO_2 electrodes prepared on FTO-coated glass substrates. Both samples were first dried in an oven at 100 °C for 18 hrs. Image (a) was taken before laser sintering, while (b) was obtained after laser sintering at 1 mm/min.

However, by utilizing the same UV source (355 nm) that is used for the transfer of the nc-TiO_2 inks, this problem can be surmounted resulting in sintered structures. Fig.13(b) shows that the laser sintering process is capable of removing the organic additives from the film and form an electrically connected network structure in the nc-TiO_2 film.

To date, the power conversion efficiency of the devices based on the laser sintered nc-TiO_2 electrodes is a factor of two greater than non-sintered films, but still less than what can be obtained from high temperature processes. However, these preliminary results indicate that the laser sintering technique is a promising approach for the fabrication of nc-TiO_2 electrodes especially for plastic DSSCs.

The merit of this approach is that it allows both to transfer the TiO_2 colloidal suspensions and in situ sinter the mesoporous nc-TiO_2 electrodes using the same UV laser, thus simplifying the processing steps required for fabricating DSSCs at low temperatures. Further optimization of the laser sintering conditions such as power intensity, laser beam spot size and translation speed leave ample room for refinement of this technique.

8. THE FUTURE OF LASER DIRECT-WRITE

The use of LDW techniques for the transfer and processing of electronic and power generating materials has been reviewed in the previous sections. Numerous examples of the types of components fabricated using this technique are described and it is clear that the LDW process is well suited to transfer or print inks or fluids containing suspensions or mixtures of the required electronic materials. The cured inks result in conformal patterns that are continuous and well adhered to the substrate. Moreover, by choosing the appropriate powder materials and ink compositions, the same technique can also be used to generate patterns of mesoporous structures with extremely high effective surface areas required for fabricating components for hybrid micropower systems including ultracapacitors, microbatteries and dye-sensitized solar cells. Each of these are indispensable elements in the fabrication of autonomous, self-powered microelectronic systems such as wireless sensor networks. However, in order to reach the full potential of the LDW process it is desired that these elements be integrated into fully functional circuits.

Work toward this goal is progressing rapidly. Laser forward transfer and laser direct-write techniques are poised to play an important role in the development of the next generation of electronic systems as work toward the goal of complete device integration steadily progresses. For example, future microelectronic designs will require integration of many different and increasingly smaller components. To achieve this, it is necessary to accurately place small-scale energy storage systems and individual electrical components such as semiconductor bare die and submillimeter discrete passive components on surfaces that may not be compatible with traditional "pick-and-place" equipment. The placement of each of these components onto virtually any type of surface is enabled by the use of laser transfer methods.

An alternative approach to device integration and miniaturization can be accomplished by incorporating electronic components, energy storage, and their interconnects within the volume of a substrate. Rather than limiting their placement to its surface, it will be possible to integrate electronic circuits within any type of form factor. The development of such "embedded" electronics is being explored using laser direct-write techniques. The remaining sections of this chapter will provide a brief overview of the latest developments taking place in device integration and embedding strategies.

8.1 Laser Transfer of Electronic Components

The use of LIFT processes for the transfer and placement of prefabricated parts or components onto a receiving substrate was first reported by Holmes et al. (Holmes, 1998). In their work, the authors describe the laser-driven release of Si-based microstructures from a UV-transparent substrate with an intermediate polymer sacrificial layer. Upon irradiation with an excimer laser pulse, a thin fraction of the sacrificial layer is vaporized, releasing the microstructure. This technique was later used to demonstrate the laser-assisted assembly of microelectromechanical devices from parts fabricated on separate substrates (Holmes, 2002). These initial results showed how to use LDW process as an alternative to conventional "pick-and-place" approaches for the placement of electronic components such as passives and semiconductor bare dies. The basic concept is quite simple: use a "sacrificial layer" such as the polymer layer used by Holmes to attach the individual components to a UV-transparent support. A laser pulse ablates the sacrificial layer generating gases that release and propel the component towards a receiving substrate placed in close proximity. This laser device transfer process is a contact-less process and thus allows the transfer of very small and very thin components, which could easily be damaged by "pick-and-place" tools.

Recently, this concept has been applied to the laser transfer of semiconductor bare dies. Karlitskaya and coworkers have developed a simple model that predicts the fluence threshold for the release of 200 x 200 μm^2 by 150 μm thick Si dies held with a PVC sacrificial layer (Karlitskaya,

2004). The model shows that the release threshold is below the damage threshold for the reverse side of the die (< 673 K) based on heat diffusion of the absorbed laser pulse through the Si substrate. In this case the authors assumed the laser transfer of devices with the active region facing opposite to the laser pulse. This configuration is not very practical since in order to establish the electrical connections between the pads on the transferred die and the acceptor substrate, extremely precise alignment is required. A better solution is to transfer the die with its active surface facing up enabling direct-write approaches to make contact to the device. The challenge however, is to be able to illuminate the active region of the die with the transfer laser pulse without damaging it. We have recently demonstrated this capability with the laser forward transfer of individual InGaN LED bare dies (250 x 350 μm^2) using a series of very low fluence UV laser pulses (Piqué, 2005b). Once laser transferred, the LED's are electrically tested and their operation verified. The fact that the devices are not damaged upon laser illumination of their active surface demonstrates the possibility to develop a uniquely versatile laser-based component placement and interconnecting system.

8.2 LDW of Embedded Electronic Circuits

The development of embedded surface mount devices (SMD's), semiconductor bare die integrated circuits (IC's), interconnects and power source elements, offers the ability to achieve levels of miniaturization beyond the capabilities of current manufacturing techniques. Given an arbitrary circuit design, significant reductions in volume and overall weight can be achieved by using embedded components. Furthermore, embedded circuits exhibit higher device density and improved electrical performance, resulting in enhanced functionality within a given form factor.

At the Naval Research Laboratory, we demonstrated the use of LDW techniques for the fabrication of embedded microbatteries and electronic circuits (Piqué, 2004a, 2005a). In this work, a simple blinker circuit comprised of six passive SMD components (4 resistors and 2 capacitors), two SMD LED's and one unpackaged IC (LM555 chipset in bare die form) was embedded in a thermoplastic polyetherimide substrate using LDW.

Laser micromachining was used to generate the pockets in the substrate wherein each component was buried. Once in place, the components were buried under a layer of polyimide. The interconnects required by the circuit were made by laser micromachining vias to expose the contact pads on each device.

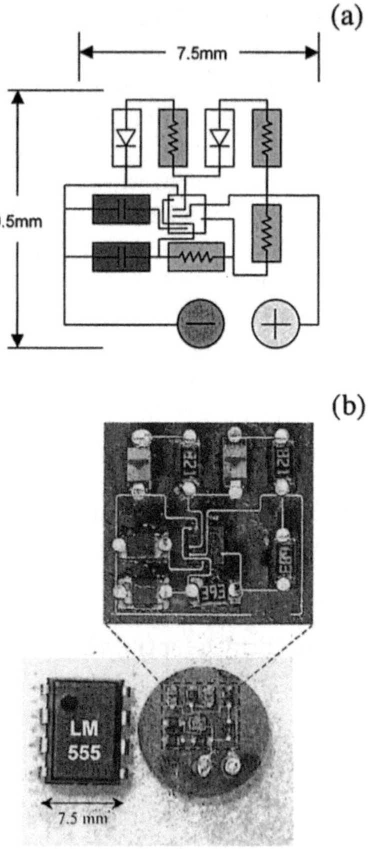

Figure 14. (a) Circuit diagram for an embedded blinker circuit. (b) Optical micrograph of the embedded circuit made by LDW, with the bottom image showing the embedded circuit next to a packaged LM555 chip for size comparison.

The metal interconnects were then generated by laser printing a conductive silver ink. The resulting embedded circuit occupied a footprint smaller than a single packaged LM555 chip. Fig. 14 shows a schematic of the circuit and photographs of the embedded blinker circuit. It is estimated that these LDW embedded circuits will occupy footprints of about 1/4 or less and require less than 1/10 of the thickness of a printed circuit board design. This shows that by using LDW processes it is possible to fabricate functional electronic circuits buried within almost any surface, with the surface or substrate serving both as circuit board and packaging.

9. SUMMARY AND OUTLOOK

Laser direct-write techniques encompass many laser-based processes, which allow the subtraction (i.e. laser micromachining), addition (i.e. laser transfer) and modification (i.e. laser annealing) of many types of materials. As this chapter has shown, LDW can be used for the generation of mesoscale patterns of electronic and power generating materials directly on a substrate without requiring lithographic processing. These patterns can be assembled to form simple structures such as metal lines for establishing the connections between devices in functional circuits, or more complex ones such as the multilayers required for a stacked Li-ion microbattery. The energy storage devices fabricated through this technique exhibit outstanding performance characteristics. In each case, the ability to transfer complex inks from a ribbon to an acceptor substrate under ambient conditions and without changing their physical or chemical properties is one of the unique attributes of the LDW process. Another key attribute of LDW processes is that they can be used for rapid prototyping applications, allowing the design, fabrication and testing of a given structure to be completed quickly.

The recent application of LDW techniques for the transfer of preformed devices such as semiconductor bare dies, and for embedding microbatteries and simple electronic circuits, opens the door for the development of a unique laser-based microelectronics fabrication tool. Such a tool would be capable of fabricating and embedding electronic circuits with the required power storage and power harvesting components within the same substrate. The resulting fully integrated systems could easily be reconfigured to fit within a desired form factor, thus allowing the placement of electronic systems in places that seem inconceivable today. In fact, the functionality of these LDW circuits can be customized for a particular application by choosing the appropriate electronic modules or building blocks from a circuit library available in the LDW tool. Such a LDW system does not yet exist, but it is just a matter of time before the various processes described in this chapter are combined into a single machine capable of making this vision a reality.

10. REFERENCES

Adrian, F.J., Bohandy, J., Kim, B.F., Jette, A.N., and Thompson, P., 1987, A study of the mechanism of metal deposition by the laser-induced forward transfer process, *J. Vac. Sci. Technol.* **B5:** 1490-1494.

Arnold, C. B., Wartena, R. C., Pratap, B., Swider-Lyons, K. E., and Piqué, A., 2002, Laser direct writing of hydrous ruthenium dioxide micro-pseudocapacitors, in: *Electroactive*

Polymers and Rapid Prototyping, D. B. Chrisey and S. C. Danforth, ed., volume 689, pages 275–280, Pittsburgh, PA. Materials Research Society.

Arnold, C. B., Wartena, R. C., Swider-Lyons, K. E., and Piqué, A., 2003 . Direct-write planar micro-ultracapacitors by laser engineering, *J. Electrochem. Soc.* **150:**A571–A575.

Arnold, C. B., Kim, H., Sutto, T.E., and Piqué, A., 2004a, Direct write laser processing for miniature electrochemical systems, *Laser Focus World*, **40**: 9-12.

Arnold, C. B., Kim, H. and Piqué, A., 2004b, Laser transferred primary alkaline microbatteries, *Appl. Phys. A*, **79:** 417-420.

Arnold, C.B., and Piqué, A., 2004c, Self-filling wet electrochemical cells by laser processing, US patent application 20040256359.

Arnold, C.B., Piqué, A., Auyeung, R. C. Y. and Nurnberger, M., 2004d, Laser-based technique for producing and embedding electrochemical cells and electronic components directly into circuit board materials, US patent application 2005000613.

Auyeung, R.C.Y., Wu, H.D., Modi, R., Piqué, A., Fitz-Gerald, J.M., Young, H.D., Lakeou, S., Chung, R., and Chrisey, D.B., 2000, Matrix-assisted laser transfer of electronic materials for direct-write applications, *Proc. SPIE* **4088:** 393-396.

Bähnisch, R., Gross, W., and Menschig, A., 2004, Single-shot, high repetition rate metallic pattern transfer, *Microelectronic Engineering* **50:** 541-546.

Barbe, C. J., Arendse, F., Comte, P., Jirousek, M., Lenzmann, F., Shklover, V., and Grätzel, M., 1997, Nanocrystalline titanium oxide electrodes for photovoltaic applications, *J. Am. Ceram. Soc.* **80**: 3157-3171.

Bates, J.D., Dudney, N. J., Lubben, D. C., Gruzalski, G. R., Kwak, b. S., Yu, X., and Zuhr, R. A., 1995, Thin-film rechargeable lithium batteries, *J. Power Sources* **54**: 58-62.

Bates, J. B., Dudney, N. J., Neudecker, B., Ueda, A., and Evans, C. D., 2000, Thin-film lithium and lithium-ion batteries, *Solid State Ionics* **135**: 33-45.

Bohandy, J., Kim, B.F., Adrian, F.J., 1986, Metal deposition from a supported metal film using an excimer laser, *J. of Appl. Phys.* **60:** 1538-1539.

Bohandy, J., Kim, B.F., Adrian, F.J., and Jette, A.N., 1988, Metal deposition at 532 nm using a laser transfer technique, *J. Appl. Phys.* **63:** 1158-1162.

Conway, B. E. 1999, Electrochemical Supercapacitors, Scientific Fundamentals and Technological Applications, Kluwer Academic, New York.

Dmowski, W., Egami, T., Swider-Lyons, K. E., Love, C. T., and Rolison, D. R., 2002, Local atomic structure and conduction mechanism of nanocrystalline hydrous RuO_2 from x-ray scattering, *J. Phys. Chem. B* **106:**12677–12683.

Esrom, H., Jun-Ying Zhang, Kogelschatz, U., and Pedraza, A.J., 1995, New approach of a laser-induced forward transfer for deposition of patterned thin metal films, *Appl. Surf. Sci.* **86:** 202-207.

Fang, Q. L., Evans, D. A., Roberson, S. L., and Zheng, J. P., 2001, Ruthenium oxide film electrodes prepared at low temperatures for electrochemical capacitors, *J. Electrochem. Soc.* **148**:A833–A837.

Fernandez-Pradas, J.M., Colina, M., Serra, P., Dominguez, J., and Morenza, J.L., 2004, Laser-induced forward transfer of biomolecules, *Thin Sol. Films* **453-454:** 27-30.

Fogarassy, E., Fuchs, C., Kerherve, F., Hauchecorne, G., and Perriere, J., 1989, Laser-induced forward transfer of high-T_c YBaCuO and BiSrCaCuO superconducting thin films, *J. Appl. Phys.* **66:** 457-459.

Greer, J. A. and Parker, T. E., 1988, Laser-induced forward transfer of metal oxides to trim the frequency of surface acoustic wave resonator devices, *Proc. SPIE* **998**: 113-125.

Herman, I. P., 1989, Laser-assisted deposition of thin films from gas-phase and surface-adsorbed molecules, *Chem. Rev.* **89**:1323-1357.

Hinsch, A., Kroon, J. M., Kern, R., Uhlendorf, I., Holzbock, J., Meyer, A., and Ferber, 2001, Long-term stability of dye-sensitized solar cells, J., *Prog. Photovolt: Res. Appl.* **9**: 425-438.

Holmes, A.S., and Saidam, S.M., 1998, Sacrificial layer process with laser-driven release for batch assembly operations, *J. of Microelectromechanical Sys.*, 7:416-422.

Holmes, A.S., 2002, Laser processes for MEMS manufacture, *Proc. SPIE* **4426**:203-209.

Humble, P. H., Harb, J. N., and LaFollettte, R., 2001, Microscopic nickel-zinc batteries for use in autonomous microsystems, *J. Electrochem. Soc.* **18**: A1357-A1361.

Ito, S., Takeuchi, T., Katayama, T., Sugiyama, M., Matsuda, M., Kitamura, T., Wada, Y., and Yanagida, S., 2003, Conductive and transparent multilayer films for low-temperature-sintered mesoporous TiO_2 electrodes of dye-sensitized solar cells, *Chem. Mater.* **15**: 2824-2828.

Kalyanasundaram, K., and Grätzel, M., 1998, Applications of functionalized transition metal complexes in photonic and optoelectronic devices, *Coordination Chemical Reviews* **77**: 347-414.

Kántor, Z., Tóth, Z., Szorenyi, T., and Tóth, A.L., 1994, Deposition of micrometer-sized tungsten patterns by laser transfer technique, *Appl. Phys. Lett.* **64:** 3506-3508.

Karaiskou, A., Zergioti, I., Fotakis, C., Kapsetaki, M., and Kafetzopoulos, D., 2003, Microfabrication of biomaterials by the sub-ps laser-induced forward transfer process, *Appl. Surf. Sci.* **208-209:** 245-249.

Kim, H., Kushto, G. P., Arnold, C. B., Kafafi, Z. H., and Piqué, A., 2004, Laser processing of nanocrystalline TiO_2 films for dye-sensitized solar cells, *Appl. Phys. Lett.* **85**: 464-466.

Kim, H., Auyeung, R.C.Y., Ollinger, M., Kushto, Kafafi, Z. H., and Piqué, A., 2005, Laser-sintered mesoporous TiO_2 electrodes for dye-sensitized solar cells, accepted for publication in *Appl. Phys. A*.

Kim, I. H. and Kim, K. B., 2001, Ruthenium oxide thin film electrodes for supercapacitors. Electrochem, *Solid-State Lett.* **5**:A62–A64.

Koeneman, P. B., Busch-Vishniac, I. J., and Wood, K. L., 1997, Feasibility of micro power supplies for MEMS, *J. Microelectromech. Sys.* **6**:355–362.

LaFollettte, R., Harb, J. N., and Humble, P., 2001, in: *Sixteenth Annu. Battery Conf. Applications and Advances*, R.S.L. Das and H. Frank, ed., IEEE, Piscatawy, NJ, pp. 349-354.

Linden, D. and Reddy, T. B., 2001, *Handbook of Batteries,* 3rd ed., McGraw-Hill, New York.

McKeown, D. A., Hagans, P. L., Carette, L. P. L., Russell, A. E., Swider, K. E., and Rolison, D. R., 1999, Structure of hydrous ruthenium oxides: Implications for charge storage, *J. Phys. Chem. B* **103**:4825–4832.

Modi, R., Wu, H.D., Auyeung, R.C.Y., Gilmore, C.M., and Chrisey, D.B., 2001, Direct writing of polymer thick film resistors using a novel laser transfer technique, *J. Mater. Res.* **16**:3214-3222.

Nakada, S., Matsuda, M., Kambe, S., Saito, Y., Kitamura, T., Sakata, T., Wada, Y., Mori, H., and Yanagida, S., 2002, Dependence of TiO_2 nanoparticle preparation methods and annealing temperature on the efficiency of dye-sensitized solar cells, *J. Phys. Chem. B* **106**: 10004-10010.

Nazeeruddin, M. K., Kay, A., Rodicio, I., Humphry-Baker, R., Muller, E., Liska, P., Vlachopoulos, N., and Grätzel, M., 1993, Conversion of light to electricity by cis-X_2Bis(2,2'-bipyridyl-4,4'-dicarboxylate)ruthenium(II) charge-transfer sensitizers (X = Cl^-, Br^-, I^-, CN^-, and SCN^-) on nanocrystalline TiO_2 electrodes, *J. Am. Chem. Soc.* **115**: 6382-6390.

Okuya, M., Nakade, K., and Kaneko, S., 2002, Porous TiO_2 thin films synthesized by a spray pyrolysis deposition (SPD) technique and their application to dye-sensitized solar cells, *Sol. Energy Mater. Sol. Cells* **70**: 425-435.

O'Regan, B. and Grätzel, M., 1991, A low-cost, high-efficiency solar cell based on dye-sensitized colloidal TiO_2 films, *Nature* **353**: 737-740.

Osgood, R. M. and Deutsch, T. F., 1985, Laser-induced chemistry for microelectronics, *Science* **227**: 709-714.

Papakonstantinou, P., Vainos, N.A., and Fotakis, C., 1999, Microfabrication by UV femtosecond laser ablation of Pt, Cr and indium oxide thin films, *Appl. Surf. Sci.* **151:** 159-170.

Pimenov, S.M., Shafeev, G.A., Smolin, A.A., Konov, V.I., and Vodolaga, B.K., 1995, Laser-induced forward transfer of ultra-fine diamond particles for selective deposition of diamond films, *Appl. Surf. Sci.* **86:** 208-212.

Piqué, A., Chrisey, D.B., Auyeung, R.C.Y., Lakeou, S., Chung, R., McGill, R.A., Wu, P.K., Duignan, M., Fitz-Gerald, J., and Wu, H.D., 1999a, Laser direct writing of circuit elements and sensors, *Proc. SPIE* **3618:** 330-339.

Piqué, A., Chrisey, D.B., Auyeung, R.C.Y., Fitz-Gerald, J., Wu, H.D., McGill, R.A., Lakeou, S., Wu, P.K., Nguyen, V., and Duignan, M., 1999b, A novel laser transfer process for direct writing of electronic and sensor materials, *Appl. Phys. A* **A69:** 279-284.

Piqué, A., Chrisey, A.D.B., Fitz-Gerald, J.M., McGill, R.A., Auyeung, R.C.Y., Wu, H.D., Lakeou, S., Nguyen, V., Chung, R., and Duignan, M., 2000a, Direct writing of electronic and sensor materials using a laser transfer technique, *J. Mater. Res.* **15:** 1872-1875.

Piqué, A., Fitz-Gerald, J., Chrisey, D.B., Auyeung, R.C.Y., Wu, H.D., Lakeou, S., and McGill, R.A., 2000b, Direct writing of electronic materials using a new laser assisted transfer/annealing technique, *Proc. SPIE* **3922:** 105-112.

Piqué, A., Weir, D.W., Wu, P.K., Pratap, B., Arnold, C.B., Ringeisen, B.R., McGill, R.A., Auyeung, R.C.Y., Kant, R.A., and Chrisey, D.B., 2002, Direct-write of sensor devices by a laser forward transfer technique, *Proc. SPIE* **4637**: 361-368.

Piqué, A., Arnold, C.B., Pratap, B., Auyeung, R.C.Y., Kim, H., and Weir, D.W., 2003, Laser direct-write of metal patterns for interconnects and antennas, *Proc. SPIE* **4977**: 602-608.

Piqué, A., Mathews, S.A., Auyeung, R.C., Ollinger, M., Kim, H., Pratap, B., Arnold, C.B., and Sutto, T.E., 2004a, Application of laser direct-write techniques for embedding electronic and micropower components, *Proc. SPIE* **5662**: 564-569.

Piqué, A., Arnold, C. B., Kim, H., Ollinger, M., and Sutto, T.E., 2004b, Rapid prototyping of micro-power sources by laser direct-write, *Appl. Phys. A*, **79**: 783-786.

Piqué, A., Pratap, B., Mathews, S. A., Karns, B. J., Auyeung, R. C., Kasser, M., Ollinger, M., Kim, H., Lakeou, S., and Arnold, C.B., 2005a, Laser direct-write of embedded electronic components and circuits, *Proc. SPIE* **5713**: 223-230.

Piqué, A., Mathews, S. A., Pratap, B., and Auyeung, R. C., 2005b, Laser forward transfer of semiconductor devices, *Laser Precision Microfabrication Conference, Williamsburg VA*.

Sarangapani, S., Tilak, B., and Chen, C., 1996, Materials for electrochemical capacitors. *J. Electrochem. Soc.* **143**:3791–3799.

Schultze, V., and Wagner, M, 1991, Laser-induced forward transfer of aluminum, *Appl. Surf. Sci. 52:303-309.*

Serra, P., Colina, M., Fernandez-Pradas, J.M., Sevilla, L., and Morenza, J.L., 2004, Preparation of functional DNA microarrays through laser-induced forward transfer, *Appl. Phys. Lett.* **85:** 1639-1641.

Sutto, T. E., Ollinger, M., Kim, H., Arnold, C.B. and Piqué, A., 2006, Laser transferable polymer-ionic liquid separator/electrolytes for solid-state rechargeable lithium ion microbatteries, *Electrochem. Solid-State Lett.* **9**:A69-A71.

Tolbert, W.A., Lee, I.-Y.S., Doxtader, M.M., Ellis, E.W., and Dlott, D.D., 1993a, High-speed color imaging by laser ablation transfer with a dynamic release layer: fundamental mechanisms, *J. Imaging. Sci. Tech.* **37:** 411-421.

Tolbert, W.A., I-Yin Sandy Lee, Xiaoning Wen, Dlott, D.D., Doxtader, M.M., and Ellis, E.W., 1993b, Laser ablation transfer imaging using picosecond optical pulses: ultra-high speed, lower threshold and high resolution, *J. Imaging. Sci. Tech.* **37:** 485-489.

Tóth, Z., Szorenyi, T., and Tóth, A.L., 1993, Ar^+ laser-induced forward transfer (LIFT): a novel method for micrometer-size surface patterning, *Appl. Surf. Sci.* **69:** 317-320.

Trasatti, S. and Kurzweil, P., 1994, Electrochemical supercapacitors as versatile energy stores, *Plat. Met. Rev.* **38**:46–56.

Vincent, C. A. and Scrosati, B., 1997, *Modern Batteries*, 2nd ed., Arnold, London.

Wartena, R. C., Curtright, A.E., Arnold, C. B., Piqué, A. and Swider-Lyons, K. E., 2004, Li-ion microbatteries generated by laser direct write, *J. Power Sources* **126**: 193-202.

Wu, P.K., Ringeisen, B.R., Callahan, J., Brooks, M., Bubb, D.M., Wu, H.D., Piqué, A., Spargo, B., McGill, R.A., and Chrisey, D.B., 2001, The deposition, structure, pattern deposition, and activity of biomaterial thin-films by matrix-assisted pulsed-laser evaporation (MAPLE) and MAPLE direct write, *Thin Sol. Films* **389-399**:607-614.

Young, D., Auyeung, R.C.Y., Piqué, A., Chrisey, D.B., and Dlott, D.D., 2001a, Time-resolved optical microscopy of a laser-based forward transfer process, *Appl. Phys. Lett.* **78**:3169-3171.

Young, D., Wu, H.D., Auyeung, R.C.Y., Modi, R., Fitz-Gerald, J., Piqué, A., Chrisey, D.B., Atanassova, P., and Kodas, T., 2001b, Dielectric properties of oxide structures by a laser-based direct-writing method, *J. Mater. Res.* **16**:1720-1725.

Young, D., Auyeung, R.C.Y., Piqué, A., Chrisey, D.B., and Dlott, D.D., 2002, Plume and jetting regimes in a laser based forward transfer process as observed by time-resolved optical microscopy, *Appl. Surf. Sci.* **197-198**:181-187.

Zergioti, I., Mailis, S., Vainos, N.A., Papakonstantinou, P., Kalpouzos, C., Grigoropoulos, C.P., and Fotakis, C., 1998a, Microdeposition of metal and oxide structures using ultrashort laser pulses, *Appl. Phys. A* **66:** 579-582.

Zergioti, I., Malilis, S., Vainos, N.A., Fotakis, C., Chen, S., and Grigoropoulos, C.P., 1998b, Microdeposition of metals by femtosecond excimer laser, *Appl. Surf. Sci.* **127-129:** 601-605.

Zergioti, I., Karaiskou, A., Papazoglou, D.G., Fotakis, C., Kapsetaki, M., and Kafetzopoulos, D., 2005, Femtosecond laser microprinting of biomaterials, *Appl. Phys. Lett.* **86:** 163902-1-163902-3.

Zhang, C., Liu, D., Mathews, S.A., Graves, J., Schaefer, T.M., Gilbert, B.K., Modi, R.,Wu, H.-D., and Chrisey, D.B., 2003, Laser direct-write and its application in low temperature Co-fired ceramic (LTCC) technology, *Microelectronic Eng.* **70**:41-49.

Zheng, J. P., Cygan, P. J., and Jow, T. R., 1995, Hydrous ruthenium oxide as an electrode material for electrochemical capacitors, *J. Electrochem. Soc.* **142**:2699–2703.

Zheng, J. P. and Xin, Y., 2002, Characterization of $RuO_2 \cdot xH_2O$ with various water contents, *J. Power Sources* **110**:86–90.

Chapter 15

DEVELOPMENT OF INERTIAL FUSION ENERGY BY LASERS

Sadao Nakai[1,2] and Kunioki Mima[1]
[1]Institute of Laser Engineering, Osaka University, Suita, Osaka, 565-0871, Japan; [2]The Gradurate School for the Creation of New Photonics Industries, Hamamatsu, Shizuoka, 431-1202, Japan

1. INTRODUCTION

The physics of the 'implosion' of a fusion fuel pellet has been well investigated since 1972 (Nuckolls et al., 1972), and achieving fusion ignition and burning, which leads to fusion energy gain, may well be demonstrated

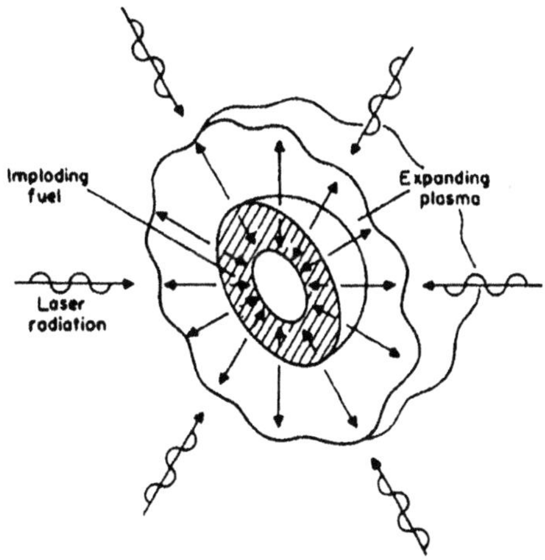

Figure 1. The concept of a laser driven implosion.

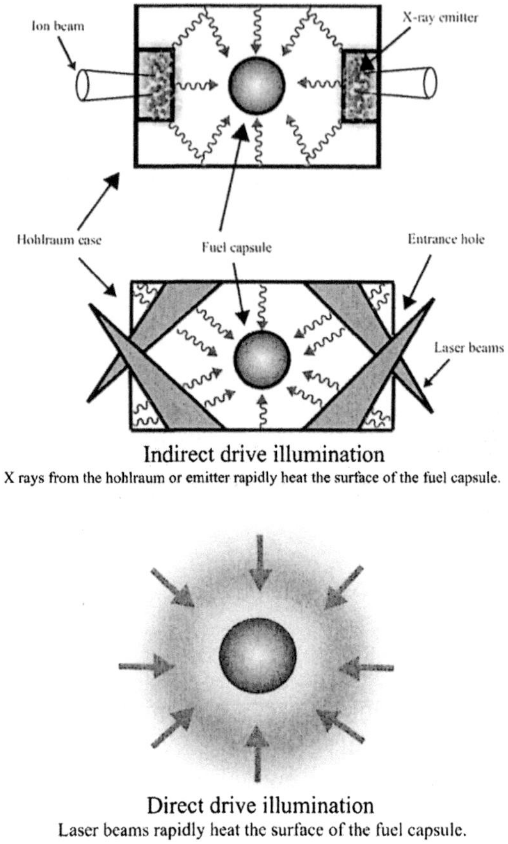

Figure 2. Basic implosion schemes 'direct drive' and 'indirect drive'.

within the next decade. The megajoule laser facilities that are now under construction in the United States and France are expected to demonstrate fusion energy gain. This will be an epoch-making achievement, which will give us the real means to solve the future energy and environment problems of the world.

When an intense laser light is uniformly impinged on a spherical fuel pellet with an intensity of the order of 10^{14}-10^{15} Wcm^{-2}, the laser energy is absorbed on the surface to generate a high temperature plasma of 2-3 keV, and an extremely high pressure of a few hundred megabars is generated. This pressure accelerates the outer shell of the target towards the target centre as schematically shown in Fig. 1. The mechanism of the acceleration is the same as rocket propulsion. When the accelerated fuel collides at the centre, compression and heating occur. If the dynamics are sufficiently spherically symmetric, the central area is heated up to 5-10 keV (called the 'central spark'), and a fusion reaction starts.

In reactor scale implosions, a fuel pellet with an initial radius of about 3 mm should be compressed to a radius of about 100 μm. For this purpose, highly precise uniformity is required for irradiation intensity distribution over the fuel pellet as well as high quality sphericity and uniformity of the fuel pellet.

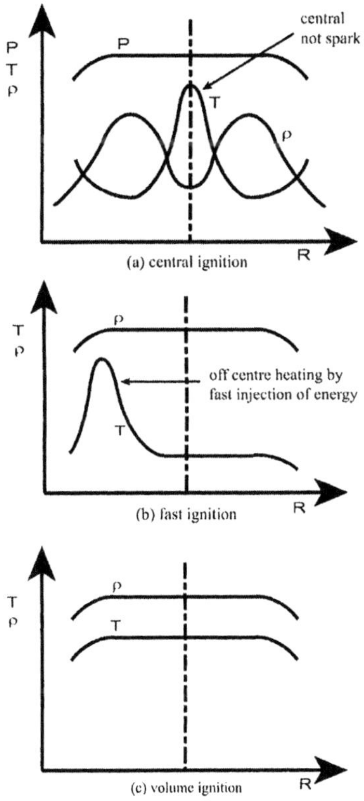

Figure 3. Ignition schemes of compressed fuel. The central ignition scheme (a) is analysed with an isobaric model, and the fast ignition (b) is analysed with an isochoric model. Uniform distribution of density and temperature is assumed for volume ignition (c).

The implosion velocity for achieving the fusion ignition temperature of 5-10 keV at the centre is required to be (3-4) x 10^7 cm s^{-1}. Therefore, the pulse length of the laser should be 10-20 ns, and in this timescale the directed megajoule energy has to be delivered to a fuel pellet of radius about 3 mm. The directed energy to drive the implosion is simply called the 'driver'.

There are two main schemes for pellet implosion: direct and/or indirect drives. The concepts of these are shown in Fig. 2. In 'direct-drive implosions', the surface of the fuel pellet is directly irradiated as described in the early part of this section. In 'indirectdrive implosions' (Lindl, 1998) the driver energy is converted to soft x-rays, which are confined in a cavity. The confined soft x-ray is absorbed on the surface of the fuel pellet to generate ablation pressure to drive the implosion.

As for the ignition of the imploded fuel, there are basically three concepts (Atzeni, 1995): central ignition (Nuckolls et al., 1972); fast ignition (Tabak et al., 1994) and volume ignition. The corresponding distribution of density, temperature and pressure are shown in Fig. 3.

When the implosion is spherically symmetric and there is no significant instability to cause mixing of the boundary, the central hot spark is formed as shown with the isobaric model (Mayer-ter-Vehn, 1982) in Fig. 3(a).

If the implosion produces a compressed fuel of uniform density, we can ignite it with local heating by introducing an intense laser of pulse length shorter than the expansion time of the compressed core. It is called fast ignition with the isochoric model as shown in Fig. 3(b). When the whole compressed core is of high density and the temperature is high enough to ignite the entire volume, it is called volume ignition as shown in Fig. 3(c).

2. GAIN SCALING OF DIRECT AND INDIRECT DRIVE, AND OF FAST IGNITION

The fusion energy gain (target gain) is defined as

$$Q = \frac{E_{\mathrm{f}}}{E_{\mathrm{l}}} \tag{1}$$

where E_{f} is the fusion output energy from the compressed fuel pellet and E_{l} is the laser energy irradiated onto the fuel pellet. It includes the entire physical process (Nakai and Takabe, 1996) of implosion by the direct drive as follows:

$$\frac{E_f}{E_l} = \eta_{abs} \bullet \eta_h \bullet G, \tag{2}$$

where η_{abs} is the absorption coefficient of the laser light incident on the fuel pellet, η_h is the hydrodynamic efficiency, which includes the transport of absorbed laser energy to the ablation front, ablative acceleration of the fuel shell and compression of the fuel at the pellet centre, and G is the fuel gain, which is the ratio of the fusion output energy to the thermal energy of the compressed fuel.

In numerical simulations, all the physical processes are included to relate the fusion energy output to the incident laser energy with the given conditions for the fusion pellet and the incident laser light. Most of the simulation codes throughout the world have been developed independently (Hogan, 1995). The validity of a simulation code has generally been verified by comparisons of the results with the experimental data. For this purpose, not only the integrated implosion but also the specified physics, which is one of the elementary processes of the implosion, is investigated with an adequately modelled experimental condition and compared with the numerical simulation. There are many databases that have been accumulated over many years by many laboratories throughout the world relating to elementary processes such as laser matter interaction, energy transport by radiation, thermal and nonthermal particles, ablation and hydrodynamics, and the instabilities related to the implosion process.

The experimental results relating to the integrated implosion and fusion reaction have also been compared with the numerical simulation results, which were reinforced by individual verification for each elementary process.

It should be noted that fusion ignition and burning front propagation through compressed fuel have not yet been demonstrated experimentally. This is really the final goal of the physics research on inertial fusion.

Fig. 4 shows the fusion energy gain scaling with incident laser energy. The gain scaling curve for indirect and direct-drive with central ignition, and the fast ignition gain curve with different fuel densities are those reported from LLNL. The gain curve from ILE is shown for fast ignition with $\rho = 300$ g cm^{-3}. Several point designs from different groups are also plotted on the same figure.

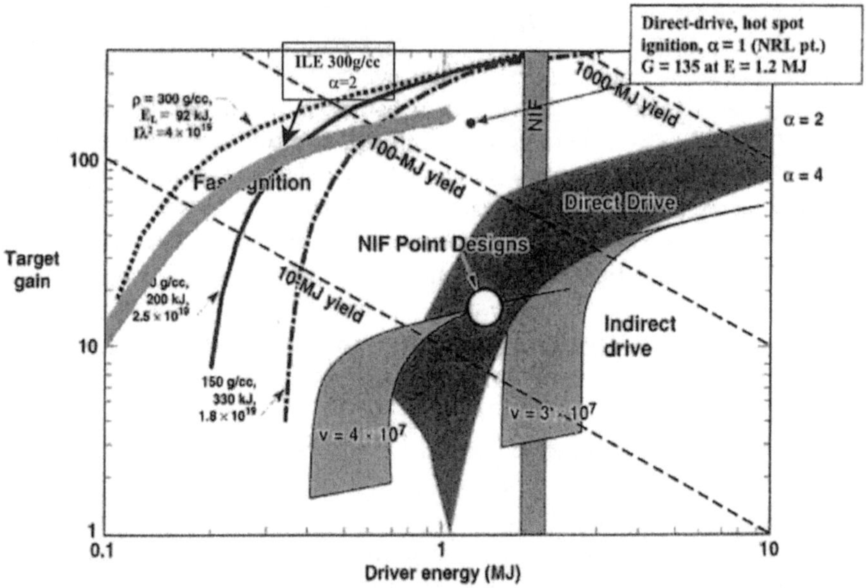

Figure 4. Target gain (fusion energy gain) as a function of the driver energy for central spark ignition with direct and indirect implosion and for fast ignition (based on the LLNL chart).

It should be noted that the completion of NIF in about 2008 will demonstrate fusion ignition and burning and energy gain. The credibility of the gain by the direct-drive central ignition target is discussed in section 3. It is strongly believed that the uniform irradiation configuration of NIF 192 beams will be able to achieve fusion ignition and burning and energy gain. If fast ignition works well, we can achieve a higher gain of 100-300 with less than 1 MJ laser energy. The recent progress in fast ignition and its credibility and uncertainty are discussed in section 4.

3. CENTRAL HOT SPARK IGNITION BY DIRECT DRIVE

3.1 Progress in direct-drive implosion

In directdrive implosion by laser, the most serious physical issue is the stability of the implosion (Bodner, 1998; Shigemori, 2002). Instability grows from the initial perturbation on the target and nonuniformity of irradiation by the laser, and leads to breakup of the shell and quenching of the hotspark at the centre, as schematically shown in Fig. 5. Concerning the stability of implosion, the experimental data on various parameters of the compressed core are compared with a one-dimensional simulation. Typical examples are shown in Fig. 6 (Nakai and Takabe, 1996), where the areal densities agree with the one-dimensional prediction up to a convergence ratio (CR; initial target radius divided by the final radius) of about 30, when the irradiation uniformity is improved by using a random phase plate (RPP). The neutron yield, however, decreased faster as CR increased, as shown in Fig. 6(b), and about 10^{-3} of the one-dimensional calculations at a CR of 30, where the high-density compression was achieved. The estimated radius of the hot spark with onedimensional simulations at the same CR was calculated to be ~5 μm (hot-spark CR ~50). The possible explanation for the discrepancy in yield would be that the remaining irradiation and target nonuniformities might cause a collapse of the hot spark region, from where most of the neutrons are generated in the one-dimensional simulation.

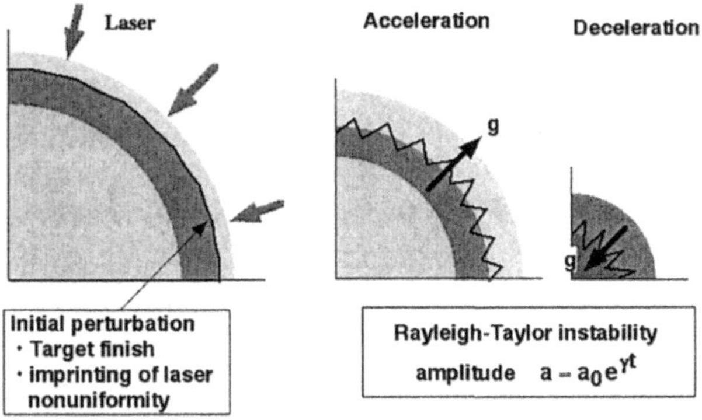

Figure 5. The biggest obstacle in directdrive implosion is shell breakup and quenching of the hot spark by hydrodynamic instability.

Recently, significant progress in directdrive implosions has been demonstrated at the LLE, University of Rochester (McCrory, et al., 2002). The LLE experiments were conducted on the OMEGA 60-beam, 30 kJ, UV laser system. The improvements in the irradiation uniformity have significantly improved the performance of implosions. A single-beam nonuniformity of 3% (averaged over 300 ps) was achieved with full implementation of 1-THz bandwidth two-dimensional smoothing by spectral dispersion (SSD) (Skupsky, et al., 1989), and polarization smoothing with birefringence wedges (Regan et al., 2000). This corresponds to an on-target nonuniformity of less than 1% rms due to beam overlap. The beam-to-beam power imbalance has been reduced to below 5% rms.

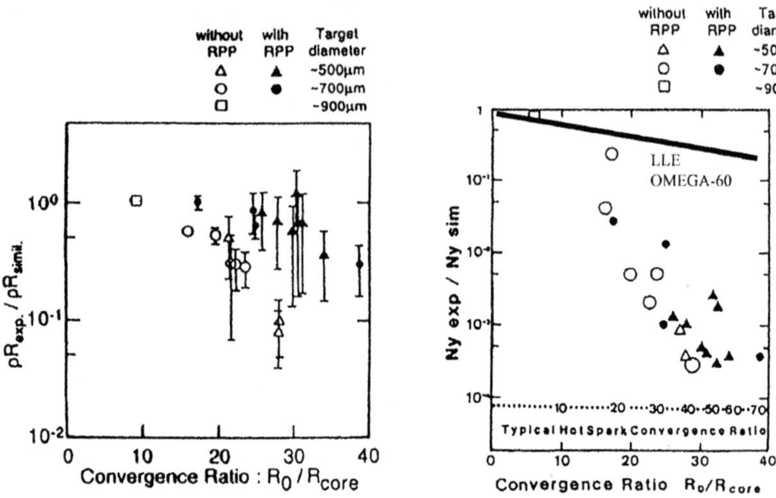

(a) (b)

Figure 6. Normalized implosion performance as a function of the computational CR or the core area. (a) The measured areal mass density normalized by the value obtained with a one-dimensional simulation code. (b) The same but for fusion neutron yield.

With these conditions, moderate CR targets (~15) produce ~30% of the neutron yield predicted by one-dimensional prediction and nearly 100% of the predicted fuel and shell areal densities. At predicted CR values close to 40, the primary neutron yield is ~20% of the one-dimensional prediction. These results are also plotted on Fig. 6(b). The remarkable progress of implosions shows the validation of directdrive inertial fusion heading towards high-gain laser IFE for power plants. It is also reasonable to be increasingly confident about achieving ignition, burning, and energy gain at the reactor scale because the size of the hotspark and surrounding cold fuel is larger at a larger pellet implosion with increased laser energy. The thickness of the mixed region and loss of the hot-spark region will become less serious, and once the ignition occurs at the centre, all the fuel will be burned to produce fusion energy. Recent simulations of high-gain implosion include the mixing effect due to R-T instability. The effect is included in predicting the gain curve in Fig. 4.

3.2 Implosion stability

The physical processes of implosion related to the stability issue are schematically shown in Fig. 7. The key element that governs the stability of the directdrive implosion was identified as being the R-T instability, which leads to the requirement of uniformity of laser irradiation and the fuel pellet.

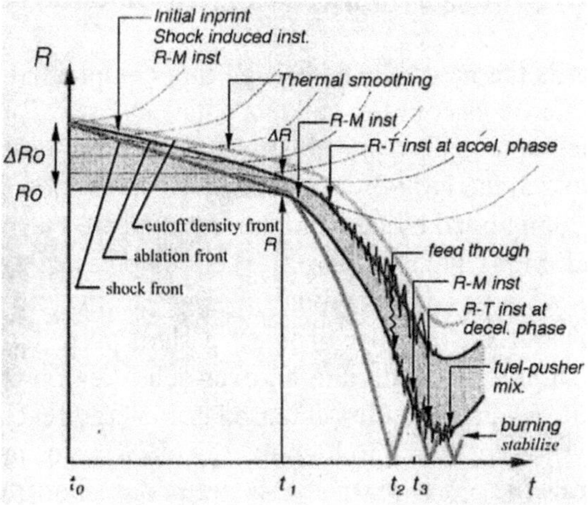

Figure 7. Implosion and physical processes related to the stability.

The growth rate of R-T instability at the ablation front (Takabe et al., 1985; Lindl, 1995; Betti et al., 1998) is given as

$$\gamma = \alpha\sqrt{\frac{kg}{1+kL}} - \beta k v_a, \tag{3}$$

where k is the wave number of the growing mode, g is the acceleration, v_a is the ablation velocity, and L is the density scale length at the ablation front. Empirical values of α = 0.9 and β = 3-4 are given for comparison of numerical and experimental results for directdrive implosions.

The concepts of stabilizing implosions aim to enlarge the characteristic scale length, L, and to increase the ablation velocity, v_a, with the reduced density, ρ_a, keeping the ablation pressure, $\rho_a v_a$, constant. They are (a) shock preheating (McKenty, et al., 2001), (b) radiation energy transport (Bodner, et al., 2000), and (c) nonlocal heat transport by hot electrons (Shigemori, et al., 2001). The gain estimations of recent numerical simulations of direct-drive implosions include a consideration of these physical processes of the instability and stabilizing mechanism. Together with the progress in fuel pellet fabrication and laser beam control technology, it can be reasonably expected that fusion ignition, burn and energy gain will be demonstrated in about 2010 by directdrive implosions with multibeam megajoule lasers if the beams are properly arranged for direct drive.

4. FAST IGNITION

4.1 Fast ignition scheme and ignition condition

Fast ignition is the new scheme for igniting compressed fuel plasmas by using an ultra-intense laser pulse, called a petawatt laser (Tabak et al., 1994). This scheme has potential advantages over central hot-spark ignition since the compression and the ignition can be separated. As shown in Fig. 8, when a fuel pellet is compressed by laser implosion, the core plasma is heated by a short-pulse and high intensity laser. The compressed core is generally surrounded by a large-scale coronal plasma at the time of maximum compression. Therefore, in order to inject a short-pulse laser into the core, the laser pulse should penetrate into an over-dense region by hole boring or cone laser guiding. When the ultra-intense laser energy reaches near the core, the laser energy is converted into high-energy electrons or ions that carry the energy and heat the core plasma. The required short-pulse energy for ignition depends upon the heating spot diameter, heating depth, transport and absorption efficiencies, electron and ion energy spectrum and so on.

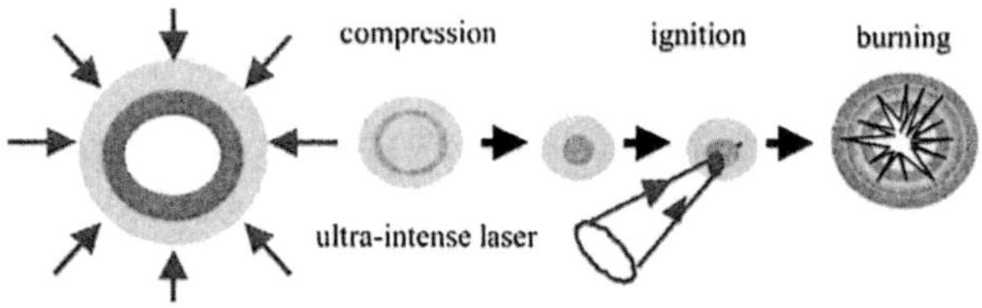

Figure 8. Fast ignition concept at the maximum compression of laser implosion: the ultra-intense short pulse laser is injected to heat the dense core plasma.

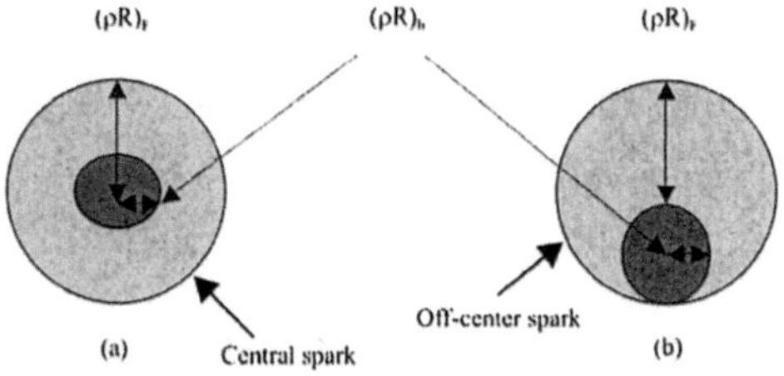

Figure 9. Structure of central hot spark and off-centre hot spark.

The fast ignition hot spark is isochoric between the spark and main fuel and off-centre. First of all, we will compare the ignition condition for an isochoric spark with that for an isobaric spark. One-dimensional burning plasma simulations are carried out to determine the ignition boundaries in ρR and T space, of which the hot-spark geometry is shown in Fig. 9 (a) (Mahady et al., 1999).

Fig. 10 shows that the ignition boundary for an isochoric hot spark is higher with respect to $(\rho R)_h$ than that for an isobaric hot spark since the main fuel layer acts as a tamper layer to suppress the expansion of the hot spark. The fast ignition is relevant to the isochoric hot spark case. In this case, the approximated ignition condition is given as

$$(\rho R)_h^3 T\left[(\text{g cm}^{-2})^3\text{keV}\right]=1.0, \tag{4}$$

when the spark temperature is around 10 keV. Since the hotspark energy is given as

$$E_g = 10.6\,(\rho R)_h^3 T\left[\,(\text{gcm}^{-2})^3\,\text{keV}\right] / \left(\frac{\rho}{\rho_s}\right)^2\ [\text{GJ}]\ , \tag{5}$$

the required spark energy from Eqs. (1)-(2) is approximately given as

$$E_g = 40\left\{\rho/(100\text{gcm}^{-3})\right\}^{-2.0}\ \text{kJ} \quad . \tag{6}$$

The ignition condition of the off-centre spark has been compared with central hot spark ignition by Mahady, et al.(1999) and Piriz (1998). For the off-centre isochoric hot spark, the two-dimensional burning simulation results in the initial geometry of Fig. 9(b). It is indicated by the triangles of Fig. 10 that the ignition condition is the same as that of the central isochoric spark case.

However, the cylindrical heat deposition geometry and finite heating time effects change the ignition condition significantly: the hot spark is produced at the edge of the main fuel by relativistic electron heating. The heated area radius and depth are r_b, and d, respectively, and the fuel density and radius are ρ and R, as shown in Fig. 11.

Note here that the stopping range of the intense electron beam is shorter than the 'thin' electron beam as discussed by Deutsch, et al. (1996). Namely, when the mean distance between beam electrons is shorter than the Debye shielding distance of the plasma, the collective interactions between beam electrons and the plasma become important and enhance the stopping power. Furthermore, the self-generated magnetic field may play important roles in the heat deposition of the intense relativistic electron beam. Since the magnetic field effects on the heat deposition processes are not clarified yet, the following discussions include only the stopping range shortening by Deutsch, et al.(1996).

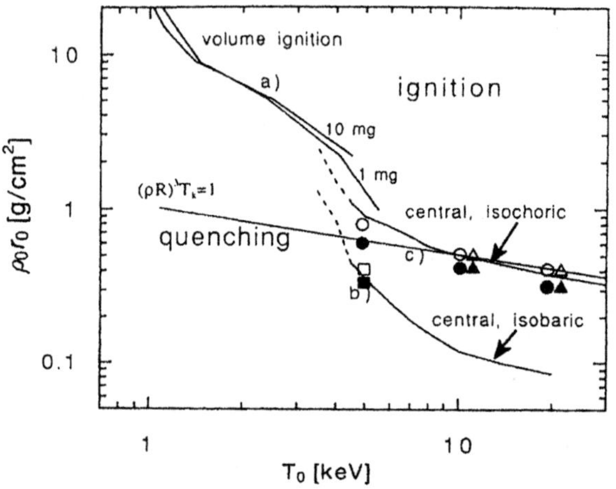

Figure 10. Ignition conditions for (a) volume ignition, (b) central isobaric spark (□ ,■), (c) central isochoric spark (△ ,▲) and off centre round isochoric spark (○ ,●).

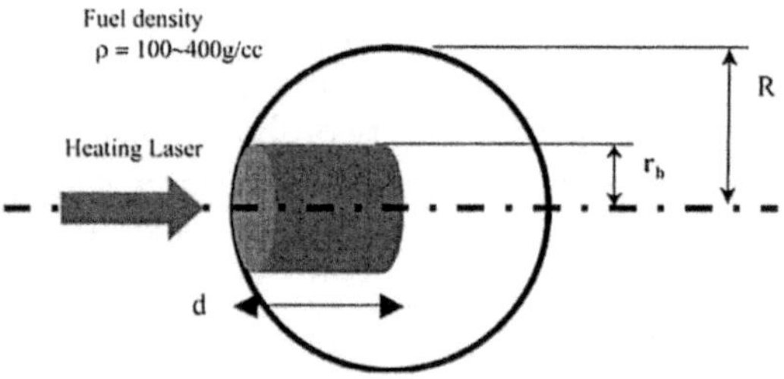

Figure 11. Hot spark-core plasma geometry. Heating pulses are injected from the left to produce a cylindrical hot spark.

In the geometry of Fig. 11, the ignition conditions have been investigated with two-dimensional burning simulations by Atzeni (1999). Once the ignition condition is met, the burning fraction is more or less the same as the central ignition. Through a heating parameter survey of two-dimensional simulations, the ignition conditions for the heating energy, power and radius, the energy deposition range, and the fuel density are explored in many papers. According to the comprehensive work by Atzeni (1999), the optimized ignition conditions are given as the following relations:

$$E_g = 140\left\{\rho/(100\mathrm{gcm}^{-3})\right\}^{-1.85}\ \mathrm{kJ} \tag{7}$$

$$P_g = 2.6\left\{\rho/(100\mathrm{gcm}^{-3})\right\}^{-1.00}\ \mathrm{PW} \tag{8}$$

$$I_g = 2.4\mathrm{x}10^{19}\left\{\rho/(100\mathrm{gcm}^{-3})\right\}^{1.95}\ \mathrm{Wcm}^{-2} \tag{9}$$

$$r_b = 60\left\{\rho/(100\mathrm{gcm}^{-3})\right\}^{-0.975}\ \mu\mathrm{m} \tag{10}$$

The range, ρ_d, is assumed to be 0.6 g cm^{-2}. Note that the ignition condition of Eq.(4) is more than three times higher than that of Eq.(3). This is due to the electron stopping range and the heating geometry. However, the fuel density dependence of the ignition energy is essentially the same in both cases. In the actual ignition, r_b, and the range, ρ_d, depend upon the petawatt laser–plasma interaction physics. The r_b depends upon the laser propagation, the electron heat transport, and the heat deposition process. The stopping range ρ_d depends on the plasma resistivity and the collision processes of relativistic electrons. The experimental and theoretical works related to those issues are reviewed in the following section.

4.2 Imploded plasma heating

As shown in Eqs.(7)-(10), the optimum compressed plasma density decreases with increasing hotspark radius, and the heating laser energy and the pulse width, τ_p, increase as $\rho/(100\ \mathrm{g\ cc^{-1}}) = (r_b/60\ \mu\mathrm{m})^{-1.025}$, $\tau_p = 55\{\rho/(100\ \mathrm{g\ cc^{-1}})\}^{-0.85}$ ps and $E_g = 140(r_b/60\ \mu\mathrm{m})^{1.9}$ kJ. When $r_b = 10\ \mu$m, $\tau_p = 10$ ps, $E_g = 4$ kJ and $\rho = 600$ g cc^{-1}, which is 3000 times the solid density of DT, and a 10–20 kJ heating pulse ignites the core plasma. These conditions may be violated because megaelectronvolt electron beams generated by the laser spread wider than a 10 μm radius and/or the compressed density is not as high as 3000 times the solid density. According to the experiments at Osaka, LLNL and so on, r_b could be 30 μm and the required laser energy will be higher than 70 kJ. So the heating spot size and coupling efficiency of short-pulse laser energy to the core plasma are the most important issues in fast ignition research.

Recently, significant theoretical and experimental research has been in progress at Osaka (Japan), LLNL (USA), Rutherford (UK), and other institutions. One such research effort concerns relativistic electron generation, transport and imploded plasma heating, in particular, in a cone guide target. Another example is efficient ion generation by petawatt lasers and its application to plasma heating. In the following section, we will discuss the implosion plasma heating research at ILE Osaka University.

In the case of spherical implosion plasma heating, the coupling efficiency of a PWM laser pulse to the core plasma is sensitive to the focus position of the heating laser because of the long-scale coronal plasma (Tanaka, 2000; Kodama, 2000). When the focus position is near the critical surface of coronal plasmas, we found that the PWM laser pulse penetrates the over-dense region and the neutron yield is significantly enhanced. However, the neutron energy spectra are broad and not isotropic as shown in Fig. 12(a). The width of the spectral peak around 2.45 MeV indicates that the reacting ion energy is of the order of 100 keV. Furthermore, the neutron spectral shape depends upon the direction of the neutron spectrometer. These results can be interpreted as the enhanced fusion reaction around the critical surface.

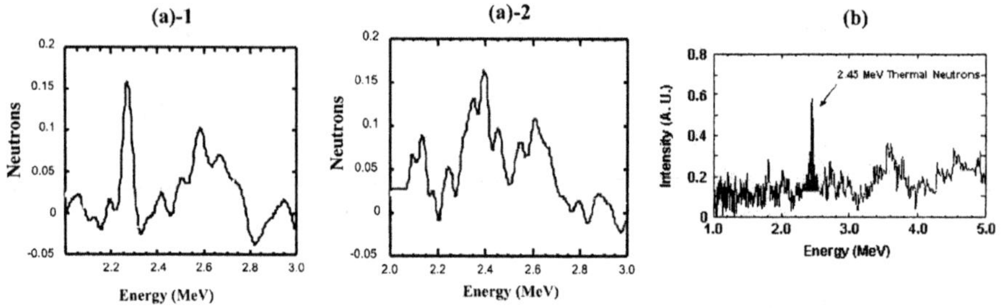

Figure 12. (a) Neutron energy spectra for direct heating of imploded CD pellet. (a)-1 and (a)-2 are the spectra in different directions with respect to the incidence laser pulse. (b) shows the neutron spectrum for a cone guided target.

According to PIC simulations, electrons are expelled from laser channels and solitons by strong laser heating and/or ponderomotive forces. As a result, ions in the laser-produced bubbles are accelerated by Coulomb explosions to a high energy (Sentoku et al., 2003). Since the fusion reaction enhancement is dominated by those high-energy ions, thermal neutron enhancement was not observed and the effect of core plasma heating was not clear in PW-module direct heating (Kitagawa, 2002). However, when the short-pulse laser energy increases to the 1 kJ level in the petawatt laser, the laser pulse penetrates the higher density region since nonlinear scattering may be saturated because of the strong plasma heating. Therefore, in petawatt laser experiments, the thermal neutrons may also increase and core plasma heating may be possible even in a coneless target.

In a cone guide target, the neutron spectral peak at 2.45 MeV is narrow, as shown in Fig. 12(b). The neutron yield is of the order of 10^5 with heating, which is ten times higher than that of nonheating cases. This indicates that thermonuclear fusion is enhanced by the temperature increase in the core plasma due to PWM laser heating. The temperature increase is estimated by the neutron yield enhancement, the observed ρ and the burning time. In the best shot, the temperature increased by 130 eV. We found from these results that 25% of the input PWM laser pulse energy is transported to the core plasma (Kodama, et al., 2000, 2001; Yamanaka, 2000; Mima, et al., 2000).

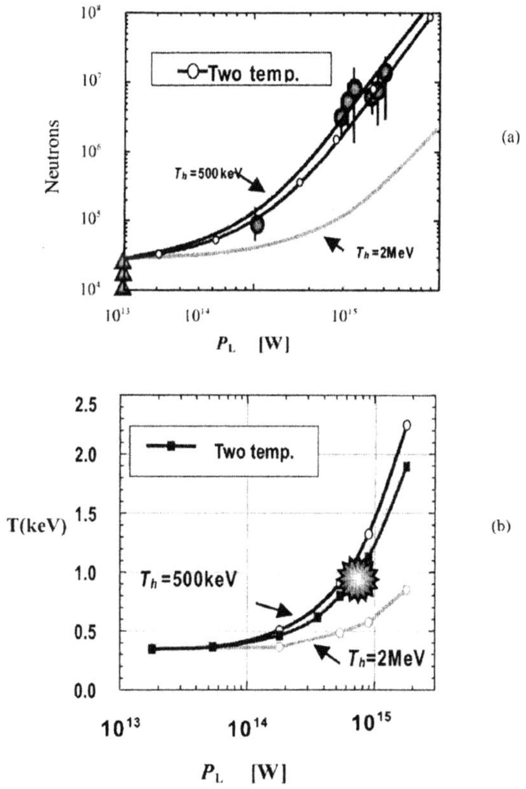

Figure 13. (a) Neutron yield dependence on the laser power (triangle, without heating; circle, with heating; curves are Fokker–Planck simulation results). (b) ion temperature dependence on the laser power (curves, FP simulation).

In petawatt laser experiments, two kinds of cone targets are imploded and heated. The cone angles are 30° and 60° degree. The heating is more efficient in the 30° cone than in the 60° cone. When the 300 J/0.6 ps CPA laser pulse is injected, the neutron yield reaches 10^7, while the neutron yield is 10^5

without heating as shown in Fig. 13 (Kodama et al., 2002). This indicates that the core plasma temperature increased by 500 eV and the energy coupling between the heating laser and core plasma is 20-25%. Since the focused laser energy included in a 5 μm diameter is less than 40%, a 20-25% coupling efficiency means that the actual coupling is higher than 50%.

Otherwise, the laser energy in the halo of the spot is collected by the cone guide. In Fig. 13, a simple scaling curve is shown, where the temperature increase is assumed to be proportional to the input short-pulse laser energy and the coupling efficiency is assumed to be the same as that of the cone guide PWM experimental results. This indicates that the coupling efficiency for a 300 J case is almost the same as for an 80 J case. This scaling law has been used for planning fast ignition experiments (FIREX). Further analysis of the experiments has been done by Fokker-Planck simulation and the neutron yield of the simulation correlates well with the experiments (Johzaki et al., 2003).

4.3 Electron heat transport research and critical issues in fast ignition

Recent 3D PIC simulation results for focused ultra-intense laser interaction with a solid plane target show that the relativistic electron current profile in the target is self-organized and confined to a small radius as shown in Fig. 14. This indicates that the small hot spark could be generated by relativistic electron heating (Sentoku, 2003).

In the cone target, an ultra-intense laser light is partially reflected on the cone surface wall and focused to the top of the cone, while the relativistic electrons are generated on the sidewall of the cone and on the top wall. Since the electrons are accelerated along the laser propagation direction, a strong current is driven along the cone axis and the relativistic electron flows are pinched to the top of the cone by the magnetic fields as shown in Fig. 15. These characteristics of laser interactions with the cone target contribute to enhancing the coupling efficiency of a short-pulse laser to a small core plasma as was indicated by the enhancement of neutron yield discussed in the previous section. However, since the PIC simulations have been limited to small scales and short time durations, it is necessary to carry out longer space scale and timescale simulations by introducing a simulation system connecting the PIC simulation with the Fokker-Planck and/or hybrid simulations in order to compare quantitatively the results of the transport experiment with the simulation results (Mima et al., 2002).

An enhanced coupling efficiency of the heating laser energy to a small core plasma is essential for achieving ignition with relatively little laser energy. Because of this point, the relativistic electron transport and energy deposition in the cone guide target and/or in the corona of the imploded plasma are the most critical issues in the fundamental physics of fast ignition, although extensive experimental and theoretical work has been conducted in an attempt to understand and control dense plasma heating by using an ultra-intense laser.

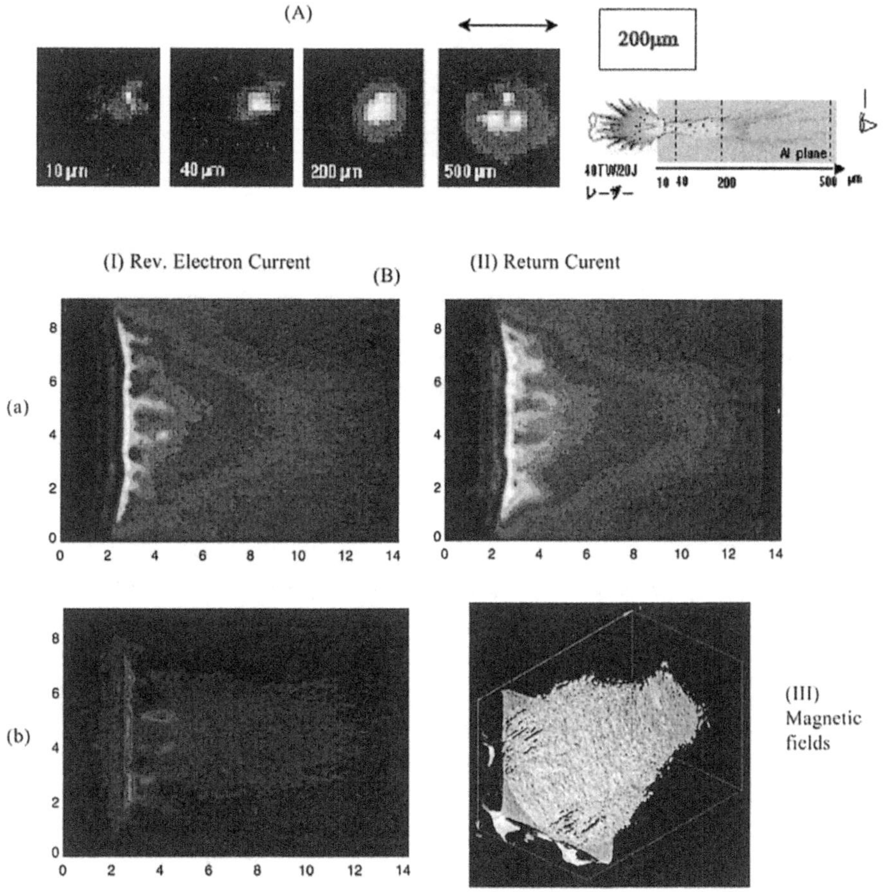

Figure 14. Relativistic electron current self-organization (self-pinch): (A) Transport experiment (B) (I)-(a) the current profile of hot electron with energy lower than 1.0MeV, (I)-(b) current profile of hot electron with energy higher than 1MeV, (II) return current profile; and (III) magnetic field iso-intensity contour (Sentoku, 2003; Kodama, 2001).

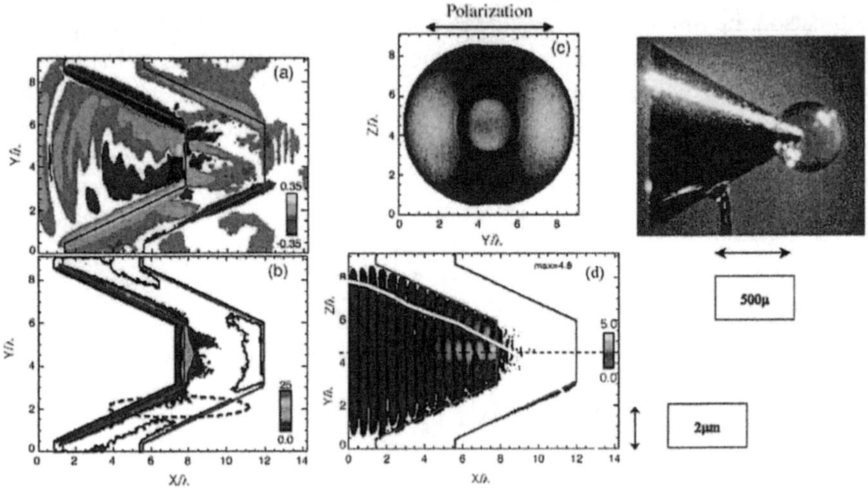

Figure 15. A photograph of a cone shell target (top right) and PIC simulation results for cone– laser interactions (a), (b), (c), (d) which are magnetic field structure (a), electron energy density distributions on the sidewall (b), electron energy density distribution at the top of the cone (c), and laser intensity distribution in the cone (d), respectively. (Mima et al., 2002; Kodama, 2000).

4.4 Petawatt laser and future fast ignition facility

A petawatt laser was recently completed, which delivers 500 J/0.5 ps (Izawa et al., 2002), at Osaka University. In PW laser experiments, imploded plasmas have been heated up to about 1 keV. The present and future PW laser target experiments at Osaka, Rutherford Appleton Laboratory, and so on (Kodama et al., 2002) will clarify the heating scaling law that determines the relations between heating laser energy, hotspark temperature, and neutron yield. Using the Fokker-Planck code, cone shell target experiments at Osaka were simulated. When a relativistic electron has the double Maxwellian energy spectrum, as shown in Fig. 16, of the cone target PIC simulation results, the predicted neutron yield and the ion temperature correlate with the experimental results as shown in Fig. 13 (Mima, et al., 2002).

The results have been extended to the larger laser energy case; the gains for the multi-10 kJ heating pulse are shown in Fig. 17. If the coupling efficiency is higher than 30%, the required heating laser energy for break-even experiments (FIREX-I) is estimated to be 10 kJ for 1500 times solid density DT plasmas, and a 30 μm spot diameter as shown in Fig. 17 (Yamanaka, et al., 2002; Mima, et al., 2002).

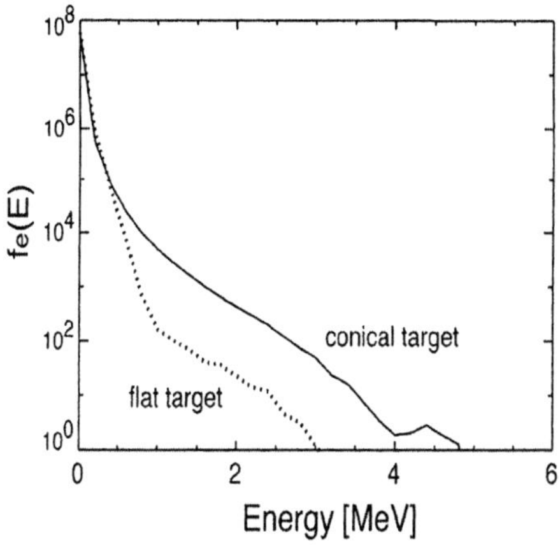

Figure 16. The number of hot electrons for the conical target is enhanced by more than one order magnitude above the flat target.

For ignition, the most critical parameter is the hot spot radius, r_b, which depends strongly upon relativistic electron transport. According to simulation and fundamental experimental results, the relativistic electron heat flow is well confined by self-generated magnetic fields. In particular, in the cone shell target, the laser energy deposited is concentrated at the top of the cone. Therefore, the hot-spark radius could be controlled by the cone top radius and the distance between the cone top and the core plasma. From the present understanding of the heating processes, ignition will be achieved with a pulse energy less than 50 kJ/10 ps for an imploded plasma ρR higher than 1.0 g cm^{-2}. The second phase of the FIREX project at Osaka University (FIREX-II) is aiming to demonstrate such an ignition.

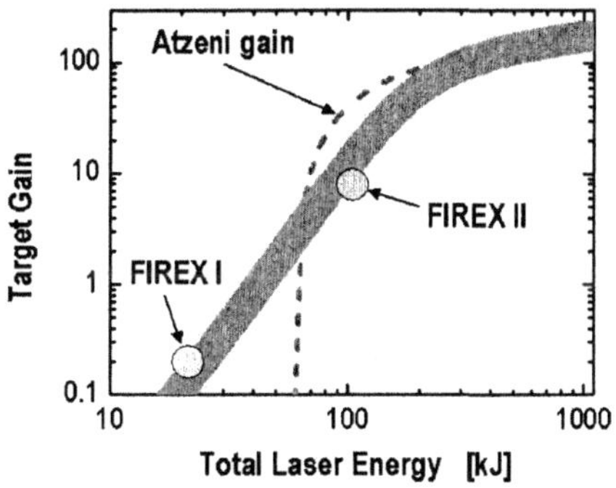

Figure 17. Gain curves for fast ignition. The core plasma is assumed $\rho = 300$ g cm^{-3}, and the isentropic factor is 2. The implosion laser energy is 10 kJ for FIREX-I, and 50 kJ for FIREX-II. The shaded area is the route from the FIREX target to the high-gain target which is essentially the same as in Atzeni's design.

5. IFE POWER PLANT DEVELOPMENT

5.1 IFE power plant systems

An IFE power plant consists of four major, separate but interconnected subsystems (elements of a power plant) as shown in Fig. 18, the functions of which are as follows:

(a) The driver, usually either a laser or particle accelerator, converts electrical power into short pulses of light or particles and delivers them to the fuel pellet in the proper spatial and temporal form to cause implosion, ignition and thermonuclear burn, i.e. fusion.

(b) In the pellet factory, fuel pellets are manufactured, filled with DT fuel and sent to the reactor and then injected into the reaction chamber.

(c) In the reaction chamber, the injected fuel pellet (target) is tracked, i.e. its position, flight direction, and velocity are measured precisely. Driver beams are directed to the target to implode it and to produce thermonuclear energy with a repetition rate of a few times a second. The thermonuclear emissions are captured in a surrounding structure called a blanket, and their energy is converted into thermal energy (heat). Tritium is also produced in the blanket.

(d) In the remainder of the plant, two major processes for material and energy are performed. Tritium and some other target materials are extracted from the recirculating blanket fluid material and from the reaction chamber exhaust gases. Then these extracted materials are recycled to the target factory. The thermal energy in the blanket fluid is converted into electricity, a portion of which is conditioned and re-circulated to power the driver.

Corresponding to the major processes, two cycles are formed in a power plant as shown in Fig. 18: the target material cycle and the power cycle. In the material cycle, the entire target disintegrates due to the implosion fusion in the reaction chamber, and a portion of the DT fuel is burned. Target debris is deposited in the chamber gas and/or in the blanket fluid. Tritium fuel is manufactured from Li in the blanket. The tritium and other target materials are extracted, conditioned, and returned to the target factory for use in the new target fabrication. In the power cycle, a series of energy conversion processes occurs, each with a characteristic efficiency and/or multiplication, as shown in Fig. 18. Electrical energy is converted to laser light or particle beam energy in the driver with an efficiency η_d. The driver energy produces thermonuclear energy by fuel pellet implosion with a target gain of Q.

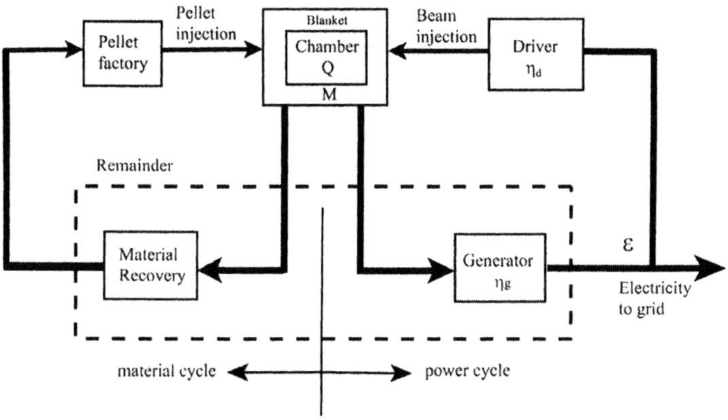

Figure 18. IFE power plant system.

The thermonuclear energy is converted into thermal energy in the blanket fluid together with the T breeding and the energy multiplication with a factor M through the neutron reactions. Finally, the heat energy is converted to electricity with efficiency η_g. A portion, ε, of the gross electricity produced must be recirculated to the power driver, completing the power cycle. In the power cycle, the basic condition of the power balance is shown as

$$\eta_d Q M \eta_g \varepsilon = 1 . \tag{11}$$

The recirculation power fraction, ε, should be less than 25% for economic reasons. If the recirculation power fraction is too large, then the cost of electricity (COE) sold rises rapidly because much of the plant equipment is used simply to generate electricity for the driver itself. The blanket gain, M, is about 1.05-1.25 depending on the design of the blanket material and structure. The efficiency of the turbine generator, η_g, ranges from 30 to 40%. Therefore, the product, $\eta_d Q$, should be

$$\eta_d Q \geq 10 . \tag{12}$$

This product determines the minimum target gain necessary for any given driver efficiency. The estimated target gain is shown in Fig. 4. The typical values are $Q \approx 30$ for the indirect drive, $Q \approx 100$ for the laser direct drive, and $Q \approx 200$ for fast ignition. Corresponding to each gain value, the necessary condition for driver efficiency is given for an indirect drive with heavy ion beam accelerator (HIB) to be 30-40%, for a direct drive with laser to be 10%, and for fast ignition, if it works well, 5%. The higher driver efficiency, η_d, with a higher target gain, Q, at a smaller driver energy, provides the condition for smaller recirculation power fraction, ε, and a lower driver cost, which gives us a competitive COE.

An IFE power plant must be safe and must have minimum impact on the environment. Ensuring these features requires that all the system components of the plant and materials used must be examined carefully for their potential negative impacts. The economic, safety and environmental (ESE) aspects should be examined in the technical design of each specific type of power plant. It should be noted that these aspects can be evaluated reasonably quantitatively with the recent developments in physics and technology related to the elements of IFE power plants.

5.2 Driver development

The specifications of the laser driver for a commercial power plant are (1) total energy (MJ/pulse), (2) intensity (10^{14}-10^{15} Wcm^{-2} on target), (3) pulse shape (tailored in 20-40 ns pulses), (4) wavelength (0.5-0.3μm), (5) spatial uniformity of irradiation (<1% rms), (6) efficiency (>10% for direct implosion, >5% for fast ignition), (7) repetitive operation (~10 Hz), (8) cost (capital and operating costs, including life and maintainability) and (9)

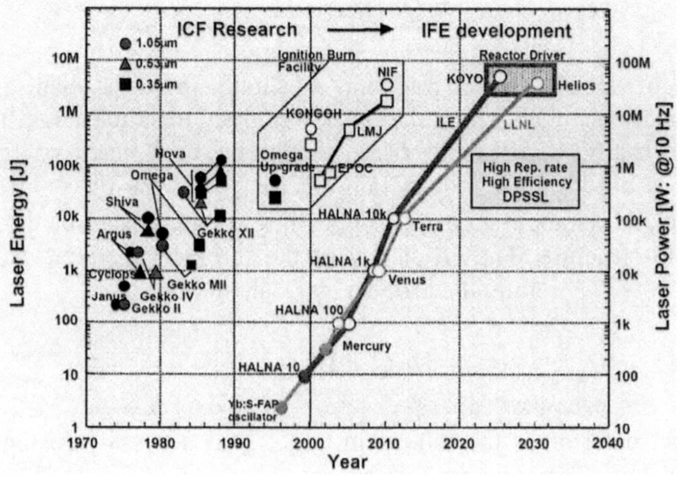

Figure 19. Glass laser system for implosion experiments and DPSSL development for power plants in the world.

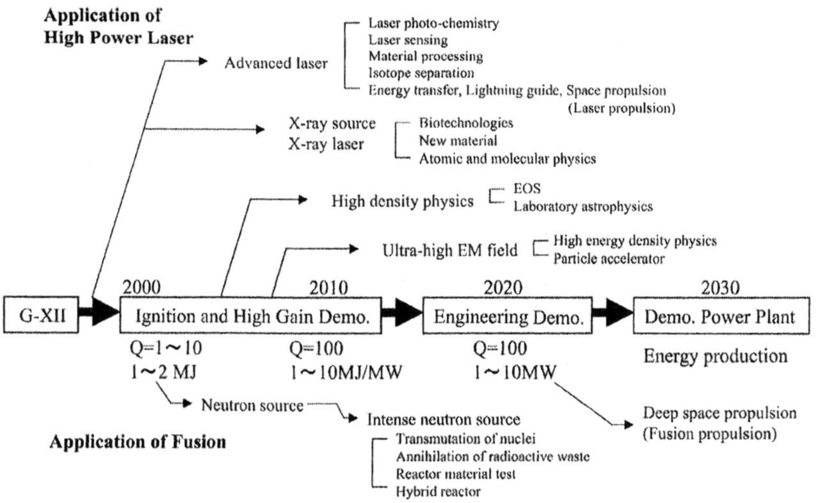

Figure 20. Application of high-power laser and spin-offs of IFE development.

reliability or availability (Hogan et al., 1995).

In the above specifications, items (1) to (5) are those required for the research phase to demonstrate ignition and fusion gain. Items (6) to (9) are those required for a driver for a laser fusion power plant. An advanced solid-state laser (Krupke, 1989) has demonstrated a breakthrough in diode laser pumping and feasible prospects of a power plant driver (Naito, et al., 1992; Orth, et al., 1996). Fig. 19 shows the progress of flash lamp pumped glass lasers on the left-hand side, which are basically single-shot lasers, and the expected development of diode pumped solid state lasers (DPSSLs) on the right-hand side.

The applicability of solid-state lasers for reactor drivers depends strongly on the capability of achieving uniform irradiation on the surface of the fuel pellet for its stable implosion. Several important concepts and technologies for better irradiation have been developed. They are: (a) RPP (distributed phase plate) to obtain a smooth envelope distribution in a far field pattern even with a beam of nonuniform intensity distribution at near-field, (b) smoothing by SSD to obtain time averaged uniformity of intensity even with a speckle structure of the instantaneous pattern, and (c) pulse tailoring in intensity (picket fence to control the shock heating of the fuel shell) in wavelength (two-colour irradiation to control nonlocal heat transport) or in coherency (partially coherent light (PCL) at the foot to reduce the perturbation by the initial imprint).

It should be noted that the Omega 60-beam achieved an on-target uniformity better that 1% rms with an improved power balance of 60 beams below 5% rms and implementation of several smoothing techniques (McCrory et al., 2002).

For better and more practical drivers of DPSSL for laser fusion power plants in commercial use, there are two major key issues. The first is high-power laser diodes for pumping and the other is solid-state laser material. Laser diodes have progressed and achieved the required specification for pumping. A cost reduction in the diode laser of 10-100 times less than the present level is required, and this is expected to be possible with the increase in demand for application of LD and/or DPSSL in industry and science as shown in Fig. 20, along with increased laser fusion power plant development (Nakai, 2000).

As for the laser material, the figure of merit is shown in Fig. 21, which shows the preferable range of the stimulated emission cross-section, σ, and the lower limit of thermal shock parameter (Matsui et al., 2000). A single crystal host material, Yb-SFAP (Bayramian et al., 2002), is utilized at LLNL

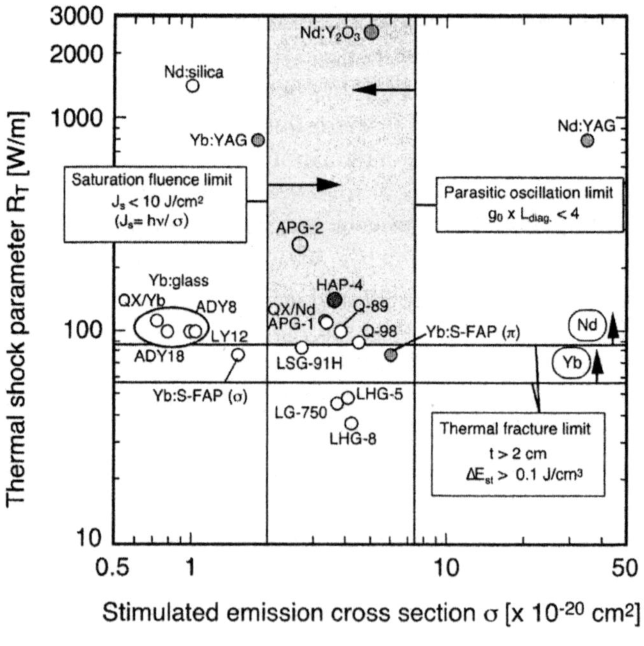

Figure 21. Laser materials for a high-power laser amplifier.

in the Mercury laser, and a reinforced glass material, Nd HAP-4, is utilized at ILE in high-average-power laser for nuclear-fusion application (HALNA) development (Matsui, et al., 1999); (Kawashima et al., 2001).

Very promising laser materials are under development in the technical field of optical ceramics. The ceramic YAG laser has already shown better performance than the single crystal YAG laser (Lu et al., 2002).

An yttria ceramic (Y_2O_3) has been made to optical quality, and its properties as a laser material have been tested (Lu et al., 2002). Fused silica is also a candidate for a largesize solid-state laser material (Fujimoto and Nakatsuka, 2002). The unique properties of high-power LD pumping such as (a) spectrum optimization being matched to the laser material, (b) high-intensity pumping with optimized pulse duration, (c) polarization control of the pumping light and (d) areal distribution control of the pumping light could lead to new optical materials for use as the laser host material. The progress of DPSSL technology, which is that of LD, laser host materials and also optical technology, is now opening many new fields in industrial technology with optical and laser processes (Krupke, 2000).

5.3 Reaction chamber

In the laser fusion power plant, small fusion targets of 10 mm diameter are successively injected and shot by driver laser beams. Following the implosion and fusion explosion of a fuel pellet, successive pulses of different energy species hit the first wall of the chamber. They are reflected or scattered laser light, x-ray from fusion plasma, neutrons and plasma debris. The energy spectrum and pulse shape of each energy species for typical high-gain implosions are estimated with burning simulation codes.

The unique feature of the IFE reaction chamber is the technical possibility of utilizing the wetted surface for the plasma-facing wall. This comes from the mitigated requirement for a vacuum in the chamber for IFE, in contrast with the MFE chamber, where a high vacuum is important for better energy confinement. The use of a liquid wall gives us flexibility in designing various kinds of chamber concepts such as a thin, liquid layer on porous solid (Booth, 1972), a woven fabric wall (Mima et al., 1992), or a thick, liquid layer of 30-50 cm that is flowing down, forming a cavity for the fusion reaction (Yamanaka, 1981; Moir et al., 1994).

On the other hand, the use of a dry wall for an IFE reaction chamber needs special care because of the high peak intensity of the wall loading due to the pulsed release of the various species of energy from the implosion fusion. The dynamic responses of the dry wall are analysed to examine the possibility of its use as the first wall of a reaction chamber (Kozaki, et al.,

2002). The dry wall seems to survive against the pulsed energy fluxes of laser light, x-rays, and neutrons that are emitted from moderate fusion explosions with reasonable separation of the first wall from the implosion fusion point. The heating and sputtering of the wall surface by the plasma debris needs more investigation to make clear the response and life span of the first wall. It should be noticed, however, that a distributed magnetic field on the surface or low-pressure gas in the cavity could mitigate the plasma flow hitting the wall surface, which could lead to a longer life of the wall. The types of reaction chamber and the features and related physics are summarized in Table 1.

Table 1. The types of laser fusion reaction chamber and the related physics.

Types	Key concept	Key phenomena and related physics	Example of conceptual design
Thick-liquid wall chamber	Magnetically guided flow or oscillating thick flow by moving vane forms a cavity.	Cavity formation and stability of thick flow Restoration of the cavity condition for target and laser injection for next cycle	SENRI HYLIFE
Wetted wall chamber	Thin liquid layer that seeps out of a porous wall or fabric tube protects the surface of the wall.	Evaporation and recondensation of wetted surface seeping out phenomenon to keep thin wetted layer	KOYO
Dry wall chamber	Solid first wall cooled from the rear by gas or liquid.	Protection of the surface from x-ray and plasma loading by using a low density gas in the chamber or a distributed magnetic field on the surface	Sombrero KOKI

In the thick liquid wall design of a chamber, most of the fusion energy, including neutrons, is absorbed in the flowing liquid. An example of a conceptual design is shown in Fig. 22 (Yamanaka, 1981).

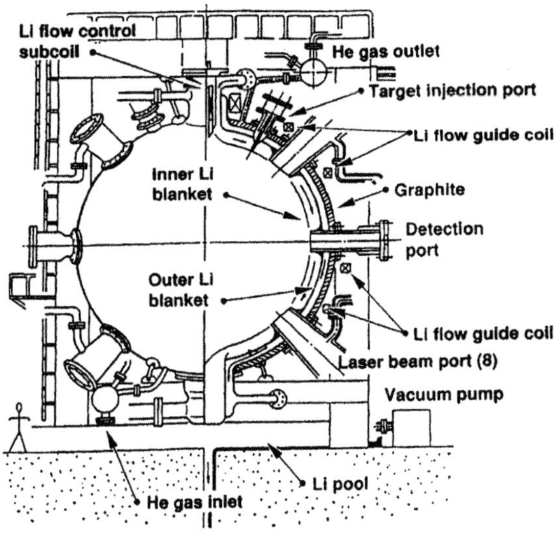

Figure 22. Thick liquid wall chamber 'SENRI'. Magnetic field guides the thick Li flow to follow the curved structure wall.

The conceptual design of the KOYO (Mima, et al., 1992) power plant is shown in Fig. 23 with the key technology and issues when utilizing a thin liquid layer for first wall protection. An example of a dry wall concept, 'Sombrero' (Sviatoslavsky, et al., 1992), is shown in Fig. 24 with some comments on the features and design principles.

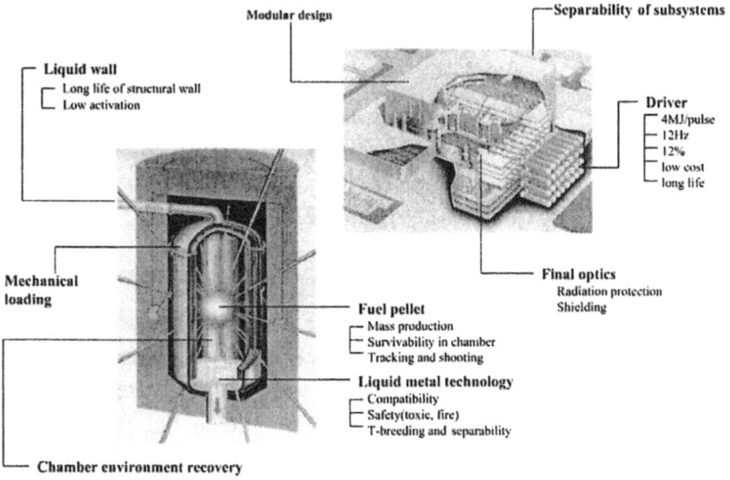

Figure 23. Key technologies and issues for IFE power plant KOYO.

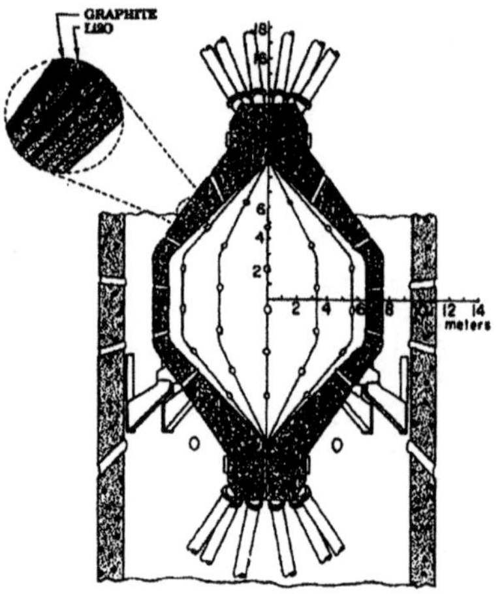

- Uses gas protection, W-armor coating, or engineered surface (e.g., brush-like) to protect first wall structure from x-ray and target ions/debris
- ~30% of fusion power must be conducted through solid FW
- FW/blanket coolant is flowing solid breeder (Li_2O) with low pressure He gas to assist in T purge and flow control
- First wall configurations under consideration are:
  - C/C composites (original design)
  - C/SiC (Sirius)
  - W/SiC, W/nanocomposited steels /stainless steel (HAPL, ARIES)

Figure 24. Sombrero is an example of a dry wall concept.

5.4 Fuel pellet

Precision fabrications of the structured fuel pellet and target for laser implosion and the related physics experiments have been well developed. Recent progress in micromachining, fine coating and characterizing techniques for microstructures in industry has interacted well with the target technology for laser fusion, in both directions.

The next step towards the power plant is to demonstrate repetitive injection, tracking, and shooting with a repetitively fired laser. This is a very challenging development for industry also. Mass production of fuel pellets at low cost is also an important issue for realistic IFE power plant development. The tritium handling technology is common with MFE, and we can share the developmental process except for the precision cryogenic fuel technology.

The effects of the residual gas in the chamber on the flying cryogenic pellet have been evaluated (Norimatsu, et al., 2002). They are the condensation effect of the chamber gas on low-temperature fuel pellets, heat flow through the gas and radiation, and disturbance to the flight trajectory by the fluctuating condition of the chamber. It can be concluded that 3-5 Hz operation of a liquid wall chamber is acceptable. The pressure of the protecting gas for the dry wall chamber could be optimized to compromise the protection effect of the first wall with the fuel pellet flight and the laser beam propagation in the chamber.

6. SUMMARY

It is expected that fusion ignition and energy gain by laser implosion will be demonstrated around the years 2010-2015 using megajoule lasers that are under construction in the USA and in France. The basic physics of implosion has been well understood, and the credibility of numerical simulations of implosion has been increased by reinforcement with physical understanding. The progress in the imploded plasma parameters achieved is remarkable, that is 10 keV high-temperature compression (Yamanaka and Nakai, 1986), and a high-density compression 1000 times over the initial density (Nakai, et al., 1989, 1990; Azechi, et al., 1991).

The recent demonstration of the formation of a hot spark surrounded by cold fuel with 1% rms uniform irradiation at Omega 60-beam at LLE of Rochester University (McCrory, et al., 2002) shows the feasibility of ignition and gain by directdrive implosion with 194 beams at NIF, together with implementation of the established techniques for uniformity improvement.

On the other hand, the fast ignition concept has been experimentally demonstrated (Kodama, et al., 2002) to be feasible for heating the compressed core efficiently. This new concept has become technically possible with the invention of new laser technologies such as chirped pulse amplification and compression for generation of ultrashort pulses (Strickland and Morou, 1985). If the fast ignition concept works well on the large scale for the compressed fuel of the reactor, it is very attractive in the design of laser fusion power plants because of the higher pellet gain with smaller driver energy. A challenging project, FIREX (Mima et al., 2002), has been kicked off, aiming to demonstrate ignition and burning through the concept of fast ignition at ILE, Osaka University.

As for power plant development, many conceptual design studies have revealed the technical issues of the key elements that constitute a power plant, such as the 'driver', 'chamber', and 'target factory'. The most critical element is the driver for a power plant, requiring rep-rate operation of the order of 10 Hz with a pulse energy of the order of megajoules in a few tens of nanoseconds with high efficiency (over 10%). This is really a big jump in the development of laser technology. On the way towards the development of the power plant driver, we can expect many fields of laser application to open up new industrial technology and scientific methodology and explore the physical parameter regions of new phenomena as shown in Fig. 20.

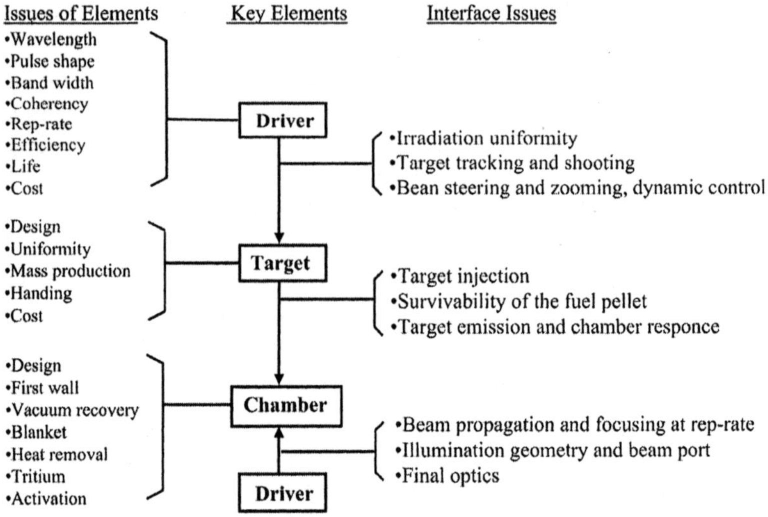

Figure 25. Elements of IFE power plant and interface issues.

It is an important feature of an IFE power plant that its key elements, the driver, target factory, and chamber, are separated from each other and softly coupled to make a system. The relation between the three key elements is shown in Fig. 25. Only the laser beam and the transport line of tiny fuel pellets connect the three areas of the driver, chamber and target factory. Even with the physical separability, there are important interface issues when all of the elements work together as a system as shown in Fig. 25.

The integrated approach to the IFE power plant is now under consideration to investigate the interface issues under the auspice of IAEA to stimulate the worldwide collaboration in the IFA development.

REFERENCES

Nuckolls J., et al., 1972, Laser Compression of Matter to Super-High Densities: Thermonuclear (CTR) Applications, *Nature*, **239**: 139.

Lindl J., 1998, *Inertial Confinement Fusion- the Quest for Ignition and Energy Gain using Indirect Drive*, Springer, New York.

Atzeni S., 1995, Thermonuclear Burn Performance of Volume-Ignited and Centrally Ignited Bare Deuterium-Tritium Microspheres, *Japan. J. Appl. Phys.* **34**: 1980-1992.

Tabak M., et al., 1994, Ignition and high gain with ultrapowerful lasers, *Phys. Plasmas*, **1**: 1626-1634.

Mayer-ter-Vehn J., 1982, *Nucl. Fusion*, **22**: 561.

Nakai S., and Takabe H., 1996, Principles of inertial confinement fusion - physics of implosion and the concept of inertial fusion energy, *Rep. Prog. Phys.* **59**: 1071-1132.

Hogan W. J., et al., ed., 1995, *Energy from Inertial Fusion,* IAEA, Vienna, ISBN 92-0-100794-9.
Bodner S. E., et al., 1998, Naval Research Laboratory, NRL/MR/6730-98-8113.
Shigemori K., et al., 2002, 19th IAEA Fusion Energy Conf. (Lyon, France, 14-19 October 2002), IAEA-CN-94/IF/P-15.
McCrory R. L., et al., 2002, 19th IAEA Fusion Energy Conf. (Lyon, France, 14-19 October 2002), IAEA-CN-94/IF-1.
Skupsky S., et al., 1989, Improved laser-beam uniformity using the angular dispersion of frequency-modulated light, *J. Appl. Phys.* **66**: 3456-3462.
Regan S. P., et al., 2000, Experimental investigation of smoothing by spectral dispersion, *J. Opt. Soc. Am. B*, **17**: 1483-1489.
Takabe H., et al., 1985, Self-consistent growth rate of the Rayleigh–Taylor instability in an ablatively accelerating plasma, *Phys. Fluids*, **28**: 3676-3682.
Lindl J., 1995, Development of the indirect-drive approach to inertial confinement fusion and the target physics basis for ignition and gain, *Phys. Plasma*, **2**: 3933-4024.
Betti R., et al., 1998, Growth rates of the ablative Rayleigh–Taylor instability in inertial confinement fusion, *Phys. Plasmas*, **5**: 1446-1454.
McKenty P. W., et al., 2001, Analysis of a direct-drive ignition capsule designed for the National Ignition Facility, *Phys. Plasmas*, **8**: 2315-2322.
Bodner S. E., et al., 2000, High-gain direct-drive target design for laser fusion, *Phys. Plasmas*, **7**: 2298-2301.
Shigemori K., et al., 2001, *Bull. Am. Phys. Soc.* **46**: 286.
Mahady A. I., et al., 1999, Pulse heating and ignition for off-centre ignited targets, *Nucl. Fusion*, **39**: 467-476.
Piriz A. R. and Sanchez M. M., 1998, Analytic model for the dynamics of fast ignition, *Phys. Plasmas*, **5**: 2721-2726.
Deutsch C., Furukawa H., Mima K., Murakami M., and Nishihara K., 1996, Interaction Physics of the Fast Ignitor Concept, *Phys. Rev. Lett.* **77**: 2483-2486.
Atzeni S., 1999, Inertial fusion fast ignitor: Igniting pulse parameter window vs the penetration depth of the heating particles and the density of the precompressed fuel, *Phys. Plasmas*, **6**: 3316-3326.
Tanaka K. A., et al., 2000, Studies of ultra-intense laser plasma interactions for fast ignition, *Phys. Plasmas*, **7**: 2014-2022.
Kodama R., et al., 2000, 18th IAEA Fusion Energy Conf. (Sorrento, Italy), IAEA-CN-77/IFP/09.
Sentoku Y., et al., 2003, Anomalous Resistivity Resulting from MeV-Electron Transport in Overdense Plasma, *Phys. Rev. Lett.* **90**: 155001-1-155001-4.
Kitagawa Y., et al., 2002, Progress of fast ignitor studies and Petawatt laser construction at Osaka University, *Phys. Plasmas*, **9**: 2202-2207.
Kodama R., et al., 2001, Fast heating of ultrahigh-density plasma as a step towards laser fusion ignition, *Nature*, **412**: 798.
Yamanaka T., 2000, 18th IAEA Fusion Energy Conf. (Sorrento, Italy), IAEA-CN-77/OV3/2.
Mima K., et al., 2000, 18th IAEA Fusion Energy Conf. (Sorrento, Italy), IAEA-CN-77/IF/1.
Kodama R., et al., 2002, Nuclear fusion: Fast heating scalable to laser fusion ignition, *Nature*, **418**: 933.
Johzaki T., et al., 2003, Analysis of Core Plasma Heating by Relativistic Electrons in Fast Ignition, *Fusion Sci. Technol.* **43**: 428-436.
Mima K., et al., 2002, 19th Fusion Energy Conf. (Lyon, France, 14-19 October 2002), IAEA-CN-94/IF3.

Izawa Y., et al., 2002, 19th IAEA Fusion Energy Conf. (Lyon, France, 14-19 October 2002) IAEA-CN-94/IF/P-04.

Yamanaka T., et al., 2002, 19th IAEA Fusion Energy Conf. (Lyon, France, 14-19 October 2002), IAEA-CN-94/OV/3-1.

Krupke W. F., 1989, *Fusion Technol.* **15**: 377.

Naito K., et al., 1992, Conceptual Design Studies of a Laser Diode Pumped Solid State Laser System for the Laser Fusion Reactor Driver, *Japan J. Appl. Phys.* **31**: 259-273.

Orth C. D., et al., 1996, A diode pumped solid state laser driver for inertial fusion energy, *Nucl. Fusion*, **44**: 75-116.

Nakai S., 2000, *AAPPS Bull.* **10**: 2.

Matsui H., et al., 2000, Conceptual Design of Laser-Diode-Pumped Water-Cooled Nd:Glass Slab Laser Driver for Inertial Fusion Energy, *Rev. Laser Eng.* **28**: 176-181 (in Japanese).

Bayramian A.J., et al., 2002, in: *IFSA 2001- Inertial fusion sciences and applications*, K. A. Tanaka, et al., ed., Elsevier, Paris, pp. 459-464.

Matsui H., et al., 1999, Conceptual design of a laser-diode-pumped Nd:glass slab laser driver for inertial fusion energy, *Fusion Eng. Des.* **44**: 401-405.

Kawashima T., et al., 2001, Design and Performance of a Diode-Pumped Nd:Silica-Phosphate Glass Zig-Zag Slab Laser Amplifier for Inertial Fusion Energy, *Japan. J. Appl. Phys.* **40**: 6415-6425.

Lu J., et al., 2002, in: *IFSA 2001- Inertial fusion sciences and applications*, K. A. Tanaka, et al., ed., Elsevier, Paris, pp 576-579.

Fujimoto Y. and Nakatsuka M., 2002, in: *IFSA 2001- Inertial fusion sciences and applications*, K. A. Tanaka, et al., ed., Elsevier, Paris, pp. 524-527.

Krupke W. F., 2000, Advanced diode-pumped solid state lasers (DPSSLs): near term trends and future prospects, *Proc. SPIE*, **3889**: 21-33.

Booth L. A., 1972, Central Station Power Generation by Laser Driven Fusion, vol. 1, Report LA-4858-MS (Los Alamos, NM: Los Alamos National Laboratory), Laser driven inertial fusion energy, 349.

Mima K., et al., 1992, 14th IAEA Fusion Energy Conf. (Wurzburg, Germany, 30 September-7 October 1992), IAEA-CN-56/G-2-3.

Yamanaka C., 1981, Report ILE-8127P Inst. of Laser Engineering, Osaka, Japan.

Moir R. W., et al., 1994, *Fusion Technol.* **25**: 5.

Kozaki Y., et al., 2002, 19th IAEA Fusion Energy Conf. (Lyon, France, 14-19 October 2002), IAEA-CN-94/FT/P-1-25.

Sviatoslavsky I. N., et al., 1992, *Fusion Technol.* **21**: 1470.

Norimatsu T., et al., 2002, 19th IAEA Fusion Energy Conf. (Lyon, France, 14-19 October 2002), IAEA-CN-94/FT/1-3Rb.

Yamanaka C., and Nakai S., 1986, Thermonuclear neutron yield of 1012 achieved with Gekko XII green laser, *Nature*, **319**: 757.

Nakai S., et al., 1989, *Bull. Am. Phys. Soc.* **34**: 2040.

Nakai S., et al., 1990, 13th IAEA Fusion Energy Conf. (Washington, DC, USA, 1-6 October 1990), IAEA-CN-53/B-I-3.

Azechi H., et al., 1991, High-density compression experiments at ILE, Osaka, *Laser Particle Beams*, **9**: 193-207.

Strickland D., and Morou G., 1985, Compression of amplified chirped optical pulses, *Opt. Commun.* **56**: 219-221.

Chapter 16

LASER SPACE PROPULSION

Applications at two extremes of laser power

Claude R. Phipps and James R. Luke
Photonic Associates, LLC, 200A Ojo de la Vaca Road, Santa Fe, NM, USA 87508

1. INTRODUCTION

1.1 Pure photon propulsion

The earliest reference to photon propulsion is in a 1953 paper by Eugen Sänger (Sänger 1955), written well before the demonstration of the laser, and perhaps before Charles Townes thought of the concept. In the paper, it is shown how one can circumnavigate the universe in 20 years using pure photon propulsion. This and related papers by Marx (1966) and Möckel (1972a) more than a decade later did not consider laser ablation propulsion, but rather rockets driven by transmission, reflection or absorption of photons. Even assuming total reflection, the momentum coupling coefficient C_m, defined as the ratio of coupled momentum $\delta m v_E$ to incident laser pulse energy W (or thrust F to power P for a continuous laser signal),

$$C_m = \delta m v_{\square}/W = F/P, \qquad (1)$$

is still only 6.7E-9 N/W, but specific impulse I_{sp} is as large as it can be, 3.1E7 seconds.

These facts mean that, as an example, to accelerate a 1-tonne object at 1 Earth gravity, 1.5 TW of optical power is required.

Möckel did not shrink from this, envisioning a 1TW, 1-km diameter x-ray laser beam with 1Å wavelength impinging on a 1-km diameter sail to propel a spacecraft to α-Centauri in 10 years.

Sänger's *Aero Digest* paper (1956) is notable for its optimistic conclusion, which provides an encouraging backdrop for this chapter:

> *No, we do not have to resign and humbly fold our hands in our laps. The infinite universe is small enough to be open to the initiative of every one of us, open to its ultimate boundaries, everything accessible to human beings.*

1.2 Ablative laser propulsion

Kantrowitz (1971) was first to suggest that it is more practical in the near term to produce thrust with costly laser power by heating a propellant to generate a vapor or plasma jet which provides the thrust. This is the principle of ablative laser propulsion. Shortly thereafter, Möckel (1972b) also considered a laser-heated hydrogen gas exhaust. Ablative laser propulsion is more practical than pure photon propulsion because the effective exhaust velocity $v_E \ll c$. Since the specific impulse for ablative laser propulsion is much smaller than that for photons,

$$I_{sp} = v_E / g_o \ll c/g_o, \tag{2}$$

the momentum coupling coefficient C_m being constrained by a constant product derived from energy conservation,

$$C_m \, I_{sp} = 2\eta_{AB}/(\psi g_o), \tag{3}$$

is orders of magnitude larger than in the case of photon propulsion, generating more thrust per watt. Then, lasers with power levels which are available today can propel useful payload masses in shortduration missions (Fig. 1). In Eq. (3), ψ is a parameter close to unity, η_{AB} is the ablation efficiency, and g_o is the standard acceleration of gravity in Eqs. (2) and (3). Both v_E and ψ are defined in section 1.2.1.

Ablation efficiency is defined as the ratio of exhaust kinetic energy W_E imparted to mass element δm to incident laser pulse energy W,

$$\eta_{AB} = W_E/W = \delta m \psi v_E^2/(2W), \qquad (4)$$

equal to 1 for pure photon propulsion. Using the two readily measurable quantities Q* and C_m, where Q* is defined as

$$Q^* = W/\delta m, \qquad (5)$$

v_E can be determined using Eqs. (1) and (5) from the product

$$v_E = C_m Q^*. \qquad (6)$$

Q* is called the specific ablation energy.

In Fig. 1, we show lines of constant Q*, in order to compare the immense energy investment ($E = mc^2$!) in a light-speed exhaust with that of, for example, water cannons (Yabe, et al. 2002).

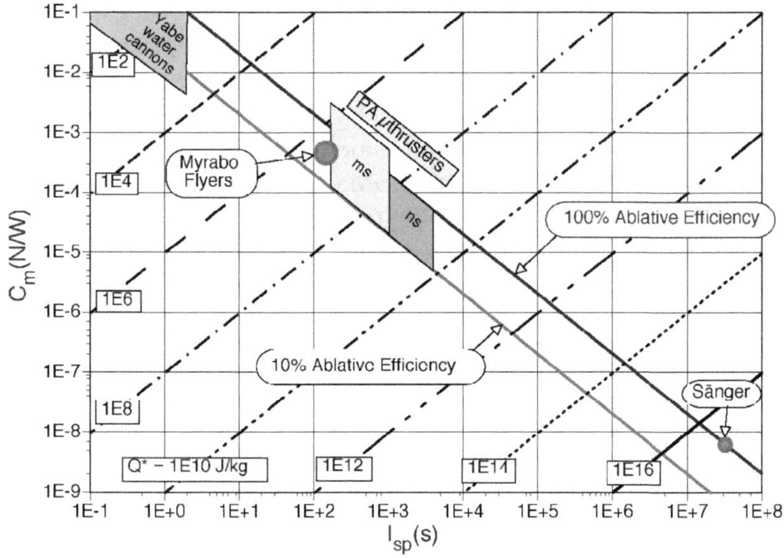

Figure 1. Specific impulse vs. momentum coupling coefficient for several types of laser-propelled devices. Lines of constant energy investment are shown. Laboratory experiment have reached much higher I_{sp}'s.

Most laser-heated exhaust streams in vacuum have $\psi \sim 1$ within 15% because of adiabatic expansion beyond the Knudsen layer at the target and electrostatic acceleration of plasma ions (see section 1.2.1).

In Fig. 1, we have illustrated the approximate parameter space occupied by Myrabo flyers in the atmosphere (Myrabo, et al. 2004), by the microthrusters which will be discussed later in this chapter (Phipps and Luke 2003, Phipps and Luke 2004), and by the Yabe water cannons, and compared them with the extreme envisioned by Sänger. Myrabo flyers in vacuum should roughly overlap the microthruster parameter space.

For ablative laser propulsion, missions are of short duration on the scale of Sänger's dreams because ablation fuel lifetime τ_{AB} of mass M with laser power P decreases quadratically with C_m:

$$\tau_{AB} = 2\eta_{AB}M/(PC_m^2) = 2\eta_{AB}M/(FC_m) . \quad (7)$$

It is noted that microthrusters can have ablative efficiency $\eta_{AB}>1$ because the definition (4) considers only laser pulse energy, excluding exothermic contributions from the ablation fuel.

1.2.1 Exhaust velocity distributions

Consider free expansion in the x-direction perpendicular to a planar target. After thermalization has taken place in the stream immediately adjacent to the thin "Knudsen layer" at the laser-heated surface, the velocity distribution of the ablated particles is a full-range three-dimensional Maxwell-Boltzmann velocity distribution with drift velocity u in the x-direction [Kelly and Dreyfus (1988)]:

$$f(v_x, v_y, v_z) = C_x C_y C_z \{\exp - \beta[(v_x - u)^2 + v_y^2 + v_z^2]\} \quad (8)$$

where $$\beta = \frac{m_E}{2kT} \quad \text{and} \quad C_x = C_y = C_z = \sqrt{\frac{\beta}{\xi}}\ , \quad (9)$$

and m_E is the mean exhaust particle mass. The definition of impulse generated by one laser pulse with energy W is

$$\delta J = \delta m <v_x> \quad (10)$$

$$C_m = \delta m\langle v_x\rangle/W \quad . \tag{11}$$

In section 1.2, we use v_E to denote $\langle v_x\rangle$. Then from Eq. (1),

$$\langle v_x\rangle = \int_{-\infty}^{\infty} dv_x v_x\, f(v_x) = C_x\sqrt{\pi/\beta}\,u = u \tag{12}$$

The amount of kinetic energy in the stream is $\delta e = \delta m\langle v_x{}^2\rangle/2$, where

$$\langle v_x{}^2\rangle = \int_{-\infty}^{+\infty} dv_x\ v_x{}^2\ f(v_x) = C_x\,[\frac{\sqrt{\check{s}}}{2\beta^{3/2}} + \sqrt{\frac{\check{s}}{\beta}}\ u^2] = [\frac{kT}{m_E} + u^2\,]\ . \tag{13}$$

To gauge the consequence of substituting $\langle v_x{}^2\rangle$ by $(\langle v_x\rangle)^2$, we calculate the ratio ψ from Eqs. (11) and (12) to find

$$\psi = \frac{\langle v_x{}^2\rangle}{(\langle v_x\rangle)^2} = \left\{\frac{u^2 + \frac{kT}{m_E}}{u^2}\right\} \quad . \tag{14}$$

If for example we consider a Mach 1 ($M = u/c_s = 1$) drift velocity with $c_s = (\gamma kT/m_E)^{1/2}$, and $\gamma = c_p/c_v = 5/3$, we have $\psi = 1.60$.

However, a preponderance of measurements summarized in Kelly and Dreyfus (1988) and in Phipps and Dreyfus (1993) show highly pronounced forward peaking which relative to the angular distribution one would obtain with $M = 1$. Where θ is the angle to the surface normal, these authors reported that the detected plume distribution varies like $\cos^{\nu}\theta$, with $\nu = 8 - 10$. Kelly and Dreyfus (1988) show that, $M = 1$ should correspond to $\nu = 4$, and $\nu = 8 - 10$ corresponds to $M \approx 2$. When $M = 2$, Eq. (14) gives

$$\psi = \frac{4\gamma + 1}{4\gamma} = 1.15\ \cdot \tag{15}$$

This situation will certainly apply in ns-pulse laser ablation, and it can also apply in ablation with longer pulses at high intensity.

$M = 2$ drift has at least two causes. These are acceleration of the stream due to unsteady adiabatic expansion beyond the Knudsen layer as discussed by Kelly and Dreyfus (1988), and electrostatic acceleration of the ions in the plasma stream by the collisionally-decoupled high-energy tail of the electron velocity distribution as discussed by Phipps and Dreyfus (1993).

2. LASER LAUNCHING – THE ULTIMATE APPLICATION

2.1 Present cost of lifting mass to low earth orbit

The way we now send things to space is very expensive. Present day costs of raising mass from the Earth's surface into low Earth orbit (LEO) with chemical rockets is more than $10,000/kg (Table 1). This cost, equivalent to the cost of gold, dominates all other considerations relating to spaceflight, limiting what we consider to be possible. But it need not be so.

Table 1. Present-day launch costs to Low Earth Orbit (LEO)

Launch System	*Minimum Cost (k$/kg)*
Rockot	*10*
Shuttle	*12*
Athena 2	*12*
Taurus	*20*
ISS, commercial	*22*
Pegasus XL	*24*
Long March CZ-2C	*30*
Athena	*41*

2.2 Expected costs with laser launching

Phipps and Michaelis (1994), taking advantage of an innovative conceptual design for a high-power laser system appropriate for launching large payloads (Phipps 1989), showed that there is an optimum set of parameters for laser space propulsion which can reduce the cost of lifting mass to LEO nearly 100-fold from its current level. Fig. 2, based on the costs derived in that work, emphasizes that rapid launch is one of the main reasons for the reduced cost. When a curve for a launch frequency typical of the Shuttle is added to the original graph, it is seen that cost becomes greater than current costs (50k$/kg for 0.01/day vs. $400/kg for 5/day). Laser launching is uniquely adaptable to high launch frequency, while chemical rocket launches, even given extensive development, have never shown that capability because of the low mobility of the associated mechanical infrastructure.

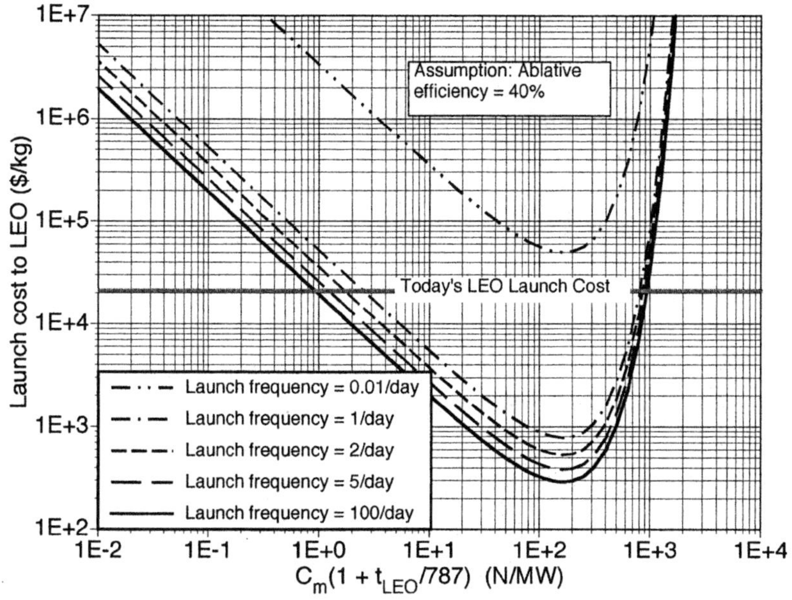

Figure 2. Cost of laser launching 10 tonnes to LEO using a repetitive-pulse laser. Figure based on Phipps and Michaelis (1994), with η_{AB} = 40% and 0.01/day launch frequency plot added to emphasize the importance of launch frequency in reducing cost.

The energy cost C in Fig. 2 is derived for flight in vacuum according to (Phipps, et al. 2000)

$$C = \frac{W}{m} = \left[\frac{(1-m/M)}{(m/M)}\right]Q^* = \frac{v_E}{C_m}\left[\exp\left(\frac{v_{LEO}+g_o t_{LEO}}{v_E}\right)-1\right] \quad , \quad (16)$$

where m/M is the ratio of mass delivered to orbit to initial mass at launch. For putting mass into LEO, this cost minimizes at about 100MJ/kg. The energy cost can, in turn, be related to dollar cost (Fig. 2) using assumptions given in Phipps and Michaelis 1994.

The parametric behavior shown in Fig. 2 occurs because the costs of personnel and facility amortization, which depend linearly on time, easily outweigh the cost of consumables and energy on the ground when launches are infrequent. The results are reasonable even for more modest laser power scenarios than the 1GW laser considered in the Phipps and Michaelis paper. Several kg can be delivered to LEO using ablative propulsion with a 1-MW laser. Present energy costs are about 3¢/MJ at retail on the ground. Accordingly, at 100MJ/kg, it ought not cost a great deal more than \$3/η per kg to reach LEO, where η is the product of all efficiencies intervening between the wall plug and the kinetic energy of the laser-ablation rocket exhaust. That this cost can be as little as \$300/kg makes sense even if η is as small as 1%.

Fig. 2 shows that cost is a sensitive function of C_m, with a relatively sharp minimum which depends on the assumed η_{AB} and time to LEO, t_{LEO}. This is because, for C_m below the optimum, more expensive laser power is needed to lift the same weight, while for C_m above the optimum, I_{sp} is smaller, and less payload mass is delivered to orbit per unit of laser energy. The optimum in Fig. 2 is what would be calculated from Möckel's optimization of the rocket equation in vacuum (Möckel 1975):

$$v_{E(opt)} = g_o I_{sp(opt)} = 0.6275\,(\Delta v + g_o \Delta t)\,. \qquad (17)$$

Where $\Delta v = v_{LEO} = 7.73$ km/s, $\Delta t = t_{LEO} = 400$s and $g_o t_{LEO} \sim 4$ km/s, Eq. (17) gives $v_{E\,(opt)} = 7.36$km/s and Eq. (3) gives $C_m = 270\ \eta_{AB}$ N/MW = 109N/MW. More careful analytics and simulations discussed in section 2.2.1 which include atmospheric effects for a launch from 30km initial altitude give a similar value

$$C_{m(opt)} = 2\eta_{AB}/v_{E(opt)} = 280\ \eta_{AB}\ \text{N/MW} \qquad (18)$$

for the same conditions.

As we will show in Section 3, C_m can be adjusted both by selecting appropriate target materials and by choosing laser intensity. The main points of this section are:

- Laser launching can be relatively inexpensive with several launches per day, a ticket to LEO costing \$50,000 or less per passenger
- Cost is a sensitive function of C_m, and
- $C_{m(opt)}$ is around 100 N/MW for fast flights from 30km initial altitude with η_{AB}=40%.

2.2.1 Flight including the atmosphere

Phipps et al. 2000 provided numerical simulations as well as analytic calculations of flight through the atmosphere into space starting from 30km altitude. Their analytic model was based on an exponential atmosphere with mass density ρ vs. altitude x in an atmosphere with scale height H:

$$\rho = \rho_o\, e^{-x/H}, \qquad (19)$$

through which the craft is flying with exponential acceleration

$$\dot{v} = \frac{v_o^2}{2H} e^{x/H} \tag{20}$$

in a profile chosen to maintain constant drag force

$$C_d A \rho v^2 = \text{const.}, \tag{21}$$

where C_d is the drag coefficient and A is the effective frontal area. This arbitrary choice for an optimum flight solution to the governing nonlinear differential equation gives mass vs. altitude

$$m = M - \int_o^f \frac{(\beta_a + 1)\, m\, g}{v\, v_E} dx \tag{22}$$

where m, v and the function β_a are all functions of position x and in particular

$$\beta_a(x) = \alpha \left\{ \frac{M}{C_d A \rho_o} \left(\frac{e^{x/H}}{2H} \right) + \frac{M}{m} \right\} . \tag{23}$$

Ultimately, a cost including an atmospheric flight portion ending in LEO is found to be given by

$$C = \frac{g_o}{v_o v_E} \left[(2MH)(1+\alpha)(e^{-x/2H} - 1) - \frac{v_o^2 M}{g} (e^{x/2H} - 1) \right] . \tag{24}$$

In Eqs. (23) and (24), α is a free parameter chosen for best m/M ratio. For the case $v_o = 0$ and $x_o = 30$km, $\alpha \sim 0.07$.

Launch from 30km altitude was emphasized because the craft was not air-breathing as in Myrabo and Raizer (1994) and therefore paid a high price for lower launch altitude. In fact, the payload mass m vanished completely at initial altitude $x_o = 2.5$km.

The results of their flight simulations for flights of a 20-kg, 1-m diameter cone-shaped craft from 30km initial altitude, using a 1-MW laser are summarized in Table 2. The optimum cost occurs at a slightly different C_m value than for the vacuum flight case, primarily because these simulations assumed $\eta_{AB} = 100\%$. The optimum cost is essentially the same as for the vacuum case (Fig. 2). The last entry in the table shows infinite cost because the fuel was depleted before any mass was delivered to orbit with $C_m = 1$mN/W. This possibility is important to keep in mind when designing laser-propelled flights.

Table 2. Energy cost for flights to LEO, including atmospheric drag, from 30km initial altitude

Assumed C_m (μN/W)	40	100	400	800	1000
Cost C (MJ/kg)	300	160	80	200	∞

Fig. 3 shows the principles of a hybrid system for laser launching from Earth which uses an electromagnetic railgun to raise the flyer to 30km altitude prior to laser ignition. From this point, laser incidence angle on the flyer need not vary more than ±10° to orbit insertion. Note that, with 20 launches per day delivering 10kg payload, parts for a 3-tonne space station can be placed on station in one month, using a 1-MW laser. Laser launching concepts which require higher power may not be practical for some time.

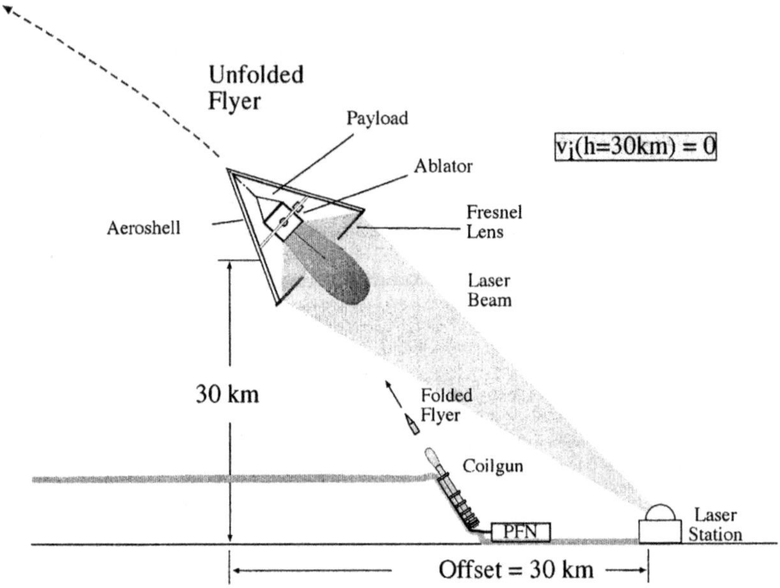

Figure 3. Hybrid laser launching system using railgun to achieve optimum altitude for laser propulsion to begin

2.3 Applications at intermediate laser power

An application of laser propulsion at the 30-kW level capable of clearing the near-Earth environment of 1-10-cm diameter space debris resulting from satellite launch activities in about two years has been reported (Phipps, 1994; Phipps, et al., 1995). In Phipps (1998) the advantages of using subnanosecond pulses were pointed out. This 1-10-cm diameter range includes about 300k objects which are threatening to space activities because they are too numerous to be tracked consistently, or avoided by spacecraft maneuvers, and yet are lethal to space assets. The much less numerous, larger ones can be tracked, and smaller ones can be absorbed using specially designed barriers. The so-called "ORION" concept uses a groundbased pulsed laser to create an ablation jet on the debris object when it is detected rising in the sky, ultimately causing its perigee to intersect the atmosphere, resulting in reentry (Fig. 4). The fluence required is much less than that which would vaporize the object. Although, at 1500km range, the beam is much larger than the debris (Fig. 4), relaxing the pointing requirements, the laser energy budget is still reasonable. Detection is possible using, e.g., a narrowband infrared laser and a heterodyne receiver, but tracking is not necessary. Laser "guidestar" and illumination lasers are also required. An alternative concept using an orbiting pulsed laser station also shows promise (Schall, 1991; Schall, 1998).

The thrust vector shown in Fig. 4 is the time-average thrust for the debris target, which will be spinning, at least after the first interaction. For this reason, a groundbased laser works as well as a spaceborne laser, because space debris are not spheres, and their surface normals cannot be specified in advance. In the figure, both radial and tangential components of the thrust vector are shown. Both components act to reduce perigee when the debris particle is rising.

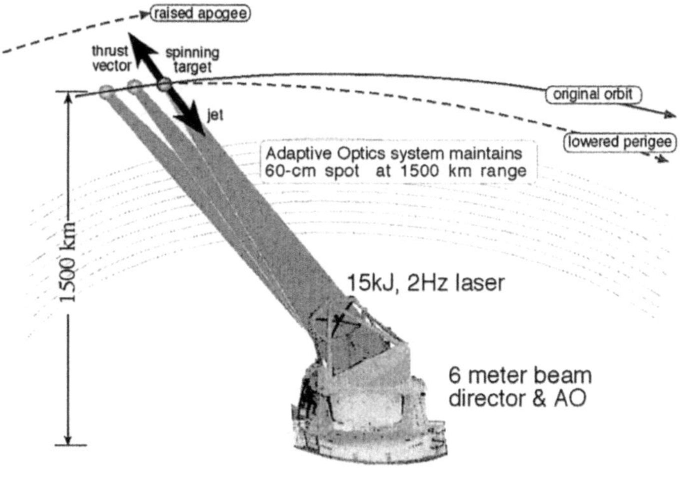

Figure 4. ORION concept for clearing near-Earth, 1-10-cm space debris

3. MICROTHRUSTERS

Throughout the early history of extra-atmospheric propulsion, emphasis was on producing engines with ever larger thrust, culminating with the 680-kN Rocketdyne F-1 engines for Apollo and engines developed for the Energiya program. Now, with the advent of micro- (≥10kg), nano- (1-10kg) and even pico-craft (<1kg), this trend is reversing. For many applications, such as pointing and positioning microsatellites, a thrust of order 100μN is desirable, together with low thrust noise and very small minimum impulse bits. This is difficult to do with conventional chemical rockets.

To meet this challenge, the field of microthrusters has evolved in the last decade, with electric propulsion as an especially interesting subset. Electric propulsion has the advantage of programmable thrust, often characterized by a minimum impulse bit (MIB) which may be as small as nN-s, and eliminates the need for storing dangerous, chemically reactive propellants on the craft. Furthermore, many electric propulsion concepts feature specific impulse, I_{sp}, which is much higher than is possible with chemistry. Table 3 gives nomenclature for this section, and Table 4 lists some comparative parameters for electric micropropulsion (Andrenucci, 2005, Phipps, et al., 2005a, Phipps, *et al.,* 2005b, Horisawa, et al., 2004, Brophy, et al., 2000, Mueller, 2000, Simon, et al., 2005]).

Table 3. Nomenclature

Symbol	Meaning
C_{ms}	System C_m = thrust/electrical power in
E	Short for "10^"
Q*	Specific ablation energy
FEEP	Field emission electric propulsion
GAP	Glycidyl azide polymer
GLYN	Polyglycidylnitrate
I_{sp}	Specific impulse
msμLPT	ms-pulse micro laser plasma thruster
nsμLPT	ns-pulse micro laser plasma thruster
NC	Nitrocellulose
μPPT	Micro pulsed plasma thruster
PVC	Polyvinylchloride
PVN	Polyvinylnitrate
η_E	Laser optical power out/electrical power in

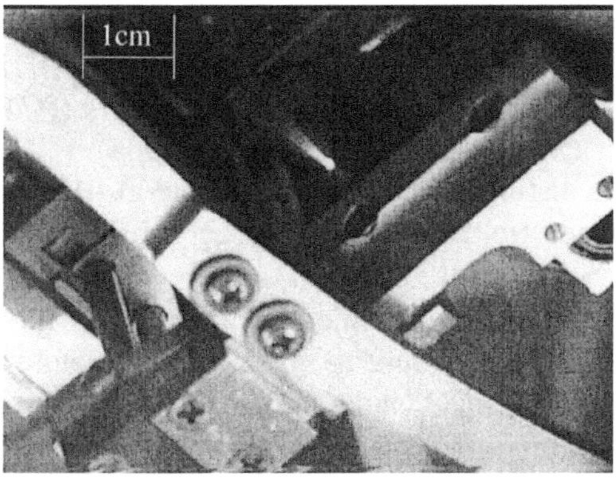

Figure 5. Microthruster jet in operation with PVC ablation fuel

We developed two laser-driven devices which occupy opposite ends of the I_{sp} spectrum (Phipps, et al. 2005b, 2005c). The nsμLPT (which uses ns-duration pulses) is competitive with ion thrusters for some applications. The msμLPT is designed to generate high thrust and must, by Eq. (3), have low I_{sp}.

Because it uses exothermic fuel tapes, the msμLPT has ablation efficiency $\eta_{AB}>1$. Highlighted cells in Table 4 represent areas in which the msμLPT and nsμLPT excel.

Table 4. Comparison of electric microthrusters

Micro-thruster	MIB (μN-s)	C_{ms} (μN/W)	I_{sp} (s)	Thrust (μN)	Engine Mass (kg)	Reference
ms-μLPT	0.5	550	200	10,000	0.4	Phipps, *et al.*, 2005a
ns-μLPT	4E-5	40	3000	100	0.8	Phipps, *et al.*, 2005b
μ PPT	2	20	1,000	30	1	Mueller, 2000e, Simon, *et al.*, 2005
LEH PPT	38	4.3	4,000	----	----	Horisawa, *et al.*, 2004
Colloid	4	180	1,000	20	0.5	Mueller, 2000d
FEEP	1	15	9,000	1,400	8.7	Andrenucci, 2000, Mueller, 2000c
Hall	----	60	1,300	30,000	1.1	Mueller, 2000b
Ion	----	40	3,100	20,000	8	Mueller, 2000a, Brophy *et al.*, 2000

Laser micro plasma thrusters are the first completely new micropropulsion technology in the last 30 years. For example, mercury ion drives and pulsed plasma thrusters (PPT's) date back to the 1960's (Sovey, et al. 1999; Guman 1968).

The main distinguishing features of the msμLPT in the electric micropropulsion field are its much larger thrust and thrust to power ratio, and its very small minimum impulse bit (MIB). Small MIB is important for positioning microsatellites with nm precision, a capability that is paramount in some system architectures such as LISA (Leach, 2003). C_m and I_{sp} can be adjusted over a modest range by changing target materials (Table 5).

The main distinguishing features of the nsμLPT are its microscopically small MIB and high I_{sp} in a low-mass engine. The energetic advantages of ns-duration and shorter pulses for manipulating objects in space via laser ablation propulsion were first pointed out in Phipps, et al., 1998 and the possibility of using onboard microchip lasers for ns-duration propulsion was demonstrated by Gonzales, et al., 2002 and 2001. In his 2002 paper, I_{sp} up to 4.9ks was reported using this technique. Other measurements going back several decades (Phipps and Michaelis, 1994, Greg and Thomas, 1966) have shown that I_{sp} values from 7 to more than 25ks can be obtained with simple desktop-scale ns-duration lasers using metallic targets in vacuum. C_m and I_{sp} can be adjusted over a significant range by changing laser intensity and pulse duration.

An especially attractive alternative is to combine both devices by using a single diode-pumped fiber laser with which, by selecting either ms-duration or ns-duration pulses, the full range of performance available in all the modern microthrusters is realized in one low-mass unit. I_{sp} and C_m can then be matched to mission requirements in real time as required by Eqs. (17) and (18).

The two types of device use different targets and target illumination geometries (Fig. 6). The laser intensity required to optimize the thrust-to-optical-power ratio C_m on opaque materials in vacuum is approximately given by

$$I = B\tau^{0.5}, \tag{25}$$

where $B = 480\ \mathrm{MW/m^2}$, for pulse durations $100\mathrm{ps} < \tau < 1\mathrm{ms}$ (Phipps, 2005).

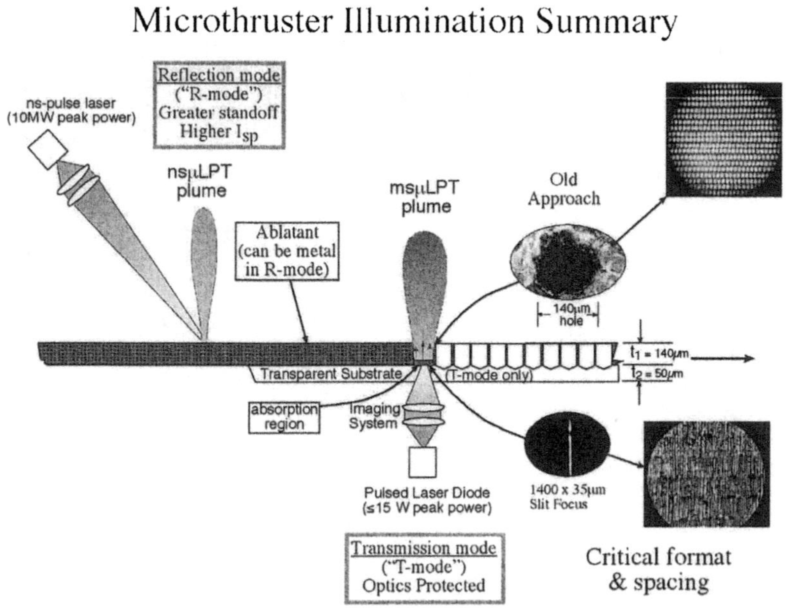

Figure 6. Illumination geometries for two types of microthruster. Targets for the ns-pulse device are metallic tapes illuminated from the same side as the plume (reflection mode). For the ms-pulse thruster, a two-layer fuel tape allows the laser optics to be protected from ablation products in so-called transmission mode. The diode laser wavelength is 920nm.

3.1 The msμLPT

High power, single-facet, 920nm diode lasers are used for brightness, low cost and durability. A "slit focus" illumination pattern on target was chosen for the best overall target fill factor (Fig. 6). Figs. 7 and 8 show the msμLPT as it will look complete.

Bare-facet, rather than fiber-coupled, diodes were chosen for optical efficiency and brightness preservation, since fibers reduce beam brightness. A simple lens pair relays a magnified image of the 1 x 100 μm laser output facet onto the target tape.

3.1.1 Ablation fuel tape: transparent layer development

The intensity that can be transmitted through the transparent layer is limited by its optical damage threshold. This threshold is further reduced compared to that for the clear material, due to UV emitted by the hot absorber layer.

Figure 7. The msμLPT. Six jets aperture on the (vertical) output face operate in matched pairs to maintain a fixed center of thrust.

A small ablation backflow exists below the damage threshold and provision (limited intensity, distance, protective windows) must be made for the protection of the illumination optics from the backstreaming material. Polyimide resin has good mechanical properties for the application, but finding materials that would adhere to it was difficult.

Figure 8. Internal view of the msμLPT

Cellulose acetate was found to have the best optical damage resistance, transparency in the 920 nm region, and adhesion, but outgassing in vacuum and vulnerability to solvents were problems. We settled on polyimide and solved the adhesion problem.

3.1.2 Ablation fuel tape: ablating layer development

At τ=2ms, no metals or metal oxides (and only some polymers) have sufficiently low thermal conductivity and specific heat to reach plasma threshold with the intensity that can be transmitted through a transparent polymer layer. Even pure graphite doesn't meet these requirements. We began with PVC as the "host" or carrier which will be heated to the temperature for plasma formation, and nanopearl carbon (typically 1 – 2% by mass) as the laser absorber. This system typically achieved C_m = 60μN/W, I_{sp} = 750s and η_{AB} = 20%. Because we wanted maximum C_m to be the leading feature in the msμLPT (maximum I_{sp} is the leading feature of the nsμLPT) and also wanted better ablation efficiency, we sought exothermic polymers for the absorbing layer. Following suggestions by and test samples from the Lippert group (see Chapter 11) as well as ideas from the Los Alamos group (Ali, et al. 2004), we tried triazene polymers, polyvinyl alcohols, nitrocellulose, PVN, GLYN and GAP, which gave progressively better results. Table 5 compares the performance of the most effective polymers.

Two different laser absorbers were used. We found that 2% nanocarbon gave less coating stickiness, but since carbon is an undesirable exhaust component, we have pursued 1% carbon and greater concentrations of the crosslinker IPDI in the GAP. We have had even better results with a tuned IR-dye as the laser absorber, since it dramatically reduced elemental carbon deposition from the exhaust. We observed lower ablation efficiency from tapes using this absorber, and some photoinstability in room light. However, IR-dye tapes have better I_{sp} and further illumination optimization may deliver equal ablation efficiency. We have also identified more stable dyes and appropriate UV-protective layers.

Ablation efficiency η_{AB} is a critical determinant of performance, since it governs the laser optical power which must be delivered to achieve a given thrust, and that parameter, ultimately, is the major factor determining C_{ms}, the "system momentum coupling coefficient," thrust per watt of input electrical power onboard the craft.

The coupling coefficient C_m = 3mN/W shown in Table 5 is sufficient to levitate a 0.15-kg object with a suitably configured 500-W laser.

Table 5. Comparative performance of ablative layer compositions

Ablatant	Absorber	C_m (μN/W)	I_{sp} (s)	η_{AB} (%)
PVC	5% nanocarbon	60	750	20
GLYN	2% nanocarbon	1280	116	73
GAP	Epolin 2057 IR dye	1300	200	125
GAP	1% nanocarbon	3000	160	235
PVN	5% nanocarbon	116	2890	164

3.1.3 Electronics

We developed DC-to-DC converters capable of delivering 25 watts at efficiencies greater than 90%. The converters will operate over a 6 to 35 V DC input range. This makes an appropriate interface to space platforms operating at 24-28 volts. The converters are controlled by a digital potentiometer, which in turn is controlled by the onboard microprocessor. This provides the precise voltage control necessary for the "super capacitor" and MOSFET switch. These converters are also used to provide power for the microprocessors and the motor.

Peak operating current can be as large as 60A during maximum thrust conditions.

These pulsed currents must come from the μLPT, since otherwise they would have to come from the host platform. In that case, interconnection inductance would degrade laser performance and could upset the platform's power systems as well. This problem was solved by the use of AVX "super capacitors" that supply the pulse current to the laser diodes.

Actual performance of the test model msμLPT is shown in Fig. 9. Predicted performance of the completed msμLPT prototype is shown in Table 6. The lifetime impulse indicated in the table is adequate for most applications involving normal attitude and position adjustment of microsatellites during their lifetime. Lifetime impulse can easily be increased to 500N-s by adding 0.5kg of fuel tape, resulting in a device with 0.9 kg mass. Such a microengine could reenter a 10kg satellite from low Earth orbit.

3.1.4 Performance

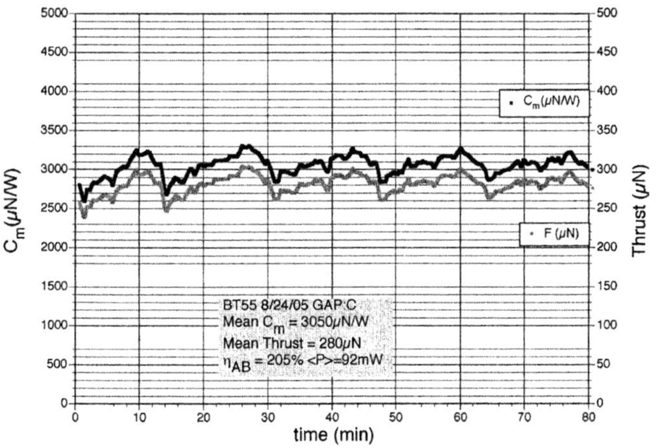

Figure 9. Thrust and C_m produced by the test model μLPT. Thrust variations were due to ablation coating thickness variations. Standard deviation relative to the mean is 4.3% for both in this case. The C_m value of 3mN/W was the largest ever measured and reported in the literature for any solid material. I_{sp} was 137s.

Table 6. Predicted performance of the msμLPT prototype

Type of laser	JDSU 6396 diode
Laser wavelength	916nm
Laser pulse duration (ms)	1.5
Tape coating material	140μm GAP:Cnanopearls
Ablatable mass (grams)	44
Tape dimensions	30m x 8mm x 190μm
Lifetime impulse (N-s)	48
Thruster dimensions (cm)	15.2 x 10.2 x 4.3
Mass (kg)	0.54
C_m (μN/W)	3,000
I_{sp} (s)	140
η_{AB} (%)	206
Normal thrust level	100 μN
Laser sequence	Sequential, in pairs
Laser average optical power (mW)	33
Laser repetition frequency (EA, Hz)	0.63
Operating lifetime (hrs)	132
Maximum thrust level	10 mN
Laser sequence	Parallel
Laser average optical power (W)	3.3
Laser repetition frequency (Hz)	63
Operating lifetime (min)	79

3.2 The nsμLPT

Two difficulties were identified in the discussion of the msμLPT in Section 3.1. One is a heavy involvement in chemistry to create polymer targets based on PVC, PVN, GLYN ,or GAP which are not sticky, resist solvents, etc. No metals could be used as targets because their thermal conductivity was too high to permit achieving the temperatures necessary for efficient propulsion.

The other problem is the requirement for transmission-mode illumination to avoid target ablation products backstreaming onto the optics (Fig. 6). Even though the transparent polymer layer in the tape targets is not penetrated by the laser-target interaction, some back-ablation does occur from the polyimide surface. In contrast, the much higher intensity available from ns-pulse illumination permits using any target material, and permits separating optics and target by a distance sufficient to eliminate the backstreaming problem.

There was some concern about losing power conversion efficiency with this change. However, ns-pulse diode-pumped fiber laser amplifiers are now capable of wallplug-to-light efficiency η_{EL} > 40% (di Teodoro, 2002), and infrared diodes have recentlly improved from 45% to 68% η_E (Alfalight, 2006), promising even better η_{EL} values in the near future. The conversion efficiency sacrifice is not prohibitive for the technology now. Also, fiber lasers are low-mass devices, unlike conventional diode-pumped solid-state lasers.

In this chapter, we report a demonstration in which thrust, specific impulse and energetic efficiency in laser propulsion with ns-duration pulses was measured for the first time using a completely self-contained microthruster suspended in a torsion-type thrust stand capable of 25nN precision in measuring thrust. Intensity on target was in the PW/m^2 range. Performance of the test stand is also described.

3.2.1 Test setup overview

Our test setup could accommodate various target materials with a target delivery system providing several hours of operating lifetime, a power supply, and infrared data link, self-contained and connected to the external world only through the torsion fiber which provided restoring force in the thrust-measurement apparatus. The vacuum thrust stand setup was identical to the device reported in Phipps and Luke (2002) except for a) 40-cm rather than 10-cm torsion fiber length, b) the critical damping attachment, and c) the interferometer element in the center of the support bar. The latter replaced the simple mirror used as a rotation readout in Phipps and Luke (2002). The entire setup mounted inside our vacuum test chamber.

The rotation sensor we developed resolved 20μrad bar rotation, giving the thrust stand 25nN thrust precision. A major design challenge was to deliver laser energy to the target illumination optics through the vacuum chamber envelope from two different benchtop laser systems while the apparatus was rotating in the test chamber with accurate focus on the spinning target (Fig. 10).

An optical link operated through the vacuum chamber wall to turn the spinning target drive on and off, and commanded the transverse illumination drive to move the lens across the disk in a radial direction at a velocity which could be adjusted from 1μm/hr to 8,000 μm/hr as necessary to avoid illumination spot overlap.

The torsion bar is suspended from a 254-μm diameter steel fiber. In Fig. 10, this attachment point is at the top of the figure. Fig. 11 shows the nsμLPT engine in operation.

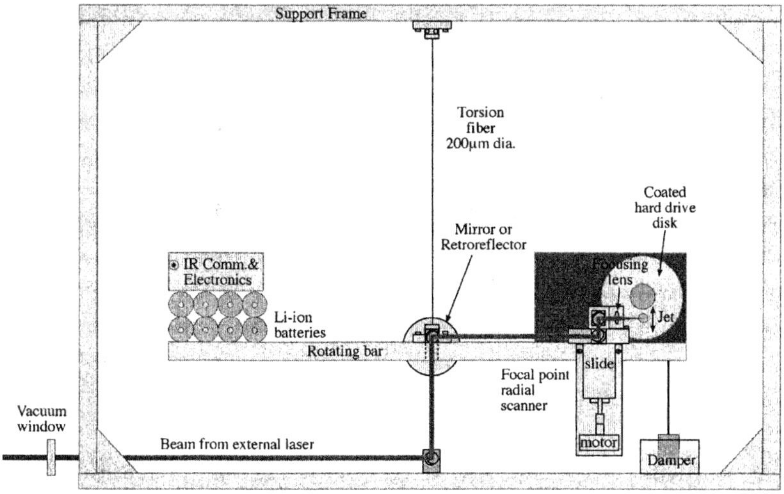

Figure 10. Vacuum thrust stand test setup. Power supply is on board the thrust measurement bar with the thruster, and command and data transfer uses an IR data link, so that the only mechanical connection with the outside world is the 254-μm diameter steel fiber supporting the bar. An interferometer based on a solid glass retroreflecting "corner cube" (described below) is the key to resolving rotation of the bar. Critical damping is provided by a flag immersed in diffusion pump oil.

Target materials were sputtered or electroplated onto an IBM 2.5-inch disk drive. The disk drive motor was operated by a bank of Li-ion batteries which also served as counterweight for the thruster. These were capable of powering the system, including the transverse drive and the optical communication link, for up to 3 hours.

A pair of shielded Langmuir/Faraday time-of-flight (TOF) detectors was installed in the chamber 5.8 and 8.6 cm distant from the target plasma.

3.2.2 Interferometer design

To measure rotation of the torsion bar, we developed a new type of optical interferometer not previously reported in the literature (Figs. 12 and 13, patent pending). The retroreflector "P" had 2.54 cm diameter aperture. The attachment point of the torsion bar is shown as "S," and thrust is applied at "F."

Probe beam (1) in Figure 12 was a 5mW, 532-nm near-diffraction-limited CW beam expanded to 15mm collimated diameter using a beam expansion telescope.

Figure 11. Jet from an operating thruster using a gold-coated nsμLPT target disk. Electroplating worked better than ion sputtering for applying the coatings, both for achieving the necessary thick (25μm) coatings and for causing less coating mechanical stress.

The retroreflector function is similar to that of a simple planar etalon, but there are important advantages in the retroreflector's favor. First, because of the complex internal optical path, the prism is about ten times as effective in creating optical path difference compared to an etalon with the same thickness as the prism depth.

Second, unlike an etalon, for which the phase shift is quadratic with incidence angle, the fringe shift in the prism is absolutely linear with rotation (Fig. 14).

Quadratic phase shift, which is found with the other etalons shown in Fig. 14, leads to calibration and sensitivity which must be specified separately with each new incidence angle. Linear phase shift is a unique advantage of the prism, making absolute calibration possible.

Table 7 summarizes the 25nN thrust stand performance. To appreciate how small the resolved forces are, consider the C_{mhv} of reflected light (no ablation), 6.7nN/W. With our balance, the pressure of a 4W collimated light beam should just be measurable. Also, the gravitational attraction of the experimenter standing 50cm from the microthruster might be observable, although about equal to the noise level of the device.

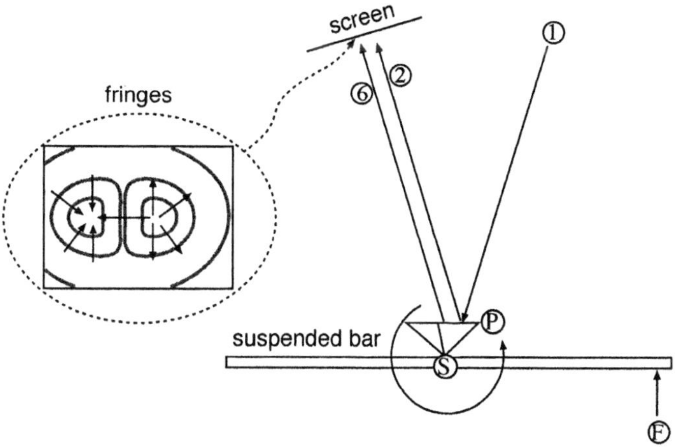

Figure 12. Schematic of torsion bar rotation interferometer viewed from above. "P" is the retroreflector. As rotation occurs, these fringes move radially outward (as illustrated by the arrows in the inset) or inward, depending on the direction of rotation of the bar. Counting the passage of the fringes, which can be done visually or using an electronic data logger, gives rotation.

3.2.3 Results

Measured thrust varied from 180nN to 2mN. Ablation efficiency was near 100% at pulse fluence Φ = 1MJ/m^2. Table 8 summarizes detailed results given in Phipps, et al. (2005b).

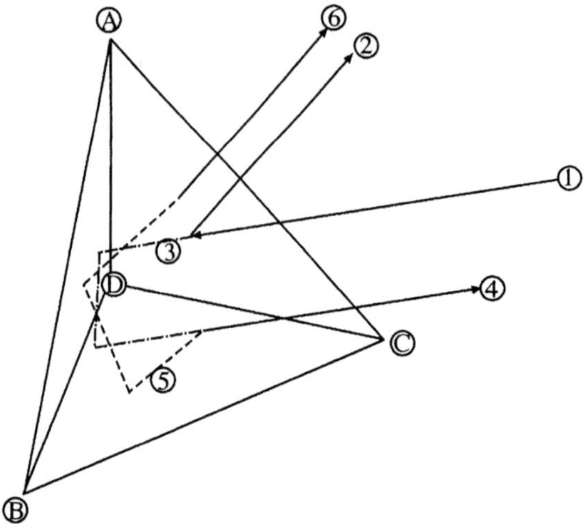

Figure 13. Illustrating the origin of beams (2) and (6) in Fig. 12. Beam (6) is the result of two internal roundtrips through the prism ABCD. Beams (2) and (6) interfere, producing the fringes shown in the inset, Fig.12.

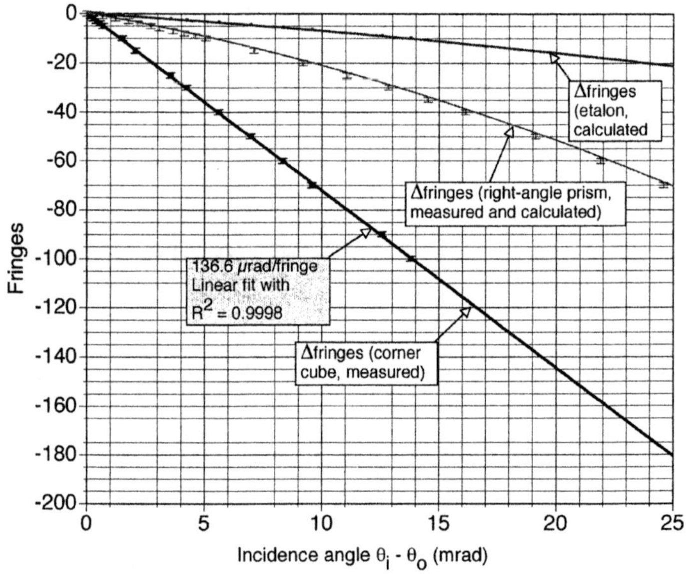

Figure 14. Calibration curve for the retroreflector cube compared to that calculated for two other types of etalon of similar depth. A zero-fringe incidence angle of 30mrad is assumed for the etalon calculation. At this angle, the prism is ten times more effective than the etalon, and its response is linear at any incidence angle. Both the etalon and the right-angle prism are distinctly nonlinear.

Table 7. Thrust stand performance

Prism diameter (cm)	2.54
Force response (nN/fringe)	171
Precision (nN)	25
Minimum thrust measurable with 25% precision (nN)	100
Maximum thrust measurable (μN)	100
Load capacity (kg)	15

Table 8. nsμLPT performance with gold target

Fluence (J/m^2)	6.36E5	2.08E7
Intensity (W/m^2)	1.40E14	4.57E15
Thrust (μN)	0.63	0.47
C_m(μN/W)	68	7.2
Q*(kJ/kg)	5.28E5	5.90E5
I_{sp}(s) from C_mQ^*	3660	212
η_{AB}(%)	100	0.75
Test duration (m)	120	16
v_i from TOF (m/s)	3.59E4	7.75E4
I_{sp} from v_i	3664	7905

The main point of Table 8 is the range of C_m and I_{sp} which is available just by changing incident fluence, although ablation efficiency suffers when it is done this way. A better way is to achieve the same goal by changing the fiber laser pulse duration from ns to ms to select the performance of Table 8, column 1 or that of Table 6 at will, both with high ablation efficiency. A secondary point is that I_{sp}, computed from ion velocities v_i determined with the TOF sensor and with mass loss, diverge strongly at the higher fluence of the two in the table, but converge near $1MJ/m^2$. This tells us that most of the thrust at that fluence is generated by ions, rather than neutrals at the lower fluence. The divergence between these two parameters at higher fluence with Au targets suggests plasma clamping (Figueira et al., 1980), a phenomenon in which, above a certain threshold, laser-produced plasma self-regulates its optical transmission in such a way that the fluence transmitted to the surface is constant, independent of incident fluence. This phenomenon will give an initial burst of high-energy ions, but a much lower energy exhaust during the rest of the laser pulse. The $1MJ/m^2$ fluence for which these two types of I_{sp} were found to agree was predicted by simulations reported in Phipps (2005b).

3.2.4 Summary

Ours are the first results reported in the literature in which I_{sp} was actually measured from mass loss, and the C_mQ^* product compared with measured ion velocities under the same conditions at appropriate background pressure. They are also the first in which an optimum fluence ($1MJ/m^2$) for high I_{sp} was shown, and this fluence was predicted by our computational model. We operated the thruster for seven hours on one gold disk target, using only 2% of the target material in the 20-μm coating. Taken together, these results demonstrate feasibility for a ns-pulse, laser-driven microthruster.

Thrust measurements for the nsμLPT were obtained with the first thrust measurement device reported in the literature that is capable of verifying that a thruster meets the thrust noise requirement of NASA's LISA program (Leach, 2003) at 0.02Hz and below.

Acknowledgments

Microthruster development reported in this chapter was supported by the U.S. Air Force under contracts FA-9300-04-M-3101 and FA9300-04-C-0030.

References

Alfalight, Madison, WI, 2006 news release at http://www.alfalight.com/press-detail.asp?articleid=52

Ali, A. N., Son. S.F., Hiskey, M.A. and Naud, D.L. 2004, Novel High-Nitrogen Propellant use in Solid Fuel Micropropulsion," *J. Prop. Power* **20** (1), 120

Andrenucci, M., Ceccanti, F., Saverdi, M., and Saviozzi, M., 2005, "Qualification status of the FEEP-5 electric micropropulsion subsystem," paper AIAA 2005-4261, 41st AIAA/ASME/SAE/ASEE Joint Propulsion Conference, Tucson

Brophy, J., Kakuda, R., Polk, J., Anderson, J., Marcucci, M., Brinza, D., Henry, M., Fujii, K., Mantha, K., Stocky, J., Sovey, J., Patterson, M., Rawlin, V., Hamley, J., Bond, T., Christensen, J., Cardwell, H., Benson, G., Gallagher, J., Matranga, M. and Bushway, D., 2000, Ion Propulsion System (NSTAR) DS1 Technology Validation Report, available from http://nmp-techval-reports.jpl.nasa.gov/DS1/IPS_integrated_Report.pdf

di Teodoro, F., Koplow J. P. and Moore, S. W. 2002, Diffraction-limited, 300-kW peak-power pulses from a coiled multimode optical fiber, *Opt. Lett.* **27**, pp. 518-520 (2002) plus private communications.

Gonzales D. and Baker, R., 2001, Microchip Laser Propulsion for Small Satellites, paper AIAA 2001-3789, 37th AIAA/ASME/SAE/ASEE Joint Propulsion Conference, Salt Lake City

Gonzales D. and Baker, R., 2002, Micropropulsion using a Nd:YAG microchip laser, *Proc. SPIE Conference on High Power Laser Ablation IV, SPIE* **4760**, pp. 752-765

Gregg, D. and Thomas, S., 1966, Kinetic Energies of Ions Produced by Laser Giant Pulses, *J. Appl. Phys.* **37**, 4313-16

Guman, W. 1968, Pulsed Plasma Technology in Microthrusters, Fairchild Hiller Corp. report AFAPL-TR-68-132

Horisawa, H., Kawakami, M., and Kimura I, 2004, Laser-electric hybrid propulsion system for microthrusters, *Proc. 5th International Symposium on High Power Laser Ablation, SPIE* **5448** pp. 918-927

Kantrowitz, A. 1971 *Astronaut Aeronaut..* **9**, no. 3, pp. 34-35

Kelly, R. & Dreyfus, R. W. 1988, *Nucl. Inst. Meth.* **B32**, 341

Leach, R., 2003, Development of a μNewton Thruster for use with a Drag Free Control system and Formation Flying Satellites, *Proc. 1st International Symposium on Formation Flying Satellites*, Toulouse

Marx, G. 1966, Interstellar Vehicle Propelled by Terrestrial laser Beam, *Nature*, **211**, pp. 22-3

Möckel, W. E. 1972a, Propulsion by Impinging Laser Beams, *J. Spacecraft and Rockets* **9**, no. 12, pp. 942-4

Möckel, W. E. 1972b Comparison of Advanced Propulsion Concepts for Deep Space Exploration, *J. Spacecraft and Rockets* **9**, no. 12, pp. 863-8

Mueller, J., 2000a, Thruster options for microspacecraft: a review and evaluation of state-of-the-art and emerging technologies, in *Progress in Astronautics and Aeronautics* **187**, M. Micci and A. Ketsdever, eds., American Institute of Aeronautics and Astronautics, Reston VA p. 80

Mueller, J., 2000b, *op. cit.,* p. 83

Mueller, J., 2000c, *op. cit.,* p. 88

Mueller, J., 2000d, *op. cit.,* pp. 96-7

Mueller, J., 2000e, *op. cit.,* pp. 102-3

Myrabo, L. N., Libeau, M. A., Meloney, E. D., Bracken R. L., and Knowles, T. B. 2004, Pulsed Laser Propulsion Performance of 11-cm Parabolic 'Bell' Engines within an Atmosphere, *Proc. High Power Ablation IV,* SPIE **5448** pp. 450-464

Myrabo, L.N. and Raizer YU. P. 1994, AIAA Plasmadynamics and Lasers Conference, Colorado Springs, Paper 94-2451

Phipps, C., Luke J., and Helgeson, W., 2005a, Giant momentum coupling coefficients from nanoscale laser-initiated exothermic compounds, paper AIAA 2005-3607, 41st AIAA/ASME/SAE/ASEE Joint Propulsion Conference, Tucson

Phipps, C., Luke J., and Helgeson, W., 2005b, 3ks specific impulse with a ns-pulse laser microthruster, paper IEPC 319, 29th International Electric Propulsion Conference, Princeton

Phipps, C., Luke J., and Helgeson, W., 2005c, Performance Test Results for the Laser-powered Microthruster, paper 16AM201, 4th International Symposium on Beamed Energy Propulsion, Nara, Japan

Phipps, C. 2005, Will your children ride a laser beam into orbit? Would you want them to?, *Proc. Third International Symposium on Beamed Energy Propulsion*, A. V. Pakhomov and L. N. Myrabo, eds., *AIP Conference Proceedings* **766**, pp.11-22

Phipps C. R. and Luke, J. R. 2004, Micropropulsion using a Laser Ablation Jet, *Journal of Propulsion and Power*, **20**, no. 6, pp. 1000-1011

Phipps C. R. and Luke, J. R. 2003, Advantages of a ns-pulse micro-Laser Plasma Thruster, in *Beamed Energy Propulsion, APS Conference Proceedings* **664**, American Institute of Physics, Melville, NY, pp. 230-9

Phipps, C. R. and Luke, J. R., 2002, Diode Laser-driven Microthrusters: A new departure for micropropulsion, *AIAA Journal* **40** no. 2, pp. 310-318

Phipps, C. R., Reilly J. P. and Campbell, J. W. 2000, Optimum Parameters for Laser-launching Objects into Low Earth Orbit", *J. Laser and Particle Beams*, **18** no. 4 pp. 661-695

Phipps, C., 1998, Advantages of using ps-pulses in the ORION Space Debris Clearing System, *Proc. International Conference on Lasers 97*, STS Press, McLean, VA, pp. 935-941

Phipps, C., Friedman, H., Gavel, D., Murray, J., Albrecht, G., George, E., Ho, C., Priedhorsky, W., Michaelis, M. and Reilly, J., 1996, "ORION: Clearing near-Earth space debris using a 20-kW, 530-nm, Earth-based, repetitively pulsed laser", *Laser and Particle Beams*, **14** (1) pp. 1-44

Phipps, C., 1994 "LISK-BROOM: A laser concept for clearing space junk," in *AIP Conference Proceedings* **318**, Laser Interaction and Related Plasma Phenomena, 11th International Workshop, Monterey, CA October, 1993, George Miley, ed. American Institute of Physics, New York pp. 466-8

Phipps, C. and Michaelis, M., 1994, Laser Impulse Space Propulsion, *Laser and Particle Beams,* **12** (1), 23-54

Phipps, C. R. and Dreyfus, R. W. (1993) Chapter 4, Laser ablation and plasma formation in *Laser Microprobe Analysis*, A. Vertes, R. Gijbels and F. Adams, eds., John Wiley, NY, pp,. 369 *et seq.*

Phipps, C. R. 1989, Conceptual Design of a 170-MJ Hydrogen Fluoride Laser for Fusion, *Laser and Particle Beams,* **7**, 835

Sänger, E. 1955, Zur Theorie der Photonenraketen, in *Probleme der Weltraumforschung (IV. Internationaler Astronautischer Kongress, Zurich, 1953)* Biel-Bienne, Laubscher, p.32

Sänger, E. 1956, Flight Mechanics of Photon Rockets, *Aero Digest*, pp. 68-73

Schall, W. 1998, "Removal of small space debris with orbiting lasers," SPIE **3343** pp. 564-574

Schall, W. 1991, "Orbital debris removal by laser radiation," Ci *Acta Astronautica*, **24.** p. 343

Simon, D., Land, B., Nedungadi A., and Cybyk, B., 2005, Instrumentation development for micro0 pulsed plasma thruster experiments, paper AIAA 2005-4264, 41st AIAA/ASME/SAE/ASEE Joint Propulsion Conference, Tucson

Sovey, J., Rawlin, V. and Patterson, J. 1999, A Synopsis of Ion Propulsion Development Projects in the United States: SERTI to Deep Space I, AIAA paper 99-2270, 35th AIAA Joint Propulsion Conference

Yabe, T., Phipps, C., Nakagawa, R., Yamaguchi M., Aoki, K., Baasandash, C., Abe, H., Yoshida, N., Ogata, Y., Nakagawa, M., Fujiwara, E., Yoshida, K., Nishiguchi, A., Ochi K. and Kajiwara, I. 2002, Numerical and Experimental Analysis of Laser-Driven Rocket and Micro-Airplane, *Proc. 2nd International Conference on inertial Fusion Sciences and Applications, Kyoto Japan,* Elsevier, Paris

Yabe, T., Phipps, C., Yamaguchi, M., Nakagawa, R., Aoki, K., Mine, H., Ogata, Y., Baasandash, C., Nakagawa, M., Fujiwara, E., Yoshida, K., Nishiguchi A. and Kajiwara, I. 2002, Microairplane propelled by laser driven exotic target, *Applied Physics Letters,* **80**, no. 23, pp. 4318-4320

Chapter 17

LASER PROPULSION THRUSTERS FOR SPACE TRANSPORTATION

W.O. Schall, H.-A. Eckel, W.L. Bohn
DLR (German Aerospace Center), Institute of Technical Physics, Pfaffenwaldring 38-40, 70569 Stuttgart

1. INTRODUCTION

The translational motion of a body in free space can only be changed by changing its momentum. Aside from photonic propulsion utilizing the photon pressure of intense electromagnetic radiation (solar or laser sailing) and a momentum change by flying through a gravitational field (gravitational assist), all other propulsion methods rely on Newton's principle action = reaction by ejecting matter in the opposite direction of the intended momentum change. This matter, the propellant, can assume any thermodynamic state, from solid to plasma, and is accelerated to an exhaust velocity, v_{ex}, either by electromagnetic or electrostatic fields or simply by a gasdynamic expansion. In the latter case the kinetic energy is acquired by addition of thermal energy to the propellant before it is exhausted. In the case of thermal laser propulsion the energy is radiated to the vehicle and concentrated into the propellant. The energy can be supplied by either continuous or pulsed radiation.

The required propulsion power and hence the laser power is proportional to the required thrust force, T, and the obtainable exhaust velocity. For thermokinetic reasons the exhaust velocity of classical chemical propulsion systems is limited by the reaction energy of the most energetic chemical reactions to about 5000 m/s. The deposition of laser energy into matter is not bound and therefore higher exhaust velocities are conceivable. The thrust is defined by the product of the exhausted mass flow rate, $\dot{m}$, and the exhaust velocity. The required thrust force is mission dependent. For a launch from the surface of the Earth the thrust must overcome the weight force of the vehicle and the air drag. In free space very small accelerations can be traded for longer mission times, thus requiring less thrust and less power over a longer time.

For propellants in the solid state, a phase change is necessary. The higher the internal energy (enthalpy, temperature) of the propellant can be raised, the higher the exhaust velocity will be and the more mass can be saved. This is expressed by the definition of the specific impulse, which is the ratio of the thrust to the weight flow rate, $\dot{w}$:

$$I_{sp} = \frac{T}{\dot{w}} = \frac{T}{\dot{m}\, g} = \frac{v_{ex}}{g} \quad (s), \tag{1}$$

where $g = 9.81\ m/s^2$ is the gravitational acceleration. From the definitions of thrust and power it is easy to show, that these are directly related to the specific impulse and the momentum efficiency, η_m, describing the conversion of incident power into thrust:

$$\frac{T}{P} = \frac{2\,\eta_m}{g\, I_{sp}} \quad . \tag{2}$$

This relation implies a decreasing thrust with increasing I_{sp} or v_{ex}. On the other hand, the importance of a high exhaust velocity, in particular for flight missions requiring a large velocity increment Δv, is seen from the fundamental rocket eqation, which relates the consumed mass, m_p, of the initial mass, m_0, to the ratio of the velocity increment to v_{ex}:

$$\frac{m_p}{m_0} = 1 - \exp\left(-\frac{\Delta v}{v_{ex}}\right) \quad . \tag{3}$$

Table 1. Characteristic data for 4 different space flight mission configuratons

Mission	Earth — LEO		LEO — GEO	
Mass at destination (kg)	10		1000	
Velocity increment (km/s)	10		4	
Initial acceleration (m/s^2)	20		0.2	
Specific impulse (s)	600	1000	400	1000
Propellant mass fraction	0.81	0.63	0.63	0.33
Thrust (N)	1059	544	544	298
Power (MW)	6.35	5.44	2.18	2.98
Burn time	243s	316s	3.5h	4.6h

With these relations we can estimate the requirement for thrust and laser power,09 for instance for placing 10 kg into low Earth orbit or transporting

1 ton from LEO to GEO. Assuming an efficiency $\eta_m = 0.5$, different I_{sp} and an appropriate initial acceleration corresponding data are given in Table 1.

A principal problem of the application of electromagnetic radiation to heat matter is the absorption of the radiation by the propellant. Gases may be transparent to the laser wavelength and require the addition of a line selective absorber. Solids may reflect a large fraction of the incident radiation. For gases without resonant absorption, radiation energy can only be taken up by few free electrons (inverse bremsstrahlung) and initiate a plasma breakdown under very high intensities (~10^9 W/cm^2 for clean air). In metals radiation energy is generally absorbed more easily by electrons in the conduction band and then distributed to the lattice (breakdown intensitity 10^6 – 10^7 W/cm^2). Plasma ignition in a gas can be enforced by impurities or by placing a metal surface in the focus. Finally, in nonmetals or organic compounds lattice vibrations may be excited and chemical bonds are broken up. In this case the material may decompose into a gas of only radicals.

In a continuously operating thruster the high intensity has to be maintained for the continuous heating of the propellant into the plasma state. From the value of the breakdown intensity it can be seen that even with megawatt powers only small plasma zones can be maintained. In contrast, high intensities can easily be provided with medium sized pulsed lasers, and in particular, in the focus of a concentrating lens or mirror. This leads to especially simple thruster configurations: in its simplest form material is ablated from the surface of a solid propellant by dumping radiative energy into the material within a very short time (< ms). The heated material leaves the surface in a gaseous or plasma state and the recoil exerts a total impulse to the propellant body equal to

$$F = \int m\, v_{ex}\, dt = \Delta m\, \overline{v}_{ex} \quad , \tag{4}$$

where again v_{ex} expresses the instantaneous exhaust velocity and corresponds in this simple case to approximately the thermal velocity. m is the ablated mass. The impulse can easily be measured, for instance by the displacement of a pendulum. From weighing the mass loss per pulse, Δm, a time averaged exhaust velocity $\overline{v}_{ex}$ can be derived and the pulse averaged specific impulse is then given by $\overline{I}_{sp} = \overline{v}_{ex}/g$. For simplicity, we now omit the dash over the symbol for the indication of a time average. A more rigorous derivation on the basis of the instantaneous values of the momentum is given by Gregory and Herren (Gregory and Herren, 2005).

By relating the impulse to the incident laser pulse energy, E, the momentum coupling coefficient, c_m, is defined as

$$c_m = \frac{F}{E} \quad \left(\frac{Ns}{J} = \frac{N}{W} = \frac{s}{m}\right) \quad , \tag{5}$$

and the specific mass consumption, μ, and its reciprocal, the deposited energy, Q^*, as

$$\mu = \frac{1}{Q^*} = \frac{\Delta m}{E} \quad \left(\frac{kg}{J}\right) . \tag{6}$$

Now the specific impulse simply is

$$I_{sp} = \frac{c_m}{g\mu} = \frac{c_m Q^*}{g} . \tag{7}$$

Applying the balance of energy between the kinetic energy of the exhaust gases and the laser pulse energy allows the definition of an ablation efficiency, η_{ab}, in analogy to equation (2), which in terms of c_m and μ yields the following relations (Phipps, et al., 2000):

$$\eta_{ab} = \frac{m v_{ex}^2}{2E} = \frac{1}{2} \mu v_{ex}^2 = \frac{1}{2} \mu v_{ex} g I_{sp} = \frac{1}{2} c_m g I_{sp} = \frac{c_m^2}{2\mu} = \frac{1}{2} c_m^2 Q^* . \tag{8}$$

Note that this definition is based on the actual laser pulse energy and not on the energy arriving on the target surface. Thus, the efficiency does include possible absorption losses in a decoupled exhausted plasma which limit the impulse generation. Finally, we can calculate the thrust of a rocket, driven by a pulsed laser with frequency ν as

$$T = c_m \nu E . \tag{9}$$

The attainable values of c_m and I_{sp} depend on the propellant itself, as well as on the laser pulse parameters. Laser pulse parameters are the wavelength, λ, pulse length, τ, the pulse energy, E, and to some extent the pulse shape. At the target and including some concentrating optics, these numbers translate into a time and location dependent intensity distribution I(x,t) in W/m^2, and the energy flux or fluence Φ in J/m^2. Experimental coupling coefficient data for surface absorbing materials and for volume absorbing materials have been collected by Phipps, et al. (Phipps, et al., 1990) and correlated to a combined quantity $q = I \lambda \sqrt{\tau}$. It turns out that c_m follows trend lines of the type $c_m = a q^{-b}$, with coefficients a and b typical for each type of absorber. In particular, for gases, c_m and I_{sp} also depend on the density of the gas and the thruster geometry due to fluiddynamic effects.

Utilizing the ablation from solid propellants a most simple rocket is conceivable, consisting of not more than a block of solid propellant with the payload attached to it. A desirable propellant should be durable and safely storable. It must be mechanically resistant to strong shocks and it should be easily machinable. The release of inner energy is favorable but the propellant must not continue to burn after the termination of the laser pulse. Finally, the ablation products should not contaminate any part of the rocket. This is particularly important for light reflecting parts of the concentrator.

There exist two types of solid absorbers for propellant. Volume absorbers are partially transparent to the laser radiation. So, the energy can be deposited to some depth of the material and a corresponding volume is vaporized and ionized. The liberated mass dominates the impulse, and thus high momentum coupling coefficients can be attained. For the same reason the specific impulse is low. Values over 1000 s are difficult to obtain (Phipps and Michaelis 1994). Depending on the wavelength, polymers and certain non-metals belong to this class of materials. In contrast, metals are surface absorbers. For metals, the penetration depth of the laser radiation is rather small and all non-reflected energy is deposited in a thin surface layer. This leads to very high plasma temperatures with higher ablation velocities. Yet, the liberated mass remains small, resulting in a low coupling coefficient but at a high specific impulse. It can be much higher than 1000 s.

Energy deposition in gases or a free liquid requires a thrust chamber to take up the momentum from the expanding gas. When pulsed energy is focused with sufficiently high intensity into a gas, it breaks down into a hot plasma ball. The expansion of this plasma ball drives a shockwave ahead, which raises the pressure in the surrounding medium. As the shockwave impinges on the walls of the thrust chamber, the pressure pulse is transferred into an impulse to the structure. If the high pressure is maintained over a certain time, a gasdynamic thrust component may add to the pressure pulse like in a conventional laval nozzle. In a multidimensional expansion geometry, the pressure behind the shock wave decreases rapidly and for short pulses the fluiddynamics can be described by Sedov's equations for a strong explosion (Sedov, 1959). In analogy to classical combustion gasdynamics one may distinguish between a laser supported detonation wave (LSD wave) and a laser combustion wave (LSC wave). For short pulses of high intensity and at high density, the temperature jump through the shockwave is sufficient to ionize the gas. All incident energy is then absorbed in this detonation wave and the radiation decouples from the original target plane. This process is also observed in the ablation of solid materials for pulse lengths longer than a characteristic flow distance away from the target. For an LSC wave the shockwave is not strong enough to absorb energy immediately behind the pressure and temperature jump. It decouples from the plasma front and remains transparent. Fig. 1 shows Schlieren pictures of evaporating material from a flat POM (PolyOxyMethylene = Polyacetal, Delrin) target 6.7 μs after the start of a 12 μs long CO_2 laser pulse of 150 J (incident intensity $8 \cdot 10^6$ W/cm^2). The ambient air pressure in picture a) is less than 50 Pa and set to 3500 Pa (= 35 mbar) in pictures b) and c). The elliptical black cloud is interpreted as the non-transparent expanding plasma. The circular shockwave produced in ambient air can be seen in the pictures b) and c). In picture c) the POM sample was seeded with about 20% aluminum powder. Obviously this constituent penetrates through the shockwave and escapes (after a delay of 2 μs) with twice the shock speed.

Figure 1. Schlieren pictures of the ablation process of POM at a fluence of 75 J/cm^2: left – vacuum; center - 35 mbar air; right – POM + 20% Al at 35 mbar air.

Since the processes described work in any gas, it is possible to use ambient air for producing thrust. Evidently, for an airbreathing laser rocket the specific impulse is infinity and the only remaining relevant performance quantity is the coupling coefficient c_m. A combination of the two processes, ablation and gas breakdown, has been conceived in the concept of the double-pulse thruster (Kare, 1990): A first, less energetic laser pulse serves to liberate a gas cloud from the surface of a solid propellant. After a few microseconds, before the gas expands into the environment, a second, much stronger pulse is placed to initiate a strong shockwave in the gas.

The impulse effect of laser ablation is enhanced, if in addition to the deposition of radiative energy an exothermic chemical reaction is initiated. Such a reaction can be the dissociation of an explosive (Liukonen, 1998, Luke, et al., 2003) or simply the combustion of the ablation products in the ambient air (Schall, et al., 2002).

2. PULSED LASER PROPULSION DEVELOPMENTS

The utilization of energy deposition in matter, transmitted by intense pulsed laser beams for rocket propulsion from ground to orbit, has first been suggested by Kantrowitz (Kantrowitz, 1972) in 1972. In the same year, substantial efforts have already been undertaken in the USA and a first report by Pirri and Weiss was published (Pirri and Weiss, 1972) on fundamental experiments with steady-state and pulsed ablation in a parabolic reflector. After a theoretical analysis by Bunkin and Prokhorov (Bunkin and Prokhorov, 1976) scaling experiments with different concentrators for air breakdown started in Russia (Ageev, et al. 1980). A considerable progress was the first free flight of a laser rocket design by Myrabo (called "Lightcraft"), jointly conducted with the U.S. Air Force in 1997 (Myrabo, et al., 1998) and resulting in a record flight height of 72 m in 2000 (Myrabo, 2001). Delrin was utilized as the ablating propellant. After Myrabo's first successful flight experiments similar experiments were reported from several other countries (Russia (Liukonen, 1998), Germany (Bohn, 1999), Japan (Urabe, et al., 2003)). Extensive analysis on the possibilities and the requirements for early introduction of laser propulsion for space flight has been carried out by Phipps and Michaelis in 1994 (Phipps, et al. 1994). Phipps has realized that the introduction of laser propulsion into a well established propulsion economy can only start in small steps, for demonstrating the credibility of this new technology. Hence he started with the development of laser ablative microthrusters for attitude control and stationkeeping (Phipps and Luke, 2002), which are prepared to fly in space soon.

A special type of pulsed ablative laser propulsion should be mentioned. In 1990, Schall (Schall, 1990), and later Kusnetsov in 1994 (Kusnetsov, et al., 1994) have suggested to de-orbit space debris by changing its orbit to one with its perigee in the decelerating upper atmosphere, utilizing the propulsive effect of ablating a fraction of the debris itself. While these proposals relied on a space based laser, Phipps et al. worked out this idea for a ground based system, called ORION (Phipps, et al., 1996). A somewhat futuristic proposal along the same idea is the deflection of near Earth asteroids on a collisional course with Earth (Campbell, et al., 2003).

Ablative laser propulsion for space flight has been promoted by attributing many conceivable advantages over conventional, i.e. chemical propulsion techniques. It is clear that large chemical boosters for carrying many tons of payload into space cannot be replaced by laser rockets in any foreseeable future due to the sheer requirement of gigawatts of power for such tasks. However, it has been realized early that laser propulsion can fill a gap between chemical rockets and electric propulsion in terms of c_m and I_{sp}. So the thrust to weight ratio is much higher than in electric propulsion, mainly because the power production source remains on the ground. A high specific impulse allows for high payload fractions without staging. The propellant can be selected to match the flight profile (Larson, et al., 2004) and for a flight within an atmosphere no propellant is required at all. Hence, no pollution can occur. In addition, easy thrust modulation is possible by changing the laser pulse energy and the pulse frequency. The laser installation and power transmission unit would constitute a considerable investment. However, as it remains at a place where it is easily serviceable, it can be kept much cheaper. For instance, on the ground it does not need space qualification. The possibility for many launches per day into Earth orbit distributes the investment and maintenance costs over many flights, thus reducing the total transportation costs per kilogram in orbit by more than an order of magnitude.

Experiments in the form of mostly lab-scale wire-guided flight modules, as well as of units for measuring mechanical impulse and specific impulse, have been made for many different thruster geometries. The most simple is of course a flat plate. This is, for instance, realized in the ablative microthruster of Phipps and Luke. (Phipps and Luke, 2002). If a concentrator is used it can at the same time work as a kind of gasdynamic nozzle. The simplest configuration is a cone or a paraboloid (Ageev, et al., 1980) or a matrix of paraboloids (Liukonen, 1998). A more complex geometry has been developed by Myrabo (Myrabo, et al., 1998). His concentrator has a ring shape around a central peak with parabolic contur, producing a ring line focus and serving at the same time as a plug nozzle. To maintain a one-dimensional blow-off the propellant target has also been placed inside a cylindical tube by some authors (Bohn, 1999, Irtuganov et al. 2005). Finally, a different concept is followed by Sasoh (Sasoh, 2001). He places the whole craft inside a tube like in a cannon to utilize the increase in gas pressure from the breakdown at the stern of the vehicle. If a long laser pulse at a high repetition frequency is radiated from behind into a thrust chamber, the light must pass through absorbing ablation products. To avoid premature absorption, Ageichik, et al. reflect the radiation sideways into the thrust chamber along its circumference (Ageichik, et al., 2005).

3. PERFORMANCE MEASUREMENTS

In the following, we will report about measurements of the coupling coefficient and the specific impulse as achieved with ablation from flat targets and inside various concentrator geometries (Schall, et al., 2000, 2002, 2005). The laser radiation was produced with an electron-beam sustained electric discharge CO_2 laser at the fundamental 10.6 μm wavelength. Maximum pulse energy was up to 450 J with an unstable resonator producing a ring mode and 250 J with a stable resonator and flat mode distribution. The nominal pulse length was between 10 and 15 μs after a leading spike of a few 100 ns. The propulsion module or material samples were suspended on long, thin strings, forming a pendulum, and placed in a vacuum chamber. The imparted impulse was derived from the displacement of the pendulum and the coupling coefficient was calculated according to equation (5). The mass loss was measured by directly weighing the propellant sample before and after one or several equal pulses. The specific impulse and the ablation efficiency were then derived according to equations (7) and (8). For most of the measurements with solid propellants and as a reference, POM has been used, which vaporizes without producing soot. Experiments were carried out with flat samples under direct and perpendicular irradiation (Schall, et al., 2005), with bell shaped parabolic concentrators of different slenderness and a toroidal concentrator of the Myrabo type (Schall, et al., 2002).

3.1 Results from flat samples of various materials

The laser beam from the stable resonator was focused (f no. = 10) to a spot size of 15.5 mm in diameter and the pulse energy was varied between 40 J and 280 J, producing a fluence between 22 and 150 J/cm^2. After the initial spike, the power rose within 2.7 μs to yield maximum pulse intensities between 2 and 15 MW/cm^2. The pulse began to decay after 6.5 μs and after about 11 to 13 μs from the ignition, the power has decayed below 5% of the maximum. Note that the maximum fluence falls within the range given by Phipps (Phipps, 2005) for peak momentum coupling. Also note that the pulse shape with its initial spike can be interpreted as a double pulse arrangement, where first material is vaporized by the initial spike and into which the main energy is subsequently dumped. During the first 4 μs the plasma front is seen in Schlieren pictures to expand with a velocity of 1.9 km/s for a fluence of 75 J/cm^2. Since the plasma does not expand further than about 12 mm from the target surface, all energy is expected to be deposited in this region (Schall, et al., 2005).

Since the blow-off from a flat sample occurs into the 3-dimensional half-space, a one-dimensional expansion was enforced for most of the experiments by placing a 41 mm long cylindrical gas guiding tube with inner diameter of 20 mm in front of the samples. Comparative measurements at atmospheric pressure without this tube showed a 40% higher mass loss for the 3D case and a comparably reduced coupling coefficient. In vacuum this difference reduces to about 15%.

As a function of the pressure the ablated mass is fairly insensitive, except for pressures below 0.1 bar. In this range up to a factor of two more mass can be ablated. However, probes of plain POM become unstable and cracked for high fluences. The coupling coefficient begins to drop significantly for pressures below 600 mbar. This is an indication that the air layer in front of the target is involved in the production of the impulse. With increasing pulse energy the amount of ablated mass has a general tendency to go down. Under atmospheric pressure this behavior is more evident and amounts to a reduction of about 30% when increasing the pulse energy from 120 J to 280

J. For the coupling coefficient a corresponding reduction is quite prominent, and in the same range of energy and in the ambient atmosphere it amounts to more than a factor of 3. Obviously, at the higher energy values a smaller fraction of energy can be deposited and turned into velocity of the ablation products. As a consequence the derived velocity in vacuum grows from 1.6 km/s at 120 J to only 2.4 km/s at 280 J. In the light of subsequent discussions it should be kept in mind, that these are pulse averaged values.

Fig. 2 and Fig. 3. show a direct comparison of the coupling coefficient and the exhaust velocity (= $g \cdot I_{sp}$) for the 1D-expansion at a pulse energy of 200 J. Twenty-five different materials have been tested in vacuum and at atmospheric pressure of air. The base materials were POM, epoxy resin, and PVN (Polyvinyl nitrate). The large amount of material that is vaporized and the consequently low specific impulse led to the idea to transfer these volume absorbers into surface absorbers by blending in other elements like aluminum, magnesium or carbon in various concentrations (Schall et al. 2005). Electron micrographs showed a spectrum of grain sizes from 15 to 30 μm for the embedded metal powders. Except for PVN the samples were squares of 22 x 22 mm with a thickness of 3 mm.

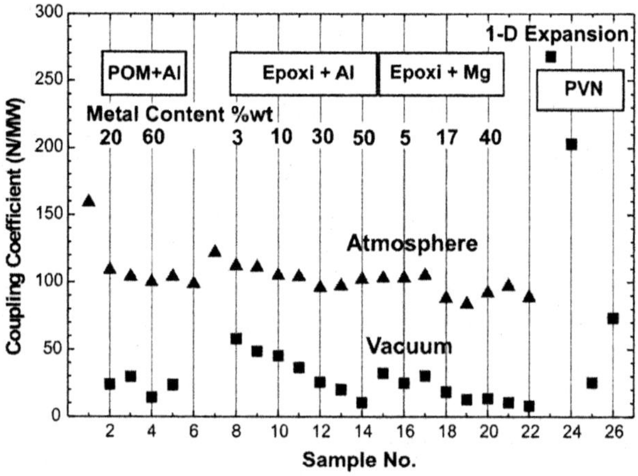

Figure 2. Comparison of the coupling coefficient for different material samples. (Sample numbers are: 1- POM; 2 – 4 - POM + Al; 5 - Butadiene+Al (unknown amount); 6 - Epoxy +Al (unknown amount); 7 – 14 - Epoxy+Al; 15 – 20 - Epoxy+Mg; 21 - POM+Fe; 22 - POM+Ti (irregular sample shape); 23 – PVN; 24 - PVN+C; 25 – 26 - PVN+Al).

As Fig. 2 shows, there were only two samples that delivered a significantly higher coupling coefficient than the rest of the materials. This was plain POM in atmosphere with over 150 N/MW, compared to an average of about 100 N/MW and PVN + C, which even in vacuum reached over 200 N/MW. This latter composition is an energetic material that releases additional energy when decomposed under laser radiation. In all investigated cases the vacuum values are much lower and only few reach 50 N/MW.

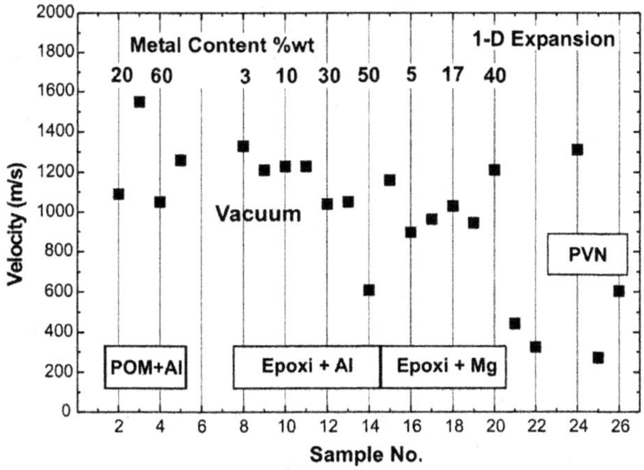

Figure 3. Comparison of the velocity of the ablation products for different material samples (The numbers are the same as in Fig. 2).

For plain POM no value is available from these measurements, but from other investigations (Bohn and Schall, 2003) it is known that it will also be reduced by 50% compared to atmospheric pressure. One reason for the much higher coupling coefficients in air can be attributed to chemical reactions of the vapor products with the oxygen of the air. From a comparison of the data of the polymeric materials with embedded metal in Fig. 2, it can be directly concluded that this approach to turn a volume absorber into a surface absorber failed. Micrographs from irradiated sample support the conjecture that the polymeric compounds easily absorb the radiation and vaporize before the metal is even melted through. Hence, this material cannot acquire high ablation speeds. On the other hand, as the Schlieren picture (Fig. 1c) shows it is remarkable that in low pressure air (35 mbar) a material cloud, presumably the metal grains, passes through the shockwave that is triggered by the expanding plasma with a velocity that can assume more than twice the original shockwave velocity. A similar observation has been made by Urech, et al. (Urech, et al., 2004). The process is not yet fully understood, but may be connected to a confined ablation (Fabbro et al. 1990).

The velocity of the ablation products is shown in Fig. 3. Only the values for vacuum are presented. Except for plane POM with a maximum credible value of 2.0 km/s, all other values range around 1 km/s. Due to the interaction with the ambient air at atmospheric pressure the use of the ablated mass only in relations (6) and (7) results in much too high values for v_{ex}. The maximum possible speed can also be found from Schlieren pictures, since the material velocity cannot be higher than the speed of the shockwave.

If it is assumed that a counterpressure of 35 mbar, for which the shockwave is still visible, does not seriously impede the blow-off, the velocity must be slower than 2.5 km/s, for instance, at a fluence of 75 J/cm^2. Applying formulas for plane shockwaves as a first approximation (isentropic exponent $\gamma = 1.4$), one finds a flow velocity behind the shock of 1.85 km/s - agreeing well with the measured plasma expansion velocity of 1.9 km/s - a pressure of 3.1 bar, and an adiabatic temperature of 3500 K for the shocked air. The values satisfy the conditions for a detonation wave (Landau and Lifshits 1959).

Although the derived velocities are remarkably high, they are by far insufficient for space propulsion requirements. Since it can be assumed that the ablation products leave the propellant surface with about thermal velocity, it is necessary either to dump much more energy into the target or try to apply gasdynamic methods for an acceleration to supersonic velocities. The first approach seems to be hampered by an increased and ineffective absorption in front of the target. The application of an expansion type nozzle structure could perhaps improve the specific impulse by accelerating the gas to a higher exhaust velocity. At the same time this nozzle structure can serve as a concentrator for a large laser beam cross-section.

3.2 Results for bell and plug nozzle shaped geometries

Extensive experiments have been made with ablation of POM in parabola-shaped and toroidal configurations (Fig. 4). These concentrator and nozzle forms had an opening diameter of 100 mm, matched to the diameter of the CO_2 laser beam. In a few experiments the slenderness of the parabola had been increased, while keeping the opening diameter constant (Fig. 5). A parabola with a length of 62.5 mm served as the reference. The experiments were conducted at atmospheric pressure and in vacuum (pressure below 0.5 mbar).

Figure 4. Parabolic and toroidal concentrator and nozzle shape.

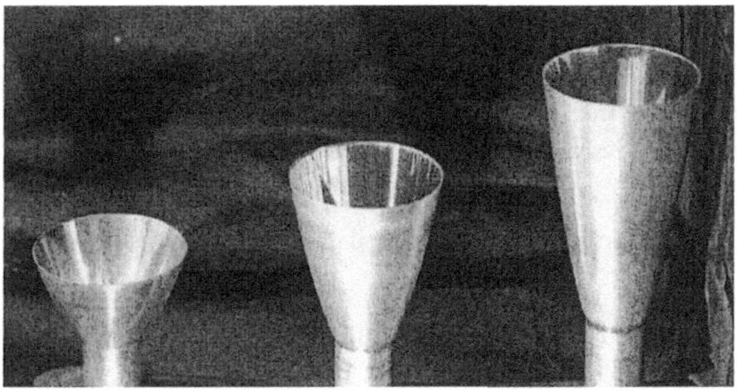

Figure 5. Investigated nozzles with parabolic concentrator and nozzle contour.

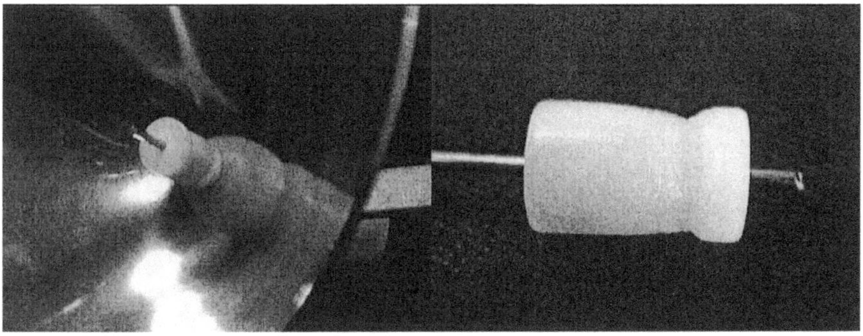

Figure 6. Propellant cylinder positioning in the parabolic nozzle (left) and a POM cylinder after irradiation.

In the parabola-shaped nozzles the propellant was placed in the range of the focus in the form of a cylinder of 8 or 10 mm in diameter (Fig. 6). The ablation occurred in radial direction along the circumference of the cylinder. Enlarging the diameter from 8 to 10 mm reduced the intensity on its surface by a factor 0.65. The peak intensity for 8 mm diameter was $3 \cdot 10^7$ W/cm^2 for a pulse energy of 100 J. In the toroidal configuration a POM ring, fitting inside the outer shroud, served as propellant. The inner ring surface coincided with the focal line.

Fig. 7 compares the coupling coefficient for the two propellant cylinders with 8 and 10 mm in diameter as a function of incident pulse energy. The peak value for the coupling coefficient in air is over 600 N/MW, while in vacuum it reaches about 400 N/MW. A few of the flat plate results are included in the diagram for comparison. These values are considerably lower than for the nozzle. An obvious tendency to decrease with increasing pulse energy, especially with increasing fluence can be observed (44% in the range from 120 to 250 J). Hence, the results for the 10 mm cylinder drop much less (12%). That the increasing fluence impedes the attaining of a high impulse is also seen in a few data where the propellant cylinder was retracted such that the radiation was centered on its front surface. The results from these experiments are considerably worse. For the same conditions, Fig. 8 shows the average vacuum exhaust velocity. Only a slight increase from 2.2 to 2.6 km/s is found. In contrast to the coupling coefficient, the exhaust velocities with the flat plates become comparable to those of the nozzles for the highest pulse energy.

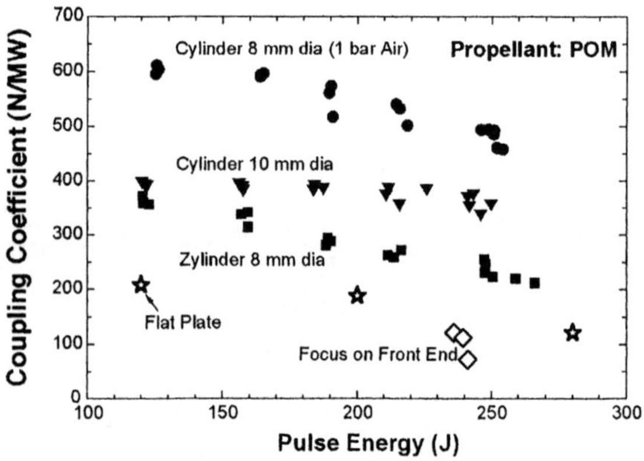

Figure 7. Coupling coefficient vs. pulse energy for different ablator configurations.

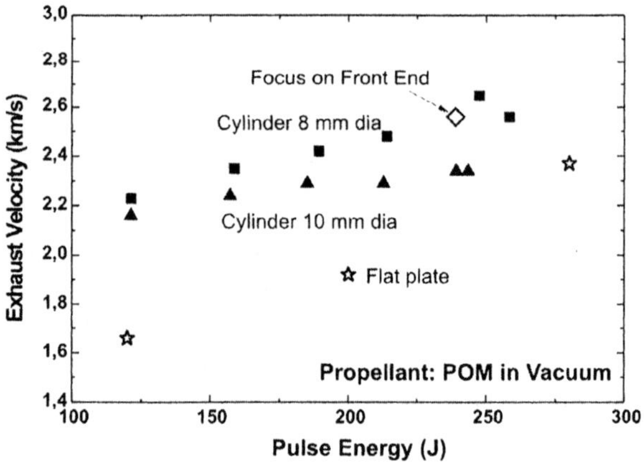

Figure 8. Average exhaust velocity vs. pulse energy in vacuum.

The toroidal form of Myrabo's Lightcraft corresponds more to so-called plug nozzles. For these the expanding gas is only guided on the inner side of the torus, while at the outer side it can expand freely. Figs. 9 and 10 compare the vacuum results with POM for the baseline bell-type nozzle and a toroidal

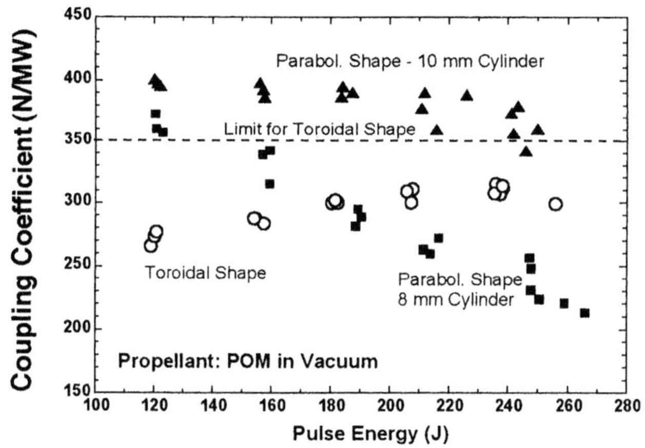

Figure 9. Comparison of coupling coefficients for parabolic and toroidal concentrator nozzles.

nozzle. The strictly linear increase of the impulse for the toroidal nozzle with the pulse energy and also of the mass loss per pulse imply a maximum coupling coefficient of 350 N/MW for this device at very large pulse energies. This limit could not be reached. Also the exhaust velocity was determined by this condition and did not go beyond 1.5 km/s.

The placement of the propellant inside a nozzle like structure obviously helps to increase the coupling coefficient but not the effective exhaust velocity, or the specific impulse. The question is now, what expansion ratio is most effective. In 1980, Ageev et al. already derived from principles of a point explosion a simple law for the increase of the coupling coefficient with the slenderness of the nozzle, considering cone as well as parabolic shapes (Ageev, et al., 1980). He predicts a strong decrease of the coupling coefficient if the half angle θ for a cone grows over 20° and for a paraboloid over 30°. For a parabolic contour the angle θ is defined as the angle enclosed by the symmetry axis and a line between the focal point and the outer edge of the contour. Ageev also predicts the existence of an optimal nozzle length.

Except for the baseline nozzle with $\theta = 43.6°$, two more nozzles with $\theta = 22.6°$ and 15.2° have been tested. For atmospheric pressure the coupling coefficient indeed rose weakly with increasing slenderness from 375 N/MW to slightly more than 400 N/MW. In vacuum, however, the highest value with 207 N/MW was measured for the intermediate nozzle, compared to 194 N/MW for the baseline contour and only 163 N/MW for the most slender type.

Myrabo et al. (Myrabo, et al., 2002) have found a decreasing coupling coefficient in air for parabolas with $\theta > 45°$. In early preliminary experiments, a 165 mm long cylindrical tube with inner diameter of 77 mm has been inserted into the base line parabola.

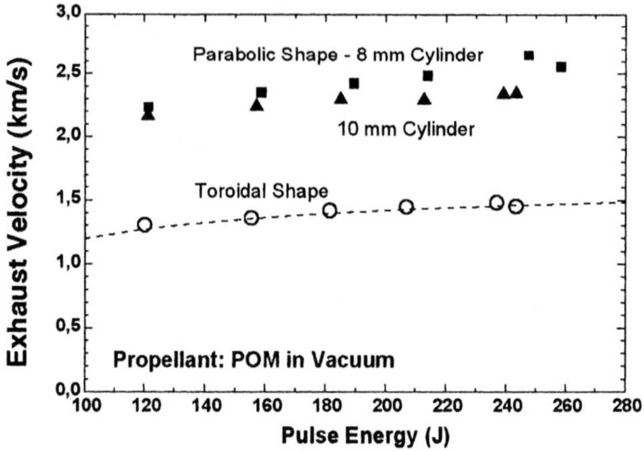

Figure 10. Comparison of average exhaust velocity for parabolic and toroidal concentrator nozzles.

This measure improved the coupling coefficient in air by almost 50% (Schall, et al., 2002). It can be concluded that a nozzle structure definitely improves the coupling coefficient. But the selection of an exact angle θ is uncritical.

3.3 Absorption losses in front of the target

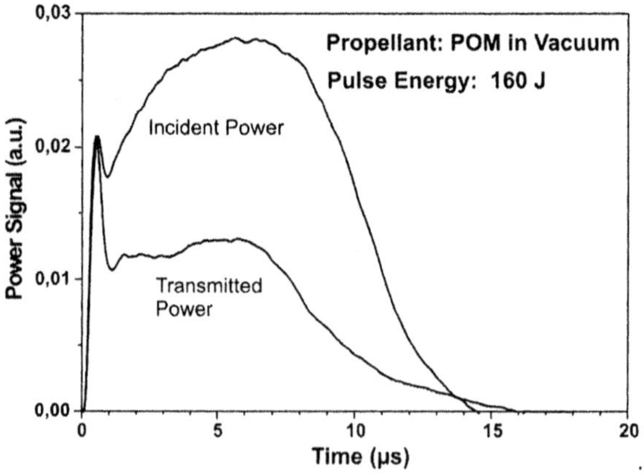

Figure 11. Time dependence of the incident and the transmitted power.

The partial absorption of a laser beam in the ablation products is a well known feature. It increases with the pulse intensity and is responsible for the performance losses that have been described above. Can these losses be quantified with respect to magnitude, time and power dependence, and location? Then, the question arises how the lost energy can be regained and turned into the production of impulse and velocity. The temporal behaviour of the absorption losses was measured by drilling a small hole (diameter 3 mm) into the flat samples and placing a powermeter behind the sample. Now the power curves of the incident laser signal can be compared with the signal measured behind the sample. This signal has penetrated through the gas and plasma layers in front of the target. Fig. 11 shows these power curves for the pulse energy of 160 J. It is seen that after the initial spike the transmitted power drops rapidly and only a fraction of the incident total pulse energy actual reaches target. With increasing pulse energy the transmitted pulse is also shortened in length. The integration over the power curves gives the fraction of transmitted energy. For POM in vacuum it drops from 55% of the incident energy of 40 J, to 36% for 280 J.

Since it is still unclear where the absorption occurs, an experiment has been started to measure the attenuation of a CO_2 probe beam, which is passed parallel to the sample surface at various distances across the ablation zone. As Fig. 12 shows qualitatively for four distances, a distinct absorption wave moves away from the sample surface. The maximum absorption strength decreases with distance but lasts as long as 10 μs after termination of the incident laser pulse.

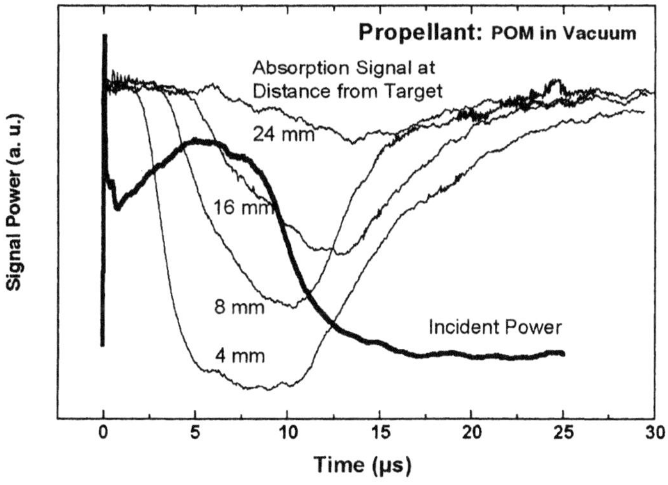

Figure 12. Absorption characteristics of a CO_2 probe beam parallel to the target surface at various distances in comparison with the laser pulse signal.

Since the continued deposition of energy into this region during the 15 μs pulse time does not improve the total mechanical impulse it must be concluded that it is not accompanied by an additional pressure rise and only supports the gas expansion velocity like the heating of a steadily flowing gas. To exert an increased impulse the release of a second strong pressure wave in the absorbing medium would be necessary. In this respect the similarity to a real double pulse arrangement does not hold.

4. CONCLUSION

Although ablative laser micropropulsion is close to introduction as an element for satellite attitude control and station keeping this is not nearly so for main thrusters of laser rockets. The major reason for this is the lack of appropriate pulsed lasers with mean output powers in the range of several hundred kilowatts and higher, short pulse lengths and wavelength. Most experiments at a large energy scale (several hundred J per pulse) have been made with CO_2 lasers at pulse lengths in the microsecond range. The resulting performance suffers from a considerable loss of energy by absorption in the exhausted gases and plasma that is ineffective for the impulse production. Technical solutions are required to constrain the absorption region and build up additional pressure that can be transformed into a mechanical impulse. The application of much shorter pulse lengths in the nanosecond regime, combined with a repetition frequency of less than several 10 kHz may also help to improve the situation. The wavelength of the CO_2 laser matches well with polymers as propellants. Some of these decompose conveniently into non-condensing gases or, as energetic fuels release additional chemical energy. This produces a high coupling coefficient, but at the same time necessarily a low specific impulse. In vacuum we have not found a specific impulse higher than 270 s. For many missions this is not high enough and by far lower of what can be achieved by chemical combustion propellants. For a single-stage-to-orbit flight, which is one of the most promising applications for ablative laser propulsion, a specific impulse of at least 550 to 600 s is demanded. Such values call for surface absorbing materials, i.e. metals, which on the other hand require wavelengths around 1 μm or less for sufficient absorption. Another, though little investigated method, has been suggested by Lo (Lo, 1977) and is the superposition of radiative laser energy on a chemical combustion of conventional propellants. Apart from this lack of knowledge it still needs a credible demonstration of laser propulsion to convince the institutional sponsors to invest more into this technology.

ACKNOWLEDGMENT

The authors wish to emphasize the merits of Jochen Tegel who has carried out many of the mentioned experiments. The authors are also grateful to the EOARD, London that has supported this research with several projects, and in particular to Dr. Frank Mead and Dr. William C. Larson for their help and advice and for the provision of one of their toroidal lightcrafts.

References

Ageev VP , Barchukov AI, Bunkin FV, Konov VI, Korobeinikov VP, Putjatin BV, Hudjakov VM (1980) Experimental and theoretical modeling of laser propulsion. Acta Astronautica 7, 79-90

Ageichik AA, Egorov MS, Ostapenko SV, Rezunkov YA, Safronov AL, Stepanov VV (2005) Model Test of the Aerospace Laser Propulsion Engine. In: Beamed Energy Propulsion: Third Intern. Symposium on Beamed Energy Propulsion, Pakhomov AV and Myrabo LN (eds.), AIP Conference Proc. 766: 183-194

Bohn WL (1999) Laser lightcraft performance. In: High-Power Laser Ablation II. Proc. of SPIE 3885: 48-53

Bohn WL, Schall WO (2003) Laser Propulsion Activities in Germany. In: Beamed Energy Propulsion: First Intern. Symposium on Beamed Energy Propulsion, Pakhomov AV (ed.), AIP Conference Proc. 664: 79-91

Bunkin FV, Prokhorov AM (1976) Use of a laser energy source in producing a reactive thrust. Sov. Phys. Usp. 19 (7), 561-573

Campbell JW, Phipps C, Smalley L, Reilly J, Boccio D (2003) The Impact Imperative: Laser Ablation for Deflecting Asteroids, Meteoroids, and Comets from Impacting the Earth. In: Beamed Energy Propulsion: First Intern. Symposium on Beamed Energy Propulsion, Pakhomov AV (ed.), AIP Conference Proc. 664: 509-520

Fabbro R, Fournier J, Ballard P, Devaux D, Virmont J (1990) Physical study of laser-produced plasma in confined geometry. J. Appl. Phys. 68 (2), 775-784

Gregory DA, Herren K (2005) Specific Impulse Definition for Ablative Laser Propulsion. In: Beamed Energy Propulsion. Third Intern. Symposium on Beamed Energy Propulsion, Pakhomov AV and Myrabo LN (eds.), AIP Conference Proc. 766: 595-604

Irtuganov VM, Kalinin VP, Sergeev VV, Smirnov AA, Sherstobitov VE, Goryachkin DA, Tul'skyi SA, Yachnev IL (2005) Experimental Investigation of Air-breathing Mode of Laser Propulsion with Elongate Cylindrical Models and CO_2 Lasers of Different Pulse Durations. In: Beamed Energy Propulsion: Third Intern. Symposium on Beamed Energy Propulsion, Pakhomov AV and Myrabo LN (eds.), AIP Conference Proc. 766: 195-204

Kantrowitz A (1972) Propulsion to Orbit by Ground-Based Lasers. Astronautics and Aeronautics 10 (5), 74-76

Kare J (1990) Pulsed Laser Propulsion for Low-cost High-volume Launch to Orbit. Space Power 9 (1), 67-75

Kuznetsov LI, Yarygin VN (1994) Laser-reactive method for disposal of small space debris. Quantum Electronics 24 (6), 555-557

Landau LD, Lifshitz EM (1959) In: Fluid Mechanics, Pergamon Press, London, New York, Paris, Los Angeles

Larson CW, Mead FB Jr, Knecht SD (2004) Laser propulsion and the constant momentum mission. In: Beamed Energy Propulsion, Second Intern. Symposium on Beamed Energy Propulsion, Komurasaki K (ed.), AIP Conference Proc. 702: 216-227

Liukonen RA (1998) Laser jet propulsion. In: XII Intern. Symposium on Gas Flow and Chemical Lasers and High-Power Laser Conference, Proc. of SPIE 3574, 470-474

Lo RE (1977) Propulsion by laser energy transmission. A New Era in Space Transportation. In: Proc. of the XXVII Intern. Astronautical Congress, Pergamon Press, 99-108

Luke JR, Phipps CR (2003) Laser Plasma Microthruster Performance Evaluation. In: Beamed Energy Propulsion: First Intern. Symposium on Beamed Energy Propulsion, Pakhomov AV (ed.), AIP Conference Proc. 664: 223-229

Myrabo LN (2001) World Record Flights of Beam-Riding Rocket Lightcraft: Demonstration of "Disruptive" Propulsion Technology. 37th AIAA/ASME/SAE/ASEE Joint Propulsion Conference, 8-11 July, Salt Lake City, UT, paper AIAA 2001-3798

Myrabo LN, Messitt DG, Mead FB Jr (1998) Ground and flight tests of a laser propelled vehicle. 36th AIAA Aerospace Science Meeting & Exhibit, 12-15 January, Reno, NV, paper AIAA 98-1001

Myrabo LN, Libeau MA, Meloney ED, Bracken RL (2002) Pulsed Laser Propulsion Performance of 11-cm Parabolic "Bell" Engines Within the Atmosphere. 33rd Plasmadynamics and Lasers Conference, 20-23 May, Maui, Hawai, paper AIAA 2002-2206

Phipps CR (2005) Will your children ride a laser beam into orbit? Would you want them to? Third Intern. Symposium on Beamed Energy Propulsion, Pakhomov AV and Myrabo LN (eds.), AIP Conference Proc. 766: 11-22

Phipps CR, Michaelis MM (1994) LISP: Laser impulse space propulsion. Laser and Particle Beams 12 (1), 23-54

Phipps C, Luke J (2002) Diode Laser-Driven Microthrusters: A New Departure for Micropropulsion. AIAA Journal 40 (2) 310-318

Phipps CR, Harrison RF, Shimada T, York GW, Turner TP, Corlis XF, Steele H, Haynes LC (1990) Enhanced vacuum laser-impulse coupling by volume absorption at infrared wavelengths. Laser and Particle Beams 8 (1-2) 281-298

Phipps CR, Albrecht G, Friedman H, Gavel D, George EV, Murray J, Ho C, Priedhorsky W, Michaelis MM, Reilly JP (1996) ORION: Clearing Near-Earth Space Debris Using a 20-kW, 530 nm, Earth-Based, Repetitively Pulsed Laser. Laser and Particle Beams 41(1) 1-44

Phipps CR, Reilly JP, Campbell JW (2000) Optimum parameters for launching objects into low Earth orbit. Laser and Particle Beams 18: 661-695

Pirri AN, Weiss RF (1972) Laser propulsion. AIAA 5th Fluid and Plasma Dynamics Conference, June 26-28, Boston, Mass., paper AIAA 72-719

Sasoh A (2001) Laser-driven in-tube accelerator. Review of Scientific Instruments 72 (3) 1893-1898

Schall W (1990) Orbital Debris Removal by Laser Radiation. 41st Congress of the Intern. Astronautical Federation, Dresden, GDR, paper IAA-90-569; also (1991) Acta Astronautica 24, 343-351

Schall WO, Bohn WL, Eckel H-A, Mayerhofer W, Riede W, Zeyfang E (2000) Lightcraft Experiments in Germany. In: High-Power Laser Ablation III, Proc. of SPIE 4065: 472-481

Schall WO, Eckel H-A, Mayerhofer W, Riede W, Zeyfang E (2002) Comparative lightcraft impulse measurements. In: High-Power Laser Ablation IV, Proc. of SPIE 4760: 908-917

Schall WO, Tegel J, Eckel H-A (2005) Ablation Performance Experiments With Metal Seeded Polymers. In: Third Intern. Symposium on Beamed Energy Propulsion, Pakhomov AV and Myrabo LN (eds.) AIP Conference Proc. 766: 423-432

Sedov LI (1959) Similarity and Dimensional Methods in Mechanics, Academic Press, New York

Urabe N, Kim S, Sasoh A, Jeung I-S (2003) Vertical Launch Performance of Laser-driven In-Tube Accelerator. In: Beamed Energy Propulsion: First Intern. Symposium on Beamed Energy Propulsion, Pakhomov AV (ed.), AIP Conference Proc. 664: 105-112

Urech L, Hauer M, Lippert T, Phipps CR, Schmid E, Wokaun A, Wysong I (2003) Designed polymers for laser-based microthrusters – correlation of thrust with material, plasma, and shockwave properties. In: High-Power Laser Ablation V, Proc. of SPIE 5448: 52-64

Chapter 18

LASER PROPULSION

Takashi Yabe, Shigeaki Uchida
Integrated Research Institute, Tokyo Institute of Technology

1. INTRODUCTION

This chapter is devoted to the description of laser propulsion. The technologies dealt with in this chapter use thermal energy or pressure of propellant fluids powered by laser radiation to generate impulse. The propellants are not necessarily in a plasma state and can be a hot fluid. In fact, this flexibility leads to the wide range of applicability of laser propulsion. Therefore, laser irradiation and propellant feeding conditions have to be optimized depending on the field of application or mission of the technology.

The concept of the laser propulsion was conceived back in the early 70's by Kantrowitz (1971). It was estimated that a one-megawatt laser system can launch a one-kilogram payload to the low earth orbits (500 to 1500 km altitude). Since then, the study of laser propulsion has evolved into a laboratory fundamental investigation on basic characterization of laser-plasma interactions in conjunction with the study of laser plasma interactions of inertial confinement fusion research (Reilly, 1979; Phipps, 1988). One typical example of these investigations was done by Phipps, where universal scaling on impulse generation in terms of laser parameters over many orders of magnitude was found. The optimization introduced above can be viewed on this scaling in terms of laser parameters such as intensity, wavelength, and pulse width.

From the early 90's, advancements of high power laser facilities, such as for materials science, accelerate the thrust measurements (Pirri, 1974) and demonstrations of so-called "Light Craft" devices (Myrabo, 2002; Sasoh,

2001; Bohn, et al., 2002; Rezunkov, 2005). Various applications have been proposed including biomedical devices (Yabe, et al., 2002b, 2003b), air plane (Yabe, et al. 2002a), satellite posture control (Phipps and Luke, 2000), orbital transfer vehicle (Uchida and Bato, 2003) as well as space debris mitigation (Phipps, 1998; Schall, 1998). Among them, laser orbital transfer vehicle (LOTV) or "laser tugboat" seems to be realized by the combination of currently available technologies and yet has significant impacts over current technology. Its feasibility will be examined in this section.

Unparalleled uniqueness of laser propulsion is that it can generate thrust remotely by beaming necessary power from distance. This characteristic leads to an advantage of eliminating the necessity of carrying a power source onboard the vehicle. However, this is not the limit of the technology when it is applied to so-called "space debris problem" (Johnson, 1991). The space debris will prohibit us from going into space if no countermeasures are taken since it will eventually cover up the near earth orbits. In principle, only laser propulsion with remote power beaming seems effective in the mitigation of a countless amount of space debris. The most effective way of using laser propulsion to mitigate space debris is to generate ablation on the surface and change its orbit so that it eventually falls to the upper atmosphere. An example of laser system for space debris removal will be described.

The first part of this chapter describes the basic elements of laser propulsion which makes the technology unique compared to others such as chemical and electrical propulsion. The uniqueness of laser propulsion technology comes from the fact that mass of propellant and laser intensity can be controlled independently and therefore, power density carried by propellant can be controlled over many orders of magnitude. Basic elements of impulse are of course mass and velocity (energy). The performance of a propulsion system can be evaluated in terms of the mass efficiency or energy efficiency, depending on the purpose of a specific application.

2. SCALING LAW IN LASER PROPULSION

2.1 Basic Concept

The most significant concept for laser propulsion is a parameter called momentum coupling coefficient, C_m, a measure of energy usage. It is defined as a ratio of impulse (momentum) generated to laser power. Different expressions are possible between pulsed and cw laser wave formats. In the case of pulsed lasers the expression becomes

$$C_m = \frac{M\Delta V}{E_{laser}} = \frac{mv_p}{E_{laser}} , \tag{1}$$

where M, ΔV , m, v_p, and E_{laser} are mass and velocity increment of projectile, mass and velocity of exhausted propellant, and laser power incident on propellant respectively. Note that the velocities are measured in the projectile coordinate. It is assumed that the propellant is in solid-state and that the propulsive force is generated in vacuum conditions and no effects from atmospheric pressure exist for simplicity. If the repetitive pulse can be expressed as averaged power, ${}_{av}P_{laser}$, the expression becomes

$$C_m = \frac{{}_{av}F}{{}_{av}P_{laser}} , \tag{2}$$

where ${}_{av}F$ is averaged force impinging on the projectile.

In the case of cw laser power source, a similar expression is possible

$$C_m = \frac{F}{P_{laser}} , \tag{3}$$

where F and P_{laser} are thrust generated and laser power incident on the propellant respectively. Expressions (1) through (3) are all equivalent to each other and are reduced to expression (1), considering that the expression (2) and (3) are he time integrated forms of (1).

The other expression of a propulsion system is a well known parameter, I_{sp}, specific impulse, and a measure of mass usage. It is defined as the ratio of impulse to the rate of propellant mass used,

$$I_{sp} \equiv \frac{\int Fdt}{g_0 \int \dot{m}dt} = \frac{v_p}{g_0} \tag{4}$$

where $\dot{m}$ and g_0 are propellant mass flow rate and the gravity constant. It has been assumed that the mass flow rate is constant in time to get the most right term that is based on the time-dependent parameter, v_p. Since the analysis of thrust characteristics can be done based on phenomena in a very short time period or instantaneous parameters, expressions (1) and (4) will be used.

The product of (1) and (4) leads to an expression

$$I_{sp}C_m = \frac{v_p}{g_0}\frac{mv_p}{E_{laser}} = \frac{2}{g_0}\frac{E_{hydr}}{E_{laser}} , \tag{5}$$

where E_{hydr} is the kinetic energy of exhausted propellant. The most right hand term is a coupling efficiency of laser pulse energy to propellant hydrodynamic motion.

Once the coupling efficiency is fixed, I_{sp} and C_m are inversely proportional to each other. This means that the efficiencies of propellant mass use and energy are not compatible with each other. In general, when low velocity propellant flow is generated, high energy efficiency and high C_m are realized at the cost of propellant mass. On the other hand, higher velocity propellant flow corresponds to better mass efficiency, high I_{sp}.

2.2 Laser Orbital Transfer Vehicle

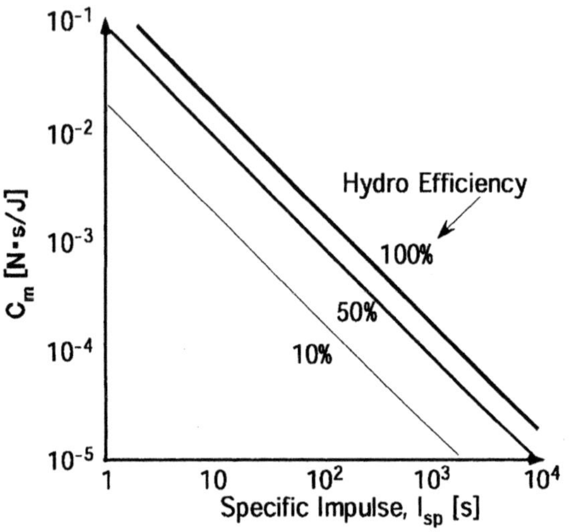

Figure 1. Relation between momentum coupling coefficient and specific impulse.

The relationship between C_m and I_{sp} is plotted in Fig. 1 with the hydrodynamic efficiency as a parameter. Each line gives the upper limit of a propulsion system performance and the target for engineering optimization. For example, when a propulsion system is applied to an orbital transfer mission, LOTV where efficient propellant mass use is important, high I_{sp} system is necessary.

The conditions of laser irradiation for specific I_{sp} or C_m values can be found in, for example, Phipps, 1998, and is reproduced in Fig. 2. It has been shown that the scaling is valid over seven orders of magnitude in $I\lambda\tau^{1/2}$, where I, λ, and τ are laser intensity, wavelength, and pulse width. The ranges of I, λ, and τ are also very wide. For example, laser intensity ranges from 10^8 to 10^{14} W/cm^2. Once wavelength and pulse width are fixed, for example to be 1 μm and 10 ns, C_m or I_{sp} is controlled by laser intensity alone. A typical value of laser intensity corresponding to a relatively high I_{sp} of 10,000 s can be deduced by using

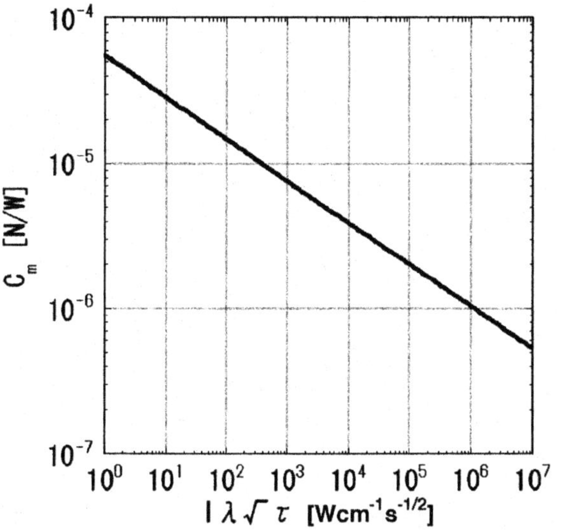

Figure 2. Laser parameter dependence of momentum coupling coefficient.

Figs. 1 and 2 and is determined to be 10^{11} W/cm^2 on the propellant. Such laser intensity can be easily realized by using a Q switching technique.

An example of the benefit of having high specific impulse is shown in Fig. 3, where payload ratio of a LOTV flying from the low earth orbit (LEO) to geo stationary orbit (GEO) is plotted (see right vertical axis) as a function of I_{sp}. (Uchida, et al., 2005). The I_{sp} of chemical (I_{sp} = 600 s) and electrical (I_{sp} = 4,000 s) propulsion systems are indicated by broken lines for comparison. A laser propulsion system with an I_{sp} of 10,000 s performs three to ten times better payload ratio.

Fig. 3 also shows the traveling periods (days) with various laser powers. For example, with a 1-kW laser system, it takes thousands of days to get to GEO while only less than 100 days is necessary when a 100-kW laser system is used. With the current state of the arts of solar cells and LD pumped solid-state lasers taken into account, two sets of solar paddle with a 1,000 m^2 area are sufficient to support such a LOTV system.

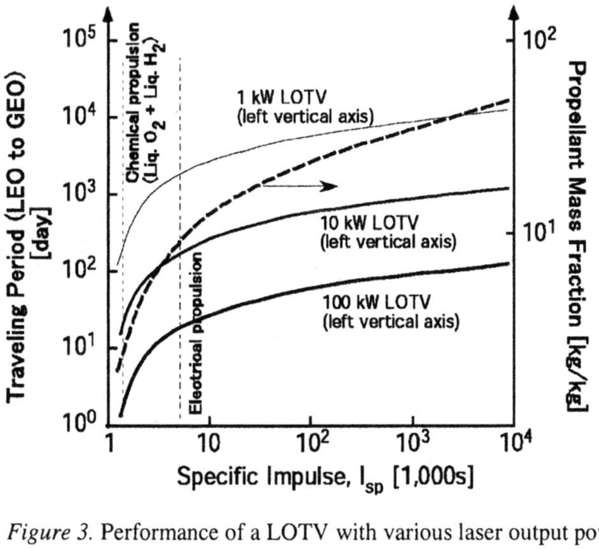

Figure 3. Performance of a LOTV with various laser output powers.

2.3 Space Debris Mitigation

In this section, an idea of mitigating relatively small space debris is described. Small debris is defined to have a mass of a few grams. This category of space debris is most hazardous since it is too small to identify its orbit from the ground, and yet its number is estimated to be enormous. Considering its relative speed of 10 km/s when colliding with operating satellites, damages to the victim could be fatal.

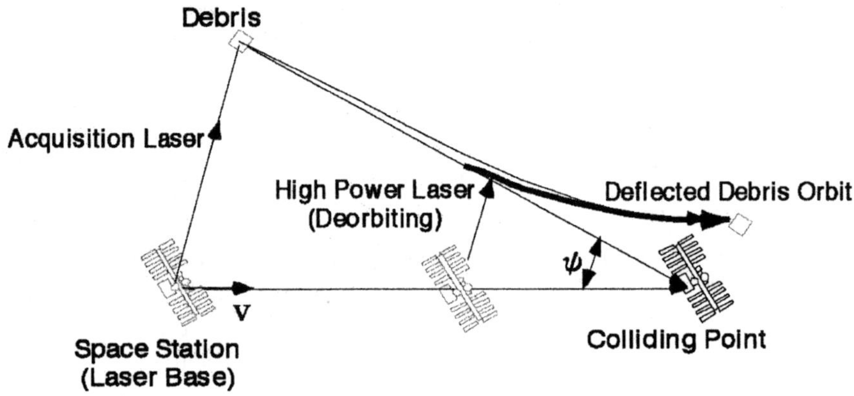

Figure 4 Concept of laser debris mitigation system

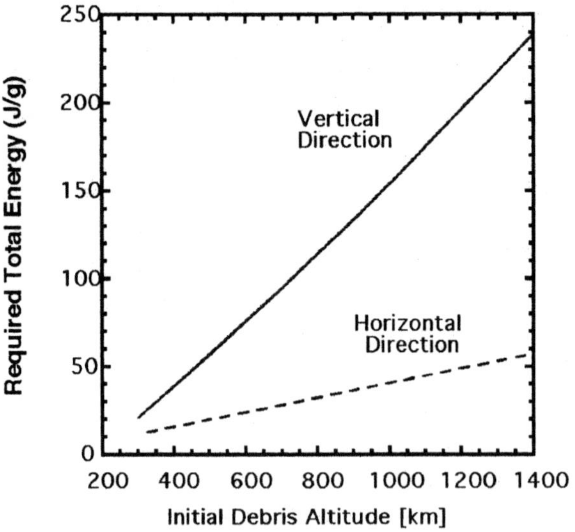

Figure 5. Laser energy required for target "reentry" as a function of initial altitude.

Since the orbits of the small debris are not known, the laser system for mitigation needs to be space borne and has to be with acquisition and tracking functions. The concept of small debris mitigation by space borne laser is depicted in Fig. 4. Before acquisition, low power laser scans the vicinity of the system for any approaching debris. Once one is detected, it seems to be approaching in a head-on collision course, since any colliding debris exhibits constant approaching angle relative to the space station velocity. Once the debris comes within range, a repetitive high-power pulsed laser is irradiated to the target. Laser ablation takes place in the direction of target surface normal leading to the deorbiting as well as avoiding collision.

The direction of the deorbiting depends on the direction of the surface at the incidence of laser pulses, however any deorbiting eventually results in the target reentry. Required laser energies for the reentry, on the other hand, depend on the direction of deorbiting and plotted in Fig. 5 as a function of the initial altitude of debris. "Vertical direction" means that the deorbiting takes place in the direction perpendicular to the earth's surface. "Horizontal direction" is parallel to the earth's surface. Laser energies are calculated so that the target acquires the velocity change for reentry and are proportional to the target weights. The momentum coupling coefficient has been taken to be 10^{-6} N/W as a typical value in Fig. 2. It can be seen that a few hundreds of laser pulses are required for the reentry of a target with unit mass.

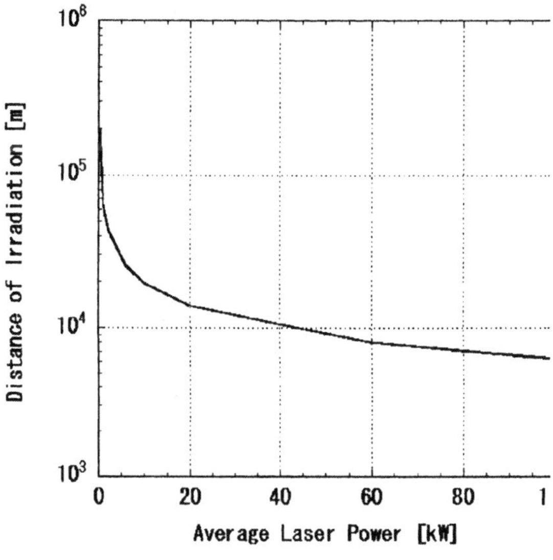

Figure 6 Laser average power for debris mitigation and irradiation distance

The average power of the laser system can be calculated if the distance of laser irradiation is determined. Fig. 6 shows the results of the estimation. A distance of 10 km leads to an average laser power of several tens of kW, about the same scale as the LOTV system. Illuminating a cm-size target from 10-km away means micro radian focusing, requiring a one-meter illuminating mirror with diffraction limited beam quality.

The case of high momentum coupling efficiency will be described in the following section.

3. HIGH-COUPLING SCHEME

3.1 Coupling Efficiency

Let us first re-examine the concept of laser propulsion using Fig. 7. When the laser illuminates the metal target, the surface layer is evaporated. This is called ablation. In reaction to this ablation, the target is accelerated in the opposite direction. The efficiency of this system is often measured by the momentum coupling coefficient.

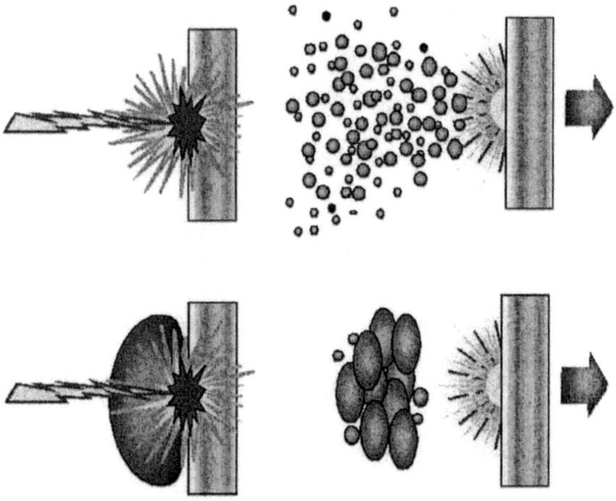

Figure 7. The schematics of laser propulsion. (Top) single-layered target. (Bottom) target with transparent overlay.

Most of the preceding studies used such a single-layered target. Yabe (1976) proposed to use the metal target covered with transparent overlay as shown in Fig. 7 (bottom). The laser penetrates through the overlay and deposits the energy in the region between overlay and metal. Such overlay can enhance the coupling as easily understood by a simple momentum conservation of two materials indicated by suffices 1 and 2:

$$M_1U_1 = M_2U_2, \tag{6}$$

where M is the total mass and U the velocity. Then kinetic energy gained by each target is

$$E_1/E_2 = M_1U_1^2/M_2U_2^2 = M_2/M_1 \; . \tag{7}$$

Therefore, if M_2 is the mass of ablated vapor and is much smaller than that of metal M_1, then most of the energy goes to vapor ($E_1 << E_2$ for

$M_1 >> M_2$). In order to increase the efficiency, we proposed to place a transparent heavier material as shown Fig. 7 (bottom).

The use of such a concept for target acceleration appeared as an application to laser-driven fusion. The efficiency enhancement is attributed to the tamped flow effect (Yabe, 1976) or cannon-ball effect (Winterberg, 1976 ; Azechi, et al., 1981). Winterberg used an enclosed configuration (like a real cannon) and Yabe employed a transparent overlay which was introduced to laser propulsion by Fabbro (1990) and Phipps (2000).

Although the enhancement is encouraging, the repetitive use of the structure is the key issue for realistic application. For this purpose, Yabe, et al., (2002a) used water overlay and proposed repetitive water supply concept(Yabe, et al., 2002b).

The experimental (symbols) and simulation (lines) results both for single layered and overlay targets are shown in Fig. 8 for various laser intensity I and pulse width τ. Among them, the Glass-P represents the experimental results by the LHMEL Nd:glass laser of 70J /25-100ns at Wright-Patterson AFB, Ohio (Phipps, et.al., 2000) with the aluminum target overcoated with glass as a transparent material, while the experiments with acrylic and water over lay were performed by 590mJ/5ns YAG laser (Yabe, et.al., 2002a). Among acrylic, glass and water overlay, water gives the most efficient thrust and it is several orders of magnitude larger than single layered target (ST).

As the simulation results by the CIP method (Yabe, 1991; Yabe, 2001) suggest, the momentum coupling is greatly enhanced by this "exotic target". Here, the momentum coupling efficiency $C_m = m\Delta u / W$ is given by the momentum $m\Delta u$ of target obtained after irradiation of laser energy W.

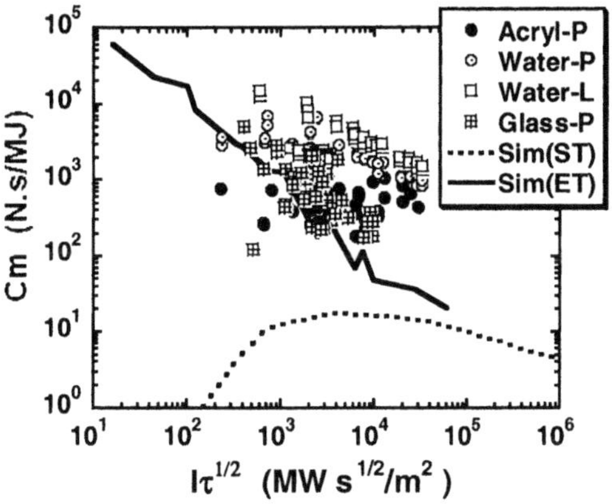

Figure 8. The momentum coupling efficiency for various laser intensity and pulse width.

Figure 9. (Left) A paper airplane with aluminum target and water droplet. (Right) The flight trajectories of a microairplane with water droplet overlay.

In the simulation of the standard target (ST) that is the single layered target, C_m has the maximum as shown in Fig. 8, because laser energy is too small to drive the aluminum target when I is less than the optimal intensity. In ST, most of the energy is transferred to low density gas and only a small fraction of energy (usually less than a few percent) is used to drive a target. On the other hand, when I is larger, strong laser generates plasma that absorbs most of the laser energy by inverse Bremsstrahlung and prevents laser from arriving.

In contrast to ST, C_m of ET increases even with decreasing intensity I, because space between two layers is filled with evaporated gas and a large amount of energy can be used to drive metal target at most, while the reduced C_m at strong laser intensity is the same as the case of ST, which is due to laser-induced cutoff by the generated plasmas.

By using such efficient propulsion with water overlay, a microairplane is driven by 590mJ/5ns YAG laser (as shown in Fig. 9), whose size is 39mm×56mm×15mm. At the rear edge of the airplane, an aluminum foil of 3.5mm×3.5mm×0.1mm-thick is pasted and the total weight is 0.2g. In addition, water droplet of 3mm-diameter, 0.014g is attached to the aluminum foil as shown in Fig. 9 (Left).

The measured velocity of the airplane for the case in which the flight path was normal to the camera was 1.4m/sec at the beginning. This gives C_m=237N.sec/MJ , in rough agreement with the scaling law in Fig. 9 for $I\tau^{1/2}$ = 1.8x10^4 MW.sec$^{1/2}$/m^2 based on the 760 μm focal spot diameter. Following this scaling law, we already achieved 5000N-sec/MJ with a larger laser on an overlay target, and much larger efficiency of about 10^5N-sec/MJ is expected from simulation results. Therefore, the use of overlay target concept for a propulsion system is attractive for practical applications. Even in the experiment of microairplane, this increase has been clearly demonstrated, and the airplane without overlay did not move and failed to gain the speed sufficient for flight.

Global warming, owing to CO_2, is internationally recognized, and the Kyoto Protocol on the restriction of CO_2 exhaust will be active, forcing most of the countries to reduce the CO_2 exhaust by a specific amount, in 5 years starting from 2008. For this purpose, we need a measuring instrument for three-dimensional distribution of CO_2.

4. PRACTICAL USE

4.1 Water Supply for Repetitive Propulsion

For practical application, the overlay structure must be repetitively constructed. Thus, a water overlay has several advantages: (1) higher efficiency, (2) easier arrangement, (3) automatic collection from air, (4) no environmental problems, and so on. Yabe, et al., (2000c) proposed a water supply in which water is contained in the cylindrical tube. The aluminum target is fixed to this cylinder. The target has a quadrilateral shape that is elongated vertically, thus giving a path of water through the gaps in both sides from the large water reservoir. After laser is irradiated through a hole shown in the front-view figure, a small fraction of water between target and hole is ejected. Then water will be quickly fed through the gap.

In a practical situation, the objects should be flying in the sky. It would be very difficult to realize this situation because we need a tracking system and large lasers. In order to realize a friction-free system and repetitive propulsion, a new system called "air-slider" was proposed. The objects are suspended by using air-flow from an air-compressor (1.57×10^{-3} m^3/s). The edge of the quadrilateral pillar is used to guide the object in one-dimensional direction so that the laser is easily irradiated. Along the surface of the quadrilateral pillar, many holes are drilled in order to supply air-flow. This system reduces the effect of friction by air-flow from the holes made in intervals of 1cm, and levitates the object so that the object can move very smoothly.

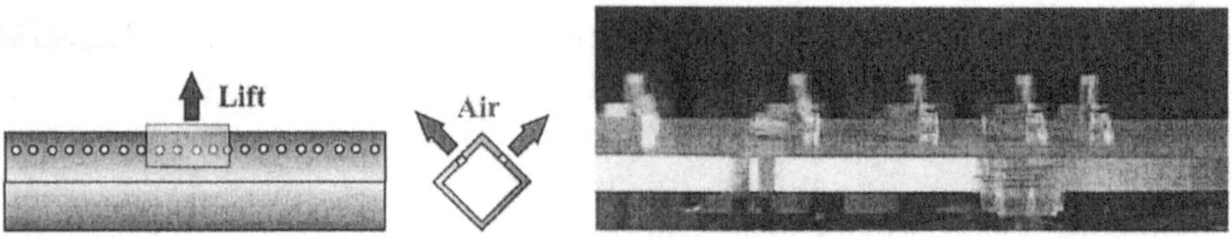

Figure 10. The tracks of water supply equipment on the air-slider. Such a levitation system can also be used for launching nuclear reactor waste and space station structural materials into space.

4.2 Posture Control of Satellite

Near future application of the laser propulsion is the posture control of a satellite. North-south posture control needs the thrust of 20mN. For this purpose, we here propose to use fiber laser (LP155R-30WDS, produced by Mitsubishi Cable Industries, Ltd.) shown in Fig. 11. This laser unit is very small (482.6 (W) ×88.1(H) ×400 (D) mm) and can be made light enough to be loaded on a satellite. This laser is robust and can be very easy to deal with because it is supplied through fiber. The input voltage to the laser is 5V and input power is 10W. Output laser is 7kW/2ns with 1-10kHz, and the

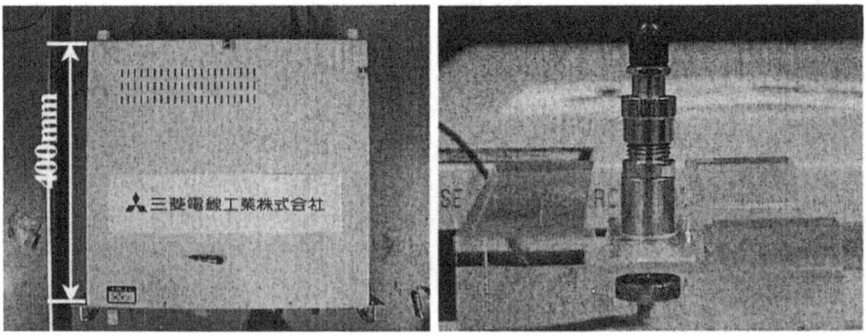

Figure 11. (Left) Photographs of fiber laser. (Right) Experimental setup for measurement of thrust force using fiber laser.

wavelength is 1550 nm. The focal spot diameter is 10mm.

The measured thrust force is shown in Fig. 12. A load cell (WMS-12B 50N, TOYODA) was used to measure the thrust force. A target was placed on the load cell. Peak thrust force of 83mN was achieved even without water overlay. We also confirmed the thrust force by an experiment using small size of pendulum (target size of 3mm×3mm×0.1mm, Al). This result means that C_m was about 16N.s/MJ using this fiber laser's peak power of 7.5kW and pulse width of 2ns.

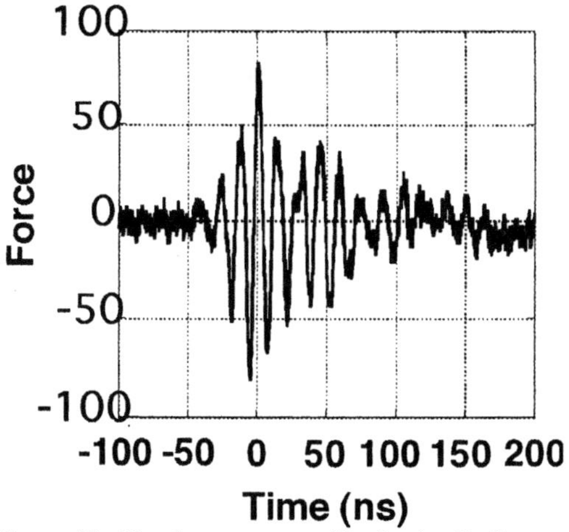

Figure 12. The thrust measured by load cell after one shot of fiber laser on pure aluminum target.

It is important to note that we can bundle a number of beams to generate much larger energy. Furthermore, we can use solar-energy-pumped laser. If we can load such laser, tracking of the target is not necessary.

The use of film target with diode lasers was proposed by Phipps and Luke (2000, 2002) and much larger thrust can be expected.

4.3 Various Propulsion for Other Vehicles

In this section, we shall discuss the possible use of laser in various vehicles. Since a laser can be made of fiber, its weight is negligibly smaller than car itself. Although a few kW is not sufficient for accelerating the car in a short period (see Table 1), it can be combined with battery that accumulates the excessive energy.

Table 1. Various vehicles that can use laser propulsion

Vehicles	*Required Power (kW)*	*Expected Surface Area (m^2)*
Assistant Robot	*0.05*	*>1*
Large Robot	*1-3*	*>1*
Solar Car(170kg) 50km/h	*0.43*	2
Solar Car(170kg) 150km/h	*6.03*	2
Automobile(1ton) 50km/h	*3.3*	2
Automobile(1ton) 150km/h	*35.4*	2
Middle-Class Tanker (100,000 ton) 14.5knot(27km/h)	*11,000*	*10,000*
Airplane 747-400	*160,000*	*300*

As for a promising method, we can generate the large amount of heat as well as hydrogen with the reaction of magnesium with water. The reaction product MgO can be deoxidized by laser. By use of magnesium, the energy can be supplied even at night and on a cloudy day (Yabe, et.al., 2005). In view of the parameters, the ships are the most promising vehicles for such energy systems that use the solar energy.

Several mechanisms are capable of converting the thrust produced by laser propulsion to wheel drive. One is the laser-driven turbine (Ohkubo, 2002). The basic principle of this system is to rotate the turbine half-submerged in water by means of the laser ablation. This system can automatically replenish water on the target by its rotation. The fin covered with water droplets is then irradiated by the laser and generates rotational powers.

Another interesting mechanism is the piston like automobile engine. In the actual system, water is injected onto the wall of the cylinder and the laser illuminates the water's inner region, just below the surface. Although the water is transparent to near infrared and visible lasers, a high intensity laser can induce the breakdown of the water if the laser is focused onto the inner region of the water, with the resulting explosion causing water to be ejected from the cylinder's surface.

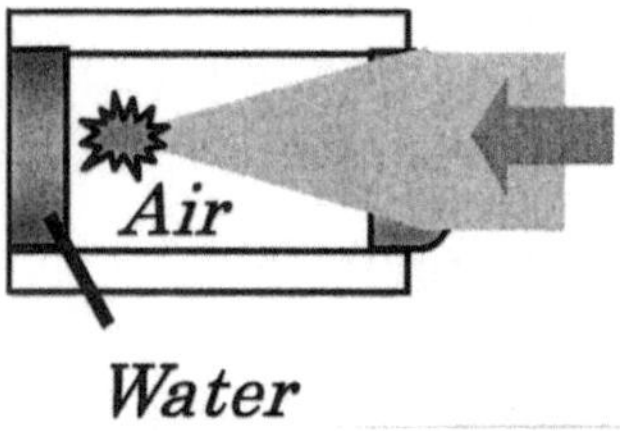

Figure 13. Schematics of water-film cannon.

In this case, we do not need metal as an ablator. We called this system "metal-free water cannon (MFWC)" (Yabe, et al., 2003) and succeeded in gaining thrust comparable to a metal target covered by water. The maximum coupling coefficient was 2444.4N.s/MJ, which is comparable to 3536N.s/MJ for the metal-water target mentioned in previous sections. With the present system, metal is not ablated and hence metal vapor which is harmful to the environment will not be exhausted. Furthermore, the metal can be used for a long time since the piston experiences no wear.

The piston filled with water has several problems. These are: (1) laser is absorbed by water in case of long tube, (2) driving force is oscillating because of bubble expansion and contraction that leads to less effective coupling and so on. We here propose a new driving concept that uses water film as shown in Fig. 13. In this system, water is repetitively supplied by a mechanism shown in the right side of the figure and laser causes breakdown in the air. This explosion drives the water film and leads to a large coupling close to 4000N-s/MJ. If the proper focusing system and window can be provided, this system can also be used for CO_2 laser.

4.4 Non-Electronic Robot (NEBOT)

In an accident at a nuclear power reactor, a lot of neutrons and radiation will be emitted and make electronic devices useless. If we need to send robots to this place, the robots must be driven by some means. We are planning to drive these robots by laser far from a long distance. We can use shape memory alloy heated by cw lasers to drive the movement of robots, as well as direct ablation process by pulsed lasers like the piston proposed earlier. These robots will be carried by the airship given below.

As shown in Table 1, the required power is an order of kW even for a large-size robot and therefore solar energy can be an important source for the robots working in the fields for a long time.

4.5 Ships

The use of water overlay is most suitable for ships because the water is automatically supplied and the coupling coefficient can be as high as possible, regardless of the specific impulse which determines the water consumption. The estimated C_m for underwater acceleration was 137 N-s/MJ for 300-mJ and 212 N-s/MJ for 560-mJ lasers, while in air it was 39 N-s/MJ for 300-mJ and 61 N-s/MJ for 560-mJ lasers (Yabe, et al., 2003b). Although the pendulum should hardly move, owing to the large drag in water, the momentum gained by the pendulum in water was three times larger than that in air.

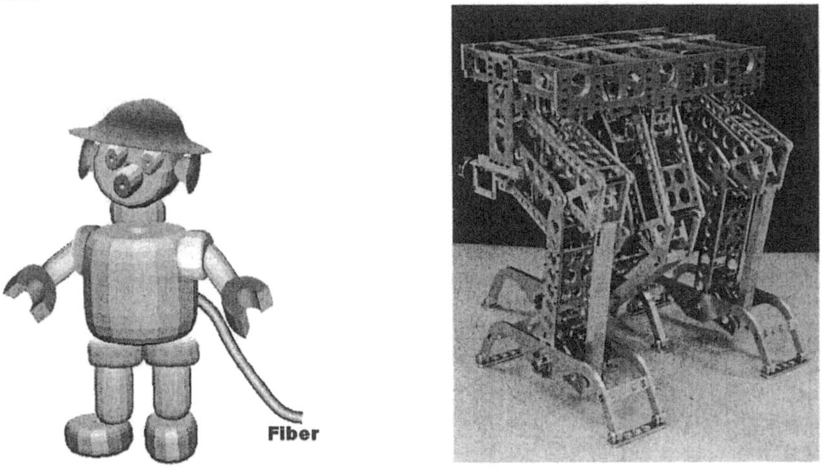

Figure 14. (Left) Robot can be driven by fiber laser. The right figure shows the assistant robot that uses only 50W poser (Size 80×60×100cm. Courtesy of Yukio Takeda, Tokyo Institute of Technology).

This is due to the efficiency enhancement by the water. This can also be used for a large ship as shown in Table 1. Those ships can provide surface area for solar energy sufficient for driving them. For example, a middle-class tanker has a surface area about $10000m^2$, and thus available solar energy is 10MW which corresponds to the present driving power 11MW. If the solar energy pumped laser can achieve efficient conversion from solar energy to laser, we do not need any fuel and energy source for driving these ships.

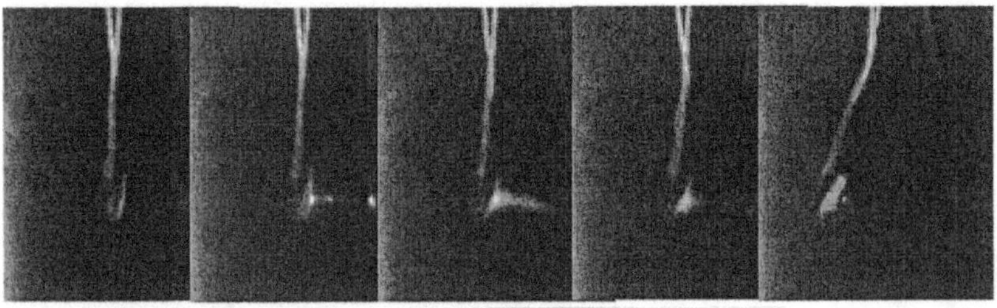

Figure 15. Pendulum driven by 300mJ laser under water.

5. REFERENCES

Azechi,H.,Miyanaga,N., Sakabe,S. Yamanaka,T. and Yamanaka,C., 1981, Model for Cannonball-Like Acceleration of Laser-Irradiated Targets, *Jpn.J.Appl.Phys.* **20**:L477

Bohn, W. L., Schall, W. O., Laser Propulsion Activities in Germany, 1st International Symposium on Beamed Energy Propulsion, 79-91 (2002).

Fabbro,F.J., Ballard,P., Devaux,D. and Virmont,J.,1990, Physical Study of Laser-Produced Plasma in Confined Geometry, *J.Appl.Phys.* **68**: 775.

Johnson,N.L.,1991, Artificial Space Debris, Orbit Book Co.

Kantrowitz,A.,1971, Propulsion to Orbit by Ground-Based Laser, *Astronaut. Aeronaut.* **10**:34-35.

Myrabo, L. N., Proc. Brief History of the Lightcraft Technology Demonstrator (LTD) Project, 1st International Symposium on Beamed Energy Propulsion, A. Pakhomov ed., AIP 49-60 (2002).

Ohkubo,T.,Yamaguchi,M.,Yabe,T. et al.,2002, Laser-Driven Micro-Ship and Micro-Turbine by Water-Powered Propulsion, *Beamed Energy Propulsion*, **AIP 664** :535-544.

Phipps,C.,1988, Impulse coupling to targets n vacuum by KrF, HF, and CO_2 single-pulse lasers, *J. Appl. Phys.*, 64(3): 1083-1096.

Phipps,C.,1998, ORION: Challenges and benefits, Proc. SPIE vol. 3343, 575-58

Phipps,C. and J. Luke,J., 2000, Micro laser plasma thrusters for small satellite, Proc. SPIE vol. 4065: 801-809.

Phipps,C.R.,Seibert II,D.B., Royse,R., King,G. and Campbell,J.W.,2000, III International Symbosium on High Power Laser Ablation, Santa Fe, SPIE Vol.4065.

Phipps,C. and Luke, J., 2002, Diode Laser-Driven Microthrusters: A New Departure for Micropropulsion, *AIAA Journal* **40:**310-318.

Pirri,A.N.,1974, *AIAA Journal*, 12(9): 1254-1261.

Reilly,J.P., 1979, Modeling of momentum transfer to a surface by laser-supported absorption waves, *AIAA Journal*, 17(10):1098-1105.

Rezunkov,Y.,2005, Performance of Laser Propulsion Engine Operating both in CW and in Repetitively- Pulsed Modes, 4th International Symposium on Beamed Energy Propulsion, K. Komurasaki ed., (to be published).

Sasoh, A.,"Laser-Driven in Tube Accelerator, *Rev. Sci. Instrum.*, **72**, 1893-1898 (2001).

Schall,W.,1998, Removal of small space debris with orbiting lasers, Proc. SPIE vol. 3343, 564-574.

Uchida,S. and M. Bato, M.,2003, Characterization of liquid propellant for improved LOTV mission, 1st International Symposium on Beamed Energy Propulsion, AIP press. 214-222.

Uchida,S.,Shimada,Y,Hashimoto,K.,Yamaura,M.,Biro,T.and Yoshida, M.,2005, 3rd International Symposium on Beamed Energy Propulsion, AIP press. 433-441, (2005).

Winterberg, F.,1976, Recoil Free Implosion of Large-Aspect Ratio Thermonuclear Microexplosion, *Lettere al Nuovo Cimento* **16** :216.

Yabe, T. and Niu, K.,1976, Numerical Analysis on Implosion of Laser-Driven Target Plasma. *J.Phys.Soc.Japan*, **40** : 863.

Yabe,T. and Aoki,T.,1991,A Universal Solver for Hyperbolic Equations by Cubic-Polynomial Interpolation I. One-Dimensional Solver. *Comput.Phys.Commun.* **66**:219

Yabe,T.,Xiao,F. and Utsumi,T.,2001, Constrained Interpolation Profile Method for Multiphase Analysis, *J. Comput. Phys.* **169**: 556-593.

Yabe,T. *et.al.*,2002a, Micro-airplane Propelled by Laser-Driven Exotic Target , *Appl.Phys.Lett.* **80**:4318-4320.

Yabe,T. *et.al.*,2002b Laser-Driven Vehicles - from Inner-Space to Outer-Space Proc.SPIE 4760 High-Power Laser Ablation IV, 21-26 April 2002,Taos, NM, USA

Yabe,T. et al.,2002c, Simulation and Experiments on Laser Propulsion by Water Cannon Target , *Beamed Energy Propulsion*, **AIP 664** :185-193.

Yabe,T. et al.,2003a, "Proposal of Liquid Cannon Target Driven by Fiber Laser for Micro-Thruster in Satellite" *Beamed Energy Propulsion*, **AIP 702** :503-512.

Yabe,T. et al., 2003b, Laser-Driven Vehicles - from Inner-Space to Outer-Space -*Appl.Phys. A* **77**:243-249.

Yabe,T.,Uchida,S.,Yoshida,K., Ikuta,K. and Okamoto,T.,2005, An Overview of Entropia Laser Initiative, *Proceedings of 4th International Symposium on Beamed Energy Propulsion (ISBEP4), Nara, Japan, November 15-18, 2005*

Chapter 19

NANO-STRUCTURING USING PULSED LASER RADIATION

Costas P. Grigoropoulos, Anant Chimmalgi and David J. Hwang
Department of Mechanical Engineering, University of California, Berkeley CA 94720-1740

1. INTRODUCTION

Fundamental understanding of microscale phenomena has been greatly facilitated in recent years, largely due to the development of high-resolution mechanical, electrical, optical, and thermal sensors. Consequently, new directions have been created for the development of new materials that can be engineered to exhibit unusual properties at submicron scales. Surface engineering is a critical subfield of nanotechnology because of the paramount importance of surface interaction phenomena at the micro/nano-machine level. Nanofabrication of complex three-dimensional patterns cannot be accomplished with conventional thermo-chemo-mechanical processes. While laser-assisted processes have been effective in component microfabrication with basic dimensions in the few-microns range, there is an increasing need to advance the science and technology of laser processing to the nanoscale. Breakthroughs in various nanotechnologies, such as information storage, optoelectronics, and biofluidics, can be achieved only through basic research on nanoscale modification and characterization of surfaces designed to exhibit special topographical and compositional features at high densities.

Since their invention in the 1980's, scanning microscopy techniques such as scanning tunneling microscopy (STM), atomic force microscopy (AFM), scanning near-field optical microscopy, and further variants thereof, have not only become indispensable tools for ultrahigh resolution imaging of surfaces and measurement of surface properties, but have also offered hitherto unexplored possibilities to locally modify materials at levels comparable to those of large atoms, single molecules, and atomic clusters. These nanometric investigation tools have been extensively used in numerous high-resolution nanostructuring studies, to manipulate single atoms, and also as effective all-inclusive nanofabrication tools. In view of rapid advances in nanotechnology, various applications and research fields are being envisioned, such as ultrahigh density data storage, mask repair, Fresnel optics for x-rays, nanoelectronics, nanomechanics, and nanobiotechnology, all of which depend on the fabrication of features only 5-10 nm in lateral dimensions (Quate, 1997). These feature sizes are more than an order of magnitude less than present industrial ultraviolet lithography standards. Due to the light diffraction that limits the size of the minimum resolvable feature to one-half of the wavelength of the radiation used, conventional optical lithography is not applicable at the nanoscale. Although other techniques, such as electron-beam, ion-beam, and x-ray lithography, provide means for high-resolution nanoprocessing, the excessive equipment cost, low processing speeds, and lack of materials suitable for fabricating various optical elements highlight the need for alternative and/or complementary process schemes.

Confinement of optical energy to small dimensions can be achieved by coupling laser radiation to near-field scanning optical microscopes (NSOMs). Historically, the concept of utilizing optical near-field was proposed by Synge in the 1920's (Synge, 1928) and experimentally examined first in the microwave range (Ash, 1972), but could not be verified initially in the visible wavelength range due to difficulty in fabricating the aperture and controlling the gap distance. Since the mid 1980's, aided by advances in scanning probe microscopy and fiber pulling technology (Lewis, et al., 1984), Scanning Near-field Optical Microscopy (NSOM) was established as a powerful nanoscale characterization tool. NSOMs were demonstrated by Pohl, et al. (1984) and Betzig, et al. (1991, 1992) as a means to achieve probing resolution far below the wavelength, yet at optical frequencies. In addition to their use as nanoscale microscopy tools, NSOMs have been applied to various materials processing tasks (Betzig et al., 1991; Wegscheider et al., 1995). Thin metal film ablation tests using optical near-field schemes have been carried out (Lieberman, et al., 1999; Nolte, et al., 1999). Since laser sources are available with different photon energies and pulse lengths, matching with the absorption properties of the substrate can be dictated in order to accomplish local chemical and structural modification with precision, accuracy and without detrimental thermomechanical side effects.

The main technical challenge stems from the difficulty in transmitting pulses of sufficient energy for nanostructuring through nanoscale openings. However, recent work has shown great potential of nanoscale laser energy deposition for precise surface modification. The surface topography, texture, crystalline structure and chemical composition could be modified with lateral feature definition in the nanometer range by coupling nanosecond and femtosecond laser beams to NSOM probes in both apertured and apertureless configurations. This work opens many possibilities for nanofabrication.

2. APERTURELESS NSOM NANOMACHINING

Surface nanostructuring of various materials, including gold, gold/palladium, poly(methyl methacrylate), and polycarbonate, by illumination of a scanning probe microscope (SPM) tip with nanosecond laser pulses has been demonstrated in the literature (Gorbunov and Pompe, 1994; Jersch and Dickmann, 1996; Huang, et al., 2002). Although laser-assisted STM-based nanoprocessing schemes have yielded resolutions on the order of ~30 nm (Lu, et al., 1999; Boneberg, et al., 1998), restrictions on the electrical conductivity of both the material to be processed and the scanning probe tip as well as the ambient working conditions make these methods unsuitable for a wide range of electrically nonconductive and biological materials. Laser nanoprocessing schemes coupled with AFM operated either in contact or noncontact mode, provide effective means for overcoming aforementioned drawbacks and enable the development of nanoscopic features of lateral dimensions less than 20 nm (Jersch, et al., 1997, Lu, et al., 2001). In spite of the increasing interest in this nanostructuring scheme, the actual processing mechanism still remains unexplained. In most cases, AFM or STM configurations were used with the tip in actual contact or in very close proximity (e.g., 3-5 Å) with the sample surface, which complicated the identification of the basic modification mechanism. Despite the high enhancement factors proposed for the electric field intensity in the vicinity of the tip (Dickmann, et al., 1997) contact of the SPM tip with the sample surface due to thermal expansion (Huber, et al., 1998; Boneberg, et al., 1998) limits the reliability and resolution of these nanostructuring techniques.

2.1 Experimental Setup

Fig. 1 shows a schematic of the experimental apparatus used by Chimmalgi, et al. (2003, 2005). Surface nanostructuring was accomplished with femtosecond pulses of 83 fs full width at half maximum (FWHM) duration, 800 nm wavelength, and pulse energy in the range of 0.125–3.7 μJ. The laser beam was focused to a $1/e^2$ beam spot diameter of ~75 μm. In the nanostructuring experiments, a constant AFM tip-sample separation distance of 3-5 nm was maintained with the aid of nanolithography software that disables tip position feedback signals and enables tip movement along complex contours. Commercially available, etched silicon and coated silicon nitride microprobes were used for both surface scanning and nanostructuring.

2.2 Enhanced Electric Field under a Scanning Probe Tip

Electric field intensity distributions were obtained with the finite difference time domain (FDTD) method using a near-field simulation software (Tempest 6.0; Pistor, 2001). A three-dimensional simulation was performed in which the four Maxwell's equations and two constitutive relationships were solved in conjunction with the current relationship.

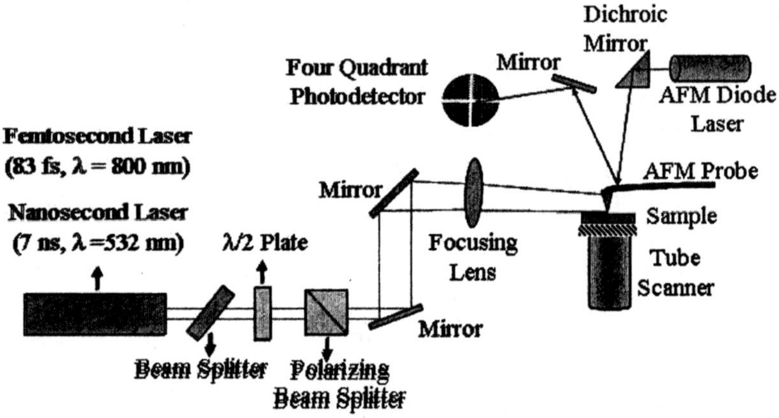

Figure 1. Schematic diagram of the apertureless near-field scanning optical microscope experimental setup for surface nanostructuring

Figs. 2a and 2b show the distribution of the electric field intensity *E* for incident uniform plane wave of unit *E* amplitude, 800 nm wavelength, *P*-state of incident polarization (i.e., polarization parallel to the incidence plane defined by the laser beam and the tip axis), at grazing angles of incidence for a tip-to-Au film surface separation distance of 2.7 nm. The total simulation volume considered for this simulation was 324 x 324 x 2025 nm^3 with a cubic discretization grid of 2.7 nm in size. An enhancement of the field intensity can be seen at the Au film surface, tip apex, and 2.7 nm from the tip apex, i.e., 5.4 nm from the film surface. The intensity enhancement is defined as the squared ratio of the scattered to the incident field amplitudes.

While an intensity enhancement as high as 150 is predicted at the Au film surface, the intensity at the tip decreases significantly [Fig. 2b]. The electric field distribution at the tip exhibits the *lightning rod effect*, which is more pronounced for longer wavelengths (Ohtsu, 1998). The lateral confinement of the field, with a FWHM of ~10 nm clearly seen from the graphs, is believed to be the main factor for the high spatial resolution observed in the nanostructuring experiments. A similar simulation carried out with the frequency-doubled (400 nm) laser yields an intensity enhancement factor of less than 40 [Fig. 2c]. The less enhanced field can be attributed to the decrease of the aforementioned *lightning rod effect* at visible wavelengths. Simulations showed a practically zero enhancement of the electric field for S-polarization. The enhancement values quoted above are theoretical estimations and the actual field strength enhancement may differ due to various influencing factors, such as approximations in the modeling of the actual tip shape, increase in the effective tip-sample separation distance

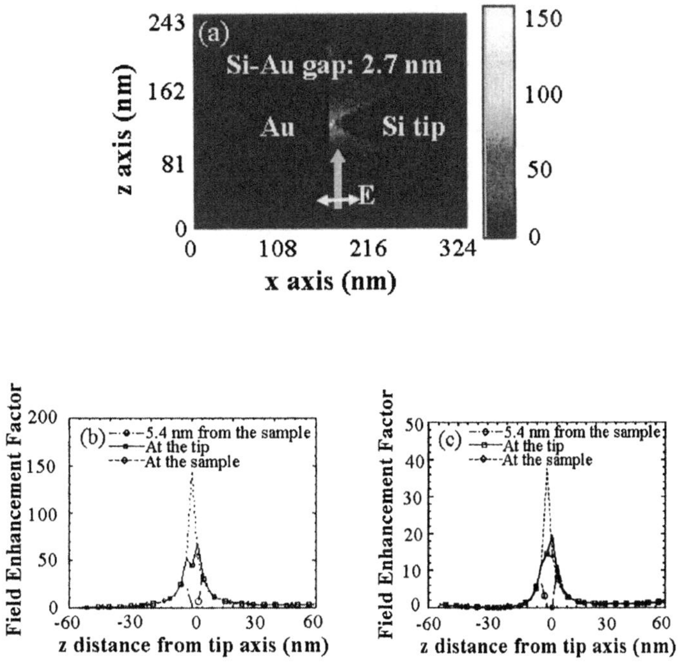

Figure 2. Normalized electric field intensity distribution underneath a Si probe tip irradiated with P-polarization femtosecond laser of wavelength equal to (a, b) 800 nm, and (c) 400 nm. The tip apex/sample surface separation is 2.7 nm

during processing due to the removal of material, surface roughness, presence of native oxide films on the tip and sample surfaces, and insitu change of the tip apex radius as a result of using the same tip for both scanning and nanostructuring.

2.3 Temperature Distribution

Heat transfer analysis of ultrathin-films subjected to ultrashort laser pulses presents significant challenges due to difficulties in modeling the dynamics of nonequilibrium heating and the lack of pertinent thermophysical data for ultrathin-films (Smith et al., 1999). The film thickness, deposition method, substrate material, and surface roughness are some of the factors directly affecting the thermophysical properties. Consequently, these properties for the thin-films vary from their corresponding bulk values. Mechanisms such as surface and grain boundary scattering tend to reduce the thermal conductivity of ultrathin metallic films used in the nanostructuring experiments (Chen and Hui, 1999).

Temporal transient temperature distributions in a film subjected to ultrashort laser pulses can be obtained from an approximate three- radiative dimensional analysis. The hyperbolic two-temperature model, simplified by neglecting lattice conduction, is used to model the energy transfer and subsequent heating by an axisymmetric heat source with Gaussian temporal and spatial distributions. Lattice conduction can be neglected because of the short timescales considered in the analysis. Temperature dependent thermophysical properties were used to improve the modeling accuracy. In addition, the size effect on the thermal conductivity of the Au thin-film was included in the analysis by using the experimental values of another study (Chen and Hui, 1999). The solution procedure involved the simultaneous solution of Eqs. (1)-(4), subject to the additional condition for the axial nodes ($r = 0$) given by Eq. (5):

$$C_e(T_e)\frac{\partial T_e}{\partial t} = -\left[\frac{q_{er}}{r} + \frac{\partial q_{er}}{\partial r} + \frac{\partial q_{ez}}{\partial z}\right] - G\left[T_e - T_l\right] + S(r,z,t), \quad (1)$$

$$\tau_e \frac{\partial q_{er}}{\partial t} + q_{er} = -k_e \frac{\partial T_e}{\partial r}, \quad (2)$$

$$\tau_e \frac{\partial q_{ez}}{\partial t} + q_{ez} = -k_e \frac{\partial T_e}{\partial z}, \quad (3)$$

$$C_l \frac{\partial T_l}{\partial t} = G\left[T_e - T_l\right], \quad (4)$$

$$\lim_{r\to 0}\left(\frac{q_{er}}{r}\right)=\frac{\partial q_{er}}{\partial r}. \tag{5}$$

The relationship for the laser source term [Eq. (6)] is based on the assumption of a spatially and temporally Gaussian beam that satisfies the conventional Beer's law of absorption. While the nominal optical absorption depth at λ = 800 nm is ~12 nm, a modified optical absorption depth z_s = 1 nm was fitted to the computed near-field absorbed energy density profile in the target obtained from the FDTD near-field simulations. An explicit finite difference method was used to solve the equations numerically:

$$S(r,z,t)=\sqrt{\frac{4\ln 2}{\pi}}\frac{(1-R)F}{t_p z_s\left(1-e^{-L/z_s}\right)}\exp\left[-\left(\frac{r}{r_s}\right)^2-\left(\frac{z}{z_s}\right)-4\ln 2\left(\frac{t-t_{peak}}{t_p}\right)^2\right]. \tag{6}$$

Adiabatic boundary conditions for all the sides and an initial temperature condition of 300 K were used in the analysis, i.e.,

$$q_{er}(0,z,t)=q_{er}(r\rightarrow+\infty,z,t)=q_{ez}(r,0,t)=q_{ez}(r,Z,t)=0, \tag{7}$$

$$T_e(r,z,0)=T_l(r,z,0)=300\text{K}. \tag{8}$$

The temperature-dependent properties were determined using the following relationships (Smith et al., 1999):

$$k_e=\xi\chi\frac{\left(\theta_e^2+0.16\right)^{5/4}\left(\theta_e^2+0.44\right)\theta_e}{\left(\theta_e^2+0.092\right)^{1/2}\left(\theta_e^2+\eta\theta_l\right)}, \tag{9}$$

$$C_e=A_eT_e. \tag{10}$$

The following constants were used for the Au film (Chen et al., 2002): T_F = 6.42 x 10^4 K, A_e = 70 J/(m^{3}K^2), η = 0.16, δ = 0.96, χ = 353 W/(mK), G = 2.6 x 10^{16} W/(m^{3}K), C_l = 2.688 x 10^6 J/(m^{3}K), and τ_e = 40 fs. The factor ξ = 0.18 was used to evaluate k_e based on experimental values quoted from the literature in order to correct for the thin-film size effect. Results of the numerical analysis for the heat transfer within the thin films for the nanostructuring experiments are discussed later in Section 2.5.1.

2.4 Experimental Surface Nanostructuring Results

In this section, representative results from various nanostructuring experiments (with typical feature sizes of 10-12 nm) performed with different metallic thin films are presented.

Fig. 3 shows the effect of the laser pulse energy on the dimensions of grid lines produced on the surfaces of 25-nm-thick Au films. It can be seen that both the depth and the width of the line features increase linearly with the laser fluence. A threshold value was observed for input laser pulse energy of ~0.5 μJ (i.e., laser fluence equal to ~0.012 J/cm^2) for which the depth of the obtained feature was almost comparable to the surface roughness. Wider and deeper features were produced by successively increasing the pulse energy. To form continuous line features, the laser was operated at 1 kHz while the sample surface was scanned at a speed of 1 μm/s. The lateral dimension of the minimum feature size in these experiments was less than 10 nm.

Fig. 4 shows nanostructuring of 15- and 25-nm-thick Au films for laser fluence of 0.069 J/cm^2. The features were produced with a linear scanning speed of 18 μm/s using a tip-sample separation distance of 3-5 nm.

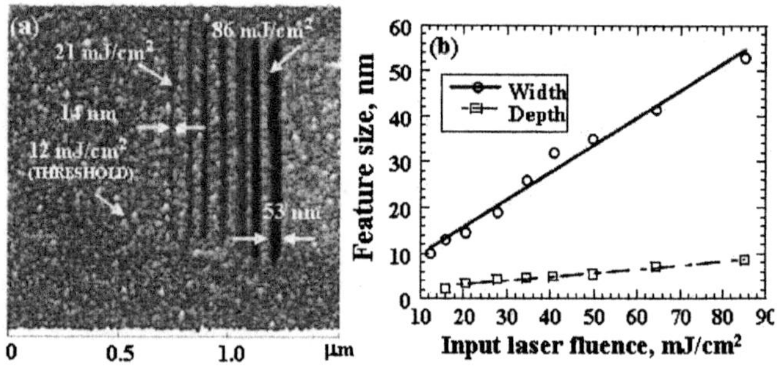

Figure 3. (a) Surface nanostructuring of 25-nm-thick Au film for laser fluence between 0.012 and 0.086 J/cm^2 and (b) dependence of lateral feature size on laser fluence.

Since the sample is mounted on the high-resolution x-y AFM scanner and the relative movement of the tip and the sample are controlled precisely using commercially available nanolithography software, features with complex contours can be produced, as demonstrated in Figs. 4(c) and 4(d).

Figure 5 shows formation of a nanocrater on 8-nm-thick In film deposited on Si substrate (Figs. 5(a) and 5(b)) and nanochannels on 13-nm-thick Au/6-nm-thick Cr dual-layer film deposited on quartz (Fig. 5(c)) using laser fluence equal to 0.035 J/cm^2 and 0.043 J/cm^2, respectively. The corresponding cross-section profiles, shown in Figs. 5(b) and 5(c), reveal feature depths larger than the thickness of the In and Au/Cr layers, respectively. Hence, nanometer-sized areas of Si and quartz were exposed in each case. Although the width of the crater and channels at the top surface is ~60 nm and ~75 nm respectively, the width of the exposed substrate region appears to be less than 10 nm in both cases. This suggests that metal thin-films can be used as masks for subsequent processing.

2.5 Mechanism of Surface Nanostructuring

Several investigators have attempted to explain the physical phenomena responsible for surface nanostructuring (Ukraintsev and Yates, 1996; Kawata, et al., 1999; Bachelot, et al., 2003). In general, material removal by ablation can be associated with photothermal, photomechanical, and photochemical effects (Hecht et al. 2000). Significant theoretical and experimental effort has been expended to study the thermal expansion of the tip and the treated sample. Indentation of the sample surface due to thermal expansion of the tip has been proposed as the main reason for the formation of nanoscopic features (Boneberg, et al., 1998; Huber, et al., 1998). Several apertureless schemes have been developed to overcome resolution limitations and significant transmission losses in aperture near-field scanning microscopy.

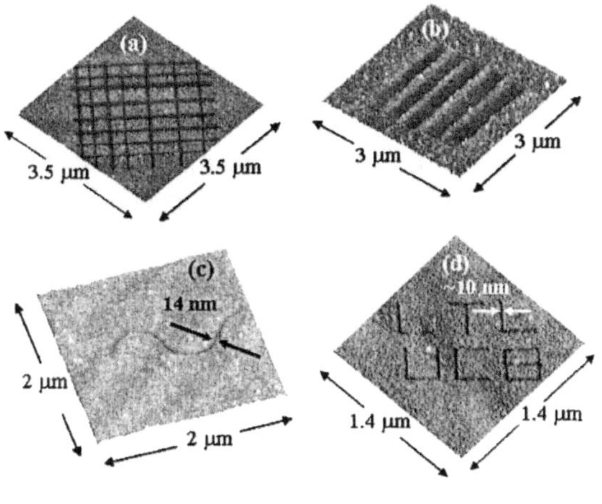

Figure 4. Surface nanostructuring of (a, c, d) ~25-nm-thick Au film and (b) 15-nm-thick Au film for laser fluence equal to 0.069 J/cm^2

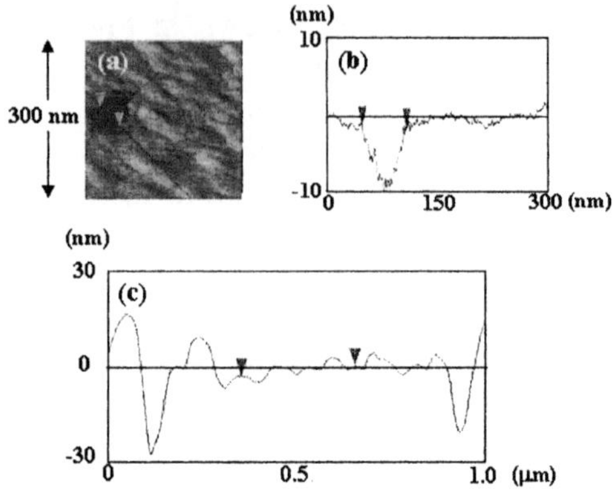

Figure 5. Formation of (a), (b) a nanocrater on In film and (c) nanochannels on 13-nm-thick Au/6-nm-thick Cr dual-layer film for laser fluence equal to 0.035 J/cm^2 and 0.043 J/cm^2, respectively. The cross-section profiles shown in (b) and (c) indicate that the feature depths in both cases exceeds the metal layer thickness exposing the underlying Si and Quartz substrate

Ultrahigh resolution fluorescence microscopy utilizing two-photon excitation and sharp metal tips for spectroscopy of single molecules have been demonstrated using similar configurations (Sanchez, et al., 1999). These schemes produce a highly localized and enhanced field underneath a sharp tip irradiated with a laser beam. Direct observations of the enhanced field using photosensitive azobenzene-containing films have confirmed the nanoscopic spatial confinement of the electric field, hence excluding the possibility of surface modification due to the tip expansion (Bachelot, et al., 2003).

A study was performed to evaluate the effect of exerting a positive force on the sample surface in the absence of laser radiation. It was found that advancing the tip into Au film did not cause plastic deformation in the film, evidently due to the low contact pressure produced by the highly compliant AFM cantilevers. Furthermore, using available calculations for the tip expansion (Huang et al., 2002), the effective temperature rise in the tip due to the effect of 7 ns laser pulses with energies between 35 and 60 μJ was determined to be in the range of 350-800 °C, which yields a tip expansion of 2-5 nm. Since the nanostructuring experiments were performed with femtosecond laser pulses, the pulse energies were an order of magnitude less than those encountered with nanosecond laser pulses, and because the tip-sample separation distance was always maintained above 3 nm, tip expansion (and hence indentation) did not contribute to the surface nanostructuring process. As indirect proof that tip expansion was not the main contributing factor in the present nanostructuring process, a complementary experiment was performed with Au films wherein the silicon AFM tip was irradiated with frequency-doubled ($\lambda = 400$ nm) femtosecond laser pulses. Although the input fluence was increased up to the limit of tip damage, only shallow features could be produced compared to the experiments conducted with fundamental-frequency ($\lambda = 800$ nm) femtosecond laser pulses. Since the visible femtosecond laser energy of $\lambda = 400$ nm is coupled more efficiently in the tip, a higher temperature rise and possible expansion of the tip would be expected.

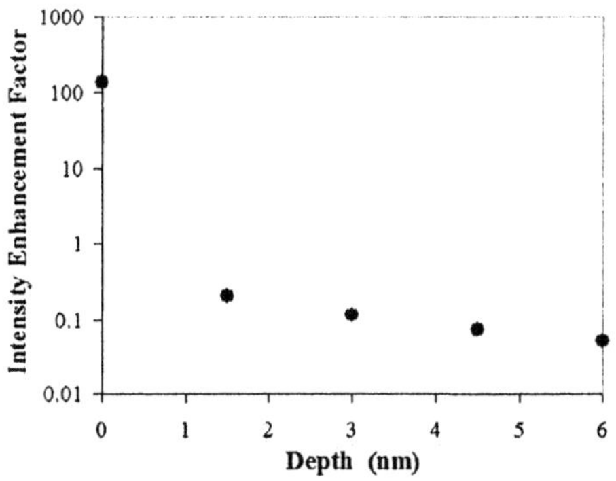

Figure 6. Electric field intensity enhancement factor underneath a Si tip irradiated with P-polarization femtosecond laser with 800 nm wavelength versus distance from the Au film surface for tip-sample separation distance and mesh grid size equal to 3 and 1.5 nm, respectively.

However, the absence of any discernible nanofeatures confirmed that tip expansion was not the underlying reason for the produced surface nanostructures.

2.5.1 Effect of Source Size on Temperature Distribution in the Modified Film

To better understand the coupling of the laser energy with the sample surface due to the enhanced electric field underneath the probe tip, simulations with a finer mesh of grid size equal to 1.5 nm were carried out for a Si tip at a distance of 3 nm from the Au film surface. Fig. 6 shows the variation of the intensity enhancement factor with distance from the surface of the Au film. For the given configuration, Beer's law of absorption predicts an exponential absorption profile of depth equal to ~12 nm. However, the much steeper near-field profile shown in Fig. 6 indicates that most of the laser energy is coupled within the top 2 nm of the film. In view of this result and considering the significant lateral confinement of the laser energy source, the temperature distribution in the Au film was determined for femtosecond laser irradiation. The tighter focusing of the radiative energy is quantified by the decrease of the $1/e^2$ radius that is fitted to the incident intensity profile on the sample surface. Thus for the laser source term, the $1/e^2$ radius fitted to the incident energy density profile on the target surface was successively reduced from ~141 to ~14 nm.

A modified absorption profile was used in which the exponential feature was maintained while the effective absorption depth was set equal to 1 nm. The simulation was carried out to evaluate the threshold laser fluence required for the center node at the surface ($r = z = 0$) to reach the melting temperature. Fig. 7 shows the transient electron and lattice temperature profiles for $r_s = 100$ nm (i.e. $1/e^2$ radius of 141 nm).

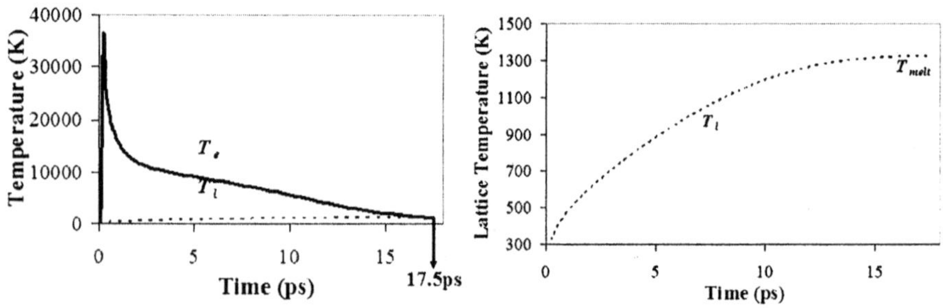

Figure 7. (a) Electron and lattice temporal temperature distributions in Au film and (b) magnified lattice temperature distribution for laser beam profile with $1/e^2$ radius equal to ~141 nm and laser fluence equal to 0.15 J/cm^2, determined at the top center node (r = z = 0).

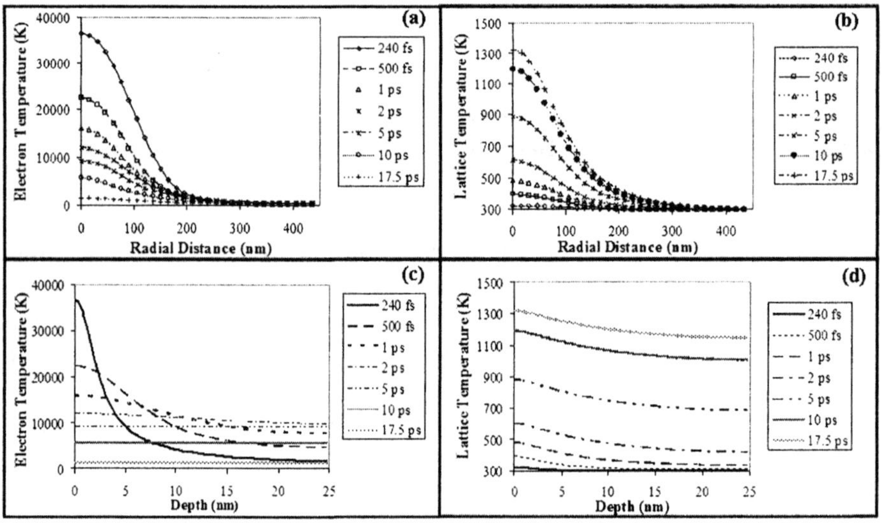

Figure 8. Electron and lattice temporal temperature distributions in the radial and depth directions in Au film for laser beam profile with $1/e^2$ radius equal to ~141 nm. (The time of maximum electron temperature and the time for the center node at the film surface to reach the melting temperature are equal to 240 fs and 17.5 ps, respectively).

The maximum electron temperature is reached at the center node at ~240 fs. Fig. 7(b) shows that the lattice temperature approaches the melting temperature significantly later, i.e., after ~17.5 ps. Fig. 8 shows temporal electron and lattice temperature distributions in the Au film along the radial and depth directions r_s = 100 nm (i.e. 1/e^2 radius of 141 nm). Fig. 9 shows that the threshold fluence for melting (the peak value of the laser energy density profile on the sample surface) increases significantly

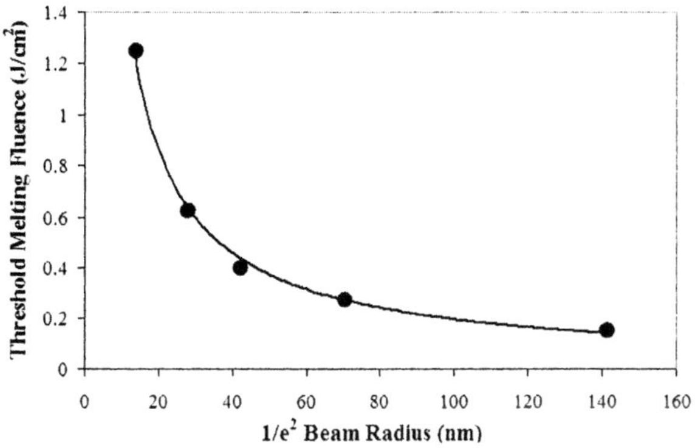

Figure 9. Threshold melting fluence vs. 1/e^2 laser beam radius.

with higher energy confinement underneath the tip, due to the rapid lateral heat diffusion in the electron system. Based on these simulations, it was observed that the threshold melting fluence increases (and thus the maximum temperature reached by the electron system also increases) and the lattice and electron systems reach the same equilibrium melting temperature faster with the decrease of the laser source size.

The favorable sharp tip geometry (leading to the lightning rod effect discussed earlier) and the multiple scattering between the resonantly coupled probe tip and the sample structure thus effectively lead to the focusing of the optical power by collecting over the larger, nearly flat probe surface and re-radiating over the smaller and sharper tip-apex structure. Indentation of the sample surface due to thermal expansion of the tip has been ruled out as a possible surface modifying mechanism. The nanostructuring results shown on various films therefore attest to a mechanism wherein the optical energy is eventually converted into thermal energy in the substrate, with the tip acting as an effective 'nanolaser source' that redistributes the energy toward the sample surface in a confined manner.

3. APERTURED NSOM NANOMACHINING

In the most common embodiment of apertured NSOMs, the optical fiber is heated locally, usually via a CO_2 laser to form a taper. A cladding layer, typically aluminum because of its high reflection at visible wavelengths, is coated on the outside of the fiber. The transmission efficiency through the fiber probe is defined as the ratio of output energy to the input coupled energy and is roughly proportional to $(d/\lambda)^n$ (n is around 2-4, depending on the probe design (Ebbesen, et al., 1998)). Because considerable input laser

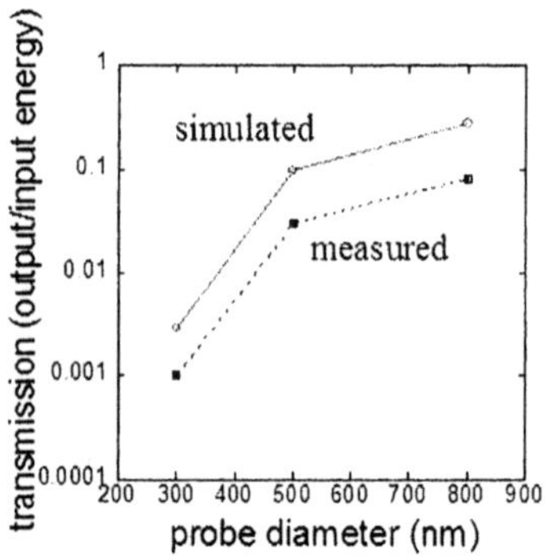

Figure 10. Calculated transmission efficiency values for different probe sizes. Measured values follow similar trend.

pulse energy is required to offset losses from apertures of decreasing size, the trade-off between minimum aperture size and minimum required light throughput should be considered in materials processing applications. The throughput of the NSOM chiefly depends on the geometry of the tapered region where the transmission efficiency drops as the fiber diameter becomes smaller than the laser wavelength. This is displayed in the FDTD simulation result shown in Fig. 10. The experimental measurement follows a similar trend, albeit indicating lower transmission, due to the inevitable losses in the system.

Fig. 11 shows the schematic of the apertured NSOM nanomachining setup employed at the Laser Thermal Laboratory. The processing laser beam is coupled to a bent cantilevered, metal-coated optical fiber, and delivered onto the sample surface. The probe laser bouncing off the upper surface is utilized for measuring the tip-sample separation. Fig. 12(a) shows the calculated energy density for the peak temperature to reach the melting temperature as a function of the aperture diameter.

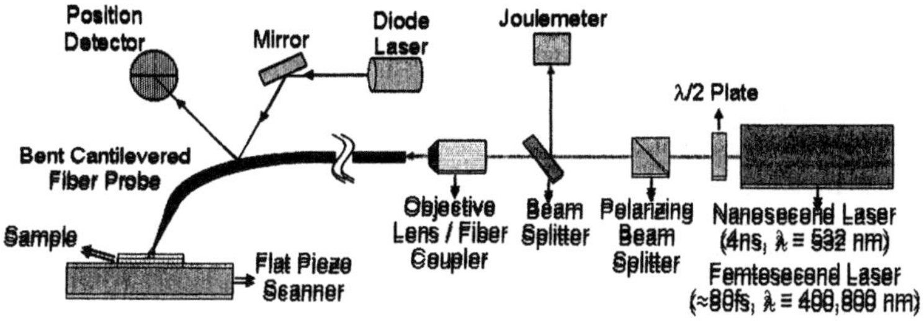

Figure 11. Schematic of the experimental setup for processing utilizing a cantilevered NSOM system.

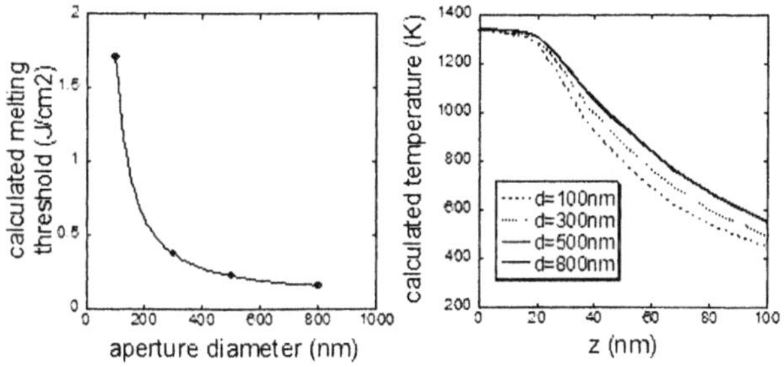

Figure 12. Thermal analysis results: (a) calculated energy density for the peak temperature to reach the melting temperature as a function of the aperture diameter and (b) temperature profile in transverse direction (metal film thickness is 20nm in this calculation).

The temperature profile in transverse direction is plotted in Fig. 12(b) while it is almost constant across the film. Processing results utilizing the 532 nm Nd:YAG laser harmonic are displayed in Fig. 13(a, b). The crater/feature profiles given in Fig. 13(b) correspond to single pulse processing. While at low laser pulse energies the surface just melts with little or no net volumetric loss, deep, clearly ablated craters are observed in the higher fluence regime. It is interesting to observe that at a certain laser fluence, the metal films can just be removed. Higher pulse energies create O(100 nm) FWHM craters in the glass substrate.

In order to investigate laser pulse width effects, ablated craters by femtosecond laser pulses (400nm wavelength and ~80 fs pulse width) are shown in Fig. 14. A fiber probe of 500nm aperture diameter was used for this study. Compared to nanosecond laser processing the ablated craters by femtosecond laser pulses are of higher aspect ratio. Ablation craters with feature width less than 200nm which penetrated through the thickness of the metal film were obtained. Clear benefits of processing with shorter laser pulses are reduction of the thermal diffusion length, tighter concentration of the deposited energy and production of sharper features. At sufficiently high fluence, the substrate can be ablated together with the metal thin film as in the crater profile depicted in Fig. 14. The fluence of 3.17 J/cm^2 produces a shallower final crater compared to the lower fluence case. This may be due to the positive deformation of glass during resolidification upon rapid heating past the glass transition temperature. Typically, the ultrashort laser pulse ablation threshold of quartz exceeds ~5 J/cm^2. Therefore, the observed substrate ablation at lower fluence has to be initiated by conductive heat transfer from the metal film.

The spectral bandwidth of ultrashort laser pulses is relatively wide. When femtosecond laser pulses propagate through the optical fiber core, they can be broadened by several mechanisms related to optical dispersion. If the optical fiber is a single mode fiber, the fs pulses primarily experience group velocity dispersion that typically broadens ~80 fs pulses at the rate of roughly one picosecond per meter length. Because the fiber probes used in the aforementioned experiment were drawn from multimode fiber, the dominant dispersion mechanism is modal dispersion. The degree of pulse broadening by modal dispersion depends on the refractive index of the glass medium as a function of laser wavelength and core size. Modal dispersion for ~100μm core diameter multimode fiber is in the range of 10's-100's picoseconds.

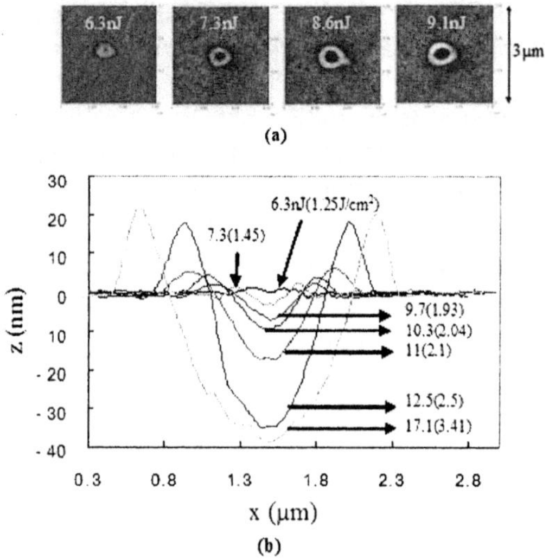

Figure 13. Sequence of features written in a 13 nm-thick gold film by 532 nm Nd:YAG laser pulses coupled to the NSOM processing apparatus, (a). AFM scans of craters generated by NSOM processing, (b). Section profiles of the generated craters. Note that the values marked are measured output pulse energies and the corresponding fluence values.

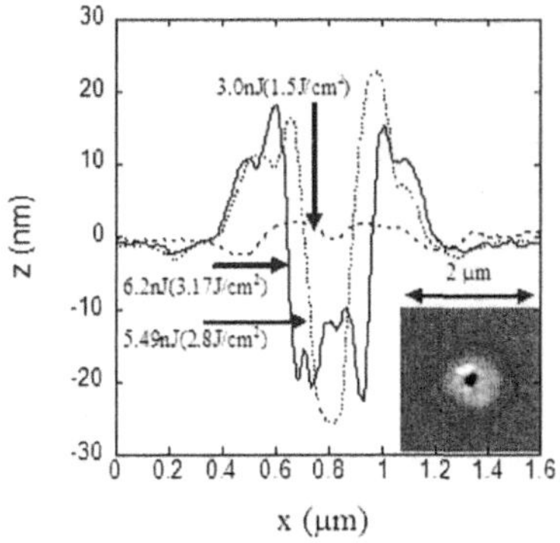

Figure 14. Ablation crater profiles by femtosecond laser (500nm probe, 400nm wavelength) as laser energy varies.

Hence, even though femtosecond laser pulses were coupled to the fiber probe, the output laser pulse was much wider. In order to maintain ultrashort laser pulses, single mode fiber based probes may be chosen, compromising light throughput, a drawback that is in part compensated by the lower ablation threshold for shorter pulses.

Special care needs to be exerted to compensate the group velocity dispersion. Nevertheless, in spite of the pulse broadening, the pulse width is still a couple of orders of magnitude shorter than nanoseconds.

Melting is difficult to avoid in laser ablation of metals, even in femtosecond laser processing where thermal diffusion is limited. The melt phase tends to produce elevated rims at the periphery of the ablation craters, hence increasing the overall feature size. This is not an issue for applications such as ablation-based mass spectroscopy (Stöckle et al., 2001).

When a solid sample is ablated using conventional lens focusing optics, the final shape of the ablated features is sensitive to the environment conditions. For example, when ultrashort laser light is focused onto the solid sample, electrons are first excited and emitted almost instantaneously (~100fs) due to high mobility and small heat capacity. If the sample is exposed to air, the emitted electrons collide with air molecules, triggering surface electron-initiated air plasma by ionization. Once the plasma is formed, it can further absorb the laser pulse energy for the remaining period of the pulse duration by inverse Bremsstrahlung, giving rise to a hot and dense state. The energy stored in the electron subsystem in the target solid is transferred to the lattice at the time scale of ~0.1-10ps, due to the much higher heat capacity of the lattice compared to electrons. Therefore, the material ejection process commences after an elapsed time of about 10-100ps. Upon interacting with the air plasma, the ejecta can bounce back and

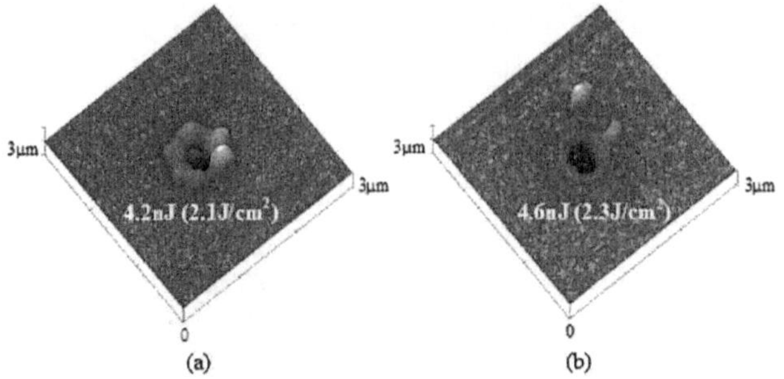

Figure 15. Ablation crater profiles (a) in ambient air and (b) moderate vacuum (~26 mtorr) environment (femtosecond laser of 400nm wavelength coupled through 500nm NSOM probe).

compress the melt. This can explain why much cleaner features can be fabricated in vacuum or non-reactive gas environments such as helium (Sun and Longtin, 2001). In nanosecond ablation, the incident beam interacts with the ejecta over a substantial fraction of the pulse duration. Consequently, the ablation threshold is not very sharp because of the more complex plasma shielding effect compared to femtosecond laser ablation. In the case of ablation through NSOM fiber probes, the laser passes through a narrow, ~nm scale gap before reaching the sample. The laser interaction with the plasma is therefore expected to be weaker compared to far-field lens focusing scheme. This argument is supported by the clear dependence of the feature size on the laser pulse energy applied for both nanosecond and femtosecond laser pulses as shown above. However, Fig. 15 shows that the topography of features ablated in ambient air exhibits qualitative difference compared with features ablated in vacuum. Further experimental study will elucidate the exact plasma formation mechanism and its evolution in the tight confines of the probe tip/sample gap.

4. NANOSCALE MELTING AND CRYSTALLIZATION

In erasable phase change optical data storage applications, a 'bit' of information is written on a crystallized region of the disk using a sufficiently strong laser pulse to induce local melting and rapid cooling leading to formation of amorphous phase (Yamada, et al., 1991). This bit can be read by using a low energy laser pulse and noting the optical response difference compared to the surrounding crystalline region. Furthermore, to erase this bit one can use intermediate laser pulse energy to raise the temperature of the film above the crystallization temperature but below the melting temperature, thus converting the amorphous region back into its corresponding crystalline phase to erase optical contrast. Understanding the nanoscale amorphization and crystallization phenomena using optical nearfield techniques is essential for ultrahigh density data storage applications.

For the study reported by Chimmalgi, et al. (2005b), two different near-field schemes were employed to explore the nanoscale melting and crystallization phenomena in a-Si thin-films. In the fiber-coupled scheme, the nanosecond laser beam was spatially confined by cantilevered near field scanning optical microscope (NSOM) fiber tip. A second approach entailed local field enhancement in the near-field of a SPM probe tip irradiated with nanosecond laser pulses in an apertureless arrangement. The experimental setups for the two schemes have already been discussed in the earlier sections (Figs. 1 and 11). A frequency-doubled (λ=532 nm) Q-switched Nd:YAG pulsed nanosecond laser operating at a maximum repetition frequency of 10 Hz and pulse duration of 4 ns was used in the nanocrystallization studies. Bent cantilevered multimode NSOM fibers, of 800nm aperture sizes, fitted onto a commercial NSOM system were used in the current experiment. The samples consisted of ~20-nm-thick a-Si films deposited onto fused quartz substrates by low pressure chemical vapor deposition (LPCVD) via a silane (SiH_4) process at a temperature of 550 °C. Characterization by an AFM showed that the films had a rms roughness value of 0.4nm over a 5x5 μm region. Unlike in the previously reported studies that mostly employed a-Si films of 50nm or greater thickness, the current study uses samples of 20nm a-Si films.

To gain insight into the spatial distributions of the enhanced electric field intensity underneath the probes (for both schemes) and to better understand the associated physical phenomena, the electric field intensity distributions were solved numerically using the finite difference time domain (FDTD) technique. Since in the current case the diameter of the tapered fiber NSOM probe is larger than the wavelength of the laser used, the direct transmission through the probe dominates over evanescent transmission modes, yielding a smooth output beam profile as clearly seen in Figs. 16(a) and 16(b) with a FWHM of ~400 nm. Also plotted in Fig. 16(b) are the intensity profiles at different z values from the tip aperture. The high enhancement factor seen in the simulation results is believed to be mainly responsible for the nanostructuring observed in the current study. It should be noted that the simulation did not consider the presence of any sample underneath the probe. For the current study only a (0, 0) mode input plane wave was assumed for simplicity, but the actual input laser beam may include multiple higher modes. Even though the length of the taper part actually affects the transmission efficiency, only limited taper length was simulated due to computational domain size limit. These approximations are reasonable as the focus of this analysis is to understand the approximate intensity distribution under the tip aperture, *i.e.* to study the shape of the output beam profile and the variation of the intensity with z. Figs. 16 (c) and 16(d) show the simulation results for a Si tip and an a-Si thin-film sample irradiated with laser pulses (of λ = 532nm), used in the current nanocrystallization experiments using the apertureless scheme. While an intensity enhancement as high as 100 is predicted at the top of the oxide layer (oxide-air interface), the intensity at the a-Si/oxide interface is greatly reduced. Nevertheless, it was found that this intensity enhancement was sufficient to induce local melting and nanostructuring of the a-Si films without damaging the tip.

Representative nanocrystallization results for the 20-nm-thick a-Si films deposited on quartz with the Nd:YAG pulsed nanosecond laser obtained with the aforementioned two schemes are now presented here. The results of a study employing single nanosecond laser pulses of varying fluence, using the fiber-coupled probe, are detailed in Fig. 17. Interesting observations about the melting transformation, the subsequent nucleation and crystallization processes were made with the gradual decrease of the fluence.

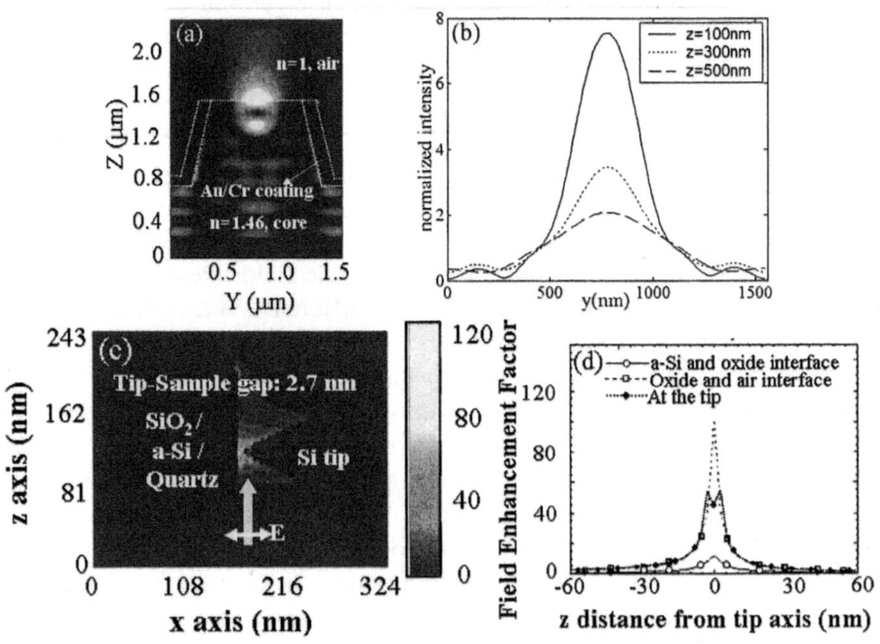

Figure 16. (a) & (b): Normalized electric field intensity distribution through a tapered NSOM probe tip for laser of wavelength 532 nm. (c) & (d): Normalized electric field intensity distribution underneath a Si SPM probe tip irradiated with P-polarization laser of wavelength 532 nm in proximity with the surface of a-Si film on quartz substrate. The tip apex/sample surface separation is 2.7 nm.

These results can be explained by considering previous findings for thin film a-Si crystallized by microscale laser pulses (Lee, et al., 2000). It should be noted that the fluence values quoted here are the actual, measured energy densities at the NSOM probe tip exit based on the pico-Joulemeter readings. Since in the present case the laser beam is confined to a FWHM of ~400nm, one can expect the threshold fluence for melting (the peak value of the laser energy density profile on the sample surface) to be considerably larger than in the corresponding micro/macroscale studies done previously.

This can be attributed to the rapid lateral heat diffusion which results in the higher observed thresholds. One can identify four different modification regimes depending on the applied laser fluence, as highlighted in Fig. 17. For the high fluence case of ~5.9 J/cm^2 [Fig. 17(a)], a clear central ablation regime with a narrow surrounding melt region can be seen. With gradual decrease in energy, but at still high enough fluence, the central region melts completely whereas the adjacent surrounding area is partially melted. Therefore, nucleation first occurs in the outer region, as evidenced by the formation of small polycrystals. Grains launched from these seeds grow laterally towards the central and completely molten region until their growth is arrested by impingement with lateral grains growing from the opposite direction or by spontaneous nucleation triggered in the severely supercooled molten silicon pool (Lee et al., 2000).

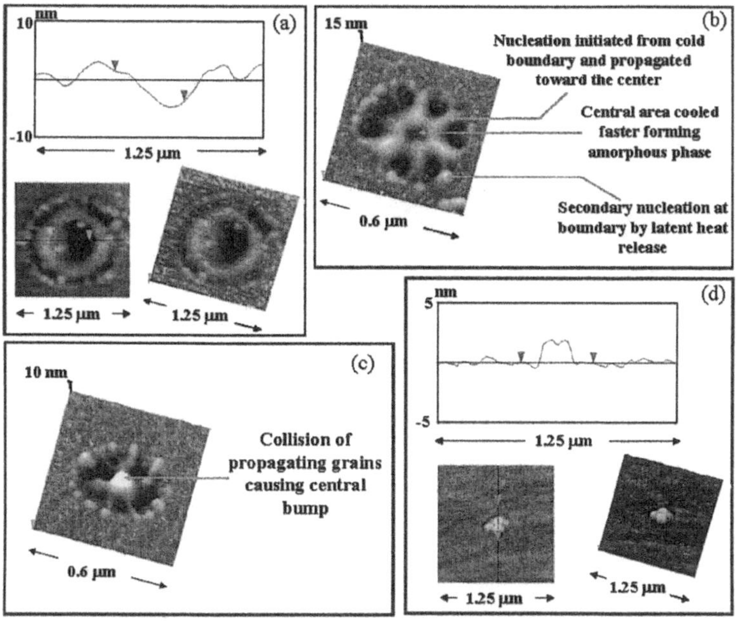

Figure 17. Limiting cases of nanocrystallization study of 20-nm-thick a-Si films based on the fiber-coupled NSOM for laser fluence equal to (a) 5.9 J/cm^2, (b) 3.3 J/cm^2, (c) 3.0 J/cm^2, and (d) 2.0 J/cm^2.

Fig. 17(b), corresponding to laser fluence of ~3.3 J/cm^2, shows such an occurrence wherein the progress of nucleation front is inhibited by the development of an ultrafine microstructure of polycrystalline/fine-amorphous phase forming a central spot structure.

Considering the solid/liquid interfacial kinetics (Stolk et al., 1993), and experimental time-resolved imaging evidence (Lee et al., 2001), the lateral crystal growth speed is around 10 m/s. Consequently, the elapsed time for the lateral crystal growth would be approximately 20-30 ns. It is estimated that the 3-D conductive loss to the non-nucleation-participating substrate drives quenching rates of the order of 10^{11} K/s at the center of the molten spot. As a result the undercooled liquid spontaneously freezes upon reaching the nucleation temperature (which is about 250 K below the nominal melting point), thereby arresting the inward propagating crystals as shown in Fig. 17(b).

The protrusions observed can be explained based on volume expansion accompanying the quenching of liquid silicon that is trapped and squeezed by the propagating crystals. For a lower fluence of ~3.0 J/cm^2, collision of the silicon crystalline grains led to the formation of a central bump structure [Fig. 17(c)]. When the fluence was further decreased to ~2.0 J/cm^2 [Fig. 17(d)], another regime, characterized by the formation of a set of individual grains was encountered. The dimensions of these nanostructures were typically in the range of ~80 nm.

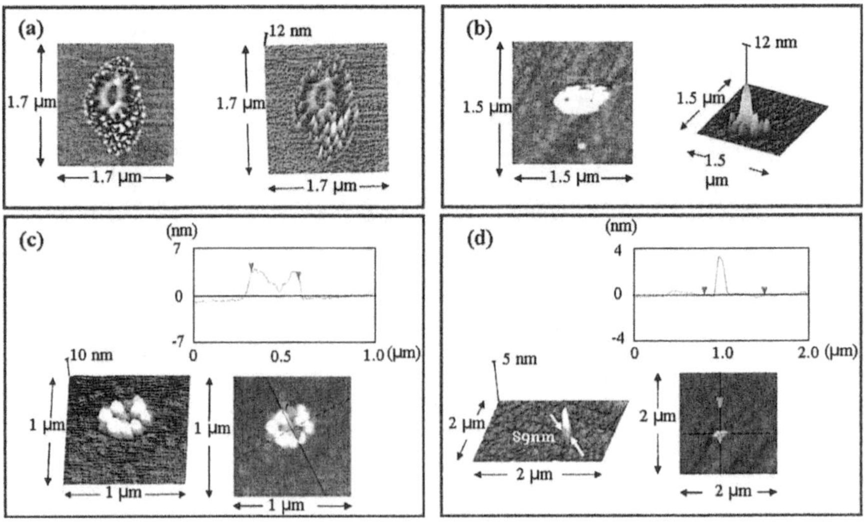

Figure 18. Nanocrystallization results of 20-nm-thick a-Si films with the apertureless NSOM for laser fluence equal to (a) 0.23 J/cm^2, (b) 0.13 J/cm^2, (c) 0.12 J/cm^2, and (d) 0.09 J/cm^2. The cross-section profiles reveal the dimensions of the produced nanofeatures

Fig. 18 shows the nanocrystallization results obtained using Scheme B with single nanosecond laser pulses of different fluence values. It is noted that the quoted fluence values in this case represent estimates of the far-field energy density at the focal spot of the laser beam near the SPM probe tip. Here again the distinct modification regimes produced as a result of the rapid melting and crystallization of the a-Si films, can be clearly identified. Due to the higher peak temperature at the center of the melt pool and because nucleation usually occurs at the solid/liquid phase boundary, polycrystalline grains growing toward the center of the supercooled melt were observed at high fluence values [Fig. 18(a)]. For a lower fluence of 0.13 J/cm^2, a bump was produced at the crystallized spot center due to collision of the silicon crystals [Fig. 18(b)]. Another regime characterized by the formation of a ring consisting of individual grains was encountered when the fluence decreased to 0.12 J/cm^2 [Fig. 18(c)]. The lateral size of these nanostructures was typically in the range of 70-80 nm. Further decrease of the input laser fluence to 0.09 J/cm^2 resulted in the formation of a single nanostructure with a lateral dimension of ~90 nm [Fig. 18(d)]. The lateral dimensions of these nanostructures could be further confined by using even lower fluence values and sharper tips, although limitations due to thermal diffusion would still exist. The ability to induce nucleation along predefined contours and the overall process control demonstrate the flexibility and effectiveness offered by the current nanostructuring schemes.

5. LASER-ASSISTED NANODEPOSITION

Laser chemical vapor deposition (LCVD) is an extremely versatile materials synthesis technique that enables the formation of technologically attractive microstructures of well-defined dimensions in a single-step maskless process (Bäuerle, 1998). A number of applications have been pursued, including the fabrication of contacts, circuit lines, interconnects, and the repair of lithographic masks. It also offers unique capabilities for the formation, coating, and patterning of non-planar, 3D objects. For example, LCVD was employed for the fabrication of micron-scale, photonic band-gap structures and the rapid prototyping of micromechanical actuators such as microtweezers and micromotors (Wanke, et al., 1993).

The decomposition of precursor molecules in LCVD can be activated either thermally (pyrolytic LCVD), non-thermally (photolytic LCVD), or by a combination of both (i.e. photophysical LCVD). In pyrolytic deposition, the heating of the substrate is employed to decompose the gases above it. In photolytic processes, one has to rely on direct dissociation of the parent gas by a bound-free transition. The type of process activation determines the morphology of the deposit while the deposition rate is a function of the laser power, intensity, pulse duration, wavelength, and substrate material. Major advantages of pyrolytic LCVD are the synthesis of high-purity materials and fast deposition rates, but an obvious drawback is that it cannot be used if the substrate softens or melts prior to gas decomposition upon exposure to laser radiation. In contrast, photolytic LCVD can be used for deposition on fragile substrates, including organic materials or complex III-V compounds. Furthermore, photolytic deposition is less sensitive to the substrate surface conditions and composition than pyrolytic deposition. However, deposition rates in photolytic processes are orders of magnitude lower than those associated with pyrolytic deposition. Photophysical LCVD achieved by a twin or single laser beam configuration seeks to combine advantages of both deposition schemes by initiating nucleation photochemically and progressing thereafter according to a conventional thermal script. For example, the deposition of Mo, W, and Pt has been shown by utilizing a multiline Ar^+ laser beam (Gilgen, et al., 1987).

Both pyrolytic and photolytic LCVD have the ability to define sub-micron features. Many possibilities arise, not only for achieving high-quality deposition of a wide range of materials on a variety of substrates, but also very significantly for fabricating 3D nanostructures.

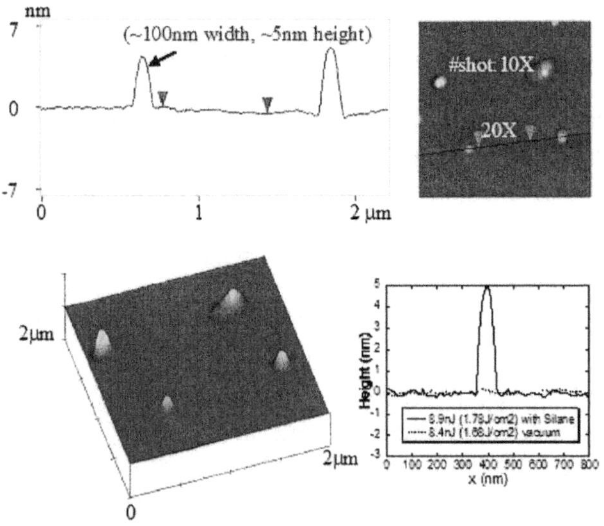

Figure 19. Nanodeposition of Si dots from silane using single nanosecond λ= 532 nm nanosecond laser shots through the apertured NSOM tip. The base diameter of the nanodeposit is ~ 100 nm.

Very recent experiments have yielded successful results for the pyrolytic NSOM-LCVD deposition of Si dots on crystalline Si by the decomposition of SiH_4. Fig. 19 shows an atomic force microscopy (AFM) image of the Si dots demonstrating that sub-100 features can be synthesized. Fabrication of three-dimensional objects of nanometric dimensions and arbitrary shape can be accomplished by extending the deposition and crystal growth through precise motion of the substrate. This nanoscale-LCVD offers the possibility of selectively controlling nucleation and then manipulating the crystal growth kinetics to drive 3D nanostructure fabrication and patterning to produce complex architectures. The fundamental thermo-chemo-physical phenomena involved must be understood. Time-resolved, in-situ spectroscopy, imaging and diagnostics can be utilized to explore and uncover the evolution of the nanostructure formation process.

6. LASER ABLATION NANOLITHOGRAPHY

Fig. 20 shows the representative schematic for the process sequence involved in the nanomachining of the metal thin-films and the subsequent wet-etching based feature transfer process wherein the features formed on the metal films are transferred to the underlying, transparent quartz substrate. The AFM scans shown in the figure are for a laser pulse energy of 6.5 nJ, output from the NSOM probe that corresponds to a fluence value of 3.2 J/cm^2 incident on the sample. To briefly summarize, the process involved the following steps: (a) the sample areas were scanned before processing to assure a clean and unaltered surface, (b) in order to facilitate locating the processed region, even after the sample is moved out of the NSOM setup, the sample was premarked using a microscale femtosecond laser ablation setup, detailed elsewhere (Chimmalgi, et al., 2005a), (c) using single laser pulses, patterns were made input laser fluence (since the ablation crater depth and width depended on the laser fluence used, a parametric study was done by gradually varying the laser fluence), (d) after cleaning the sample with DI water, the sample was moved to a normal AFM and high resolution scans were performed on the processed regions to map out the ablation crater profiles, (e) using the patterned metal layers as mask, wet-etching was performed on the sample by immersing it in 10% HF solution (diluted in DI water) for a duration of 2 minutes to transfer the feature into the underlying quartz substrate, (f) the sample was then thoroughly rinsed in DI water to remove any leftover etchant, (g) the sample was once again moved to an AFM and the etched regions were scanned to note the increase in the depth of the craters (this step is depicted in Fig. 20(b)), (h) now the masking layers were stripped off by immersing the sample in commercial Au and Cr etchants sequentially, for time durations of 5 minutes each which were long enough to completely etch away the entire thickness of the metal layers (~13nm in case of Au and ~6nm for Cr), (i) the sample was once more

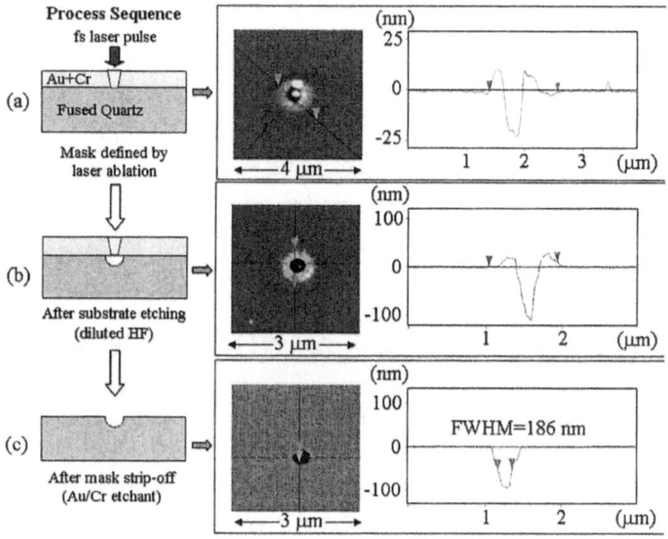

Figure 20. Process sequence for NSOM based nanomachining of thin metal films and subsequent feature transfer onto transparent quartz substrate. Also shown in the figure are the AFM scans corresponding to each of the steps in the process sequence, for a single case of input laser fluence of 3.2 J/cm^2.

thoroughly cleaned with DI water and transferred to an AFM to perform high resolution scanning for the last time to map out the final transferred feature profiles on quartz (refer to Fig. 20(c)).

Selected results from the parametric study are shown in Fig. 21, wherein input laser fluence was varied as a parameter. AFM crater profiles for the cases of input laser fluence of 2.8 J/cm^2, 2.9 J/cm^2, and 3.2 J/cm^2 are shown in Figs. 21 (a), (b), and (c) respectively. For the current study, the smallest crater profile obtained had a Full Width Half Maximum (FWHM) lateral dimension of ~90nm. It is noted here that though the top lateral dimension of the inner negative region of the ablated crater in Fig. 20(a) is ~300nm, the exposed region on quartz is much smaller due to the Gaussian-like crater profile which replicates the laser beam profile and is characteristic of laser ablation. Thus, it is possible to expose a narrowly confined region of quartz by closely controlling the input energy fluence. Further, the use of HF based wet-etching technique in this study was done only to demonstrate the feasibility of using these metal thin-films as effective masking layers for

NSOM ablation lithography. HF based etching being an isotropic etching technique effectively enlarges the lateral dimensions of the final obtained crater. However, the issue of resolution loss could be remedied by using alternative dry-etching techniques such as Deep Reactive Ion Etching (DRIE), which allows straight transfer of the pattern from the mask onto the underlying substrate. Further, it is also worth noting that as seen from Figs. 20 and 21, the elevated rim structures, that are inevitable for laser ablated craters, were restricted only on top of the metal masking layer and hence after strip-off, a clean rimless transferred crater profile was obtained. Use of NSOM probes of smaller aperture dimensions and thinner metal films could also further reduce the lateral dimensions of the features. The throughput from the current 500nm probes was high enough not only to ablate the entire thickness of the two metal film layers, but also to damage the underlying quartz substrate for the cases of high input laser fluence. Thus by employing probes of even smaller aperture sizes (which correspond to lower throughput), it should still be feasible to go through the metal film layer and expose smaller regions on the quartz substrate. Also by choosing suitable etchants, the same process could be used for a variety of film/substrate combinations, thus not just limited to transparent substrates.

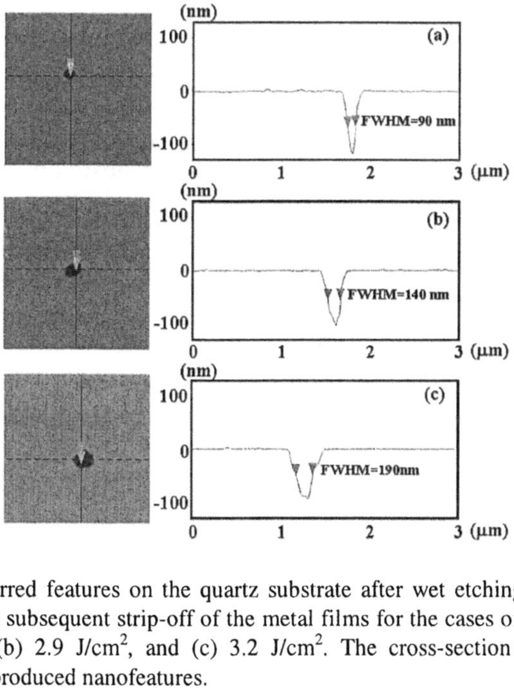

Figure 21. Transferred features on the quartz substrate after wet etching through the metal masking layers and subsequent strip-off of the metal films for the cases of laser fluence equal to (a) 2.8 J/cm^2, (b) 2.9 J/cm^2, and (c) 3.2 J/cm^2. The cross-section profiles reveal the dimensions of the produced nanofeatures.

The choice of femtosecond laser for this study is justified on account of the fact that the short duration pulses minimize the thermally mediated diffusion of the metal and subsequent compound formation with the underlying substrate material (in this case, silicide formation with the quartz substrate).

The same study done with nanosecond laser pulses led to serious compound formation between the Au/Cr metal layer and the quartz substrate that prevented the use of the patterned films as masks since the silicides acted as etch-stops making the HF based wet-etching process ineffective. Further, femtosecond lasers with few hundred nanojoule pulse energy and high repetition rates of ~10MHz are commercially available which ensures considerably fast machining rates limited only by the probe scanning speeds.

7. CONCLUSIONS AND OUTLOOK

Surface nanostructuring of various films at high spatial resolutions was performed with great accuracy and reproducibility by using two near-field optical methods. Nanostructures were defined with minimum lateral feature dimensions of ~10 nm and various complex nanopatterns were produced on metal thin-film samples. Experimental ablation results on Au thin-films and numerical simulation results for the electric field intensity and temperature distribution in the films provided useful insight into the laser thin-film interaction dynamics at the nanoscale. Nanoscale melting and rapid crystallization of a-Si films using a nanosecond laser source was demonstrated using the both apertured and apertureless NSOM schemes. The ability to nucleate and produce these single nanostructures in a controlled fashion could open up a number of potential applications. Further, LCVD based nanodeposition study results were presented using the apertured NSOM scheme which provides us with a useful tool to produce arbitrary shaped three-dimensional nanostructures. Use of thin metal films as effective masking layers was demonstrated wherein ablated nanopatterns generated on metal masking layers were effectively transferred on to the underlying transparent quartz substrate.

Using arrays of scanning microprobe tips with integrated actuator and sensor mechanisms could lead to increased throughput of these surface nanostructuring schemes. Furthermore, by incorporating improved probe designs with dedicated waveguide structures or using switching devices like digital micromirror arrays, precise optical delivery schemes could be devised for coupling the beam with the microprobe tips. Possible applications of these nanostructuring processes are envisioned in high-resolution nanolithography, controlled nanodeposition, ultrahigh density data storage, mask repair, nanoelectronics, nanophotonics, and various nanobiotechnology applications.

ACKNOWLEDGMENT

Support to this work by the U.S. National Science Foundation is gratefully acknowledged. AC was also supported by a UC Berkeley Nanoinitiative Fellowship.

NOMENCLATURE

C	volumetric heat capacity
d	absorption penetration depth
d_{th}	thermal penetration depth
d_{eff}	heat-affected-zone depth (HAZ)
D_b	diameter of the input laser beam
E	electric field
E_p	laser pulse energy
G	electron-lattice coupling coefficient
F	laser fluence
k	thermal conductivity
L	film thickness
T	temperature
q	heat flux
r	radial coordinate
R	surface reflectivity
r_s	laser spot radius (1/e position)
t_p	FWHM laser pulse duration (= 83 fs)
t_{peak}	time instant of maximum incident energy density
z	depth coordinate
Z	thickness of metal film
z_s	optical penetration depth
S	volumetric laser heat source

Greek

$\theta_e = T_e/T_F$	normalized electron temperature
$\theta_l = T_l/T_F$	normalized lattice temperature
λ	wavelength
τ	relaxation time

Subscripts

e	electron system
l	lattice system
r	radial coordinate
z	depth coordinate
F	Fermi temperature

REFERENCES

Ash, E. A., Nicholls, G, 1972, Super-resolution Aperture Scanning Microscope, *Nature,* **237**:510-512

Bachelot, R., H'Dhili, F., Barchiesi, D., Lerondel, G., Fikri, R., Royer, P., Landraud, N., Peretti, J., Chaput, F., Lampel, G., Boilot, J., and Lahlil, K., 2003, Apertureless Near-Field Optical Microscopy: A Study of the Local Tip Field Enhancement using Photosensitive Azobenzene-containing Films, *Journal of Applied Physics*, **94**:2060-2072.

Bäuerle, D., 1998, Laser-Induced Fabrication and Processing of Semiconductors: Recent Developments, *Physica Status Solidus (a)*, **166**:543-554.

Bethe, H. A., 1944, Theory of Diffraction by Small Holes, *Physical Review*, **66**:163-182.

Betzig, E., Trautman, J. K., Harris, T. D., Weiner, J.S., and Kostelak, R.L., 1991, Breaking the Diffraction Barrier: Optical Microscopy at Nanometric Scale, *Science*, **251**:1468-1470.

Betzig, E., and Trautman, J. K., 1992, Near-Field Optics: Microscopy, Spectroscopy and Surface Modification Beyond the Diffraction Limit, *Science*, **257**:189-195.

Boneberg, J., Munzer, H. J., Tresp, M., Ochmann, M., and Leiderer, P., 1998, The Mechanism of Nanostructuring upon Nanosecond Laser Irradiation of a STM Tip, *Applied Physics A,* **67**:381-384.

Chen, G., and Hui, P., 1999, Thermal Conductivities of Evaporated Gold Films on Silicon and Glass, *Applied Physics Letters*, **74**:2942-2944.

Chen, J. K., Latham, W. P., and Beraun, J. E., 2002, Axisymmetric Modeling of Femtosecond-pulse Laser Heating on Metal Films, *Numerical Heat Transfer B,* **42**:1-17.

Chimmalgi, A., Choi, T. -Y., Grigoropoulos, C. P., and Komvopoulos, K., 2003, Femtosecond Laser Apertureless Near-Field Nanomachining of Metals Assisted by Scanning Probe Microscopy, *Applied Physics Letters*, **82**:1146-1148.

Chimmalgi, A., Grigoropoulos, C. P., and Komvopoulos, K., 2005a, Surface Nanostructuring by Nano-/femtosecond Laser-assisted Scanning Force Microscopy, *Journal of Applied Physics*, **97**:104319(1)-104319(12).

Chimmalgi, A., Hwang, D.J., and Grigoropoulos, C.P., 2005b, "Nanoscale Rapid Melting and Crystallization of Semiconductor Thin Films," to appear in *Nanoletters*, published on-line at http://pubs.acs.org/journals/nalefd/.

Dickmann, K., Jersch, J., and Demming, F., 1997, Focusing of Laser Radiation in the Near-field of a Tip (FOLANT) for applications in Nanostructuring, *Surface and Interface Analysis*, **25**:500-504.

Ebbesen, T. W., Lezec, H. J., Ghaemi, H. F., Thio, T., and Wolff, P. A., 1998, Extraordinary Optical Transmission through Sub-wavelength Hole Arrays, *Nature*, **391**:667-669.

Gilgen, H. H., Cacouris, T., Shaw, P. S., Krchnavek R. R., and Osgood, R.M., 1987, Direct Writing of Metal Conductors with near-UV Light, *Appl. Phys. B*, **42**:55-66.

Gorbunov, A. A., and Pompe, W., 1994, Thin Film Nanoprocessing by Laser/STM Combination, *Physica Status Solidus A*, **145**:333-338.

Hecht, B., Sick, B., Wild, U. P., Deckert, V., Zenobi, R., Martin, O. J. F., and Pohl, D. W. 2000, Scanning Near-Field Optical Microscopy with Aperture Probes: Fundamentals and Applications, *Journal of Chemical Physics*, **112**:7761-7774.

Huang, S. M., Hong, M. H., Lu, Y. F., Lukyanchuk, B. S., Song, W. D., and Chong, T. C., 2002, Pulsed-Laser Assisted Nanopatterning of Metallic Layers combined with Atomic Force Microscopy, *Journal of Applied Physics*, **91**:3268-3274.

Huber, R., Koch, M., and Feldmann, J., 1998, Laser-induced Thermal Expansion of a Scanning Tunneling Microscope Tip Measured with an Atomic Force Microscope Cantilever, *Applied Physics Letters*, **73**:2521-2523.

Jersch, J., and Dickmann, K., 1996, Nanostructure Fabrication using Laser Field Enhancement in the Near Field of a Scanning Tunneling Microscope Tip, *Applied Physics Letters,* **68**:868-870.

Jersch, J., Demming, F., and Dickmann, K., 1997, Nanostructuring with Laser Radiation in the Nearfield of a Tip from a Scanning Force Microscope, *Applied Physics A*, **64**:29-32.

Kawata, Y., Xu, C., and Denk, W., 1999, Feasibility of Molecular-resolution Fluorescence Near-field Microscopy using Multi-photon Absorption and Field Enhancement near a Sharp Tip, *Journal of Applied Physics*, **85**:1294-1301.

Lee, M., Moon, S., Hatano, M., Suzuki, K., and Grigoropoulos, C. P., 2000, Relationship between Fluence Gradient and Lateral Grain Growth in Spatially Controlled Excimer Laser Crystallization of Amorphous Silicon Films, *Journal of Applied Physics*, **88**:4994-4999.

Lee M., Moon S., and, Grigoropoulos C.P., 2001, In Situ Visualization of Interface Dynamics during the Double Laser Recrystallization of Amorphous Silicon Thin Films, *J. Cryst. Growth*, **226**:8-10.

Lewis, A., Isaacson, M., Harootunian, A., and Muray, A., 1984, Development of a 500 Å Spatial Resolution Light Microscope I. Light is Efficiently Transmitted through lambda /16 Diameter Apertures, *Ultramicroscopy,* **13**:227-231.

Lieberman, K., Shani, Y., Melnik, I., Yoffe, S., and Sharon, Y., 1999, Near-field optical photomask repair with a femtosecond laser, *J. of Microscopy*, **194**:537-541.

Lu, Y. F., Mai, Z. H., Qiu, G., and Chim, W. K., 1999, Laser-induced Nano-oxidation on Hydrogen-Passivated Ge (100) Surfaces under a Scanning Tunneling Microscope Tip, *Applied Physics, Letters,* **75**:2359-2361.

Lu, Y. F., Hu, B., Mai, Z. H., Wang, W. J., Chim, W. K., and Chong, T. C., 2001, Laser-Scanning Probe Microscope based Nanoprocessing of Electronics Materials, *Japanese Journal of Applied Physics,* **40**:4395-4398.

Nolte, S., Chichkov, B. N., Welling, H., Shani, Y., Lieberman, K., and Terkel, H., 1999, Nanostructuring with Spatially Localized Femtosecond Laser Pulses, *Opt. Lett.*, **24**:914-916.

Ohtsu, M., 1998, *Near-Field Nano/Atom Optics and Technology*, Springer-Verlag, Tokyo, Japan.

Pistor, T. V., 2001, *Electromagnetic Simulation and Modeling with Applications in Lithography*, Memorandum No. UCB/ERL M01/19.

Pohl, D. W., Denk, W., and Lanz, M., 1984, Optical Stethoscopy: Image Recording with Resolution $\lambda/20$, *Applied Physics Letters*, **44**:651-653.

Quate, C. F., 1997, Scanning Probes as a Lithography Tool for Nanostructures, *Surface Science,* **386**: 259-264.

Sanchez, J., Novotny, L., and Xie, X. S., 1999, Near-field Fluorescence Microscopy based on Two-Photon Excitation with Metal Tips, *Physical Review Letters*, **82**:4014-4017.

Smith, A. N., Hostetler, J. L., and Norris, P. M., 1999, Nonequilibrium Heating in Metal Films: An Analytical and Numerical Analysis, *Numerical Heat Transfer A,* **35**:859-873.

Stöckle, R., Setz, P., Deckert, V., Lippert, T., Wokaun, A., and Zenobi, R., 2001, Nanoscale atmospheric pressure laser ablation-mass spectrometry, *Anal. Chem.*, **73**:1399-1402.

Stolk P. A., Polman A., and, Sinke W.C., 1993, Experimental Test of Kinetic Theories for Heterogeneous Freezing in Silicon, *Phys. Rev. B*, **47**:5-13.

Sun, J., and Longtin, J. P., 2001, Inert gas beam delivery for ultrafast laser micromachining at ambient pressure, *J. Appl. Phys.*, **89**:8219-8224.

Synge, E. H., 1928, A Suggested Method for Extending Microscopic Resolution into the Ultra-microscopic Region, *Phil. Mag.*, **6**:356-362.

Ukraintsev, V. A., and Yates, J. T., 1996, Nanosecond Laser Induced Single Atom Deposition with Nanometer Spatial Resolution using a STM, *Journal of Applied Physics*, **80**:2561-2571.

Wanke, M. C, Lehmann, O., Muller, K., Qingzhe W., and Stuke, M., 1997, Laser Rapid Prototyping of Photonic Band-gap Microstructures, *Science*, **275**:1284-1286.

Wegscheider, S., Kirsch, A., Mlynek, J., and Krausch, G., 1995, Scanning Near-field Optical Lithography, *Thin Solid Films*, **264**:264-267.

Yamada N., Ohno E., Nishiuchi K., Akahira N., and, Takao M., 1991, Rapid-phase Transitions of GeTe-Sb2 Te3 Pseudobinary Amorphous Thin-films for an Optical Disk Memory, *J. Appl. Phys.*, **69**:2849-2856.

Chapter 20

SOFT LASER DESORPTION IONIZATION – MALDI, DIOS AND NANOSTRUCTURES

Akos Vertes
Department of Chemistry, Institute for Proteomics Technology and Applications, The George Washington University, Washington DC 20052

1. INTRODUCTION

In 1948, Arne Tiselius won the Nobel Prize in chemistry for the discovery of electrophoresis and its application to analyze large molecules, specifically complex serum proteins. During the following decades, gel electrophoresis and its derivative techniques became a central method for molecular biology. They allowed for the identification of molecular mass with an accuracy of ~1 kDa, i.e., a thousand atomic mass units. Although this method enabled the separation of thousands of proteins from biological samples, the determination of the actual molecular weight, and with it finding the true identity of many substances, remained unachievable.

When it came to accurate mass determination of small molecules, mass spectrometry was the method of choice. However, for the accurate mass analysis of large molecules, their low volatility presented a fundamental obstacle. Typically, the larger the molecule, the harder it is to transfer it into the gas phase without degradation. Depositing the energy necessary for volatilization initiates other competing processes that elevate the internal energy of the adsorbate and results in its fragmentation. This competition between evaporation (or desorption) and fragmentation was recognized early on and the method of rapid heating was proposed to minimize the latter (Beuhler, et al., 1974). Lasers with submicrosecond pulse length were appealing tools for rapid heating but the initial results indicated limitations with respect to the ultimate size of the biomolecules ($m/z < 1{,}500$) that could be successfully analyzed (Posthumus ,et al., 1978). The goal of probing large molecules with laser desorption mass spectrometry remained elusive for a decade.

After a frustrating period of near misses in the early 1980's, in 1988 the quest to desorb and ionize large peptides and proteins using pulsed laser radiation succeeded (Tanaka, et al., 1988; Karas and Hillenkamp, 1988). As a result, during the past decade soft laser desorption ionization (SLDI) methods have become indispensable in the mass spectrometric analysis of biomolecules. With the analytical needs in the biomedical field—most notably in genomics, proteomics and metabolomics—multiplying, there is a continuous impetus to enhance the existing methods and explore new ones.

The first, and thus far most successful, SLDI technique is matrix-assisted laser desorption ionization (MALDI). In this method, the sample is mixed with an organic matrix that efficiently absorbs the laser energy. Upon exposure to pulsed laser radiation, the ensuing phase transition of the matrix volatilizes the large molecules. A small fraction of these molecules are also ionized enabling the mass spectrometric determination of their accurate mass. The combination of MALDI with mass spectrometry (MALDI-MS) is an enabling technology for the emerging discipline of proteomics. For the discovery of SLDI, in 2002 Koichi Tanaka became a corecipient of the Nobel Prize in chemistry.

The improvement in the accuracy of molecular mass determination brought about by MALDI was monumental. While for 10 kDa proteins the mass accuracy of gel electrophoresis was at best ~1%, MALDI routinely provided ~10 ppm. This thousand-fold improvement enabled the identification of post-translational modifications, a feature responsible for determining the biological function of proteins.

The utility of MALDI-MS for the analysis of large molecules inspired related efforts to use SLDI for the investigation of low molecular mass compounds. Due to the overwhelming matrix signal, MALDI itself had limited capabilities in the low mass region. In 1999, based on a nanoporous silicon substrate Siuzdak and coworkers introduced an efficient SLDI method termed desorption ionization on silicon (DIOS) (Wei, et al., 1999).

It was assumed that the substrate used in this technique derived its utility from the nanoporous structure. This hypothetical link between nanostructured surfaces and their enhanced desorption ionization properties initiated a host of laser desorption studies on nanomaterials. The considered systems included quasi-one-dimensional structures, such as nanopores (Wei, et al., 1999), nanotubes (Xu, et al., 2003), nanowires (Go, et al., 2005) and nanogrooves (Okuno, et al., 2005), nanoparticles (McLean, et al., 2005) and their aggregate phases, e.g., nanoparticle films, (Chen, et al., 2005; Luo, et al., 2005b) and, more recently, ordered arrays of silicon nanocavities (Finkel, et al., 2005) and microcolumns (Chen and Vertes, 2005; Chen, et al., 2005). Ion production from all of these surfaces was demonstrated for small to midsize molecules, including peptides, up to ~6 kDa.

There are four processes that determine the abundance and properties of the generated ions in SLDI: energy deposition, phase change, ionization and energy redistribution. As all the events are driven by the energy of the laser pulse, energy deposition, i.e., coupling of the laser energy into the target is an important factor. The efficiencies of phase change and ionization control the ion yields, whereas the increase in the internal energy of the produced ions due to energy redistribution promotes their fragmentation. The objective of this chapter is to explore the contribution of these processes to MALDI and to the other emerging SLDI methods. While there is a growing body of knowledge on the mechanistic aspects of MALDI, the nature and role of the decisive factors in SLDI from nanostructures is in a nascent state. Beyond the straightforward surface area effects, the potential role of phenomena specific to nanomaterials, such as quantum confinement and nanoscopic liquid confinement, as well as near-field effects, is considered.

2. MATRIX-ASSISTED LASER DESORPTION IONIZATION

2.1 Energy deposition – nonlinear effects

It was clear early on that matching the laser pulse parameters (wavelength and fluence, and to a certain degree pulse duration) with the optical and thermal properties of the matrix was critically important to the success of the MALDI experiment. Matrix materials were selected, in part, based on their linear optical properties at the wavelength of the laser radiation. Solid phase optical absorption coefficients of the most successful matrix materials in the near-ultraviolet (UV) are in the 10^4 to 10^5 cm^{-1} range (Allwood, et al., 1996, 1997). For example, the linear (low fluence) absorption coefficient of 2,5-dihydroxybenzoic acid (2,5-DHB) at the nitrogen laser wavelength of 337 nm is $\alpha = 7.95\times10^4$ cm^{-1}. Similarly, water ice is a MALDI matrix in the infrared (IR) at ~2.94 μm, where the OH vibration band gives rise to efficient linear absorption, $\alpha = 1.2\times10^4$ cm^{-1}.

Nonlinear absorption effects are generally considered negligible in MALDI (Dreisewerd, 2003). While this might be true for UV excitation at the ion generation threshold, at other wavelengths and/or higher laser fluences nonlinear absorption can be expected. In practice, experiments are not always carried out at the threshold fluence. A possible cause of nonlinearity is the high linear absorption coefficient of the matrix, as it helps to populate the excited states. Thus, if the excited state lifetime is sufficiently long the matrix material can behave as a saturable absorber and with increasing fluence become more and more transparent.

The nonlinear optical properties of the UV matrix materials are not known. There is data, however, for matrixes in the mid-IR that supports the hypothesis that nonlinear absorption plays a role. Careful optical absorption measurements as a function of laser fluence indicate that at 2.94 μm wavelength the absorption coefficient of water declines as the energy density of the material increases (Shori, et al., 2001). This effect can be explained by the disruption of the hydrogen bond network due to heating by the laser pulse (Bencsura and Vertes, 1995). This weakening of the non-covalent water structure gives rise to a blue shift of the absorption resonance and, thus, to reduced absorption at 2.94 μm. For example, the transmission of a 4.26 μm water layer increases from ~0.5% in the low fluence limit (linear absorption) to 11% at 0.8 J/cm^2 laser fluence, a value only slightly higher than the fluence threshold for IR-MALDI (see Fig. 2 in Shori, et al., 2001).

A similar blue shift was observed in IR-MALDI experiments. For certain matrixes (phloroglucin, triethanolamine, and glycerol), the ionization threshold fluence as a function of wavelength in the 2.7–3.6 μm range showed a minimum that was blue shifted compared to the absorption maximum of the low intensity (linear) optical absorbance (Menzel et al., 2001). In the same study, the nonlinear behavior for glycerol was confirmed through fluence dependent optical absorption measurements.

There are two major consequences of the nonlinear optical response regarding energy deposition. First, the laser light penetrates deeper into the target than what one would expect from the linear theory. Second, the temperature distribution in the target exhibits a maximum at a finite depth beneath the surface. As it is shown in section 2.2, these two factors have important implications for material removal and plume dynamics.

Sample morphology also plays a special role in the efficiency of energy deposition. It is known that upon repeated laser exposure of a polycrystalline MALDI target, the smaller crystallites and/or the thinner crystal features evaporate first. This preferential ablation dramatically accelerates when the thermal diffusion length and the smallest dimension of a crystallite are comparable in size (Sadeghi and Vertes, 1998). In these systems, energy dissipation through heat conduction is limited. This leads to rapidly escalating temperatures, superheating and eventually phase explosion in the entire crystallite. Sample morphology (texture) was also found to influence the laser fluence threshold necessary to produce matrix ions. Studying pressed pellet samples, it was shown that with increasing the pressure used to produce the pellet structural coarsening occurred and, as a result, the threshold fluence increased (Chen and Vertes, 2003). While these morphology effects are significant in MALDI, they are expected to be even more important, perhaps even dominant, in SLDI from nanostructured surfaces.

Current models of MALDI do not incorporate nonlinear absorption and morphology effects. For a broad class of conditions, realistic estimates of target heating cannot be given without accounting for these aspects of energy deposition.

2.2 Surface evaporation vs. phase explosion

In MALDI, the molecules of interest are embedded into the matrix crystals. Volatilization of the analyte relies on a phase transition induced by fast laser heating. Due to the extreme heating rates produced by the short laser pulse (>10^6K/s), the surface layers of the matrix become superheated. Once the spinodal decomposition temperature is reached, these layers undergo phase explosion. Before and after the phase explosion, vigorous surface evaporation takes place. The relative importance of phase explosion and surface evaporation depends on laser fluence, pulse length, and matrix material parameters.

Experimental evidence for the presence of both of these processes was found in ion and neutral velocity measurements in vacuum (e.g., Juhasz, et al., 1997), and for atmospheric pressure MALDI (AP-MALDI) conditions, in shock front velocity measurements (Apitz and Vogel, 2005). The velocities in vacuum are typically in the few hundred m/s range with variations detected for different matrixes and analytes.

To follow the plume dynamics stemming from surface evaporation in vacuum, a fluid dynamics model was used (Vertes, et al., 1993). The model enabled the description of density, velocity and temperature distributions in and above the target for a period of 100 ns. It gave good predictions for neutral velocities and for the desorbed flux as a function of laser irradiance.

In the fast imaging studies of Apitz and Vogel (2005), the atmospheric pressure IR-laser ablation of water (70 ns Er:YAG laser pulse at 5.4 J/cm^2 fluence) clearly exhibited two phases: surface evaporation followed by a massive phase explosion. Initially, the rapid evaporation at the surface created a shock front that propagated away from the target. The advancing front was further accelerated by a delayed phase explosion. Quantitative analysis of the shock front position vs. time data in Fig. 9 of their publication is presented below to reveal this time delay, t_{pe}, and calculate the energy expended on surface evaporation, E_{se}, and on phase explosion, E_{pe}.

The similarity model, originally developed by Taylor for the description of nuclear explosions (Taylor, 1950), describes the time dependence of the shock front position, *R(t)*, as:

$$R(t) = S(\gamma)\rho_0^{-\frac{1}{5}} E^{\frac{1}{5}} t^{\frac{2}{5}}, \qquad (1)$$

where $S(\gamma)$ is a weak function of the specific heat ratio, γ, ρ_0 is the background gas density and *E* is the energy of the explosion. Replotting the Apitz–Vogel data in the $t^{2/5}$ variable clearly demarcates the two phases (see Fig. 1).

Using the explosion analogy for the data in Fig. 1, one might presume that the two phases of the plume expansion correspond to the compounded effect of two consecutive "explosions." The first one corresponds to the abrupt onset of vigorous surface evaporation, whereas the second describes the phase explosion.

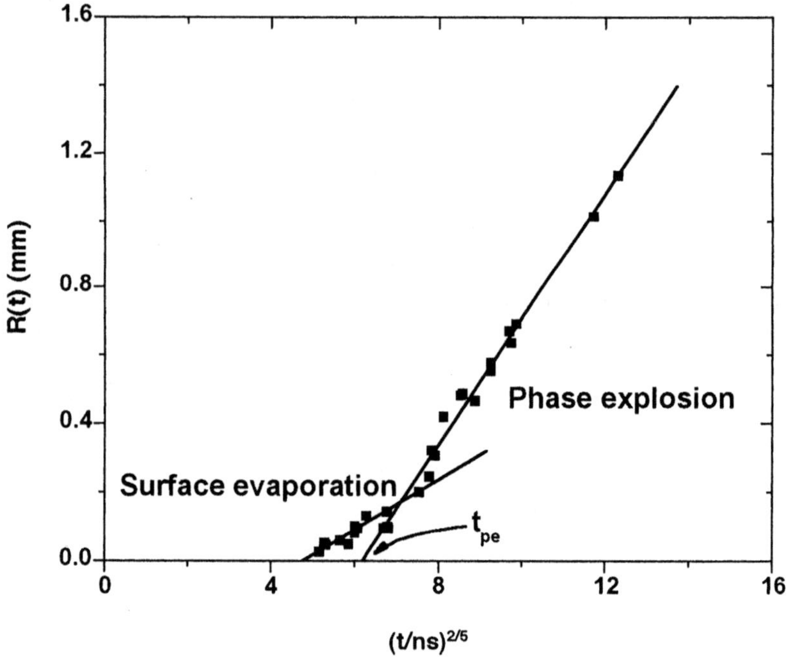

Figure 1. Shock front position as a function of time for Er:YAG laser ablation of water. Experimental data for 5.4 J/cm^2 fluence from Apitz and Vogel (2005) (■) replotted in $t^{2/5}$ variable. Solid lines indicate our linear fit using Eq. (1) for the surface evaporation and phase explosion.

Indeed, it is shown in Fig. 1 that Eq. (1) gives excellent fits to the data with a slope of 0.072 for surface evaporation, a slope of 0.186 for phase explosion, and with 0.89 and 0.99 regression coefficients, respectively.

From Eq. (1) and from the ratio of the slopes for the two phases of the expansion, q, one can determine the ratio of energies for the surface evaporation, E_{se}, and phase explosion, E_{pe}, processes:

$$q = \left(1 + \frac{E_{pe}}{E_{se}}\right)^{\frac{1}{5}}. \qquad (2)$$

Because the laser pulse energy is primarily spent in the surface and volume processes, $E_l = E_{se} + E_{pe}$. From the total fluence and the slope ratio in Eq. (2), one can calculate the fluence fractions used in the two phases. From the slopes of the fitted curves in Fig. 1, $q = 2.57$ and the ratio of the two energies is, $E_{pe}/E_{se} = 111$. Thus, one can conclude that the overwhelming majority of the laser pulse energy is utilized in the volume process. In particular, only 0.05 J/cm^2 of the 5.4 J/cm^2 laser fluence drives the surface evaporation, whereas the remaining 5.35 J/cm^2 is used for the volume process. Extrapolating the second linear region back to $R(t_{pe}) = 0$ yields $t_{pe} = 94\ ns$, i.e., compared to the start of the laser pulse and even to the onset of the surface process the phase explosion is substantially delayed. This approach to extract the phase explosion delay and energy is valuable because there is no other method (experimental or theoretical) available to determine them.

Molecular dynamics simulations of the desorption process gave insight into the first 1 ns of the desorption process with exceptional detail. Early results revealed the fast disruption of hydrogen bond network in matrix solids due to laser heating, described the desorption of the analyte molecule through surface evaporation and followed the matrix guest energy transfer in MALDI (Bencsura and Vertes, 1995; Bencsura et al., 1997; Wu et al., 1998). Further investigations based on a simplified model potential that enabled the modeling of larger systems for a longer period of time found a transition from surface evaporation to phase explosion at elevated fluence values (Zhigilei et al., 1997). More detailed work indicated that longer laser pulses created thermal confinement conditions that resulted in a phase explosion, whereas shorter pulses induced stress confinement and photomechanical disintegration of the surface layers (Zhigilei and Garrison, 2000).

It is very likely that under typical MALDI conditions both surface evaporation and phase explosion contribute to the plume formation. Similar to the above analyzed water ablation scenario, the deposited laser energy initiates a rapid surface evaporation and, in case of nonlinear absorbers, creates a superheated subsurface layer. After a certain incubation period, t_{pe}, this metastable phase decomposes through phase explosion.

2.3 Ionization

Depending on the chemical nature of the matrix and the analyte, the ionization in MALDI can follow several paths. Often one can separate the formation of matrix ions (primary ions) in the condensed phase from analyte ionization (secondary ions) in the plume (Zenobi and Knochenmuss, 1998). Among the various mechanisms suggested for primary ion formation, perhaps the best supported is the one based on exciton pooling (Knochenmuss, 2002). This model is unique in the sense that it predicts several quantitative aspects of the MALDI process, such as the laser fluence threshold, the fluorescence yield for the most studied matrix, 2,5-DHB, and the ion yield. A refined version of the model (Knochenmuss, 2003) incorporates secondary ion formation through ion-molecule reactions in the plume and provides quantitative predictions for the ion suppression effects, for analyte ion yields and for the results of fast pump-probe experiments (Knochenmuss and Vertes, 2000).

In order to elucidate the relationship between matrix ionization and fragmentation, pumping rate studies were conducted using nanosecond and picosecond laser excitation (Chen and Vertes, 2003). Fig. 2 shows the dramatic difference in the fluence dependence of the molecular ion yields of 2,5-DHB for nitrogen laser (4 ns @ 337 nm) and for mode locked 3×ω Nd:YAG laser (22 ps @ 355 nm). This difference can be rationalized by the ladder climbing and ladder switching pathways outlined in the scheme below:

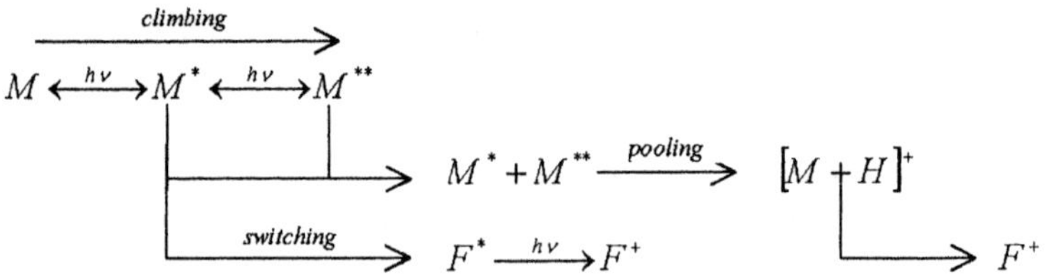

In the ladder climbing scheme, the singlet excited state, M^*, of the matrix molecule, M, is further photoexcited to M^{**}. The created mobile excitons combine to produce the protonated matrix species, $[M + H]^+$. Depending on its internal energy, this ion can thermalize or decompose into an F^+ fragment ion. In the ladder switching mode, the excited matrix molecules can directly break up into excited fragments, F^*, and fragment ions, F^+.

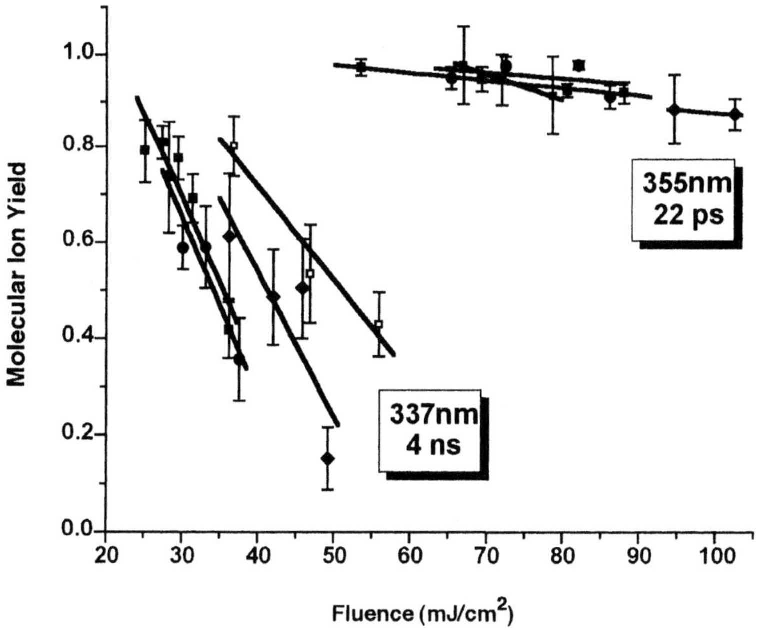

Figure 2. Molecular ion yields of 2,5-DHB pellets made with 100 MPa (■), 200 MPa (●) and 300 MPa (◆) pressure, and of dried droplet samples (□) measured with ns and with ps laser. Reprinted with permission from (Chen and Vertes, 2003). Copyright 2003 American Chemical Society.

Although both channels are active in MALDI, for ps excitation ladder climbing is dominant, whereas for ns laser pulses the relative contribution of ladder switching is larger. This mechanism is an extension of the photophysical model that can explain the large difference in molecular ion yields in Fig. 2. It also incorporates matrix fragmentation, a ubiquitous process in MALDI that clearly cannot be neglected.

The production of protonated analyte ions, $[A + H]^+$, and their fragments, AF^+, competes with adduct formation:

$$[M+H]^+ + A \rightarrow [A+H]^+ + M$$
$$[A+H]^+ \rightarrow AF^+$$

$$F^+ + A \rightarrow [A+F]^+$$

The analyte ions are produced with significant internal energy resulting in metastable decay and fragmentation. These processes are further discussed in the following section.

2.4 Internal energy and fragmentation

In SLDI, there are two factors that determine the energy transfer and, correspondingly, the internal energy content of the analyte ions. The first one is associated with the desorption process (e.g., the phase transition of the matrix), whereas the second one is linked to the thermochemistry of ion formation.

In MALDI, the first component is controlled by the details of the phase transition (surface evaporation vs. phase explosion), the thermodynamic properties of the matrix (melting, boiling, spinodal decomposition and critical temperature, heats of melting and vaporization, and heat capacities in the different phases) and the transport properties (thermal conductivity) of the target (Luo, et al., 2002). Additional changes in the internal energy of ions can be induced by in-plume collisions with neutrals in the ion source (Gabelica, et al., 2004). This latter process occurs on a longer timescale and it is dependent on the characteristics of the ion source.

For protonated analytes, the energy transfer associated with the ionization primarily scales with the difference in gas phase basicities between matrix and analyte. A larger difference in gas phase basicities typically means high-

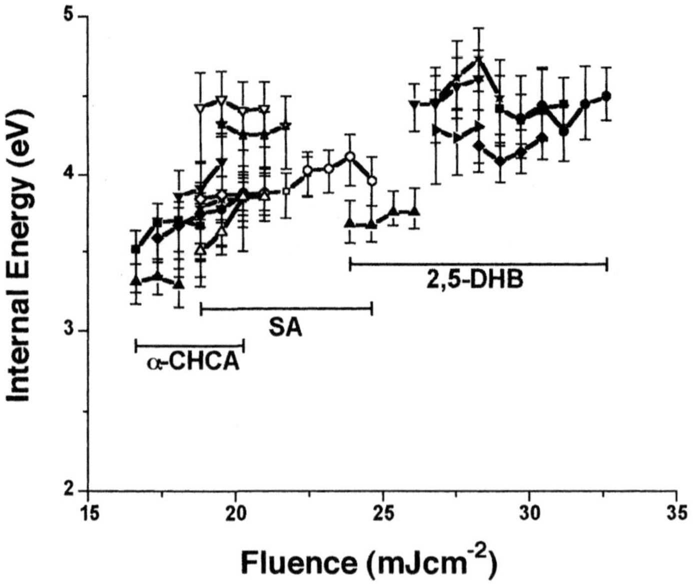

Figure 3. Fluence dependence of internal energies of preformed ions used as thermometer molecules in MALDI. Three different matrixes produced gas phase ions with internal energies increasing as α-CHCA < SA < 2,5-DHB. Data is from (Luo, et al., 2002). The horizontal bars indicate the utilized fluence ranges for the three matrixes. Different symbols correspond to the various thermometer ions. The symbols for SA matrix are hollow for better distinction.

er internal energy for the analyte ions (Stevenson, et al., 2000; Konn, et al., 2005).

Determining the internal energy transfer associated with the desorption process alone requires that the ionization step is eliminated. This was achieved by using the cations of benzyl-substituted benzylpyridinium salts as thermometer ions in MALDI (Luo et al., 2002). The ions of these salts are already preformed in the solid phase. Following their fragmentation in the MALDI process and measuring the survival yields for three matrixes (α-cyano-4-hydroxycinnamic acid (CHCA), 3,5-dimethoxy-4-hydroxycinnamic acid or sinapinic acid (SA) and 2,5-DHB) enabled the determination of the associated internal energies (see Fig. 3). The conversion from survival yields to internal energy was based on the Rice-Ramsperger-Kassel-Marcus (RRKM) theory of unimolecular decomposition.

The average internal energy values found for the different matrixes are 3.69 ± 0.21 eV, 4.04 ± 0.27 eV, and 4.30 ± 0.29 eV for CHCA, SA, and 2,5-DHB, respectively. Compared to the previous qualitative assignment based on fragmentation patterns of small peptides (Stimson, et al., 1997), the CHCA<SA<2,5-DHB sequence of internal energies found for the thermometer ions shows the opposite trend.

This apparent contradiction is the consequence of the different processes involved. While the peptides in the earlier study undergo both desorption and protonation, and thus reflect the internal energy transfer associated with both, the thermometer ions only experience desorption. The opposite sequence of energy transfer from the same set of matrixes to neutral and ionic analytes highlights the comparable role desorption and ionization play in determining internal energy transfer. The internal energy content of the produced ions is the primary factor that determines their fragmentation and, in turn, our ability to infer structural information from SLDI mass spectra.

3. DESORPTION IONIZATION ON SILICON

The mass analysis of large molecules has been greatly facilitated by MALDI. For small molecules (<700 Da), however, the spectral interferences created by the ionic matrix species in this region present a key obstacle. Other limitations in MALDI stem from the solubility and cocrystallization requirements introduced by the sample preparation. Based on the SLDI approach, other methods are developed to alleviate these difficulties.

In the past fifteen years, the photoluminescent properties of porous silicon (pSi) fuelled a meticulous exploration of this low dimensional semiconductor. The application of pSi as an efficient matrix-free SLDI surface was discovered in 1999 (Wei, et al., 1999). This new approach was not only appealing because it did not produce interferences in the low mass region, but it also had the potential to be integrated into lab-on-a-chip devices.

The nanoporous structure of pSi is produced by photoelectrochemical etching of silicon wafers. For the typical n-type DIOS surfaces, the pore diameter distribution spans from nano- (<2 nm) to meso-scale (2-50 nm) and the porosity is 30 to 40% (Shen, et al., 2001). The pores can be several μm long; they are, on the average, perpendicular to the surface and are not interconnected. During the etching process, various surface species, such as hydrides, oxyhydrides, and oxides are formed (Cullis, et al., 1997).

Energy deposition in pSi exhibits some unique characteristics. The linear optical absorption coefficient of bulk silicon at the nitrogen laser wavelength of 337 nm is 10^6 cm^{-1} and between 300 and 972 K it is independent of temperature (Jellison and Modine, 1982). For pSi, at the same wavelength an effective absorption coefficient of up to 10^5 cm^{-1} can be observed (Kovalev, et al., 1996). The energy deposition, therefore, is very efficient in both systems. Due to limited dissipation through heat conduction in the thin pore walls (quasi-one-dimensional system), however, the temperature on the wall surfaces of pSi rises more rapidly and to a higher value than at the surface of bulk silicon.

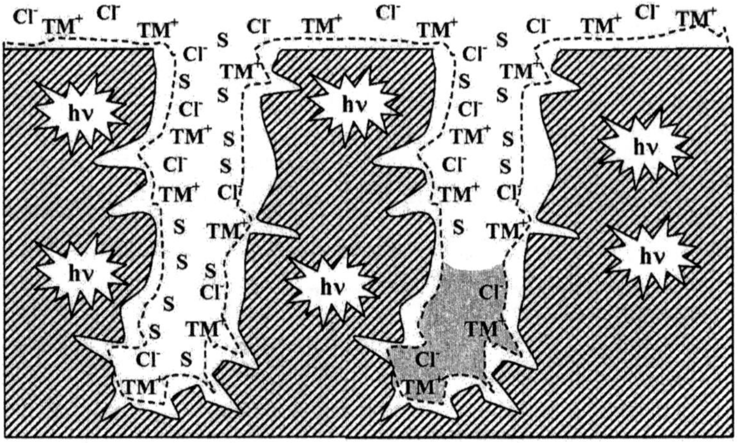

Figure 4. Adsorption of TM ions (TM^+ and Cl^-) and solvent molecules (S) in nanoporous silicon. Pore walls are sufficiently thin to induce thermal confinement and consequently high wall temperatures. **Left pore** only contains various adsorbates ("dry" desorption), whereas **right pore** shows trapped solvent (gray area) formed as a result of capillary condensation or retained due to inefficient pumping through nanopore ("wet" desorption). Reprinted with permission from (Luo, et al., 2005a). Copyright 2005 American Chemical Society.

This confinement of the deposited energy in the pore walls means that in pSi a particular surface temperature can be achieved at significantly lower laser fluence than in bulk silicon. Experimental results that are specific only to the desorption phase of the DIOS process are not yet available. Some conclusions on possible desorption mechanisms can be drawn from parametric studies and from ion yield measurements.

Investigations of the effect of laser wavelength, crystallographic orientation, preparation conditions, and other physical properties of the pSi (pore morphology, and thermal and optical properties) on ion generation were conducted to optimize the desorption conditions (Kruse, et al., 2001). There was no significant difference between DIOS mass spectra obtained at 266, 355 and 532 nm indicating much less sensitivity to wavelength than in MALDI. These experiments also proved that the porosity and wetting properties of the pSi substrate played a major role in determining DIOS ion yields but some of the other unique optical properties of the surface, i.e., photoluminescence, had no effect. Residual gases were identified as important factors in DIOS. When acetic acid or water vapor was introduced at a partial pressure of 2×10^{-6} Torr, it resulted in a 10 to 40-fold ion yield enhancement (Alimpiev, et al., 2001).

This observation highlights the importance of the trapped solvent and residual gas accumulated in the pores. Fig. 4 depicts two possible scenarios for the distribution of adsorbates in the pores. The left pore in Fig. 4 shows a "dry" environment, where the solvent and adsorbate molecules (in this case, thermometer cations and anions) compete for the adsorption sites. The right pore is "wet" due to the liquid solvent (gray area) either produced by capillary condensation or retained from sample preparation due to inefficient pumping through the nanopore.

The desorption of thermometer ions can proceed through either the dry or the wet pathway. As a result of energy deposition by the laser pulse, in the dry pores the rapid heating of the pore walls results in desorption of the adsorbates. In the confined space of the pores, a plume develops that undergoes one-dimensional expansion toward the mouth of the pore. During the expansion, the pressure is kept relatively high due to the confined space. In the wet pores the rapidly heated trapped solvent reaches the boiling and possibly the spinodal temperature. This gives rise to a high density solvent plume expanding with the released adsorbate.

Ionization of the adsorbed molecules can proceed through charged surface states (e.g., hydrides) or through ion-molecule reactions in the plume. As mentioned above, protons from the dissociation of water or acidic compounds can play a major role. The plume reactions, including

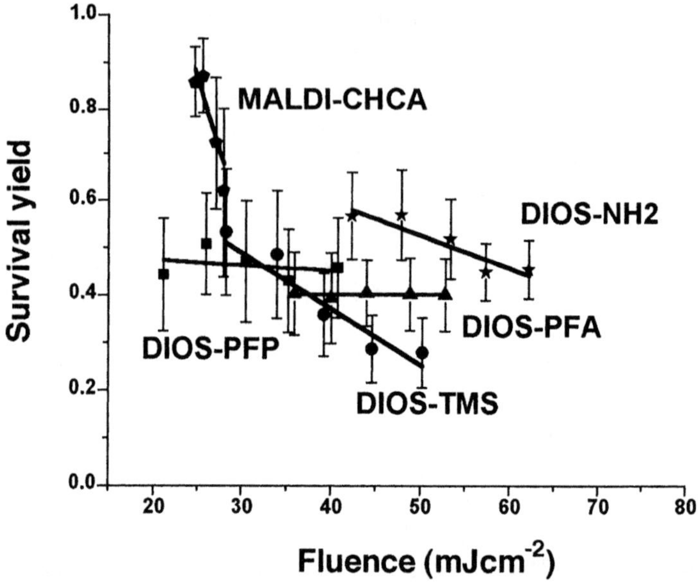

Figure 5. Comparison of survival yields for 3-methoxy-benzylpyridinium thermometer ions desorbed from four different DIOS surfaces with a nitrogen laser. Results for MALDI experiments using CHCA matrix are also included. Reprinted with permission from (Luo et al., 2005a). Copyright 2005 American Chemical Society.

protonation, are facilitated by the high density of the plume sustained by the confined space.

Using the survival yield of thermometer ions as a measure of internal energy, large differences between MALDI and DIOS were observed (Luo, et al., 2005a). Fig. 5 shows the fluence dependence of survival yields of guest ions for MALDI from CHCA matrix and for DIOS, from four derivatized surfaces. Trimethylsilyl- (TMS), amine- (NH2), perfluoroalkyl- (PFA), and perfluorophenyl (PFP) modification of the porous silicon was carried out to gauge the role of the hydrophilic character.

The survival yields in DIOS are uniformly lower than in MALDI with CHCA matrix. This is an indication that, perhaps as a result of plume confinement in the pores, DIOS imparts significantly more internal energy.

Indeed, comparing the dynamics of a quasi-one-dimensional and a three-dimensional expansion, the displacement of the shock front, *R(t)*, as described by the generalized form of Eq. (1) exhibits different kinetics (Sedov, 1993):

$$R(t) = \lambda_d \left(\frac{E}{\rho_1} \right)^{\frac{1}{2+d}} t^{\frac{2}{2+d}} . \quad (3)$$

In this expression, d is the dimension of the expansion, E is the deposited energy, ρ_l is the density of the environment, and λ_d is a dimension-dependent constant. Assuming similar conditions in MALDI and DIOS, the density of the background gas is expected to be higher in the latter due to the long and narrow pores. This results in a slower expansion. Furthermore, comparing the 1D and 3D expressions reveals a crossover in the position of the shock front at time

$$t_{co} = \left(\frac{\lambda_3}{\lambda_1} \right)^{\frac{15}{4}} \left(\frac{\rho_1}{E} \right)^{\frac{1}{2}} . \quad (4)$$

Initially, the radius of the 3D expansion exceeds the displacement in the 1D case, which, in turn, means lower density in the 3D expansion. Thus, t_{co} characterizes the time of confinement in a one-dimensional expansion. This confinement period also brings about faster energy transfer and higher reaction rates in the 1D case.

In Fig. 5, increasing the laser fluence within a narrow fluence range (~5 mJ/cm^2) leads to a sharp decline in the survival yields in MALDI, whereas, even in a wider fluence range (~20 mJ/cm^2), it produces very little change in DIOS. This has important implications for the stability of fragment ion production. While fragmentation in MALDI is strongly fluence dependent and fluctuates from laser shot to laser shot, the production of fragments shows much less variability in DIOS.

It is also apparent in Fig. 5, that the non-polar surface derivatized with PFP requires the lowest fluence, whereas the polar NH2 surface produces ions at significantly higher laser fluence. Considering the ionic nature of the adsorbed species, this behavior is expected. Doubling or even tripling the laser fluence, however, does not result in appreciable difference in the internal energy content of the desorbed ions. This unusual response is specific to DIOS and requires further investigation.

4. OTHER NANOSTRUCTURES

There is a large variety of other nanostructures that facilitate the SLDI of small to moderately sized biomolecules. Among the possible morphologies, quasi-one-dimensional systems such as nanopores in DIOS, nanowires, nanotubes, and nanogrooves stand out. They offer the lowest fluence threshold for ion production (nanowires) and the greatest versatility in terms of analytes. Nanoparticles, including quantum dots and nanoparticle films as well as ordered nanostructures such as nanocavities and silicon microcolumn arrays are being explored for SLDI applications. In the following subsections a few examples of these emerging nanostructures are discussed.

4.1 Nanowires

Semiconductor nanowires are of great interest for applications in nanoelectronics and as biological and chemical sensors. These quasi-one-dimensional systems can be produced by vapor-liquid-solid (VLS) methods using nanometer scale catalyst clusters (Yazawa, et al., 1992).

Gold clusters of 40 nm diameter as catalysts were deposited on silicon wafers and processed at 480°C in SiH_4 environment to produce the silicon nanowires shown in Fig. 6. To achieve the desired polarity, these nanowires were oxidized and silylated.

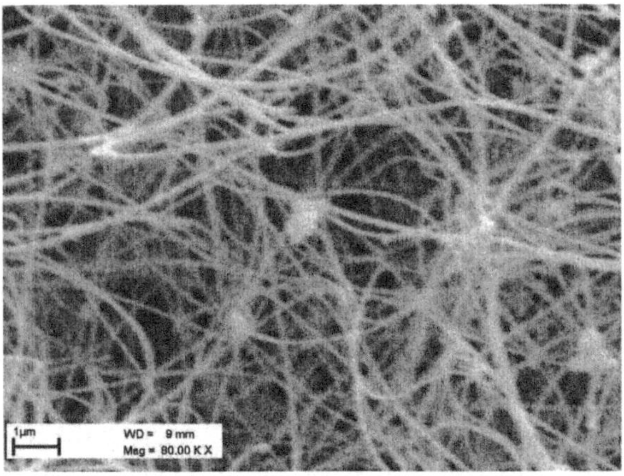

Figure 6. Scanning electron microscope (SEM) image of entangled silicon nanowires used in laser desorption experiments.

The ion yields of the species desorbed from these nanowires were investigated in SLDI mass spectrometry experiments (Go et al., 2005). A wide selection of small molecules and peptides were ionized with good efficiency. The nanowires exhibited significantly lower ionization fluence thresholds than DIOS surfaces or common MALDI matrixes. Correspondingly, the energy transfer during desorption from nanowires was limited. Fig. 7 shows the survival yields for 3-methoxy-benzylpyridinium thermometer cations in SLDI experiments. The fluence threshold for the nanowire surfaces was ~3 mJ/cm^2, whereas for DIOS it was ~21 mJ/cm^2. The lowest MALDI ion generation fluence was observed in CHCA matrix to be at ~25 mJ/cm^2.

The survival yields from PFP-derivatized nanowires were significantly higher than from PFP-DIOS surfaces and comparable to the ions desorbed from CHCA matrix in MALDI. The low fluence required for ion generation in combination with the high thermometer ion survival yields indicates that laser desorption from silicon nanowires is one of the gentlest SLDI methods we know.

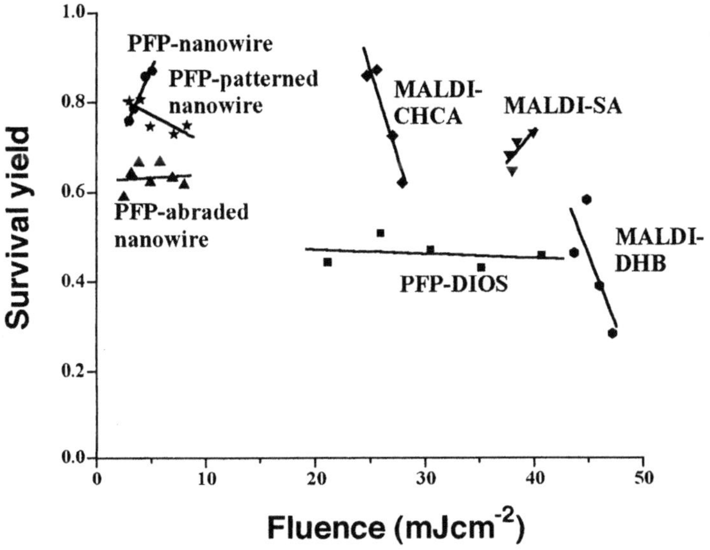

Figure 7. Comparison of survival yields of 3-methoxy-benzylpyridinium thermometer cations desorbed from native, patterned and abraded silicon nanowires, a perfluorophenyl-derivatized DIOS surface and from three different MALDI matrixes. Based on data in (Go et al., 2005). Copyright 2005 American Chemical Society.

4.2 Nanoparticles and nanoparticle films

Nanoparticles were first used in SLDI in 1988 (Tanaka, et al., 1988) but the efficiency of organic matrix materials displaced them for the following decade (Karas and Hillenkamp, 1988). In 1999, particle suspension matrixes were revisited (Schurenberg, et al., 1999) and it was noticed that the ion yields improved with decreasing particle size. Insulator, semiconductor, and metal particles were tested for their efficacy in facilitating ion production (Crecelius, et al., 2002; Turney, et al., 2004).

Recently, size selected gold nanoparticles were used in SLDI experiments (McLean et al., 2005). When the particle size was reduced from 10 nm to 2 nm, a significant improvement in peptide ion yield was observed. At the same time, the abundance of gold cluster ions diminished. This was interpreted as a consequence of quantum confinement in <3 nm gold particles. These results were obtained at relatively high laser pulse energy, 120 μJ/pulse, which raised the possibility that the experiments were carried out in a mixed ablation-desorption regime.

To further explore the laser fluence requirements, gold nanoparticle films were produced with the vertical colloidal deposition (VCD) technique (Diao, et al., 2003). First, 12 nm colloidal gold nanoparticles were produced by solution phase Na_2S reduction of $HAuCl_4$. Then, evaporation-driven VCD was used to produce the nanoparticle films with uniform particle distribution. When these films were used as SLDI substrates, compared to earlier nanoparticle studies (McLean et al., 2005) ion yield measurements on small peptides indicated a significantly reduced threshold at 3 to 4 μJ/pulse (Chen et al., 2005; Luo et al., 2005b). The produced mass spectra were completely free of gold cluster ions. This served as an indication that on the nanoparticle film surfaces ablation did not play a role in ion generation.

The nanoparticle films offer additional advantages over individual particles. Unlike the particles that tend to aggregate and coalesce, the VCD surfaces are very stable and require 30 to 40 times reduced laser pulse energy for ion generation.

4.3 Silicon microcolumn arrays

Silicon microcolumn arrays are very robust structures that spontaneously develop under a wide range of conditions when a silicon wafer is exposed to repeated laser pulses at a certain fluence level. Initially, cone array formation was observed when aluminum oxide (Al_2O_3) target in vacuum was exposed to a few hundred 6 ns laser pulses at 266 nm wavelength and 0.5 J/cm^2 fluence (Rothenberg and Kelly, 1984). In 1998, silicon microcolumn arrays were produced by femtosecond laser processing (Her, et al., 1998). Low resistivity silicon wafers in 500 Torr SF_6 environment were exposed to 500 pulses of 800 nm laser radiation with 100 fs pulse length at 1 J/cm^2 fluence.

At these fluences periodic surface structures seem to develop irrespective of the environmental pressure (from vacuum to atmospheric pressure), laser wavelength (from near IR to near UV), and laser pulse length (from 100 fs to 6 ns). In a pulse length study, however, it was noted that 100 fs pulses gave a column separation of 10 μm and increasing the pulse length to 10 ps resulted in column arrays with 5 μm periodicity (Her et al., 2000). At 250 ps pulse length no columns were observed.

To explore the ps pulse length range, 22 ps pulses from a mode-locked 3×ω Nd:YAG laser (355 nm) were used to process low resistivity silicon wafers (Chen and Vertes, 2005; Chen, et al., 2005). Microcolumn arrays were produced under a wide range of environmental conditions, including ambient air, 10 Torr SF_6 and liquid water (see Fig. 8). The formation of the microcolumn structures required 1000 to 1200 laser pulses with 0.4–1.0 J/cm^2 fluence.

Figure 8. Tilted (45°) SEM view of silicon microcolumn array produced in water with 1000 pulses of a mode locked 3×ω Nd:YAG laser at 355 nm wavelength. The 22 ps pulses delivered 0.4 J/cm^2 fluence.

These results indicate that near UV laser pulses of tens of ps in duration can successfully create silicon microcolumn arrays. Of the tested preparation methods, the aqueous environment produced the columns with the sharpest tips, i.e., with less than 500 nm in diameter.

Silicon microcolumn arrays were used in their native state as desorption surfaces for SLDI of small peptides, including, angiotensin I, substance P, insulin, and their mixtures. In the spectra, up to m/z ~6000 predominantly protonated ions were observed. The submicron tip diameter microcolumn arrays produced in water demonstrated the highest ion yields with a threshold fluence of ~30 mJ/cm^2 that was comparable to values observed in MALDI. Fig. 9 shows the SLDI spectrum of substance P desorbed from a silicon microcolumn array substrate produced in 10-Torr SF_6 gas environment.

Although other silicon nanostructures, such as nanopores and nanowires, are also used to produce peptide ions in SLDI experiments, the silicon microcolumn surfaces are significantly simpler to produce and can be more readily integrated into microchemical chips. Due to the uniform light absorption characteristics of the surface from near UV to mid-IR, it is expected that a wide variety of lasers can be used for SLDI experiments.

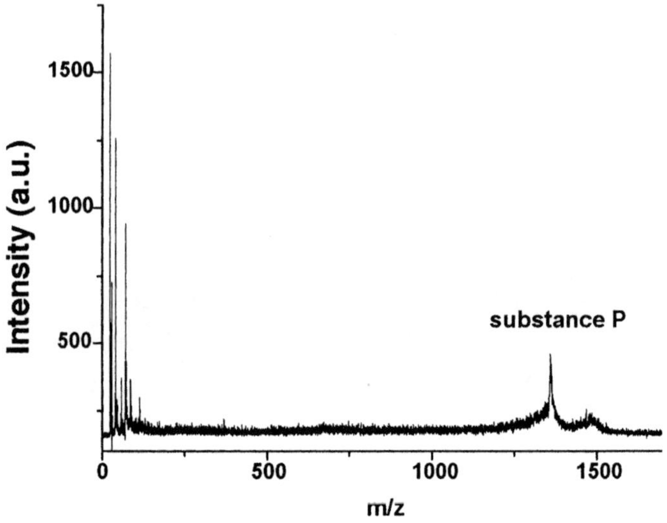

Figure 9. Mass spectrum of substance P with SLDI at a laser fluence of ~75 mJ/cm^2. Silicon microcolumn array substrate was produced in 10 Torr SF_6 gas environment using 1200 pulses of 1 J/cm^2.

It is unclear if the SLDI experiments reported here for silicon microcolumn arrays utilize the specific properties of the array. The effect of the periodicity, the column length and the incidence angle of the laser radiation has not yet been studied. Exploring the effect of these parameters is necessary to improve our understanding of how these ordered systems facilitate SLDI.

5. OUTLOOK

In the past one and a half decades, the analysis of biomolecules has greatly benefited from the improved mass accuracy of MALDI mass spectrometry. Based on the achievements, MALDI has become a mainstream method. Further refinements of the method are primarily expected in particular applications and in its novel combinations with separation techniques.

For the analysis of small to medium size molecules, matrix-less desorption surfaces, such as DIOS and other nanostructures are being explored. These emerging methods exhibit broad applicability in the analysis of small biomolecules and pharmaceuticals.

The interaction of laser light with nanostructures also presents new challenges in our understanding how coherent electromagnetic radiation interacts with systems that exhibit quantum confinement, nanoscopic liquid confinement and confined reactions. The dimensions of nanostructures can be commensurate to the wavelength of the radiation and thus give rise to near field effects in desorption and ionization. Based on the improved understanding of these phenomena, new "designer" surfaces can be envisioned that are specifically created to enhance SLDI.

Other potential developments include the integration of SLDI nanostructures into chip-based analytical methods. Combining the separation and the spectroscopic steps in a single device promises improved efficiency and higher throughput in biomedical analysis.

6. ACKNOWLEDGMENTS

Funding for some of the research described in this chapter was provided by grants from the U.S. Department of Energy (DE-FG02-01ER15129), the W. M. Keck Foundation (041904) and the George Washington University Research Enhancement Fund (GWU-REF). Support from the Department of Energy does not constitute an endorsement of the views expressed in the article. The author is grateful for the Academy Award Fellowship of the Royal Flemish Academy of Belgium for Science and the Arts. Discussions with A. Vogel were helpful in clarifying the experimental details of the water ablation data.

7. REFERENCES

Alimpiev, S., Nikiforov, S., Karavanskii, V., Minton, T., and Sunner, J., 2001, On the mechanism of laser-induced desorption–ionization of organic compounds from etched silicon and carbon surfaces, *J. Chem. Phys.* **115**(4):1891–1901.

Allwood, D. A., Dreyfus, R. W., Perera, I. K., and Dyer, P. E., 1996, UV optical absorption of matrices used for matrix-assisted laser desorption/ionization, *Rapid Commun. Mass Spectrom.* **10**(13):1575–1578.

Allwood, D., Dreyfus, R., Perera, I., and Dyer, P., 1997, Optical absorption of matrix compounds for laser-induced desorption and ionization (MALDI), *Appl. Surf. Sci.* **109/110**:154–157.

Allwood, D., and Dyer, P., 2000, Quantitative fluorescence measurements performed on typical matrix molecules in matrix-assisted laser desorption/ionisation, *Chem. Phys.* **261**(3):457–467.

Apitz, I., and Vogel, A., 2005, Material ejection in nanosecond Er:YAG laser ablation of water, liver, and skin, *Appl. Phys. A* **81**(2):329–338.

Bencsura, A., Navale, V., Sadeghi, M., and Vertes, A., 1997, Matrix-Guest Energy Transfer in Matrix-Assisted Laser Desorption, *Rapid Commun. Mass Spectrom.* **11**(6):679–682.

Bencsura, A., and Vertes, A., 1995, Dynamics of hydrogen bonding and energy transfer in matrix-assisted laser desorption, *Chem. Phys. Letters* **247**(1-2):142–148.

Beuhler, R. J., Flanigan, E., Greene, L. J., and Friedman, L., 1974, Proton transfer mass spectrometry of peptides. A rapid heating technique for underivatized peptides containing arginine, *J. Am. Chem. Soc.* **96**(12):3990–3999.

Chen, Y., Luo, G., Diao, J., Chornoguz, O., Reeves, M., and Vertes, A., 2005, Laser desorption/ionization from nanostructured surfaces: nanowires, nanoparticle films and silicon microcolumn arrays, *J. Phys.: Conf. Ser.* in press.

Chen, Y., and Vertes, A., 2003, Pumping rate and surface morphology dependence of ionization processes in matrix-assisted laser desorption ionization, *J. Phys. Chem. A* **107**(46):9754–9761.

Chen, Y., and Vertes, A., 2005, Black silicon as matrix-free laser desorption ionization substrate, *Abstracts of Papers, 230th ACS National Meeting*, Washington, DC, United States, Aug. 28-Sept. 1, 2005, **2005**(111):ANYL.

Crecelius, A., Clench, M. R., Richards, D. S., and Parr, V., 2002, Thin-layer chromatography–matrix-assisted laser desorption ionisation–time-of-flight mass spectrometry using particle suspension matrices, *J. Chromatogr. A* **958**(1-2):249–260.

Cullis, A. G., Canham, L. T., and Calcott, P. D. J., 1997, The structural and luminescence properties of porous silicon, *J. Appl. Phys.* **82**(3):909–965.

Diao, J. J., Qiu, F. S., Chen, G. D., and Reeves, M. E., 2003, Surface vertical deposition for gold nanoparticle film, *J. Phys. D Appl. Phys.* **36**:L25–L27.

Dreisewerd, K., 2003, The desorption process in MALDI, *Chem. Rev.* **103**(2):395–425.

Finkel, N. H., Prevo, B. G., Velev, O. D., and He, L., 2005, Ordered silicon nanocavity arrays in surface-assisted desorption/ionization mass spectrometry, *Anal. Chem.* **77**(4):1088–1095.

Gabelica, V., Schulz, E., and Karas, M., 2004, Internal energy build-up in matrix-assisted laser desorption/ionization, *J. Mass Spectrom.* **39**(6):579–593.

Go, E. P., Apon, J. V., Luo, G., Saghatelian, A., Daniels, R. H., Sahi, V., Dubrow, R., Cravatt, B. F., Vertes, A., and Siuzdak, G., 2005, Desorption/ionization on silicon nanowires, *Anal. Chem.* **77**(6):1641–1646.

Her, T.-H., Finlay, R. J., Wu, C., Deliwala, S., and Mazur, E., 1998, Microstructuring of silicon with femtosecond laser pulses, *Appl. Phys. Lett.* **73**(12):1673–1675.

Her, T.-H., Finlay, R. J., Wu, C., and Mazur, E., 2000, Femtosecond laser-induced formation of spikes on silicon, *Appl. Phys. A* **70**(4):383–385.

Jellison, Jr., G. E., and Modine, F. A., 1982, Optical absorption of silicon between 1.6 and 4.7 eV at elevated temperatures, *Appl. Phys. Lett.* **41**(2):180–182.

Juhasz, P., Vestal, M. L., and Martin, S. A., 1997, On the initial velocity of ions generated by matrix-assisted laser desorption ionization and its effect on the calibration of delayed extraction time-of-flight mass spectra, *J. Am. Soc. Mass Spectrom.* **8**(3):209–217.

Karas, M., and Hillenkamp, F., 1988, Laser desorption ionization of proteins with molecular masses exceeding 10 000 Daltons, *Anal. Chem.* **60**(20):2299–2301.

Knochenmuss, R., 2002, A quantitative model of ultraviolet matrix-assisted laser desorption/ionization, *J. Mass Spectrom.* **37**(8):867–877.

Knochenmuss, R., 2003, A quantitative model of ultraviolet matrix-assisted laser desorption/ionization including analyte ion generation., *Anal. Chem.* **75**(10):2199–2207.

Knochenmuss, R., and Vertes, A., 2000, Time-delayed 2-pulse studies of MALDI matrix ionization mechanisms, *J. Phys. Chem. B* **104**(23):5406–5410.

Konn, D. O., Murrell, J., Despeyroux, D., and Gaskell, S.J., 2005, Comparison of the effects of ionization mechanism, analyte concentration, and ion "cool-times" on the internal energies of peptide ions produced by electrospray and atmospheric pressure matrix-assisted laser desorption ionization, *J. Am. Soc. Mass Spectrom.* 16(5): 743–751.

Kovalev, D., Polisski, G., Ben-Chorin, M., Diener, J., and Koch, F., 1996, The temperature dependence of the absorption coefficient of porous silicon, *J. Appl. Phys.* **80**(10):5978–5983.

Kruse, R.A., Li, X., Bohn, P. W., and Sweedler, J. V., 2001 Experimental factors controlling analyte ion generation in laser desorption/ionization mass spectrometry on porous silicon, *Anal. Chem.* **73**(15):3639–3645.

Luo, G., Chen, Y., Siuzdak, G., and Vertes, A., 2005a, Surface modification and laser pulse length effects on internal energy transfer in DIOS, *J. Phys. Chem. B* **109** (51): 24450-24456

Luo, G., Diao, J., Chornoguz, O., Reeves, M., and Vertes, A., 2005b, Laser desorption/ionization mass spectrometry from nanoparticle films, *Abstracts of Papers, 230th ACS National Meeting*, Washington, DC, United States, Aug. 28-Sept. 1, 2005, **2005**(109):ANYL.

Luo, G. H., Marginean, I., and Vertes, A., 2002, Internal energy of ions generated by matrix-assisted laser desorption/ionization, *Anal. Chem.* **74**(24):6185–6190.

McLean, J. A., Stumpo, K. A., and Russell, D. H., 2005, Size-selected (2-10 nm) gold nanoparticles for matrix assisted laser desorption ionization of peptides, *J. Am. Chem. Soc.* **127**(15):5304–5305.

Menzel, C., Dreisewerd, K., Berkenkamp, S. and Hillenkamp, F., 2001, Mechanisms of energy deposition in infrared matrix-assisted laser desorption/ionization mass spectrometry, *Int. J. Mass Spectrom.* **207**(1/2): 73–96.

Okuno, S., Arakawa, R., Okamoto, K., Matsui, Y., Seki, S., Kozawa, T., Tagawa, S., and Wada, Y., 2005, Requirements for laser-induced desorption/ionization on submicrometer structures, *Anal. Chem.* **77**(16):5364–5369.

Posthumus, M. A., Kistemaker P. G., Meuzelaar H. L. C., and Ten Noever de Brauw M. C., 1978, Laser desorption–mass spectrometry of polar nonvolatile bio-organic molecules, *Anal. Chem.* **50**(7):985–991.

Rothenberg, J. E., and Kelly, R., 1984, Laser sputtering. Part II. The mechanism of the sputtering of Al_2O_3, *Nucl. Instrurn. Meth. Phys. Res. B* **229**(2-3):291–300.

Sadeghi, M., and Vertes, A., 1998, Crystallite size dependence of volatilization in matrix-assisted laser desorption ionization, *Appl. Surf. Sci.* **127/129**:226–234.

Schurenberg, M., Dreisewerd, K., and Hillenkamp, F., 1999, Laser desorption/ionization mass spectrometry of peptides and proteins with particle suspension matrixes, *Anal. Chem.*, **71**(1):221–229.

Sedov, L. I., 1993, *Similarity and Dimensional Methods in Mechanics*, 10-th ed., CRC Press, Boca Raton.

Shen, Z. X., Thomas, J. J., Averbuj, C., Broo, K. M., Engelhard, M., Crowell, J. E., Finn, M. G., and Siuzdak, G., 2001, Porous silicon as a versatile platform for laser desorption/ionization mass spectrometry, *Anal. Chem.* **73**(3):612–619

Stevenson, E., Breuker, K., and Zenobi, R., 2000 Internal energies of analyte ions generated from different matrix-assisted laser desorption/ionization matrices, *J. Mass Spectrom.* **35**(8):1035–1041.

Stimson, E., Truong, O., Richter, W. J., Waterfield, M. D., and Burlingame, A. L., 1997, Enhancement of charge remote fragmentation in protonated peptides by high-energy CID MALDI-TOF-MS using "cold" matrices, *Int. J. Mass Spectrom. Ion Processes* **169/170**:231–240.

Tanaka, K., Waki, H., Ido, Y., Akita, S., Yoshida, Y., and Yoshida, T., 1988, Protein and polymer analyses up to m/z 100 000 by laser ionization time-of-flight mass spectrometry, *Rapid Commun. Mass Spectrom.* **2**(8):151–153.

Taylor, G., 1950, The Formation of a Blast Wave by a Very Intense Explosion. I. Theoretical Discussion, *Proc. Roy. Soc. London A* **201**(1065):159–174.

Turney, K., Drake, T. J., Smith, J. E., Tan W., and Harrison, W. W., 2004, Functionalized nanoparticles for liquid atmospheric pressure matrix-assisted laser desorption/ionization peptide analysis, *Rapid Commun. Mass Spectrom.* **18**(20):2367–2374.

Vertes, A., Irinyi, G., and Gijbels, R., 1993, Hydrodynamic model of matrix assisted laser desorption mass spectrometry," *Anal. Chem.* **65**(17):2389–2393.

Wei, J., Buriak, J. M., and Siuzdak, G., 1999, Desorption-ionization mass spectrometry on porous silicon, *Nature* **399**(6733): 243–246.

Wu, X., Sadeghi, M., and Vertes, A., 1998, Molecular dynamics of matrix-assisted laser desorption of leucine enkephalin guest molecules from nicotinic acid host crystal, *J. Phys. Chem. B* **102**(24):4770–4778.

Xu, S., Li, Y., Zou, H., Qiu, J., Guo, Z., and Guo, B., 2003, Carbon nanotubes as assisted matrix for laser desorption/ionization time-of-flight mass spectrometry, *Anal. Chem.* **75**(22):6191–6195.

Yazawa, M., Koguchi, M., Muto, A., Ozawa, M., and Hiruma, K., 1992, Effect of one monolayer of surface gold atoms on the epitaxial growth of InAs nanowhiskers, *Appl. Phys. Lett.* **61**(17):2051–2053.

Zenobi, R., and Knochenmuss, R., 1998, Ion formation in MALDI mass spectrometry, *Mass Spec. Rev.* **17**(5):337–366.

Zhigilei, L. V., and Garrison, B. J., 2000, Microscopic mechanisms of laser ablation of organic solids in the thermal and stress confinement irradiation regimes, *J. Appl. Phys.* **88**(3):1281–1298.

Zhigilei, L. V., Kodali, P. B. S., and Garrison, B. J., 1997, Molecular dynamics model for laser ablation and desorption of organic solids, *J. Phys. Chem. B* **101**(11):2028–2037.

Chapter 21

MATERIALS MODIFICATION WITH INTENSE EXTREME ULTRAVIOLET PULSES FROM A COMPACT LASER

M. E. Grisham,[1] G. Vaschenko,[1] C. S. Menoni,[1] L. Juha,[2,3] M. Bittner,[2] Yu. P. Pershyn,[4] V. V. Kondratenko,[4] E. N. Zubarev,[4] A. V. Vinogradov,[5] I. A. Artioukov,[5] and J. J. Rocca[1]

[1]NSF ERC for Extreme Ultraviolet Science and Technology and Department of Electrical and Computer Engineering, Colorado State University, Fort Collins, CO 80523; [2]Institute of Physics, Academy of Sciences of the Czech Republic, Na Slovance 2, 182 21 Prague 8, Czech Republic; [3]Center for Radiochemistry and Radiation Chemistry, Czech Technical University in Prague, Brehová 7, 115 19 Prague 1, Czech Republic; [4]Metal and Semiconductor Physics Department, National Technical University "KhPI", Kharkov, Ukraine; [5]P. N. Lebedev Physical Institute, Moscow 117942, Russia

1. INTRODUCTION

The demonstration of compact high power extreme ultraviolet (EUV) lasers (Benware, et al., 1998; Macchietto, et al., 1999) has enabled the investigation of the interaction of intense EUV laser light with materials. Two main characteristics of the EUV light differentiate this part of the electromagnetic spectrum (10–50 nm wavelengths). First, the absorption length in materials at EUV wavelengths is typically much shorter than that for light in other regions of the spectrum. The short penetration depth allows for the deposition of large irradiation doses in very thin, ~ 10 – 20 nm, subsurface layers of materials. Second, the high photon energy associated with EUV light allows single photons to break bonds in materials such as common polymers. This feature of EUV light results in an ablation mechanism that is qualitatively different from that corresponding to visible and infrared (IR) irradiation. To be able to take full advantage of the unique characteristics of EUV light in materials processing and patterning (Capeluto, et al., 2005) it is essential to gain a good understanding of the interaction mechanisms. In this contribution, we summarize our findings resulting from the study of the interaction of several materials with 46.9 nm wavelength light produced by a compact laser. In the following section, we describe the methods and instrumentation used in the EUV ablation experiments. This includes a description of the compact EUV laser used. The second section discusses the main findings resulting from EUV laser ablation of polymer materials. The third section deals with the experimental determination of EUV laser-induced optical damage in EUV Sc/Si multilayer mirrors and the corresponding damage mechanism.

2. EXPERIMENTAL TECHNIQUES

The experimental setup used to irradiate different samples is shown in Fig. 1 (Benware, et al., 1999). Its central component is a compact laser producing pulses of approximately 1 ns duration at a wavelength of 46.9 nm (26.4 eV photon energy). The laser beam propagates in vacuum into a 0.5 m diameter interaction chamber containing the focusing optics and irradiated sample.

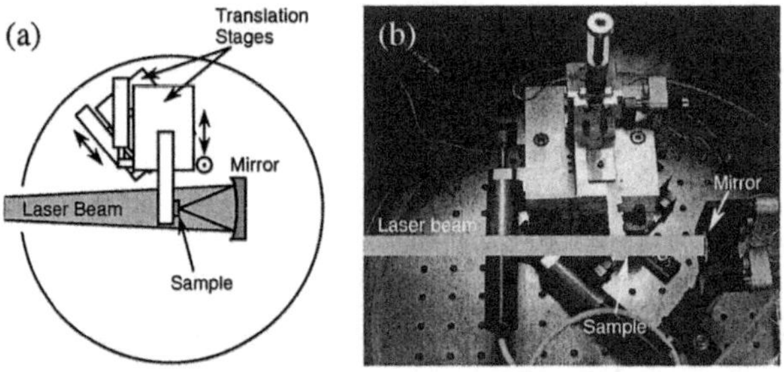

Figure 1. Schematic diagram (a) and a picture (b) of the experimental setup for irradiation with focused EUV laser beam. The laser emission is focused on the sample surface with a spherical (R = 10 cm) Sc/Si multilayer-coated mirror positioned normal with respect to the laser beam for minimum aberrations. Samples are attached to XYZ translation stage that is mounted on a two-axis mount with 50° orientation of one of the platforms. This provides a simultaneous variation of the fluence and position of the irradiated spot on the sample.

The EUV laser employed in the experiment is a compact table-top device that uses a fast discharge in argon filled capillary tube to create and collisionally excite Ne-like Ar ions, producing a population inversion and amplification in the 46.9 nm transition of these ions (Benware, et al., 1998; Macchietto, et al., 1999). The fast current pulse creates a magnetic force that rapidly compresses the plasma to form a dense and hot plasma column with

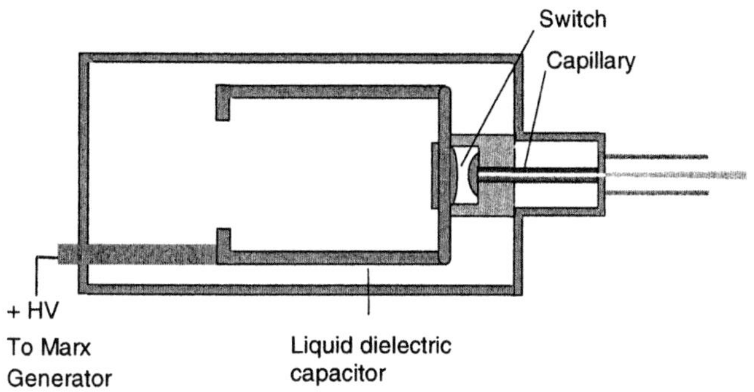

Figure 2. Schematic diagram illustrating the compact EUV capillary-discharge laser design.

a large density of Ne-like ions and a length to diameter ratio of up to ~ 1000:1. Collisional electron impact ionization of the ground state Ne-like ions produces a population inversion between the $3p^1S_0$ and $3s^1P_1^0$ levels resulting in gain-saturated amplification at 46.9 nm (Rocca et al., 1994; Rocca et al., 1996).

Fig. 2 shows a schematic of the laser. It consists of an argon filled alumina capillary that is excited by discharging a compact water dielectric capacitor. The water in the capacitor serves as both the dielectric material for the capacitor and also as a cooling medium for the capillary. The water capacitor is pulse charged using a four stage Marx generator, and is discharged through a self-breaking spark-gap switch pressurized with SF_6 gas. The laser output pulse characteristics depend on the length of the capillary discharge channel that can be selected to have a length of up to 36 cm. To conduct the experiments described herein capillaries of 18.4 cm in length were used to produce pulses with an energy of ~ 0.13 mJ and a duration of ~1.2 ns FWHM (Benware, et al., 1998). The alumina capillaries had an internal diameter of 3.2 mm, and were filled with 440 mTorr of argon gas. The fast current pulse had a peak amplitude of ~ 25 kA, a 10 % - 90 % rise time of ~ 25 ns, and a first half cycle duration of the pulse of ~ 110 ns. Refraction of the rays in the plasma column amplifier creates a laser beam with annular beam profile and a divergence of about 4.6 mrad (Fig. 3). To conduct the ablation experiments the laser was operated at a repetition rate of 1 - 2 Hz. Approximately 5000 laser shots were obtained from a single capillary before the laser output energy started to decrease significantly due to degradation of the capillary walls.

Careful positioning of the target samples was accomplished by a set of two coupled translation stages (Fig. 1). The laser emission was focused onto the target with a 2.5 cm diameter spherical mirror of 10 cm radius of curvature.

The mirror was coated with a Sc/Si multilayer stack deposited by magnetron sputtering to achieve a normal incidence reflectivity of ~ 30 % at 46.9 nm. The mirror was placed in the interaction vacuum chamber at ~ 1.44 m from the capillary-discharge amplifier. The sample was attached to a 1.6 mm thick support that intercepted a small fraction of the laser beam (Benware, et al., 1999), which at the location of the sample has a diameter of ~ 13 mm.

(a) (b)

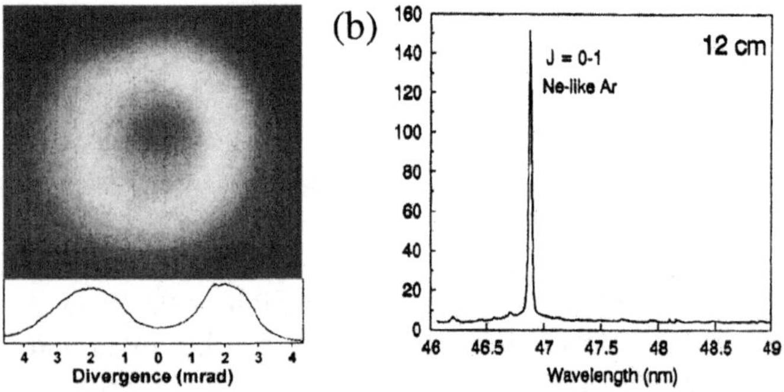

Figure 3. (a) Image of the laser beam at 1.38 m from exit of the capillary (Benware, et al., 1998). The annular beam pattern is caused by refraction in the plasma column. (b) Spectrum of the laser output emission shows high monochromaticity of this source ($\lambda/\Delta\lambda > 10^4$) (Rocca, et al., 1994).

Motorized translation stages were used to allow for the motion of the sample along an axis that forms an angle of ~ 50° with respect to the laser beam and for its accurate positioning in the horizontal and vertical directions (Grisham et al., 2004). The displacement of the samples at 50° allowed us to vary the distance between the sample and the focal plane (focal spot ~ 5 μm in diameter) to select EUV radiation fluences from ~ 0.01 to > 10 J/cm^2 range, while simultaneously changing the target area irradiated by the beam after each shot. In some samples individual spots were irradiated by single laser shots, whereas in others a large number of shots were used to uniformly irradiate ~ 2 × 2 mm^2 areas with a fixed EUV fluence to allow for x-ray diffraction studies of the damaged films.

Early work (Benware, et al., 1999) explored the focusing of the EUV laser beam with 40 % reflection mirror, and demonstrated its use for the ablation of metal samples. The laser was focused onto thin brass strips (1.6 x 10 mm wide). The focused laser beam readily ablated metals within several hundred micrometers of the focal plane. A series of imprints were obtained by continuously moving the translation stage and operating the laser at a repetition rate of 1 Hz.

Fig. 4 shows scanning electron microscopy (SEM) micrographs with a progression of the ablation patterns as the target was moved away from the mirror toward the focus. Each of these patterns is a result of a single laser shot. Within a few hundred micrometers of the focus the ablated pattern has an annular shape. The central discontinuity in the ablation pattern is caused by the shadow of the sample holder that blocks a portion of the incoming beam.

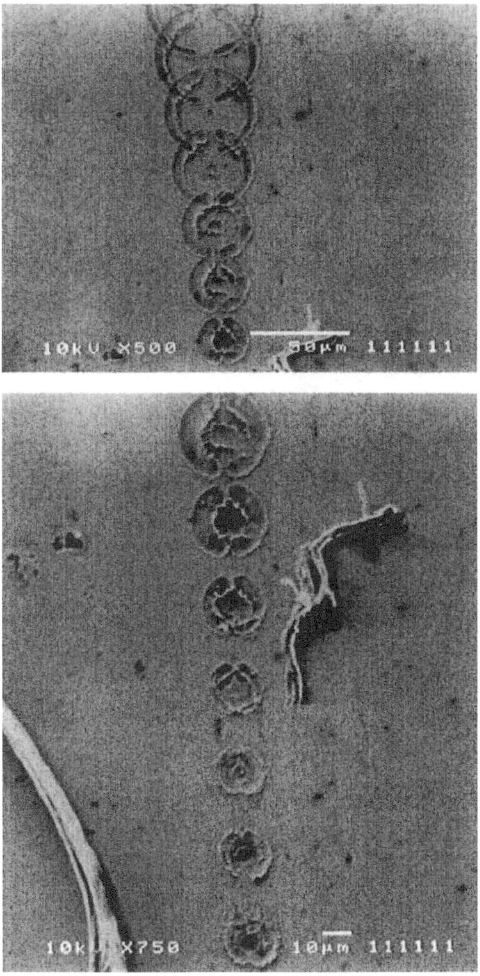

Figure 4. SEM images of EUV laser ablation patterns on the surface of a brass sample, illustrating the convergence of the focused beams and the characteristic shapes of the ablation craters (Benware, et al., 1999).

As the focal region is approached the depth of the ablation craters is observed to increase, and an additional ablated spot develops at the center. Finally, in the focal plane the ablation region develops into a single spot with a deep crater on axis. The smallest spot ablated in the series measures ~ 17 μm and contains a deep crater with ~ 3 μm diameter on axis. Ray-tracing computations were conducted to analyze the results and to obtain an estimate of the power density achieved.

The beam was approximated as originating from a point source located 263 cm from the mirror and the angular distribution was selected to closely match the measured laser beam intensity in the near field as well as the far field. The trajectories of 13,000 rays were computed to simulate the EUV beam propagation.

Fig. 5 shows the computed cross section of the beam intensity distributions in the focal region. The filled circles in Fig. 5 represent the experimental values.

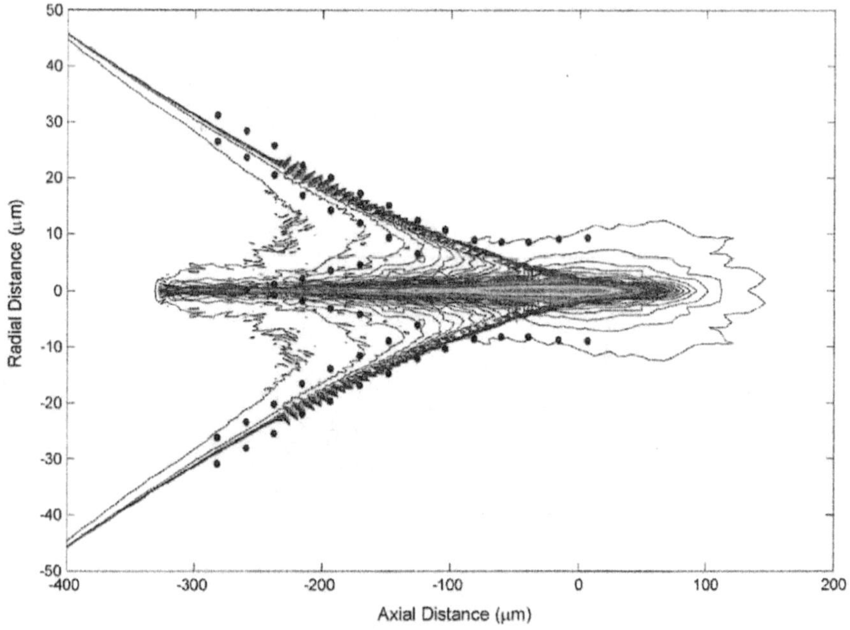

Figure 5. Computed cross section of the 46.9 nm laser beam in the focal region. The filled circles are the measured boundaries of the patterns ablated on a brass sample (Benware, et al., 1999).

All the major features of the observed ablation profiles shown in Fig. 4 are well described by the ray-tracing computations. The computations illustrate that a few hundred micrometers from the focus, the majority of the rays form thin annular disks. As the focal region is approached, the computation shows a high peak intensity along the axis which is in concordance with the experimental data. These features are caused by spherical aberrations and not the annular shape of the beam. Spherical aberrations would cause similar shapes with any beam profile with sufficient intensity in the periphery. Near the plane of minimum confusion the intensity profile is computed to be dominated by a sharp central peak. This sharp central peak is responsible for deep holes that are observed in the ablation patterns. The average intensity within 2 μm of the axis can be estimated from the computed fraction of rays that intersect this region to be ~ 1×10^{11} W/cm^2. This corresponds to a 100 J/cm^2 fluence.

3. EUV ABLATION OF POLYMERS

Ablation of organic polymers with optical radiation from visible and UV lasers has been a subject of extensive investigations (see for example, Srinivasan and Braren, 1989; Srinivasan, 1994; Lippert and Dickinson, 2003, and references therein). In contrast, polymer ablation induced by EUV radiation was investigated only in a smaller number of studies. Poly (butene-1 sulphone) (Yaakobi, et al., 1983), PMMA (Juha, et al., 2002a; Fiedorowicz, et al., 2004), poly (ethylene terephthalate) (Juha, et al., 2002b) and PTFE (Anderson, et al., 1994; Juha, et al., 2002a) have recently been ablated by incoherent, non-monochromatic laser-produced plasma's emitting within the EUV spectral region. The use of a Z-pinch plasma source with emission in the EUV has also yielded ablation of PMMA (Juha, et al., 2002a) and PTFE (Juha et al., 2002b; Deno, et al., 1996). Very recently PMMA was efficiently ablated by sub-100-fs pulses of 86-nm-radiation provided by a free-electron laser (Juha et al., 2002b; Sobierajski et al., 2002). Also, a number of studies on direct photoetching of organic polymers induced by EUV synchrotron radiation have been published (Y. Zhang, et al., 1995; X. Zhang, et al, 1995; Katoh and Zhang, 1998). However, according to Haglund's criterion (Haglund, 1996), due to the low fluence of the synchrotron radiation, photoinduced material removal is more similar to laser desorption than to laser ablation. In this section we discuss the EUV ablation behavior of three common organic polymers: PTFE, PMMA, and PI, irradiated with an intense focused 46.9 nm laser beam at a pulse repetition rate of 1 Hz.

The samples were exposed with the EUV laser at Colorado State University and were analyzed at the Institute of Physics of the Czech Academy of Sciences. The ablation processes induced by nanosecond pulses of 46.9 nm laser radiation are compared with those occurring in the polymer materials irradiated with conventional, longer wavelength laser sources. The

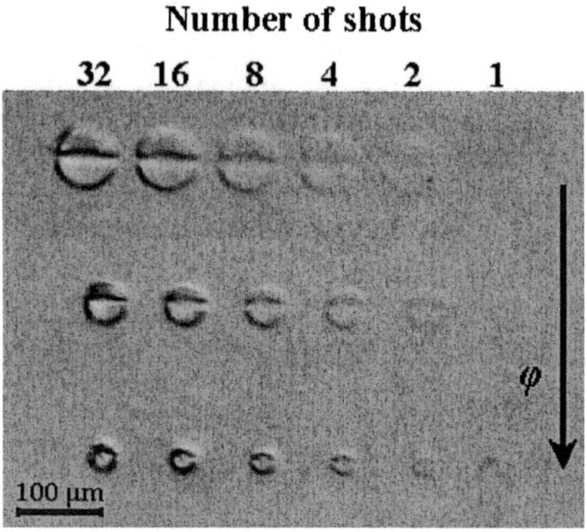

Figure 6. Optical micrograph of a PMMA sample irradiated with 46.9 nm laser light at three fluences (~1, 2, and 4.5 J /cm^2) with an increasing number of laser shots from 1 (right) to 32 (left). The fluence increases from top to bottom and the number of shots increases from the right to the left (Juha, et al., 2005a)

samples consisted of 1-mm-thick sheets cut into 2.0 x 5.0 mm^2 chips. The PTFE and PI samples were polished, while the PMMA was used without any prior surface treatment.

Fig. 6 shows an optical micrograph of a PMMA sample exposed to three different fluences, with increasing value in the direction of the arrow (Juha et al., 2005a). The six craters in each row were formed upon accumulation of 1, 2, 4, 8, 16, or 32 laser shots at each particular fluence. The horizontal ridges observed in the ablated craters are again the result of the shadow produced by the sample holder, which blocks part of the laser beam (Benware, et al., 1999). The PTFE and PI samples were irradiated in the same way as PMMA. Cross-sections of the ablated craters were measured with a profiler (Alpha Step 500) in the direction crossing the ridges of the unirradiated zones. The irradiation fluence was determined from the laser pulse energy and the ablated area defined as the region where the ablation depth becomes distinguishable from the unexposed surface after a single-shot exposure.

Fig. 7 shows a series of ablation profiles for PMMA created by a various number of laser shots at a fluence of 2.0 J /cm^2. The ablation rates evaluated from the measured profiles for all three polymers ablated by either 4 or 32 accumulated laser pulses are compared in Fig. 8.

Numerous previous ablation experiments conducted with UV and VUV excimer laser pulses of nanosecond duration have shown that PTFE, PMMA, and PI differ significantly in their ablation behavior.

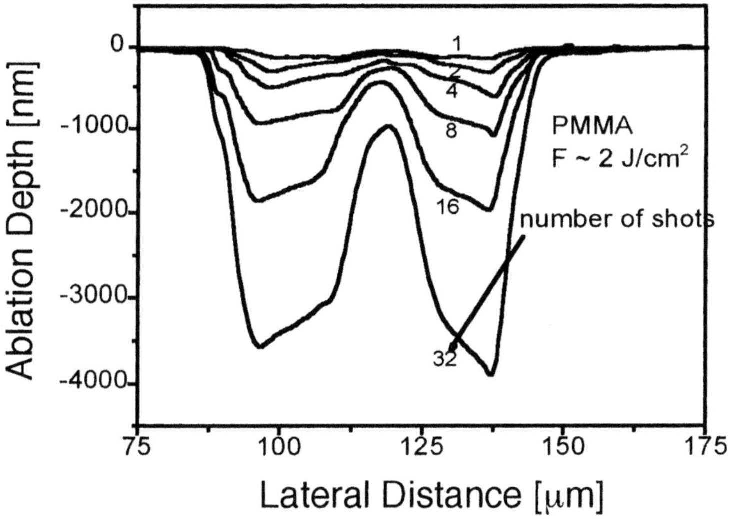

Figure 7. Profiles of the craters ablated in PMMA by 46.9 nm laser light at a fluence of 2.0 J/cm^2 (Juha, et al., 2005a).

PI and PMMA belong to groups of strongly and weakly absorbing materials in the UV region, respectively, while PTFE has the first absorption band at ~160 nm.

Experiments (Costela et al., 1995) with even shorter wavelength emission (157-nm-F_2 laser) have shown efficient ablation of all the polymers of our interest, but the ablation rate as well as the attenuation length, which is generally considered (Bauerle, 1996; Lippert and Dickinson, 2003) to be the parameter that controls the ablation process, still differ from one polymer to another (see Table 1).

In contrast, the 46.9 nm results summarized in Fig. 8 and in Table 1 show similar ablation rates within the experimental error for PTFE, PMMA, and PI. This result can be explained by the values of the attenuation lengths for the 46.9 nm and 157 nm radiation (Table 1). While the attenuation lengths are different in all three materials at 157 nm, they are very similar in PMMA and PI at 46.9 nm. The absorption of 46.9 nm radiation in PTFE is stronger than in PMMA and PI.

However, polymer chain scissions by high-energy photons are induced more efficiently (Holmes-Siedle and Adams, 2002) in PTFE than in PMMA and PI. This is likely to even out the ablation rate for PTFE with other polymers. This uniformity of ablation rates could be very beneficial for using the EUV laser radiation in direct polymer nanostructuring.

Long wavelength ablation with a laser fluence ϕ is a threshold process in which the polymer ablation rate d is controlled by a threshold fluence ϕ_{th} (Lippert and Dickinson, 2003):

$$d = 1/\alpha \ln (\phi / \phi_{th}), \quad (1)$$

where α is the absorption coefficient of the material.

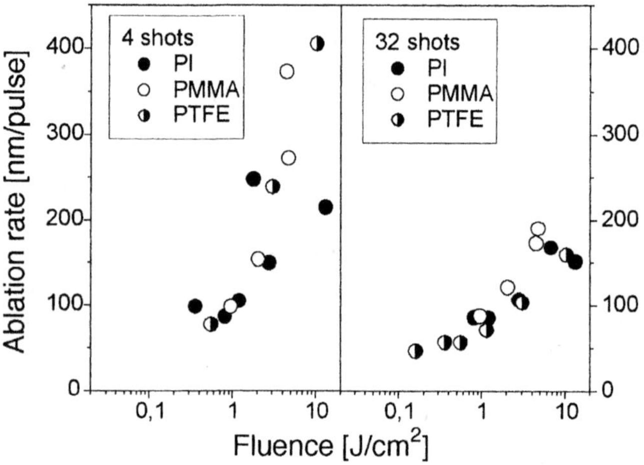

Figure 8. Ablation rates of PTFE, PMMA, and PI irradiated by 46.9 nm laser beam as a function of fluence (Juha, et al., 2005a).

Table 1. Attenuation lengths and ablation rates of PTFE, PMMA, and PI irradiated by 46.9-nm and 157-nm light (Juha, et al., 2005a). The attenuation lengths at 46.9 nm were calculated from (Henke et al, 1993) and the ablation rates were taken from Fig. 8 for 32 laser shots. The values related to the F_2-laser irradiation (λ = 157 nm) were published in (Costela, et al., 1995).

Polymer	Ablation rate [nm/pulse] at 46.9 nm $\phi \sim 1$ J/cm^2	Ablation rate [nm/pulse] at 157 nm $\phi \sim 300$ mJ/cm^2	Attenuation length [nm] at 46.9 nm	Attenuation length [nm] at 157 nm
PTFE	83	370	12	172
PMMA	87	260	19	117
PI	88	150	16	79

In contrast with the long wavelength case, at EUV wavelengths the energy of a single photon is sufficient to break chemical bonds and to overcome a cohesive energy of the polymer. Therefore, a question arises of whether a threshold fluence concept is valid for ablation at 46.9 nm.

It is evident from Fig. 8 that the measured ablation rates at this wavelength do not fulfill Eq. (1). The slope is varying with the fluence and there is no well-defined threshold. This suggests the need for establishing a new model describing the ablation rate of polymer ablation by the EUV radiation, especially in the low fluence regime.

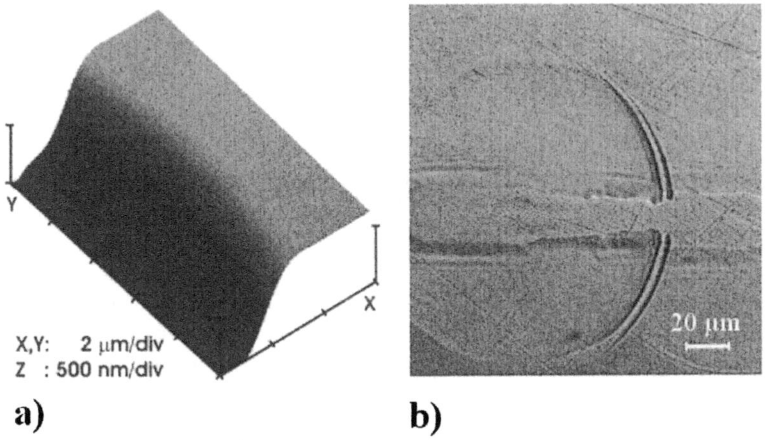

Figure 9. (a) AFM image of the edge of a crater ablated in PMMA by 16 accumulated 46.9 nm laser pulses at a fluence of 2.5 J /cm^2; (b) optical micrograph of the craters ablated in PI at 0.25 J /cm^2 with 32 (left) and 16 (right) laser pulses (Juha et al., 2005a).

As illustrated by the AFM image in Fig. 9(a) and optical micrograph in Fig. 9(b), all ablation craters produced in this work for the three polymers were observed to have smooth, high-quality surfaces and cleanly cut walls with very well developed sharp edges that are not affected by thermal damage. This confirms the previous observations of the improved quality of ablated surfaces as the wavelength of radiation is reduced from the visible-UV range to 157 nm (Costela, et al., 1995; Lapczyna and Stuke, 1998) and further down to 125 nm (Riedel and Castex), and that at the shortest wavelength high quality surfaces are obtained even upon irradiation with relatively long (nanosecond) pulses (Juha et al., 2005b). This improvement is caused by a strong localization of the absorbed energy, i.e. both the attenuation and the thermal diffusion lengths are here very short (~ 10 nm; see Table 1 and (Dyer, 2003)). The direct, radiation-chemical action of 26.4 eV photons on the structure of molecular solids can be partly responsible for this observation, i.e. only a portion of absorbed energy is thermalized.

Experiments conducted on organic polymers with synchrotron radiation (X. Zhang, et al., 1995; Maida, et al., 2001; Beetz and Jacobsen, 2003) and plasma-based XUV sources (Juha, et al., 2002a; Juha, et al., 2002b) indicate that the chain scissions are dominant processes in materials irradiated by short wavelength radiation.

Summarizing this section, we have observed an efficient ablation of three common polymers by a focused intense nanosecond pulse from a 46.9 nm laser. The ablation rates (50 – 400 nm/pulse) are similar for all polymers investigated. There is no clear threshold ablation fluence detectable in the range of fluences used in the experiment. The ablation craters are clean, with well-defined edges and no bubbles or thermal damage. This testifies in favor of a key role of the non-thermal effects in EUV laser ablation with nanosecond duration pulses.

4. EUV LASER DAMAGE OF Sc/Si MULTILAYER MIRRORS

Numerous practical applications of the EUV sources require reflective optics for steering and focusing the beam. In the 35 - 50 nm wavelength range Sc/Si multilayer mirrors show the best reflectivity values, reaching as high as 54 % (Uspenskii, et al., 1998). However, the damage to these mirrors when exposed to high peak powers of EUV light has not been studied. This problem is now of particular relevance because the peak power and fluence of EUV laser sources have recently increased significantly (for example, the radiation fluence at the exit of a capillary-discharge Ne-like Ar laser operating at 46.9 nm can exceed 1 J/cm^2 (Macchietto, et al., 1999)) and is soon expected to achieve unprecedented values with the commissioning of EUV and soft x-ray free electron lasers (Yurkov, 2003; Yu, et al., 2003). In this section we describe the investigation of the optical damage of the Sc/Si multilayer coatings performed using the 46.9 nm capillary-discharge laser. The study was conducted at Colorado State University using the experimental setup shown in Fig.1 with the samples manufactured at the Metal and Semiconductor Physics Department, National Technical University in Ukraine. The damage to the multilayer coatings resulting from the exposure to the EUV beam was analyzed with scanning electron microscopy (SEM), transmission electron microscopy (TEM), selected-area electron diffraction, and small-angle x-ray diffraction (λ = 0.154 nm) techniques. As a result of the study, we have determined the damage threshold and the main damage mechanisms taking place in the irradiated areas.

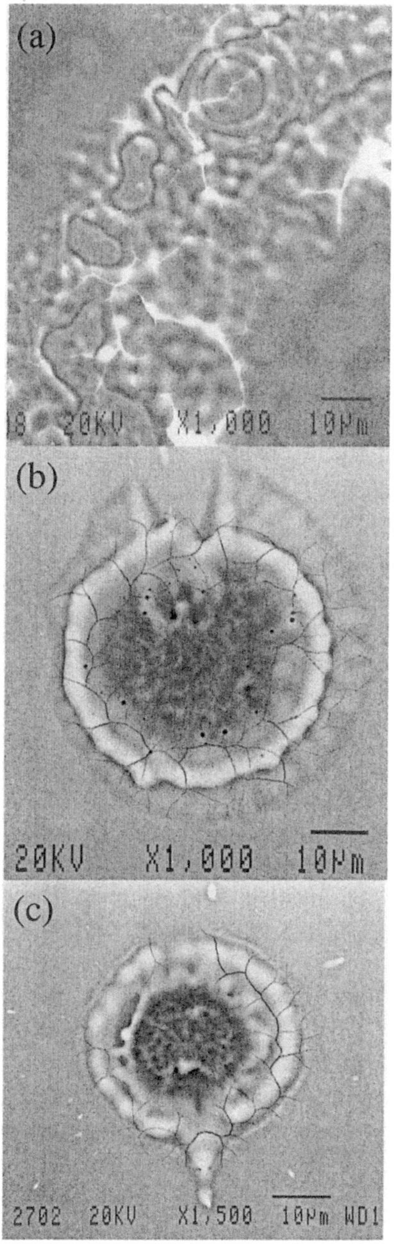

Figure 10. SEM micrographs of the damaged area of the Sc/Si coatings exposed to EUV laser beam fluences of (a) 0.13 J/cm^2, (b) 1.5 J/cm^2, and (c) 2.8 J/cm^2. Only a segment of the entire damaged area is shown in (a) (Grisham, et al., 2004).

The Sc/Si multilayers were deposited by dc magnetron sputtering at 3 mTorr of Ar pressure on superpolished borosilicate glass (surface roughness of ~ 0.4 nm) and on Si wafers (surface roughness of ~ 0.6 nm). The multilayers on borosilicate glass consisted of ten periods of Sc/Si layers, each with a thickness of ~ 26.7 nm and a ratio of layer thickness of H(Sc)/H(Si) ~ 0.7. A top 5-nm-thick Si protection layer capped the multilayers. The multilayer coatings deposited on Si consisted of 33 periods of Sc/Si pairs with the same parameters as those deposited on borosilicate glass. In these structures the crystalline Sc layers were always separated from the amorphous Si layers by ~3 nm of amorphous ScSi interface layers formed by interdiffusion (Fedorenko et al., 2001). Fig. 10 shows typical SEM images of damaged areas in coatings deposited on a Si wafer resulting from average EUV fluences of 0.13, 1.5, and 2.8 J/cm^2. At 0.13 J/cm^2 (Fig. 10(a)) we observe large areas with apparent discoloration. These areas are most likely produced by heat-triggered interdiffusion in the upper layers of the coatings. This surface modification, which already appears at fluences of ~0.08 J/cm^2, establishes the damage threshold for the Sc/Si multilayers defined in this work. In comparison, the onset of damage in bare Si substrates measured in this work occurs at a significantly larger irradiation fluence of 0.7 J/cm^2. The areas with larger local fluences (Fig. 10(a)) are molten and covered with cracks resulting from significant mechanical tensile stress generated by thermal expansion and the following cooling down process (Lee et al., 1999). At larger fluences (1.5 J/cm^2), some of the molten material is displaced toward the periphery, forming a crown (Fig. 10(b)). Also, a substantial concentration of cracks and micrometer-sized pits resulting from boiling are observed at this fluence. At an even larger fluence of ~3 J/cm^2 the coating is fully evaporated from the center of the irradiated spot and the Si substrate is also damaged (Fig. 10(c)). Electron microanalysis data reveal that Sc is absent in the center part of the dark spot shown in Fig. 10(c).

Small-angle X-ray diffraction analysis of the samples with 2 × 2 mm^2 area irradiated with ≥ 0.1 J/cm^2 emission fluence shows a noticeable drop in the intensity of the diffraction peaks with respect to the unexposed areas. Thus, the sample irradiated with a fluence of ~ 0.13 J/cm^2 loses 20 – 30 % in diffraction intensity, whereas that where the fluence was ~ 0.21 J/cm^2 (Fig. 11) loses 75 – 85 %. However, the diffraction peak's position remains approximately the same, indicating that the coating is only partially destroyed. This evidence suggests that while at these fluences the top layers of the coating are molten, the layers adjacent to the substrate remained unchanged. This interpretation of the X-ray diffraction data was confirmed by cross-section TEM imaging of the sample exposed at 0.21 J/cm^2. The TEM image of Fig. 12(a) shows that the top 700 nm of the coating are molten, while ~180 nm (7 periods) adjacent to the substrate are not destroyed. The top molten layer constitutes an alloy of Sc_3Si_5 and crystal Si as determined from electron diffraction data. Analysis of the surviving multilayer coating beneath the molten layer indicates that changes in layer thickness have occurred within a distance of less than 2 periods from the molten region (Fig. 12). Over the whole irradiated area, the heat affected zone (HAZ) did not exceed ~100 nm (4 periods).

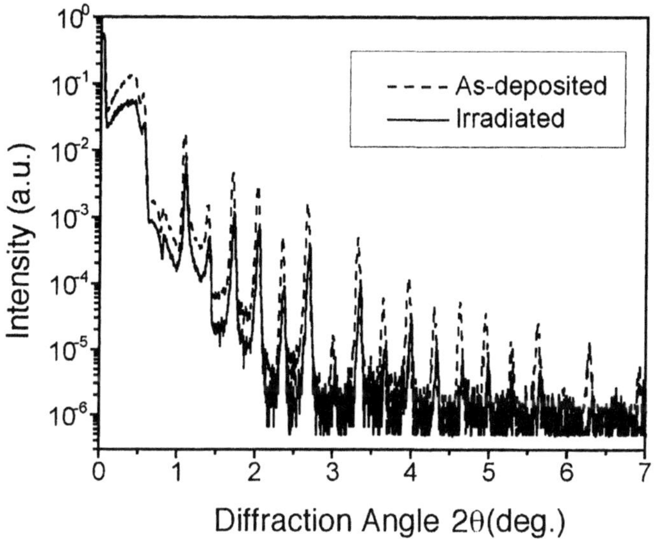

Figure 11. Small-angle X-ray diffraction patterns in the as-deposited and EUV-irradiated coatings. Laser fluence is 0.21 J/cm^2 (Grisham, et al., 2004).

Comparison of the layer structure in the heat affected zone with that of isothermally annealed samples indicates that the various stages of structural and phase transformations observed within a few periods of the coating under laser irradiation are the same as in samples annealed at different temperatures. Both structural and phase transformations in Sc/Si multilayers at temperatures as high as 970 °C have been previously studied in detail (Voronov, et al., 2002). The changes taking place in the Sc-containing layer nearest to the molten region (indicated by I in Fig. 12(b)) correspond to a stage of formation and crystallization of Sc_3Si_5 silicide that have been observed in isothermally annealed coatings at 430 °C after 1 h.

In the next Sc-containing layer (indicated by II in Fig. 12(b)), only minor expansion of the ScSi silicide interface layers is observed, which is a result of the reaction of solid-state amorphization. Similar effects have been observed at annealing temperatures of less than 200 °C (Voronov, et al., 2002). The insertion of antidiffusion barriers at Sc–Si interfaces might help to increase the damage threshold of these multilayers, as suggested by the increased heat resistance observed in the isothermally heated structures with such barriers (Voronov et al., 2002).

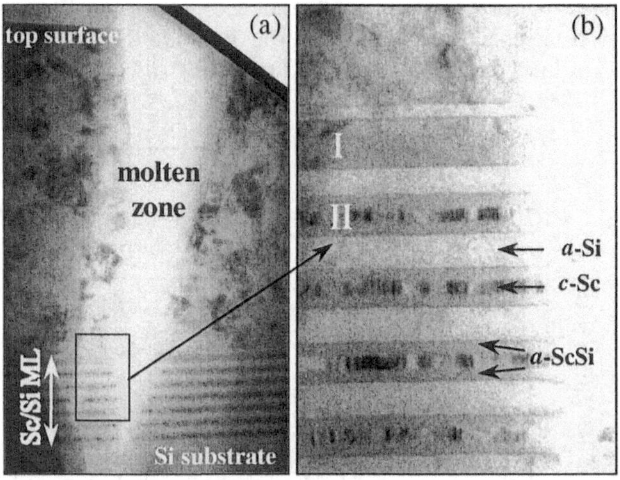

Figure 12. Cross-sectional TEM image of (a) the molten zone and (b) survived layers (magnified) of a sample irradiated with 0.21 J/cm^2 pulses of 46.9 nm laser radiation (Grisham, et al., 2004).

The damage threshold determined in this work for Sc/Si multilayer mirrors (~ 0.08 J/cm^2) is close to the values reported at significantly shorter wavelengths for other multilayers based on different material combinations, such as Mo/Si, W/C, and W/Si (MacGowan, et al., 1992; Leguern, et al., 1997). At the same time, this value is appreciably lower than the damage threshold of 0.7 J/cm^2 found for a single crystal silicon. On one hand such discrepancy looks natural and is related to the fact that the single crystal damage requires melting and evaporating homogeneous material, whereas for phase-nonequilibrium Sc/Si multilayer system the degradation starts with interlayer diffusion interaction that requires significantly lower energy. On the other hand, the melting threshold for Sc/Si multilayers (~0.13 J/cm^2) is also much lower than that for the single crystal. Two main effects are responsible for this disparity. The first effect is a result of the exothermic nature of scandium silicide formation from pure components, which leads to a significant local heat concentration (Lukashenko, et al., 1992). The second effect is a significant phonon scattering at the interfaces between different materials, which results in a poor heat-conducting ability of layers (Pugachev, et al., 1982; Balandin, et al., 1999; Chen and Neagu, 1997; Chen, 1997) compared to a bulk material. The abrupt heat-affected zones observed in the ablated multilayers in this work also evidence this low heat conductivity. To reduce this effect and therefore increase the damage threshold of the multilayer coatings, one may consider replacing *c*-Sc layers with the diffusion-formed silicide interlayers by Sc_3Si_5 layers being in thermodynamic equilibrium with Si layers. This would effectively reduce the number of the interfaces in the structure of the coating and would improve the heat conductivity of the structure.

5. CONCLUSION

In summary, we have realized a series of experiments with a compact 46.9 nm wavelength laser that produces intense pulses of nanosecond duration to study the ablation behavior of metals, common polymers, and Sc/Si multilayers. The key ablation process in polymers is likely to be a radiolysis of the polymer chains by EUV photons, resulting in the formation of numerous small molecular fragments that are subsequently removed from the surface of the samples. The EUV ablation rates for different polymers were found to be almost material independent, ~ 50 – 400 nm/pulse. In each material EUV irradiation was observed to leave smooth craters with well defined edges and without signs of thermal damage. No threshold behavior was detected in the EUV ablation of the polymers in the range of fluences used in the experiment. In contrast to polymers the irradiation damage in metals and in Sc/Si multilayers is thermal in nature. A damage threshold of 0.08 J/cm^2 was measured in the multilayer mirror coatings deposited on Si or borosilicate glass substrates, compared with a measured value of 0.7 J/cm^2 for bare Si substrates. These results are relevant to the use of these mirrors with newly developed high-power EUV laser sources and provide a benchmark for their further improvement. In combination, the experiments demonstrate that compact extreme ultraviolet lasers are new tools available for surface modification studies and patterning.

ACKNOWLEDGMENTS

This work was supported by the Engineering Research Centers Program of the National Science Foundation under NSF Award Number EEC-0310717, and in part by the U.S. Department of Energy, Chemical Sciences, Geosciences and Biosciences Division of the Office of Basic Energy Sciences. The part of this work carried out in the Czech Republic was partially funded by the Czech Ministry of Education, from program of the National Research Centers (Project LC528) and program INGO (Grants LA055 and 1P2004LA235), and by the Academy of Sciences of the Czech Republic (Grant Z10100523).

REFERENCES

Anderson, A. T., Tobin, M. T., and Peterson, P. F., 1994, X-ray response of National-Ignition-Facility first surface materials, *Fusion Technol.* **26**: 804–808.

Balandin, A., Khitun, A., Liu, J. L., Wang, K. L., Borca-Tasciuc, T., and Chen, G., 1999, Optimization of the thermoelectric properties of low-dimensional structures via phonon engineering, *Proc. International Conference on Thermoelectrics*, IEEE, Piscataway, pp. 189–192.

Bauerle, D., 1996, *Laser Processing and Chemistry*, 2nd ed., Springer, Berlin.

Beetz, T., and Jacobsen, C., 2003, Soft X-ray radiation-damage studies in PMMA using a cryo-STXM, *J. Synchrotron Radiat.* **10**: 280–283.

Benware, B. R., Macchietto, C. D., Moreno, C. H., and Rocca, J. J., 1998, Demonstration of a high average power tabletop soft X-ray laser, *Phys. Rev. Lett.* **81**: 5804–5807.

Benware, B. R., Ozols, A., Rocca, J. J., Artioukov, I. A., Kondratenko, V. V., and Vinogradov, A. V., 1999, Focusing of a tabletop soft-x-ray laser beam and laser ablation, *Opt. Lett.* **24**: 1714–1716.

Capeluto, M. G., Vaschenko, G., Grisham, M., Marconi, M. C., Ludueña, S., Pietrasanta, L., Lu, Y., Parkinson, B., Menoni, C. S., and Rocca, J. J., 2005, Nanopatterning with interferometric lithography using a compact $\lambda = 46.9$ nm laser, *IEEE Trans. Nanotechnol.*, in press.

Chen, G., and Neagu, M., 1997, Thermal conductivity and heat transfer in superlattices, *Appl. Phys. Lett.* **71**: 2761–2763.

Chen, G., 1997, Size and interface effects on thermal conductivity of superlattices and periodic thin-film structures, *J. Heat Transfer* **119**: 220–229.

Costela, A., Garciamoreno, I., Florido, F., Figuera, J. M., Sastre, R., Hooker, S. M., Cashmore, J. S.,and Webb, C. E., 1995, Laser-ablation of polymeric materials at 157 nm, *J. Appl. Phys.* **77**: 2343–2350.

Deno, H., Sugiyama, S., Kakudate, Y., Yoshida, M., and Fujiwara, S., 1996, VUV ablation of polymers by emission from gas-puff Z-pinch plasmas, *Appl. Surf. Sci.* **96-8**: 563–568.

Dyer, P. E., 2003, Excimer laser polymer ablation: twenty years on, *Appl. Phys. A: Mater. Sci. Process.* **77**: 167–173.

Fedorenko, A. I., Pershin, Yu. P., Poltseva, O. V., Ponomarenko, A. G., Sevryukova, V. S., Voronov, D. L., and Zubarev, E. N., 2001, Structure of Sc/Si multilayer mirrors in as-deposited state and after annealing, *J. X-Ray Sci. Technol.* **9**: 35–42.

Fiedorowicz, H., Bartnik, A., Bittner, M., Juha, L., Krasa, J., Kubat, P., Mikolajczyk, J., and Rakowski, R., 2004, Micromachining of organic polymers by direct photo-etching using a laser plasma X-ray source, *Microelectron. Eng.* **73-74**: 336–339.

Grisham, M., Vaschenko, G., Menoni, C. S., Rocca, J. J., Pershyn, Yu. P., Zubarev, E. N., Voronov, D. L., Sevryukova, V. A., Kondratenko, V. V., Vinogradov, A. V., and Artioukov., I. A., 2004, Damage to extreme-ultraviolet Sc/Si multilayer mirrors exposed to intense 46.9-nm laser pulses, *Opt. Lett.* **29**: 620–622.

Haglund, Jr., R. F., 1996, Microscopic and mesoscopic aspects of laser-induced desorption and ablation, *Appl. Surf. Sci.* **96-8**, 1–13.

Holmes-Siedle, A., and Adams, L., 2002, *Handbook of Radiation Effects*, 2nd ed., Oxford University Press, Oxford, p. 569.

Juha, L., Krasa, J., Präg, A., Cejnarova, A., Chvostova, D., Rohlena, K., Jungwirth, K., Kravarik, J., Kubes, P., Bakshaev, Yu. L., Chernenko, A. S., Korolev, V. D., Tumanov, V. I., Ivanov, M. I., Bernardinello, A., Ullschmied, J., and Boody, F. P., 2002a, Ablation of poly(methyl methacrylate) by a single pulse of soft X-rays emitted from Z-pinch and laser-produced plasmas, *Surf. Rev. Lett.* **9**: 347–352.

Juha, L., Präg, A. R., Krasa, J., Cejnarova, A., Kralikova, B., Skala, J., Chvostova, D., Krzywinski, J., Andrejczuk, A., Jurek, M., Klinger, D., Sobierajski, R., Fiedorowicz, H., Bartnik, A., Pína, L., Kravarik, J., Kubes, P.,. Bakshaev, Yu. L, Chernenko, A. S., Korolev, V. D., Tumanov, V. I., Ivanov, M. I., Scholz, M., Ryc, L., Tomaszewski, K., Viskup, R., and Boody, F. P., 2002b, Ablation of Organic Polymers and Elemental Solids Induced by Intense XUV Radiation, in: *X-ray Lasers: 2002*, J. J. Rocca, J. Dunn, and S. Suckewer, eds., AIP Conf. Proc. **641**, pp. 504–509.

Juha, L., Bittner, M., Chvostova, D., Krasa, J., Otcenasek, Z., Präg, A. R., Ullschmied, J., Pientka, Z., Krzywinski, J., Pelka, J. B., Wawro, A., Grisham, M. E., Vaschenko, G., Menoni, C. S., and Rocca., J. J., 2005a, Ablation of organic polymers by 46.9-nm-laser radiation, *Appl. Phys. Lett.* **86**, 034109.

Juha, L., Bittner, M., Chvostova, D., Krasa, J., Kozlova, M., Polan, J., Prag, A. R., Rus, B., Stupka, M., Feldhaus, J., Letal, V., Otcenasek, Z., Krzywinski, J., Pelka, J. B., Andrejczuk, A., Sobieraiski, R., Ryc, L., Boody, F. P., Fiedorovicz, H., Bartnik, A., Mikolajczyk, J., Rakowski, R., Kubat, P., Pina, L., Horvath, M., Grisham, M. E., Vaschenko, G. O., Menoni, C. S., and Rocca, J. J., 2005b, Short-wavelength ablation of molecular solids: pulse duration and wavelength effects, *J. Microlithography, Microfabrication, and Microsystems* **4**: 033007.

Katoh, T., and Zhang, Y., 1998, High aspect ratio micromachining by synchrotron radiation direct photo-etching, *Microsyst. Technol.* **4**: 135–138.

Lapczyna, M., and Stuke, M., 1998, Direct fabrication of micro mesas by VUV laser ablation of polymers: PMMA (polymethylmethacrylate), *Appl. Phys. A: Mater. Sci. Process.* **66**: 473–475.

Lee, S. K., Chang, W. S., and Na, S. J., 1999, Numerical and experimental study on the thermal damage of thin Cr films induced by excimer laser irradiation, *J. Appl. Phys.* **86**: 4282–4289.

Leguern, F., André, J.-M., Lebreton, J.-P., Dutrannoy, J.-L., Chauvineau, J.-P., Krastev, K.,. Larcade, J.-L., Friart, D., Nazet, C., and Barchewitz, R., 1997, Experimental study and simulation of the damage induced to various multilayer interferential mirrors by soft X-ray plasma-laser sources, *J. X-Ray Sci. Technol.* **7**: 271–283.

Lippert, T., and Dickinson, J. T., 2003, Chemical and spectroscopic aspects of polymer ablation: special features and novel directions, *Chem. Rev.* **103**: 453–485.

Lukashenko, G. M., Polotskaya, R. I., and Sidorko, V. R., 1992, Thermodynamic properties of scandium, lanthanum, neodymium, and gadolinium silicides and germanides, *J. Alloys Compounds* **179**: 299–305.

Macchietto, C. D., Benware, B. R., and Rocca, J. J., 1999, Generation of millijoule-level soft-x-ray laser pulses at a 4-Hz repetition rate in a highly saturated tabletop capillary discharge amplifier, *Opt. Lett.* **24**: 1115–1117.

MacGowan, B. J., Mrowka, S., Barbee, Jr., T. W., DaSilva, L. B., Eder, D. C., Koch, J. A., Pan, L. S., Turner, J. A., Underwood, J. H., and Young, P. E., 1992, Investigation of damage to multilayer optics in x-ray laser cavities: W/C, WRe/C, WC/C, stainless-steel/C, and Cr_3C_2/C mirrors, *J. X-Ray Sci. Technol.* **3**: 231–282.

Maida, O., Kohma, N., Ueno, M., Shibuya, A., Kanashima, T., Okuyama, M., and Ohashi, H., 2001, Evaporation and expansion of poly-tetra-fluoro-ethylene induced by irradiation of soft X-rays from a figure-8 undulator, *Jpn. J. Appl. Phys., Part 1* **40**: 2435–2439.

Pugachev, A.T., Churakova, N.P., and Volkov, Yu. A., 1982, Heat conduction of gold thin films in the temperature interval of 80 to 300 K, Poverkhnost. Fizika, khimiia, mekhanika **9**: 149.

Riedel, D., and Castex, M. C., 1999, Effective absorption coefficient measurements in PMMA and PTFE by clean ablation process with a coherent VUV source at 125 nm, *Appl. Phys. A: Mater. Sci. Process.* **69**: 375–380.

Rocca , J. J., Shlyaptsev, V. N., Tomasel, F.G., Cortazar, O. D., Hartshorn, D., Chilla, J. L. A., 1994, Demonstration of a discharged pumped table-top soft-x-ray laser, *Phys. Rev. Lett.* **73**: 2192–2195.

Rocca, J. J., Clark, D.P., Chilla, J. L. A., Shlyaptsev, V. N., 1996, Energy extraction and achievement of the saturation limit in a discharge-pumped table-top soft x-ray amplifier, *Phys. Rev. Lett.* **77**: 1476–1479.

Sobierajski, R. J., Krzywinski, J., Andrejczuk, A., Faatz, B., Felten, F., Jacobi, S., Juha, L., Jurek, M., Kauch, A., Klinger, D., Pelka, J. B., Saldin, E., Schneidmiller, E., Sikora, M., Steeg, B., and Yurkov, M., 2003, *Free Electron Lasers 2002*, Argonne, edited by K.-J. Kim, S. V. Milton, and E. Gluskin, Elsevier, Amsterdam, p. II–77.

Srinivasan, R., and Braren, B., 1989, Ultraviolet laser ablation of organic polymers, *Chem. Rev.* **89**: 1303–1316.

Srinivasan, R., 1994, *Laser Ablation: Principles and Applications*, J. C. Miller, ed., Springer, Berlin, p. 107.

Uspenskii, Yu. A., Levashov, V. E., Vinogradov, A. V., Fedorenko, A. I., Kondratenko, V.V., Pershin, Y. P., Zubarev, E. N., and Fedotov, V. Y., 1998, High-reflectivity multilayer mirrors for a vacuum-ultraviolet interval of 35-50 nm, *Opt. Lett.* **23**: 771–773.

Voronov, D. L., Zubarev, E. N., Kondratenko, V. V., Penkov, A. V., Pershin, Y. P., Ponomarenko, A. G., Artioukov, I. A., Vinogradov, A. V., Uspenskii, Y. A., and Seely, J. F., 2002, Thermoresistive multilayer mirrors with antidiffusion barriers for work at the wavelengths 40-50 nm, in: *X-ray Lasers 2002*, J. J. Rocca, J. Dunn, and S. Suckewer, eds., AIP Conf. Proc. **641**, pp. 575–582.

Yaakobi, B., Kim, H., Soures, J. M., Deckman, H. W., and Dunsmuir, J., 1983, Submicron x-ray lithography using laser-produced plasma as a source, *Appl. Phys. Lett.* **43**: 686–688.

Yu, L. H., DiMauro, L., Doyuran, A., Graves, W. S., Johnson, E. D., Heese, R., Krinsky, S., Loos, H., Murphy, J. B., Rakowsky, G., Rose, J., Shaftan, T., Sheehy, B., Skaritka, J., Wang, X. J., and Wu, Z., 2003, First ultraviolet high-gain harmonic-generation free-electron laser, *Phys. Rev. Lett.* **91**: 074801.

Yurkov, M. V., 2003, Development of free-electron x-ray lasers on the TESLA test accelerator (DESY, Germany), *At. Energy* **94**: 108–112.

Zhang, X., Jacobsen, C., Lindaas, S., and Williams, S., 1995, Exposure strategies for polymethyl methacrylate from *in situ* x-ray absorption near edge structure spectroscopy, *J. Vac. Sci. Technol. B* **13**: 1477–1483.

Zhang, Y., Katoh, T., Washio, M., Yamada, H., and Hamada, S., 1995, High aspect ratio micromachining Teflon by direct exposure to synchrotron radiation, *Appl. Phys.Lett.* **67**: 872–874.

Chapter 22

LASER RESTORATION OF PAINTED ARTWORKS

Fundamentals, Modeling and Advances

Giannis Bounos, Austin Nevin, Savas Georgiou, Costas Fotakis
Institute of Electronic Structure and Laser, Foundation for Research and Technology – Hellas, 71110, Heraklion, Crete, Greece

1. INTRODUCTION

Lasers have by now found important and various applications in the preservation and restoration of artworks, as their use offers a number of distinct advantages over conventional methods. A most important aspect of the application of lasers in art conservation lies in analysis. However, because of the particular emphasis of this volume on laser ablation, we focus here exclusively on the use of lasers for the cleaning of works of art, in particular of paintings, parchments, etc. Indeed, laser-based restoration closely corresponds to a wide spectrum of highly successful laser ablation-based intervention applications, such as etching of polymers [Srinivasan R., et al., 1989; Bäuerle D., 2003] and photorefractive keratectomy and laser-based excision of tissue in medicine [Vogel A., et al., 2003]. In all these cases, laser irradiation is exploited in order to effect highly precise and controlled material removal (either for the purpose of eliminating unwanted degraded material/pathogenic tissue or for the purpose of appropriately shaping the substrate (microstructuring)) with minimal collateral damage to the substrate predominantly from organic materials. Consequently, similar principles underline all these applications. In this chapter, we consider how concepts and methodologies from these other applications have been adapted to the special requirements of restoration and how, in turn, the development of laser restoration methodologies has resulted in the reinvestigation of processes in model systems.

Cleaning of works of art may be desirable for maintaining their aesthetics, but often it is absolutely imperative for prolonging their lifetime

[Wolbers R. et al., 1990]. However, the efficacy of traditional cleaning methods is often compromised by the high complexity of the substrates/painting Traditional methods rely on the use of mechanical or chemical means. Mechanical means include abrasive methods, jet spraying, or use of scalpel, etc. However, even in the case of extensive experience, human errors can lead to inadvertent material removal from the substrate and to the destruction of its texture. On the other hand, chemical solvents and gels may penetrate into the substrate and result in dissolution or leakage of pigments, etc. Furthermore, the employed chemicals (e.g., methyl ether ketone, methylene chloride, phenol, etc.) are often highly aggressive, thereby introducing serious health hazards.

In comparison with conventional techniques, laser processing offers some well demonstrated advantages (e.g. automation, selectivity, versatility, a high degree of online control by interfacing with a variety of in situ laser-based analytical techniques etc.). These are most important criteria to be met in the treatment of complex substrates. However, the determining advantage of laser processing is the nearly submicron resolution structuring of the substrate with minimal collateral damage. Even for the most experienced operator, removal of layers with sub-micron resolution is simply impossible.

However, the extension of laser based techniques to the restoration of artworks poses a number of specific problems due to the complexity of paintings [Asmus, J. F., 1986; Georgiou, S., et al., 1997]. Besides the apparent issues of efficiency and effectiveness, the most important one concerns the nature and extent of any side effects that may be induced to the substrate. This is crucial for artworks, since minor chemical or structural modifications may result in their accelerated aging/deterioration.

In the following, we first provide an overview of the methodologies that have been developed for dealing with the various restoration problems of painted artworks (Sections 2 and 3). Particular emphasis is placed on the issue of chemical effects that may be induced (Section 4) and any dependence on laser and material properties (Section 5). Addressing this question, in fact, touches upon fundamental aspects of laser/material interactions.

2. STRUCTURE OF PAINTINGS

Paintings, which range from miniature pendants to large-scale wall paintings, are characterised by their technique - the specific order in which they were constructed and painted - and are found on various kinds of substrates (including earth and straw, stone and lime, metal, glass, ceramics, wood, pottery, fabric and paper). Independently of any differences in composition, paintings are composed of pigments usually dissolved/dispersed in some kind of binding medium [Reed, R., 1970]. The colour/appearance of paintings depends not only on the type of pigment/colorant used, but also on the type of medium which is employed for its application, on the type of surface on which it is applied, as well as on the varnish that may be applied subsequently; all components are critical for the physical and chemical properties of paintings and have direct relevance to their conservation.

Artists have turned to naturally occurring materials for both pigments and binding media. Broadly, pigments are either inorganic or organic (from plants and animals). The earliest used pigments are represented by silicate based earth pigments (ochres) found in prehistoric cave art, but with technological advances: synthesis of blue and purple glass pigments has been documented in archaeological findings (Egyptian blue and Han Blue (China)). Although many pigments are derived from minerals, this does not imply long-term stability; some are particularly sensitive to alterations, both chemical and physical, which can be accompanied by changes in colour (e.g. the alteration from red to brown-black of cinnabar (HgS), which is due to a physical change in crystal structure), whereas the darkening of silicate earth pigments upon exposure to heat is due to reduction of the state of hydration. Organic pigments are also very common in artworks. These are generally more light- sensitive than mineral based materials [Gettens, R., et al., 1966].

The application of pigments to any surface requires some kind of binding medium; in fresco paintings, this is provided by the substrate of calcium hydroxide (lime) which reacts to form calcium carbonate. However, in the vast majority of paintings, naturally occurring materials are added to pigments to function as adhesives; these range from plant gums and tree resins to seed oils (linseed, walnut, and poppy) and protein-based materials of animal origin (eggs, milk and glues). The use of oil-based media is generally limited to European paintings, whereas other media are widespread elsewhere. In addition, waxes (encaustic) have been used for paintings. In recent centuries, synthetic polymers, including various acrylics and silicate mixtures, have been used as media because of flexible working properties than of the natural polymers [Casadio, F., et al., 2004].

Regardless of origin or chemical composition, all media share a common function; they serve to adhere the pigment to a substrate. The film-forming capability of media used in works of art can be the result of oxidative cross-linking reactions (as in the case of oil and egg films) or through the loss of water from polypeptides or polysaccharides. In addition to the use of polymers for the application of pigments, their use for the varnishing and finishing of paintings is widespread; varnishes, often based on naturally occurring resins (mastic and dammar, for instance) can also contain oil-based media. Varnish layers (typically 50-100 μm) are applied to the painting for a variety of reasons, mostly for the improved optical effects which they can impart; not only is the appearance of some colours more "saturated" following the addition of a varnish, but further changes in the surface texture can be achieved. However, with time, changes in the chemical composition of varnish occur, as well as darkening effects from the deposition of dirt and further cross-linking reactions necessitating the removal and replacement of the uppermost layers of varnish.

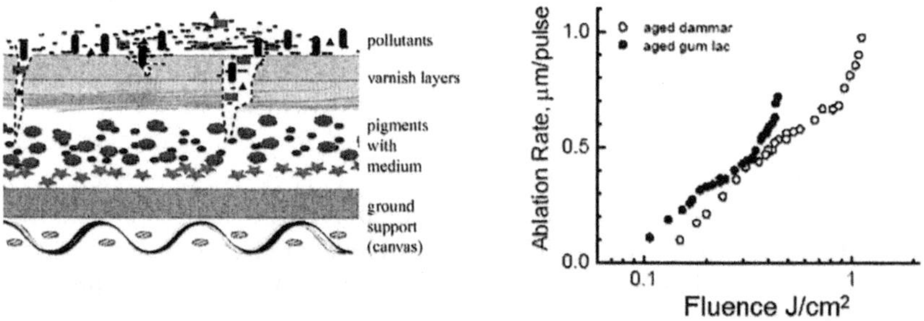

Figure 1. (a) Schematic of painted artwork. (b) Etching efficiencies for two types of varnishes.

3. OVERVIEW OF RESTORATION PROBLEMS

Broadly, the typical problems in restoration of painted artworks are:

- Cleaning of the surface (varnishes) and of the support material (canvas, wood, etc.) from accumulated pollutants (degradation preoducts, etc.). The degraded layers may be so hard and/or chemically inert that conventional mechanical or chemical methods become highly inefficient. It is also not uncommon for the 'contaminants' to be of similar chemical and physical properties of the original material, hence the use of solvents can compromise the underlying materials and therefore mechanical alternatives are traditionally adopted. The purpose of the laser here is to effect layer-by-layer material removal with minimal influence to the pigment medium.
- Removal of overpaintings in order to recover the original painting; this is very much like the previous problem expect for the much higher caution that may be needed in ensuring accurate delineation between the painting

to be removed from the lower one, especially as areas of overpaint can be confused with restorations.

- In many cases, the offending/degrading agent is not of a uniform, relatively large thickness, but instead in the form of small, isolated dots adhered to the surface or even partly entrapped to the substrate. In this case, a more refined mode of laser interaction must be employed, characterized by localized action only on the units to be removed.

4. LASER CLEANING METHODOLOGIES

4.1 Cleaning of Varnish/Overpainting

Lasers are now routinely employed to effect removal of coatings including polymers from a variety of substrates [Bäuerle, D., 2003; Luk'yanchuk, B. S., 2002]. In these cases, cleaning relies on the exploitation of the massive, unselective material ejection, i.e. ablation, that is effected upon irradiation with intense laser pulses. In these applications, the objective is the removal of the coating with no major collateral effect to the morphology or structure of the substrate.) In the case of artworks, the application is considerably more demanding, since the substrates are comprised of stratified media of varying chemical and structural composition and variable fragility.

The first aspect to consider concerns the efficiency and accuracy of material removal that can be attained with laser irradiation. To this end, Fig. 1(b) illustrates the dependence of etching rates for two types of varnishes on laser fluence [Fig. 1(b)]. Generally, depending on varnish optical and chemical properties, the features of the etching curves are described by either of two simple phenomenological models, corresponding to the "steady-state" and the "blow-off" models employed in polymer ablation. In few cases, the dependence of the etching depth (δ) is well described by:

$$\delta = \frac{F_{LASER} \check{S} F_{thr}}{\rho E_{abl}} \quad , \qquad (1)$$

where F_{thr} and E_{abl} represent, respectively, the minimum fluence (threshold) and the minimum energy density for material ejection ρ: density of material. This formula describes with some success the etching curves for longer laser pulses ($\tau > 1\mu s$-$1 ms$). However, as usual for polymers for nanosecond pulses, the "blow-off" (or "layer-by-layer" removal) model is generally more applicable. The basic premise is that for an incident fluence, all material within a depth such as $F_{transmitted} \geq F_{thr}$ is removed. Assuming Beer's law for the absorption process, the dependence of the etching depth (δ) on the incident laser fluence is given by:

$$\delta = \frac{1}{\alpha_{eff}} \ln \left(\frac{F_{LASER}}{F_{thr}}\right) \text{ for } F_{LASER} \geq F_{thr} \quad , \tag{2}$$

where α_{eff} is the (effective) absorption coefficient . Detailed discussion of the principles underlying these phenomenological models can be found in [Bäuerle, D., 2003; Georgiou, S., 2004].

Based on Eq. (2), removal requires the use of wavelength that is relatively strongly absorbed by the coating so as to achieve good removal rate at moderate laser fluences. Generally, paint-coating removal has been attempted by a variety of laser systems, (e.g. CO_2, Nd:YAG (1064 nm and its harmonics), excimer and high-power diode lasers). However, in the case of the restoration of painted artworks, excimer lasers have proven the most useful, since varnishes generally absorb well in the UV. (e.g., for varnish and its degradation products in painted artworks, $\alpha \approx 105$ cm^{-1} at 248 nm [de la Rie, 1988]), thereby ensuring efficient and "clean" etching at moderate fluences. On the other hand, because of the very small etching depth for very high α, processing at the highly absorbed by varnishes 193 nm wavelength turns out to be very time-consuming (despite the fact that otherwise it produces excellent results). From the etching rate curves (λ=248 nm) in Fig. 1(b), it is seen that material removal can be effected with a resolution of 0.1 to 1 μm per pulse. Clearly, the accuracy of material removal by laser irradiation far surpasses that afforded by either mechanical or chemical techniques. Given a laser repetition rate of 10-50Hz, contaminated surface layers of 20 to 300 μm thickness can be removed in a reasonable amount of time by Excimer laser. Illustrative examples for the restoration of painted artworks have been described previously [Georgiou S. et al 1997, Zafiropoulos V. et al 1997, Fotakis C. et al 1997]. Laser cleaning of canvas, support material (wood, silk, etc) has also been achieved.

Generally, the optimal fluences for painting cleaning are found to be in the range of 200 - 600 mJ/cm^2, while the removal of overpaintings requires ≈ 2 to 3 times higher fluences. The important point, of course, is to effect material removal with minimal, if any, influence to the sublayers. (i.e. paintings, illuminated manuscripts, etc.). To this end, the detailed studies described in Section 2 demonstrate that this objective can be satisfied if a thin layer of varnish (usually, 5 to 10 μm thick) is left intact on the pigment medium. Essentially this layer acts as a filter, preventing light from reaching the pigments and binding media.

Even in the case of well-defined polymers, simple equations like (1) or (2) may fail in describing the full range of the etching curves and their dependency on irradiation parameters (e.g. wavelength) and on properties (optical, physical, etc.) of the substrates. A detailed discussion of this problem has been presented by [Bityurin, N. et al., 2003]. In the case of varnishes, which are characterized by a high degree of ill-defined polydispersity, and/or chemical variability, deviations from the simple functional dependences can be highly pronounced. [In addition to these problems, the number of pulses employed in establishing the threshold and etching rates is unfortunately not often specified, so that comparisons between different laboratories are not directly possible.] Incidentally, rather high emphasis in the field of laser restoration is placed on the "ablation efficiency" of the process, i.e. the energy necessary to remove a given mass $\eta_{abl} = \rho\delta/F_{LASER}$, where ρ is the density of the material and F_{LASER} represents the energy absorbed. For the "blow-off" model, η_{abl} reaches a maximum at moderate fluences above the ablation threshold decreasing subsequently [Vogel, A., et al., 2003; Georgiou, S., 2004]. For applications, it would appear best to work at fluences close to the maximum efficiency. However, as already mentioned, there are other crucial factors that must be taken into account. Thus, the criterion of maximizing efficiency can be highly misleading (let alone that its determination is prone to experimental inaccuracies and its interpretation questionable).

Figure 2. Painting of the Flemish School, National Gallery of Athens, Greece before and after cleaning/restoration from oxidized varnish layers, using both lasers and solvents.

4.2 Selective Material Removal/ 'Dry' Laser Restoration

In the previous case/problem, the etching depth is smaller than the film thickness that needs to be removed. Thus, the concept of layer-by-layer removal implied by Eq. (2) suffices for the purposes of the application. In many cases, however, the impurity film has a thickness comparable or smaller than the etching depth, or it consists of isolated pigments on the surface of the substrate. This situation is very common in treating parchments [Reed, R., 1970; Kautek, W., et al., 1998]. It is important to note that conventional removal techniques such as ultrasonic, jet spraying, etc. are ineffective or altogether fail in dealing with micron and submicron particles [Luk'yanchuk, 2002]. On the other hand, this objective may also be attained by laser irradiation, albeit only under certain conditions. The solution illustrates in the most direct way the versatility of laser processing and how these can be exploited to deal with different problems.

The selective removal of impurities may be effected via essentially thermal evaporation processes induced by irradiation at fluences below the ablation threshold. Indeed, for a number of molecular systems, there is now strong evidence by experiments, simulations and theoretical analysis that at intermediate laser fluences, a thermal desorption/evaporation process operates. In contrast, the ablative regime, as defined above, entails the massive, (volume), non-selective ejection of material by other mechanisms (explosive boiling, stress-wave spallation, photochemical pressure generation, etc.). For a number of polymers, similarly at low fluences, mass loss, as measured by sensitive quartz crystal balances, is observed. The loss is due to the desorption of fragments/species weakly bound to the polymer matrix that are formed upon laser irradiation.

A satisfactory removal rate via evaporation can be attainted for fluences:

$$F_{laser} \geq F_{\min} = \rho\alpha^{-1}\Delta H_{sub}(1-R)^{-1}, \tag{3}$$

where ΔH_{sub} represents the sublimation or evaporation enthalpy, ρ the density, and R the reflectivity. Of course, the threshold for damage of substrate should be much higher than the fluence required for the removal of contaminants ($F_{s,min}>F>F_{cont,min}$). Therefore, for this approach to be effective, contaminants must have a higher absorption coefficient than the substrate, and/or a much lower sublimation enthalpy. Alternatively, the substrate must exhibit a high reflectivity. The ideal situation would be for the contaminant layer to be highly absorbing and the substrate to be highly reflective with minimal, if any, absorptivity. In such a case, the laser beam is completely reflected by the substrate once the contaminant layer has been removed. Such "*self-limiting*" cases are encountered in the removal of inorganic artefacts, but they are not common in the treatment of organic materials.

The efficiency of this approach has been examined [Kautek, W., et al., 1998; Kolar, J., et al., 2002] for parchments with carbonaceous contamination. For graphite, at λ= 308 nm, R_{sub}=0.35, $\alpha_{cont} \approx 2x10^5$ cm^{-1} whereas for parchment, $\alpha_{sub} \leq 400$ cm^{-1}. In agreement with Eq. (3), selective removal of dirt without any morphological changes is attained by irradiation at 308 nm. However, chemical alterations to the substrate are induced, including a ≈30% drop in the polymerization degree of cellulose.

In microelectronics, a particularly effective way of particle removal from surfaces, including polymers [Luk'yanchuk, B. S., 2002], is attained via laser-absorption by the substrate resulting in its thermal expansion and the subsequent acceleration and ejection of the particles. For evident reasons, this approach has not been examined in the case of artefacts.

Understandably, the range of fluences between the cleaning threshold and surface modification of the polymer substrate is quite narrow. In fact, because of plausible field enhancement under the particles [Mosbacher, M., et al., 2001], local ablation of the substrate and damage to the substrate may be effected at fluences much lower than in the absence of particles. In addition to these problems, for polymeric substrates and particles, further issues may become important (e.g., photochemical modifications).

4.3 Liquid Assisted Material Removal

Laser-induced material removal and particle removal is much enhanced when a liquid film (typically a few tenths to several μm thick) is applied on top of the substrate ("steam laser cleaning") [Tam, A. C., et al., 1998; Dongosik, K., et al., 2001]. For sufficient deposited energy, vaporization of the liquid (more precisely, explosive boiling) takes place. The high-amplitude pressure wave that is generated by the fast-growing bubbles results in a significant enhancement of the etching efficiency and in the sufficient accelerations (10^8 to 10^9 m/s^2) of any surface-adhered particles for their detachment and their ejection. Most importantly, laser steam cleaning efficiencies are much higher than those in dry cleaning. Furthermore, in contrast to dry laser cleaning, the cleaning efficiency is largely independent of the particle size, thereby permitting efficient cleaning even of collections of particles with a wide size distribution. Although most work has focused on particle removal from silicon wafers, few studies have also been reported on steam-laser cleaning of polymer substrates [Lee, Y. P., et al., 1998].

On the basis of the above, processing of paintings with the application of a thin layer of alcohol has been suggested [de Cruz, A., 2000]. OH-rich solvents absorb very well at 2.94 μm, so that laser irradiation at this wavelength is preferentially absorbed by the solvent with reduction of the heat diffusion to the substrate. However, applying of liquid on the highly sensitive organic/painted surfaces may have deleterious side effects. Furthermore, because of the high pressures developed in the bubbles, the potential for mechanical collateral damage is much higher than that for ablation in air. This is not a major concern in industrial applications, but it can be of detrimental influence in the processing of fragile artworks.

5. EXPERIMENTAL SETUPS AND TECHNIQUES

There are well-established setups/designs that have evolved in industrial applications of lasers and have been adapted for the purposes of laser cleaning schemes [Zafiropoulos, V., et al., 1997; Fotakis, C., et al., 1997]. However the most important issue for the successful restoration of artworks is the use of techniques for the online monitoring of the process, therefore adjusting appropriately processing parameters. Monitoring can be achieved by the variety of optical and spectroscopic techniques. Furthermore, these techniques can be employed for the detailed examination of the processes involved in laser restoration. To this end, a brief description (by no means, exhaustive) of methods that have been employed is given below.

Direct analytical information, mainly about atomic constituents of the ablated volume, can be obtained through the spectrally and temporally-resolved analysis of the spectrum radiated by the plasma produced during ablation (LIBS). The independence of the process from charging effects (thus, from substrate conductivity) in combination with the high sensitivity for pigment constituents makes LIBS powerful for monitoring laser restoration of painted artworks. Detailed description of the work in this direction can be found in [Anglos, D., 2001].

Concerning the examination of irradiated areas, chromatography is extensively used by restorers in the examination of the chemical changes induced to artworks. However, as the amount of extracted material cannot be controlled very well and generally it is larger than the effective optical penetration depth such measurements can be in error, regarding quantification of chemical changes. Much more sensitive chemical characterization is afforded by laser-induced fluorescence [Anglos, D., et al., 1996], Raman [Castillejo, M., et al., 2000] and FTIR [Perez, C., et al., 2003]. The application of these techniques for a variety of diagnostic as well as restoration problems has been described.

Structural information about the painting can be derived from reflectography [Balas, C., 1997; Georgiou, S., et al., 1997]. This relies on the fact that UV absorptivity of varnish increases strongly upon degradation/oxidation, and it is characterized by a much lower reflectivity than non-degraded/"clean" varnish. Thus, reflectography provides a direct tool for optimizing restoration materials, including laser-based ones, by detecting the fresh layers that are exposed as dirt and debris is removed. Furthermore, multispectral imaging techniques can be used for mapping the composition and coloration of a painted artwork.

Holographic methodologies are a powerful means for the examination of photomechanical (structural) effects over extended areas of the irradiated objects. The versatility of the technique allows its adaptation to the different detection requirements posed by a large range of movable and immovable works of art [Athanassiou, A., et al., 2002].

5.1 Assessment of the Laser Induced Effects

Besides the issue of efficiency, the most important consideration in the adoption of any restoration technique concerns the nature and extent of side effects that may be induced to the substrate. To this end, much of the following discussion relies on studies on appropriately 'designed' polymeric doped systems, serving essentially as model systems. The necessity of using model system derives first from the fact that detailed case-by-case studies of effects in realistic samples becomes clearly tedious and impractical. Most importantly, even if a detailed examination were possible, this would not specify the responsible mechanisms, as necessary for the systematic/rational optimization of the techniques. By comparison the simplicity of the reactivity patterns of the dopants enables the systematic characterization of the induced modifications as a function of laser fluence. Most importantly, the doped systems constitute a good, even if idealized, model of the paint layer in artworks, which essentially consists of chromophores dissolved or dispersed within an organic medium. Certainly, there are still a number of observations on realistic artefacts whose generality or importance cannot be yet specified (i.e., are they specific to the chemical composition of the material or not).

5.2 Thermal effects

Thermal side effects (decomposition as well as any induced morphological changes) are of major concern in the optimization of laser restoration applications, since the processed substrates (e.g., painted artworks, parchment, etc.) tend to be thermally sensitive. Analytical approaches describing heat diffusion effects upon ns irradiation would be most useful. For simple polymers, analytical approaches have been developed relying on the consideration of heat conduction coupled with thermal decomposition/desorption of the material (photofragmentation neglected) leading to a differential equation for the evolution of the temperature in the substrate as a function of time and depth. In particular, the so-called volume model [Bityurin, N., et al., 2003] has succeeded in explaining experimental results on the ablation of polymers and organic materials. However, such detailed models cannot be reliably used for real-life artefacts.

For the typical absorptivity of varnishes, temperatures as high as 700°C-800°C may be reached at fluences close to their 248nm ablation threshold.

The physical properties of varnishes are not well-known, but it is likely that these temperatures much exceed their melting point. Melting induced upon laser irradiation can have deleterious effects on the appearance and the integrity of the irradiated substrate, e.g. for varnishes being mixtures of various oligomers and polymers enhanced segregation may occur within the molten zone; preferential desorption of weakly bound species may occur, resulting in changes of the chemical composition of the substrate; the back-pressure exerted by the ejected material may result in irregular surface morphology. However, the depth of the molten material and the duration of melting decreases much with absorption coefficients. Furthermore, [Tokarev, V. N., et al., 2004] have shown that the flow velocity of material is very much reduced with decreasing melt depth. Thus, even a simple thermal mechanism may result in "clean" processing. Thus, contrary to common arguments, lack of a melting zone around the irradiated area does not necessarily indicate a "non-thermal" (photochemical) mechanism of ablation

The detailed analytical studies above demonstrate that the heat penetration depth shows a complicated temporal dependence. However, for moderate to strong α at sufficiently high fluences, $z_{thermal}(t) \approx (Dt)^{1/2}$, where t is the time of energy removal (t ~1-10 μs). Typically, D for polymers and amorphous organic materials is $\approx 10^{-3}/10^{-4}$ cm^2/s. Thus, the thermal diffusion length for typical ns pulses is estimated to be ≈100 nm - 500 nm, i.e., smaller or at most comparable to the typical optical penetration depth (1-10μm) of varnishes and of typical pigments in the UV. Thus, the extent of "damage" is mainly specified by the substrate absorptivity at the irradiation wavelength. For instance, in the case of painted artworks, optical microscopic examination demonstrates "clean" etching for ablation at 248 nm, a strongly absorbed wavelength (α~10^5 cm^{-1}). In contrast, irradiation at weakly absorbed wavelengths (e.g., at 308 nm, α ~10^2 cm^{-1}) results in pronounced collateral damage and/or morphological changes. Similarly, thermal conduction clearly limits the efficiency of laser-cleaning techniques when the separation between the absorbing units (impurities) and of the substrate is comparable to the thermal diffusion length. Thus, upon IR irradiation of parchments in the presence of absorbing impurities, it has been shown that H_2O loss and cross-linking of the cellulose takes place [Kolar, J., et al., 2002], as a result of the heat conducted from the absorbing units.

5.3 Influence of Molecular Weight

Molecular weight is a fundamental parameter which directly affects the physical parameters of polymers. In addition, this parameter can be crucial in laser applications both in medicine as well as in the restoration of artwork, since for substrates in these applications it varies greatly from case to case.

Indeed, work on PMMA (at 308 nm and 248 nm) demonstrates that the ablation threshold as well as the etching efficiency may decrease much with polymer molecular weight. This has been confirmed by more recent studies. To a first-order approximation, this increase can be described to the fact that in Eq. (2), F_{thr} depends on the cohesive energy of the system. For polymeric systems, a high MW implies that a higher number of bonds must be broken for the chains to decompose into small units and oligomers, as necessary for desorption/ejection in the gas phase. Because of the difference in etching efficiencies, a higher amount of chemical modification/product formation in the substrate is observed for substrates with increasing MW.

The consequences of this dependence of the etching threshold on molecular weight have been very little explored, although preliminary reports in the case of painted artworks indicate correspondence with the previous results on well-defined polymers. Yet, some implications can be drawn. First, the ablation threshold and material removal efficiency will vary greatly with degree of polymerization of the varnishes (in practice, this reflects the empirical finding of restorers that "harder" material requires higher fluences). Second, since the degree of polymerization varies substantially with increasing depth (usually decreases from the upper, superficial layers to the lower, non-oxidized layers, due to enhanced exposure of the outer layers to light, humidity, etc.), the ablation threshold and etching efficiency may vary as the cleaning procedure continues. Therefore, it is imperative that laser fluence is appropriately adjusted during the procedure. Hence, elucidation of the importance of this polymer parameter will allow the refinement and further optimization of previously developed methodology.

5.4 CHEMICAL PROCESSES AND EFFECTS

The most crucial question in laser processing of paintings, parchments, etc. concerns the nature and extent of chemical modifications induced to the substrate. The processed molecular substrates include a wide variety of chromophores, which, upon UV excitation, may dissociate into highly reactive fragments [Mills, J. S., 1956]. Additional species may be formed by the thermal or stress-induced breakage of weak bonds. These species may form by-products (e.g., oxidation products) in the short or long-term with detrimental effects for the integrity of the substrate. Thus, minimization of their formation is crucial for the optimization of the laser restoration applications.

5.4.1 The photochemical mechanism

It is worth considering the issue of photothermal vs. photochemical mechanisms. The photochemical mechanism is a somewhat overly used explanation/justification for the use of lasers in both medical applications and for the restoration of works of art. According to the "photochemical" model, the formation of a high number of photofragments with high translational energies and/or the formation of gaseous photoproducts that exert a high pressure may result in material ejection. Because the photon energy is largely "consumed" in bond dissociations, the heat "load" to the substrate is minimal with a consequent improvement in the morphology of the processed surface.

However, even for simple, well defined molecular systems, it has proven very difficult to assess the contribution of such a mechanism. In fact, the issue of photochemical vs. thermal mechanisms has been the most hotly debated one in the field of ablation. After numerous studies it now appears that for a number of common organic and polymeric systems, overall, a thermal mechanism is dominant for ablation in UV. Only, for few specifically designed polymers, which upon photolysis produce a very high number of gaseous (N_2, CO, etc.) species, there is evidence for the operation of a photochemical mechanism [Fujiwara, H., et al., 1995; Lippert, T., et al., 1997; Lippert, T., 2003]. Vogel, A., et al., (2003) suggest that although the extreme view of "exclusive" photochemical mechanism is unlikely, photochemical processes can still result in a reduction of the substrate's integrity (e.g. via bond decomposition, etc.), thereby facilitating material ejection. A similar possibility has also been suggested by molecular dynamics simulations [Yingling, Y. G., et al., 2001].

In view of all these controversies, it is very difficult to ensure if a photochemical mechanism is applicable in the irradiation of the complex materials encountered in artworks. Certainly, varnishes do include a number of photolabile units that could result in some production of gaseous products. However, as compared to the case of specifically doped polymers, the extent of any gaseous product formation is not well-defined. At any rate, the following studies demonstrate that even if a largely thermal mechanism is dominant, appropriate irradiation conditions can be found for minimizing both thermal and chemical effects to the substrates/artworks.

5.4.2 Factors affecting chemical effects in the substrate

Generally, varnishes and pigments are photolabile with relatively high quantum dissociation yields (e.g. for common organic chromophores

$\eta \sim 0.2\text{-}1$). Thus, upon irradiation, the number of dissociated/photomodified molecules scales with depth z (from the surface),

$$N_D(z) = \eta \, \frac{\sigma N F_{LASER}\, e^{-az}}{h\nu}, \tag{4}$$

where σ is the absorption cross-section, N the number density of the chromophores, and $h\nu$ the photon energy. However the removal of a depth δ via the etching process implies that the number of fragments/products remaining in the substrate is lowered. This description is, however, correct only to a first approximation, because (1) the temperature changes will much affect the evolution of the chemical processes and the extent of the products formed, and (2) excitation of molecules to higher electronic states may occur that react differently from that assessed upon 1-photon excitation.

The interplay of the various factors affecting chemical processes and modifications is best illustrated in studies employing photosensitive organic compounds, etc. dispersed within polymer films (dopants) [Lippert, T., et al., 1997]. For some dopants, no particular difference in their decomposition is found when irradiating above vs. below the ablation threshold. For instance, in the UV ablation (248 nm or at 266 nm, 60 ps pulses) of PMMA films doped with 5-diazo Meldrum's acid [Fujiwara, H., 1995], no significant change was found in the dopant decomposition yield and in the product signal intensity above vs. below the ablation threshold. In the examination of the 1,1,3,5- tetraphenylacetone (TPA) dopant within PMMA matrix upon excitation at 266 nm with nanosecond pulses, TPA was found to decompose into two diphenylmethyl radicals and CO [Arnold, B. R., et al., 1992]. At low fluences, the diphenylmethyl radical concentration grows linearly with fluence, while at higher fluences, excited radicals are also observed, formed via a two-photon excitation process.

The previous dopants dissociate to stable compounds (CO or N_2) with a decomposition quantum yield close to unity. Thus, it is rather understandable that no particular change in their photolysis occurs as the fluence is raised above the threshold. On the other hand, many other chromophores, such as encountered in paintings, dissociate into reactive radicals. Since the reactivity of such radicals depends sensitively on the temperature, it may be much affected by heat diffusion effects. Thus, for such compounds/chromophores, considerable deviations may appear in the subsequent reactivity of their fragments/radicals (even in the case that the photoexcitation/dissociation steps do not differ in the ablative vs. sub-ablative regime).

Because of the above considerations, we have chosen to study/employ, the photolabile iodonaphthalene or iodophenanthrene as dopants in PMMA due to their chemical similarity to pigments and varnishes [Mills, J. S., 1988]. Upon excitation, these compounds/chromophores dissociate to aryl radical and iodine. (Thermal decomposition for these compounds is insignificant because of the high C-I bond energy (2.6 eV)). The aryl (naphthyl-Nap, and phenanhthrenyl-Phen) radicals react with nearby polymer units by abstraction reactions to form ArH product (Scheme 2). [Athanassiou, A., et al., 2000] Indeed, for low concentrations of NapI or PhenI (ArH) (<1% wt), the main dopant-deriving product is ArH (detectable in pump-probe experiments by the aromatic $^1B_{3u} \rightarrow {}^1A_{1g}$ emission (at ≈330 nm for NapH and ≈375 nm for PhenH)), [Lassithiotaki, M., et al., 1999].

$$\text{ArX (Ar=Nap, Phen)} \xrightarrow{h\nu} \text{Ar} + \text{X}$$

$$\text{Ar} + \text{polymer} \rightarrow \text{ArH}$$

$$\text{Ar} + \text{Ar} \rightarrow \text{Ar}_2$$

Scheme 1. Pathways of aryl product formation in the irradiation of ArI doped polymers

To this end, the fluence dependence of the ArH product emission intensity following a single pump pulse on virgin polymer is plotted as a function of laser fluence (Fig. 3(b)) [Athanassiou, A., et al., 2002]. (This intensity is nearly proportional to the amount of the product that remains in the substrate after irradiation). At low fluences, product yield/amount scales linearly with F_{LASER}, consistent with 1-photon photolysis of these compounds. However, at higher fluences (but still below the threshold), the dependence of the product amount becomes supralinear. Increase cannot be ascribed to saturation of the absorption process [Andreou, E., et al., 2002], but instead, reflects an enhancement in ArH product formation per unit volume. The increase in ArH is due to the increased substrate temperatures attained with increasing F_{LASER}, as discussed in detail subsequently. On the other hand, at fluences close to the threshold, the ArH product amount in the substrate is found to reach a limiting value.

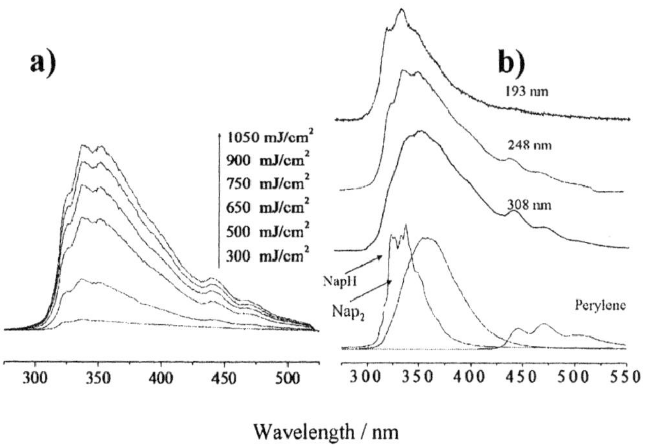

Figure 3. (a): Laser-induced fluorescence spectra recorded from NapI-doped PMMA samples (1.2% wt) after their irradiation with a single ''pump'' pulse at 248 nm with increasing laser fluence (For this system, ablation is at ~750 $mJcm^{-2}$, as established by profilometric measurements). (b) The figure also illustrates the approximate deconvolution of the probe spectrum into the emission bands of the suggested species (recorded from PMMA films doped with the indicated compounds).

At least for organic molecules decomposing via a photothermal mechanism, reactivity of species/fragments/radicals in the melt may differ much from that in the glassy condition. Further, diffusion in molten zone much increases, so that formation of by-products via recombination reactions can occur. This possibility is illustrated in the irradiation of NapI/PMMA samples with a dopant concentration higher than 1% wt. Following pump irradiation at low fluences (at 248 nm and 308 nm), NapH-product is the exclusive dopant-deriving product (Fig. 3(a)). However, for pump irradiation at higher fluences, formation of Nap_2 and perylene is demonstrated by the broad emission at ≈ 375 nm and the double peak structure around 450 nm, respectively (Fig.3(b)). At the employed dopant concentrations, NapI aggregation is minimal, as confirmed by a number of techniques. Thus, the bi-aryl products are formed through the reaction of diffusing Nap radicals (Scheme 1). A more detailed consideration of the dependence of these effects on laser parameters is presented below.

5.5 Dependence on Absorptivity

Wavelength is the most important parameter that needs specification in implementations of UV ablation [Srinivasan, R., et al., 1989; Georgiou, S., et al., 2003]. It follows from Eq. (2) that irradiation at relatively strongly absorbed wavelengths must be selected for ensuring efficient etching and good surface morphology. In fact, these are usually the only criteria employed for selection of the optimal wavelength. In addition, the operation of a photochemical mechanism is often invoked for irradiation of organics at 248 nm and 193 nm.

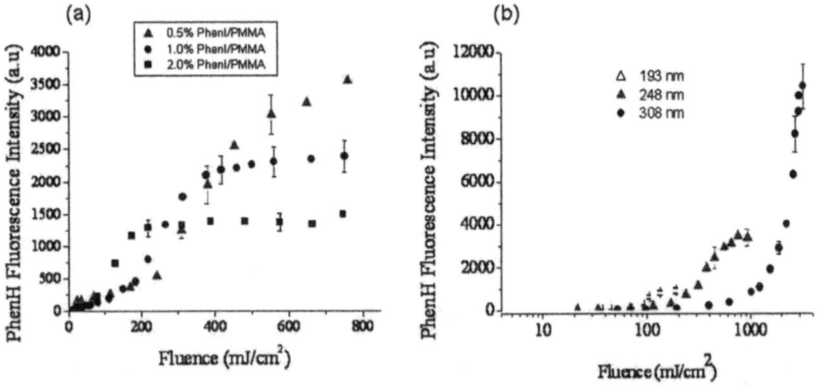

Figure 4. a) F_{LASER}-dependence of aryl (PhenH) product remaining in the substrate following irradiation (each time with a single pulse) of PMMA films doped with PhenI at 248 nm for three different PhenI concentrations. At this wavelength, PMMA absorbs very weakly and thus absorption is almost exclusively due to the PhenI dopant. (b) Product laser-induced fluorescence recorded after the ablation of PhenI-doped PMMA samples (0.5% wt) at the indicated excimer laser wavelengths.

To illustrate the importance of absorptivity on the extent of induced chemical modifications, we consider the efficiency of ArH formation (Ar = Nap or Phen) yield in the substrate upon irradiation of ArX doped polymers as a function of laser fluence for irradiation at 3 different concentrations of the dopant (Fig. 4(a)) and at 3 different excimer wavelengths (Fig. 4(b)). In all cases, the dependence is qualitatively the same, but the quantitative features differ much. Most importantly, Fig. 4(a) shows that although the concentration of the photolabile chromophore increases, the product remaining in the substrate at fluences above the threshold is much reduced with increasing substrate absorptivity [Athanassiou, A. et al., 2001]. At the simplest level, this dependence can be understood from the fact that the absorption coefficient determines the relative ratio of etching depth vs. optical penetration depth and thus the depth over which products remain in the substrate. In parallel, with increasing substrate absorptivity, the ablation threshold also decreases, thereby resulting in a decrease of the photon flux.

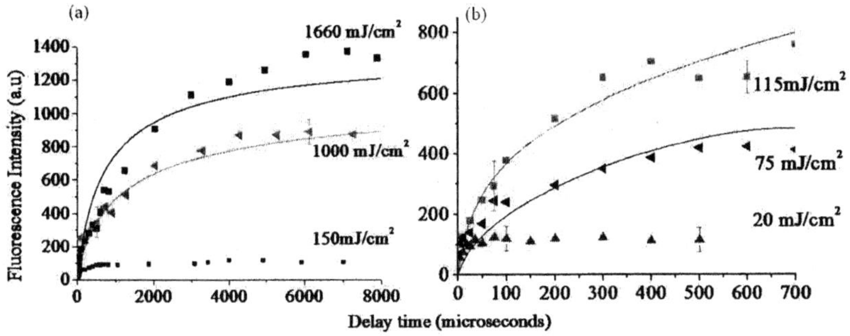

Figure 5. PhenH product fluorescence intensity as a function of the delay time between the pump and the probe beams in irradiation of PhenI/PMMA films (0.5% wt) at indicated laser fluences (representing a fluence well below the threshold, a fluence close to the corresponding threshold, and a fluence well above the threshold) at (a) 248 nm and (b) 193 nm.

A more fundamental understanding of the involved factor(s) derives from the comparison of the kinetics of the product formation at the different wavelengths (Fig. 5). For irradiation at weakly absorbed wavelengths (e.g. 248 nm for PMMA, 308 for PS), at high laser fluences (above the swelling onset) ArH-product formation is characterized by a high rate and it continues for well over ≈1 ms, whereas at low fluences it is quenched after ≈100 μs. In contrast, for irradiation at strongly absorbed wavelengths (193nm), ArH formation is quenched very fast (Fig 6.) In fact, the differences in the timescale of product kinetics correspond well to the different thermal relaxation times $t_{th} \approx 1/(\alpha^2_{eff} D_{th}) \approx 10^{-3}\text{-}10^{-2}$ s at the weakly absorbed wavelength and $\approx 10^{-5}\text{-}10^{-4}$ s for the strongly absorbed wavelengths. These differences in kinetics largely account for the different F_{LASER}- dependences observed for the amount of product remaining in the substrate upon irradiation at the corresponding wavelengths.

This difference in product formation kinetics can be understood within the framework of the volume photothermal model [Bityurin, N., et al., 2003]. Reactions of radicals within polymers usually follow a "pseudo-unimolecular" Arrhenius equation, i.e. $[Product]=[R](1-e^{-kt})$, where [R] represents the concentration of the radicals produced by the photolysis and k the reaction constant $k=Ae^{-Ea/RT}$. Detailed simulations and modelling of the observed kinetics can be found in [Bounos, G., et al., 2004]. Hydrogen atom abstraction reactions by small aromatic radicals are characterized by an activation energy E_{act} in the 40-60 kJ/mol range and a preexponential factor A of $10^5\text{-}10^7$ s^{-1}[Haselbach, E., et al., 1990]. For modeling ArH formation, the temperature evolution in the substrate upon irradiation is estimated by:

$$T(z,t) = T_0 + \frac{\alpha_{eff} F_{LASER}}{2C_P} \times \exp(\alpha_{eff}^2 D_{th} t) \times$$
$$[\exp(-\alpha_{eff} z) erfc(\alpha_{eff} \sqrt{D_{th} t} - \frac{z}{2\sqrt{D_{th} t}}) + \exp(\alpha_{eff} z) erfc(\alpha_{eff} \sqrt{D_{th} t} + \frac{z}{2\sqrt{D_{th} t}})] \quad (5)$$

where z is the depth from the film surface (erfc: complementary error function). [Kuper, S., et al., 1993] The limitations of the formula are discussed in Bounos, G., et al., 2004. T_0: initial temperature, ρ: polymer density, Cp: higher-temperature heat capacity, D_{th}=4x10^{-8} m^{2}s^{-1} [Bäuerle, D., 2000] and α_{eff}:effective absorption coefficients determined by transmission measurements at selected "pump" laser fluences below the thresholds. The simulation reproduces nearly quantitatively the F_{LASER}-dependence of the ArH product formation at various excimer wavelengths for a number of polymer systems and for different dopant polymer concentrations (for $E_a \approx$ 55 kJ/mol). At low laser fluences, only a small percentage of the photoproduced Ar radicals react to ArH. Because of the low temperatures, reactivity is "quenched" (i.e. $e^{-Ea/RT}$ becomes very small), already on μs scale. With increasing laser fluence, the reaction efficiency increases sharply and in parallel a higher percentage of the radicals in the sublayers react as a result of heat diffusion. This accounts for the obervation of ArH formation being limited by the heat diffusion time. Accordingly, the effective "reaction depth" can be approximated by:

$$\sqrt{\frac{1}{\alpha_{eff}^2} \ln[\frac{(\alpha_{eff} F_{LASER} / \rho C_P)[\ln A - \ln(\alpha_{eff}^2 D_{th})]}{(E_{act}/R)}]} \approx 7 - 10\ \mu m$$

for weakly absorbing systems and ≈ 1 μm for the strongly absorbing ones). Therefore, for the same absorbed energy $\alpha_{eff} \cdot F_{LASER}$, the ArH yield decreases with increasing α_{eff} (assuming all other factors being equal) as a result of the much shorter t_{th}. Close to the ablation threshold, Eq. (5) fails. At these fluences, energy removal by material ejection and the sharp increase of the polymer C_p (due to the decomposition into smaller fragments) limit the attained temperatures, thereby resulting in the observed leveling-off of product formation in the remaining substrate. The observation of a plateau is in accordance with the "blow-off" model, according to which the remaining substrate is subject to F_{thr}, with the additional radicals/products formed with increasing F_{LASER} being removed by the etching process.

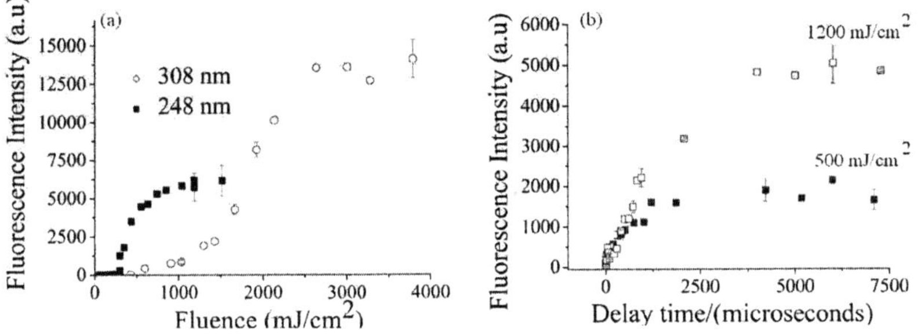

Figure 6. (a) Intensities of Nap_2 as a function of F_{LASER} in the irradiation of 1.2%wt NapI/PMMA at 248 nm and 308 nm. (b) Nap_2 formation kinetics in the irradiation of 1.2 wt% NapI/PMMA at 248 nm at the indicated fluences (ablation threshold » 1000 mJ/cm²).

We next consider recombination product formation via diffusion-limited reactivity and its dependence on substrate absorptivity. To this end, Fig. 6(a) illustrates the extent of Nap_2 formation in the irradiation of NapI/PMMA at 248 nm and 308 nm. Clearly, formation of the bi-aryl species is much reduced with decreasing wavelength i.e. polymer absorptivity, to the point that at 193 nm irradiation hardly any bi-aryl species are detected.

As previously demonstrated by a number of spectroscopic examinations, at the employed NapI concentrations (≤ 1.2% wt), dopant aggregation is insignificant . [Lassithiotaki, M., et al., 1999]. Thus, Nap_2 and perylene must form exclusively via diffusion-limited reaction(s) (Scheme I). In view of this conclusion, we have modeled the Nap_2 formation by a 2nd-order reaction, with a Smoluchowski-type rate constant $K=8k_BT/300\eta$ (where η the medium viscosity, η: Pa-s, and k_B: Boltzmann constant). At temperatures above the glass transition, the temperature dependence of polymer viscosity η is usually approximated by:

$$\eta = \eta_0 \exp\left(\frac{-C_1(T-T_{ref})}{C_2+(T-T_{ref})}\right) \tag{6}$$

where η_0: the viscosity at T_{ref}; T_{ref}: a reference temperature; C_1, C_2: constants characteristic of the polymer. The simulation reproduces, at least semi-quantitatively, the observations of Nap_2 formation. With increasing F_{LASER}, the Nap concentration increases and, in addition, viscosity further decreases, so that a much higher percentage of radicals reacts to Nap_2. However, the melt depth (approximately, $\eta \leq 10^6$ Pa-s) scales as:

$$h_{melt} \approx \frac{1}{\alpha_{eff}} \ln \frac{\alpha_{eff} F_{LASER}}{\rho C_p (T_m - T_0)}, \quad \text{for } F_{LASER} > \frac{(T_m - T_0)\rho C_p}{\alpha}, \tag{7}$$

where T_m is the melting temperature and T_0 the ambient temperature. In fact, it also depends on heat diffusion and any material and energy desorption/removal. As a result of the nearly inverse dependence of h_{melt} on α_{eff} and the shorter melt condition for the higher α_{eff}, radical mobility is much reduced, with a consequent reduction or minimization in the extent of by-product formation. Evidently, as compared to morphological examinations or even the more elaborate interferometric method developed [Tokarev, V. N., et al., 2004], Nap_2 formation provides a highly sensitive probe of the polymer viscosity changes upon laser irradiation, permitting direct quantitative assessment of the viscosity changes upon laser irradiation. For instance, the approach has been used to quantitatively monitor the viscosity changes of PMMA upon irradiation at 308 nm, 248 nm, and 193 nm. [Bounos, G., et al., 2004].

In all, with increasing absorptivity, both the extent of formation of by-products deriving via either thermally-activated reactions or via diffusion limited ones is highly reduced. It is generally assumed that the necessity for irradiation at strongly absorbed wavelengths is due to the need for attaining efficient etching and good morphology of the treated area. However, as shown above, with a high substrate absorptivity, the extent of induced chemical modifications and of side-product formation in the substrate is highly reduced, i.e., a high degree of "photochemical protection" to the remaining material is afforded. This high "protection" provides a strong rational/justification for the use of lasers for the processing of even highly thermally and photolabile substrates.

Based on the previous results, it is concluded that for varnishes with a high absorptivity ($\alpha \approx 10^5$ cm^{-1}) at 248 nm, chemical modifications effected by the ablation should be localized within $\delta \sim 1$ μm from the etched surface. Thus, if during material removal, a thin layer of varnish is left unprocessed, it is ensured that influence to the underlying painting medium will be negligible. This has, indeed, been demonstrated in studies on KrF laser processing of samples of cinnabar red pigment (HgS) [Castillejo, M., et al., 2002; Pouli, P., et al., 2000] in seed oil covered with varnish. Examination of the remaining pigment/medium by a number of analytical techniques demonstrates that no oxidation products are detected when at least a thin varnish layer is left intact. In contrast, when all varnish is removed and the laser has directly irradiated the pigment medium, black oxidation products are observed in the remaining material.

As indicated above, the "blow-off model" suggests that the extent of photofragmentation and product formation remaining in the substrate beyond the ejected depth is constant with laser fluence (for $F \geq F_{thr}$). However, this neglects the possibility that fragments weakly bound to the matrix and/or products that are formed within the thermally affected zone below the ejected layer may diffuse to the surface and thermally desorb ("post-ablation" desorption). For instance, in the KrF laser ablation of varnishes doped with photolabile compounds (e.g. C_6H_5Cl), the examination shows that for products such as HCl accumulation in the substrate is minimal. HCl, interacting relatively weakly with the matrix, desorbs even from the non-ejected layers. With increasing substrate absorptivity, products are formed "confined" closer to the surface and the probability of weakly bound species desorbing increases. This effect may be an important factor for the success of laser-material processing, since it indicates that small (usually the most reactive) species formed even below the etched depth may be removed efficiently.

5.6 Dependence on number of laser pulses

In practice, a number of laser pulses is used to remove the overlayer. For irradiation above the ablation threshold, an equilibrium is expected to establish between new species produced and species removed, with the degree of accumulation depending on the etching depth vs. optical penetration depth. For a high absorptivity, accumulation of products occurs only to small depth. In contrast, at moderate fluences, at which an essentially thermal desorption mechanism may operate, desorption of fragments/products weakly bound to the matrix may occur; however, removal of the strongly-bound side products is inefficient. This effect is particularly pronounced for irradiation at weakly absorbed wavelengths.

For instance, in irradiation of ArI/PMMA or varnishes with successive laser pulses, the probe LIF product spectra become progressively broad and red-shifted, as a result of the formation and accumulation of the polymer and dopant decomposition species (e.g. for PMMA the species can be assigned to the conjugated polymer-derivatives also detected by IR by [Larciprette, R., et al., 1987].) However, *at the same total photon dose*, accumulation of by-products is higher for irradiation at low fluences [Andreou, E., et al., 2002, Athanassiou, A., et al., 2002].

A similar trend is also observed for chromophores which are part of the polymer chain. For instance, in the irradiation of PS polymer films with successive laser pulses at low laser fluences at 248 nm, the PS fluorescence peak at ~ 320 nm decreases gradually to very low values, due to the polymer fragmentation to benzyl and/or phenyl radicals (Fig. 7). The decrease per pulse becomes more pronounced with increasing laser fluence.

In parallel, a growing broad emission band at ≈440 nm is observed due to the formation of polyene structures. In contrast, with irradiation above the threshold, the ratio of degradation/PS emissions never reaches the extreme case found at low laser fluences, due to efficient removal of conjugated/degraded species.

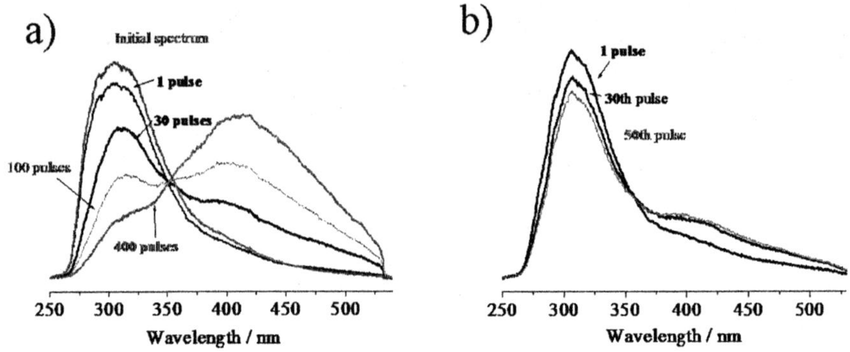

Figure 7. Pulse evolution of the laser-induced fluorescence probe spectra of polystyrene recorded following irradiation at 248 nm with the indicated number of pulses (a) at ≈200 mJ/cm^2 i.e., a fluence below the ablation threshold and (b) at ≈1000 mJ/cm^2 i.e., a fluence above the ablation threshold. It is clear that accumulation of degradation byproducts with successive laser pulses is significant at low laser fluences.

In practice, in the implementations of UV ablation, the number of required pulses is specified by the etching efficiency as compared to the amount of material that must be removed. Thus, a compromise may have to be drawn between efficient material removal and minimization of photochemical effects.

6. FEMTOSECOND ABLATION

Ablation with femtosecond pulses has attracted significant attention because of the several advantages that it has indicated to provide for material processing. In most of the work thus far, three main advantages of fs laser irradiation have been emphasized. First, heat diffusion effects are minimal. Second, because of material ejection occurring well after the laser pulse, there is no plasma shielding; thus, maximum coupling of the incident laser energy into the substrate is effected. Third, the efficient energy use (i.e. negligible loss due to heat diffusion) enables processing at much lower fluences than possible with nanosecond pulses. Most importantly, the high intensities attained with femtosecond pulses result in a highly efficient multiphoton process. In addition, due to the much enhanced possibility of multiphoton processes, the ablation threshold is much reduced. This enables processing of substrates that are transparent or weakly absorbing at the irradiation wavelength (eg. the threshold for ablation of poly(methyl methacrylate) with ~ 500 fs pulses at 248 nm is ~ 5 times lower than that for nanosecond pulses). It is generally demonstrated that quality of structuring with fs pulses far surpasses that attained in ns ablation. Ablation with femtosecond pulses is reviewed by [Baudach, S., et al., 2001].

However, in addition to the above factors, novel features are involved that have not yet been elucidated (for instance, in the irradiation of ArI-doped systems with 500 fs pulses at 248 nm). The F_{LASER}-dependence of ArH yield in the fs irradiation differs markedly from the ns one (Fig. 8). Most interestingly, recombination (e.g., Nap_2 type) or other by-products are not observed even for very high dopant concentrations [Athanassiou, A., et al., 2000]. Furthermore, the amount of any ArH product in the substrate remains highly limited as fluences are close to the ablation threshold.

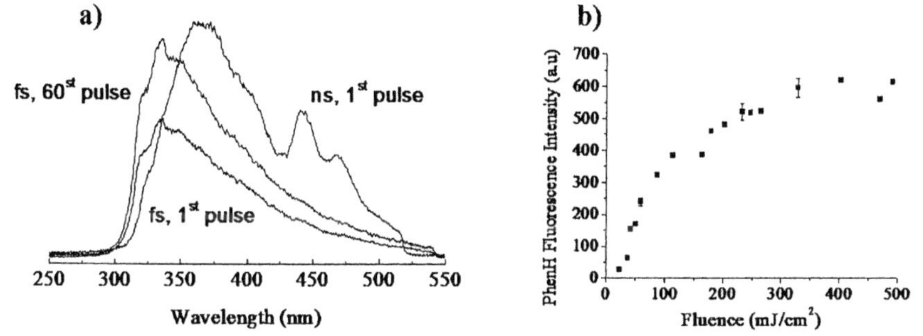

Figure 8. (a) LIF spectra recorded from 4% NapI-doped PMMA after irradiation with the indicated number of pulses with 500 fs, 248 nm pulses, and for comparison purposes, the corresponding spectrum recorded following irradiation with 1 ns pulse at 248 nm. (b) Yield of PhenH following irradiation of PhenI/PMMA (0.5% wt) virgin substrate with a single 500 fs "pump" pulse as a function of the "pump" fluence.

The much lower ArH product formation for fs irradiation can be easily explained on the basis of multiphoton processes much reducing the "effective" optical penetration depth. Indeed, Larciprete, et al., (1987) indicated that at least a 2-photon process is indicated in the 248 nm irradiation of PMMA with 500 fs pulses. However, this argument is in itself insufficient to account for the highly selective chemical modifications observed with fs irradiation. Likely, because of the limited thermal dissipation in the ablation with fs pulses, the mobility of any radicals remains highly restricted. It is also tempting to argue that with the shorter pulses, material ejection occurs too fast for formation of by-products to compete. Studies are underway for establishing the responsible factors.

Independently of mechanistic considerations, it is clear that besides the well-acknowledged advantages (i.e, limited heat load, highly efficient and localized energy deposition, etc.), fs ablation also affords a high degree of control over the induced chemical modifications. Thus, processing with fs lasers is expected to be highly potent in the treatment of molecular substrates. In particular, the highly reduced optical penetration depth suggests the treatment of a wide range of artworks, even in the near absence of "protecting" varnish layer.

7. CONCLUSIONS

In all, laser technology is well demonstrated to afford a highly effective method of removal of unwanted material from molecular solids/systems. Furthermore, the operation of different mechanisms of laser-induced material removal can be exploited in resolving different restoration problems. Certainly, several experimental parameters must be carefully optimized in order to achieve proper cleaning. Furthermore, online monitoring is essential for safeguarding against damage. With these precautions, laser cleaning can be a highly accurate and versatile method, by far surpassing the degree of selectivity and control afforded by traditional cleaning methods. It is worth noting that the versatility of the interaction implies the potential for other applications besides restoration. For instance, the same phenomenon/ablation may be exploited for the microetching of holograms for authentication and security purposes.

Particular emphasis is placed on the side effects of the procedures and, in particular, on the minimization of any chemical modifications induced to the substrate, which are discussed from a fundamental/mechanistic standpoint and are exemplified in the case of the laser-based restoration of painted artworks. Certainly, the interaction of intense laser pulses with organic substrates is highly complex. Yet, deleterious influences have been shown, empirically, that can be avoided to a large extent. Studies on model systems provide further insight into the factors that are involved in these interactions. In particular, these studies provide a fundamental understanding and establish criteria for the systematics for guiding development of laser processing cleaning schemes.

APPENDIX: ABBREVIATIONS

α	Absorption coefficient	λ	Wavelength
c_p	Heat capacity at constant pressure	MW	Molecular Weight
c_v	Heat capacity at constant volume	N_{pulse}	Number of laser pulses
δ	Ablation (etched) depth per pulse	η	Photodissociation quantum yield
$E_{binding}$	Binding energy to the substrate	PI	Polyimide
D	Thermal Diffusivity	PMMA	Polymethyl-methacrylate
D_{sp}	Species Diffusion constant ≈	PS	Polystyrene
E_a	Activation energy	R	Reflectivity
E_{cr}	Critical energy density for ablation	R_B	Gas constant
F_{LASER}	Laser fluence	ρ	Density
F_{thr}	Threshold fluence for material removal	r_p	Particle Radius
ΔH_{vap}	Evaporation enthalpy	σ_p	Particle absorption coefficient
ΔH_{sub}	Sublimation enthalpy	T	Temperature
I	Laser intensity	t_{th}	Thermal diffusion time
k(T)	Reaction rate constant	τ_{pulse}	Laser pulse duration
κ_B	Boltzmann constant	η	Expansion velocity

REFERENCES

Andreou E., Athanassiou A., Fragouli D., Anglos D., Georgiou S. Laser and Material Parameter Dependence of the Chemical Modifications in the UV Laser Processing of Model Polymeric Solids. Laser Chem 2002; 20:1-21

Anglos D., Solomidou M., Zergioti I., Zafiropulos V., Papazoglou T.G., Fotakis C. Laser-Induced Fluorescence in Artwork Diagnostics: An Application in Pigment Analysis. Appl Spectrosc 1996; 50: 1331-1334

Anglos D. Laser-Induced Breakdown Spectroscopy in Art and Archaeology Appl Spectrosc 2001; 55: 186A-205A

Arnold B.R., Scaiano J.C. Laser ablation of doped polymers: transient phenomena as the ablation threshold is approached. Macromolecules 1992;25: 1582-1587

Asmus J.F. More light for art conservation. IEEE Circuits Dev. Magaz. 1986;2: 6-

Athanassiou A., Andreou E., Anglos D., Georgiou S., Fotakis C. UV laser ablation of halonaphthalene-doped PMMA: chemical modifications above versus below the ablation threshold. Appl Phys A 1999;86: S285-S289

Athanassiou A., Lassithiotaki M., Anglos D., Georgiou S., Fotakis C. A comparative study of the photochemical modifications effected in the UV laser ablation of doped polymer substrates. Appl Surf Sci 2000;154-155: 89-94

Athanassiou A., Andreou E., Fragouli D., Anglos D., Georgiou S., Fotakis C. A comparative examination of photoproducts formed in the 248 and 193 nm ablation of doped PMMA. J Photochem Photobiol A 2001;145: 229-236

Athanassiou A., Andreou E., Bonarou A., Tornari V., Anglos D., Georgiou S., Fotakis C. Examination of chemical and structural modifications in the UV ablation of polymers. Appl Surf Sci 2002;197-198: 757-763

Balas C. An imaging calorimeter for noncontact tissue color mapping. IEEE Trans Biomed Eng 1997; 44: 468-474

Baudach S., Krüger J., Kautek W. Femtosecond Laser Processing of Soft Materials. Rev Laser Eng 2001;29: 705-709

Bäuerle D., Laser processing and chemistry. Berlin Heidelberg New York: 3rd edn. Springer, 2000.

Bityurin N., Luk'yanchuk B.S., Hong M.H., Chong T.C. Models for Laser Ablation of Polymers. Chem Rev 2003;103: 519-552

Bounos G, Athanassiou A, Anglos D, et al. Product formation in the laser irradiation of doped poly(methyl) methacrylate) at 248nm: Implications for chemical effects in UV ablation. J Phys Chem B 2004;108 (22): 7050-7060

Casadio F., Giangualano I., Piqué, F. Organic materials in wall paintings: the historical and analytical literature. Reviews in conservation 2004; 5: 63-80

Castillejo M., Martin M., Silva D., Stratoudaki T., Anglos D., Burgio L., Clark R.J.H. Analysis of pigments in polychromes by use of laser induced breakdown spectroscopy and Raman microscopy. J Mol Struct 2000;550-551:191-198

Castillejo M., Martin M., Oujja M., Silva D., Torres R., Manousaki A., Zafiropulos V., van den Brink O. F., Heeren R. M. A., Teule R., Silva A., Gouveia H. Analytical Study of the Chemical and Physical Changes Induced by KrF Laser Cleaning of Tempera Paints. Anal. Chem. 2002; 74(18): 4662-4671.

de Cruz A. Laser removal of contaminants from painted artworks. J. of Cultural Heritage 2000; 1: S173-S180

Dongsik K., Park H.K., Grigoropoulos C.P. Interferometric probing of rapid vaporization at a solid–liquid interface induced by pulsed-laser. Int J Heat Mass Trans 2001;44: 3843-3853

Fotakis C., Anglos D., Balas C., Georgiou S., Vainos N.A., Zergioti I. and Zafiropulos V. In: Tam AC (ed) OSA TOPS on Lasers and Optics for Manufacturing, vol. 9 Optical Society of America, Washington DC, USA 1997; pp 99-104

Fujiwara H., Nakajima Y., Fukumura H., Masuhara H. Laser Ablation Dynamics of a Poly(methyl methacrylate) Film Doped with 5-Diazo Meldrum's Acid. J Phys Chem 1995;99: 11481-11488

Georgiou S., Zafiropulos V., Anglos D., Balas C., Tornari V., Fotakis C. Excimer laser restoration of painted artworks: procedures, mechanisms and effects. Appl Surf Sci 1997;127-129: 738-745

Georgiou S., Hillenkamp F. Introduction: Laser Ablation of Molecular Substrates.Chem Rev 2003; 103: 317-319

Georgiou S. Laser cleaning methodologies of polymer substrates.Adv. Polym Sci 2004; 168: 1-49

Gettens R., Stout G. Painting materials: a short encyclopaedia, Dover, London, 1966.

Haselbach E., Rohner Y., Suppan P. Photodissociation of Halonaphthalenes in Solution: Comparative Photochemistry of 1-Iodo-, 1-Bromo-, and 1-Chloronaphthalenes. Helv Chim Acta 1990; 73: 1644-1652

Kautek W., Pentzien S., Rudolph P., Krüger J., König E. Laser interaction with coated collagen and cellulose fibre composites: fundamentals of laser cleaning of ancient parchment manuscripts and paper. Appl Surf Sci 1998; 127-129: 746-754

Kolar J., Strlič M. IR pulsed laser light interaction with soiled cellulose and paper. Appl Phys 2002 A 75: 673-676

Kuper S., Brannon J., Brannon K. Threshold behavior in polyimide photoablation-single-shot rate measurements and surface-temperature modeling. Appl Phys A 1993; 56:43-50

Larciprete R., Stuke M. Direct observation of excimer laser photoablation products from polymers by picosecond-UV-laser mass spectroscopy. Appl Phys B 1987; 42: 181-184

Lassithiotaki M., Athanassiou A., Anglos D., Georgiou S., Fotakis C. Photochemical effects in the UV laser ablation of polymers: Implications for laser restoration of painted artworks. Appl Phys A 1999; 69: 363-367

Lee Y.P., Lu Y.F., Chan D.S.H., Low T.S., Zhou M.S. Steam Laser Cleaning of Plasma-Etch-Induced Polymers from Via Holes. Jpn J Appl Phys 1998;37: 2524-2529

Lippert T., Stoutland P.O. Laser–material interactions probed with picosecond infrared spectroscopy. Appl Surf Sci 1997;109-110: 43-47

Lippert T., Dickinson T.J. Chemical and Spectroscopic Aspects of Polymer Ablation: Special Features and Novel Directions. Chem Rev 2003;103: 453-485

Luk'yanchuk B.S, Laser Cleaning. Singapore: World Scientific,2002

Mosbacher M., Dobler V., Boneberg J., Leiderer P. Universal threshold for the steam laser cleaning of submicron spherical particles from silicon.Appl Phys A 2000; 70(6): 669-672

Mosbacher M., Mónzer H-J., Zimmermann J., Solis J., Boneberg J., Leiderer P. Optical field enhancement effects in laser-assisted particle removal. Appl Phys A 2001;72: 41-44

Mills J.S. "Photochemical and thermal degradation of films of dammar resin" Studies in Conservation 33: 53, J Chem Soc 2196 (1956)

Pouli P. and Emmony DC. The effect of Nd:YAG laser radiation on medieval pigments Journal of Cultural Heritage 1 August 2000; Volume 1, Supplement 1, S181-S188

Perez C., Barrera, M. and Díez, L. Positive findings for laser use in cleaning cellulosic supports, Journal of Cultural Heritage, 2003; Volume 4, Supplement 1: 194s-200s

Reed R., Ancient Skins, Parchments and Leathers. London New York: Seminar Press,1970

de la Rie ER (1988?89) "Old master paintings: A study of the varnish problem" Anal Chem 61: 1228A

Srinivasan R., Braren B. Ultraviolet laser ablation of organic polymers. Chem. Rev 1989;89: 1303-1316

Tam A.C., Park H.K., Grigoropoulos C.P. Laser cleaning of surface contaminants. Appl Surf Sci 1998;127/129: 721-725

Tokarev V.N., Lazare S., Belin C., et al. Viscous flow and ablation pressure phenomena in nanosecond UV laser irradiation of polymers. Appl Phys A-Mater 2004; 79: 717-720

Vogel A., Venugopalan V. Mechanisms of Pulsed Laser Ablation of Biological Tissues. Chem Rev 2003;103: 577-644

Wolbers R., Sterman R.N, Stavroudis C "Notes for Workshop on new Methods in the Cleaning of Paintings" The Getty Conservation Institute, Marina del Rey, CA (1990)

Yingling Y.G., Zhigilei L.V., Garrison B.J. The role of the photochemical fragmentation in laser ablation: a molecular dynamics study. J Photochem Photobiol A 2001;145: 173-181

Zafiropulos V., Fotakis C. Chapter 6. In Laser Cleaning in Conservation: an Introduction, Cooper M. ed. Butterworth Heinemann, Oxford, 1997

Index

Springer Series in

OPTICAL SCIENCES

Volume 1

1 Solid-State Laser Engineering
By W. Koechner, 6th revised and updated ed. 2006, 447 figs, 45 tabs, xvi, approx 747 pages

Published titles since volume 80

93 Fundamentals of Semiconductor Lasers
By T. Numai, 2004, 166 figs, XII, 264 pages

94 Photonic Crystals
Physics, Fabrication and Applications
By K. Inoue, K. Ohtaka (Eds.), 2004, 209 figs, XV, 320 pages

95 Ultrafast Optics IV
Selected Contributions to the 4th International Conference on Ultrafast-Optics, Vienna, Austria
By F. Krausz, G. Korn, P. Corkum, I.A. Walmsley (Eds.), 2004, 281 figs, XIV, 506 pages

96 Progress in Nano-Electro Optics III
Industrial Applications and Dynamics of the Nano-Optical System
By M. Ohtsu (Ed.), 2004, 186 figs., 8 tabs., XIV, 224 pages

97 Microoptics
From Technology to Applications
By J. Jahns, K.-H. Brenner, 2004, 303 figs., XI, 335 pages

98 X-Ray Optics
High-Energy-Resolution Applications
By Y. Shvyd'ko, 2004, 181 figs., XIV, 404 pages

99 Mono-Cycle Photonics and Optical Scanning Tunneling Microscopy
Route to Femtosecond Ångstrom Technology
By M. Yamashita, H. Shigekawa, R. Morita (Eds.) 2005, 241 figs., XX, 393 pages

100 Quantum Interference and Coherence
Theory and Experiments
By Z. Ficek and S. Swain, 2005, 178 figs., XV, 418 pages

101 Polarization Optics in Telecommunications
By J. Damask, 2005, 110 figs., XVI, 528 pages

102 Lidar
Range-Resolved Optical Remote Sensing of the Atmosphere
By C. Weitkamp (Ed.), 161 figs., XX, 416 pages

103 Optical Fiber Fusion Splicing
By A.D. Yablon, 2005, 137 figs., XIII, 306 pages

104 Optoelectronics of Molecules and Polymers
By A. Moliton, 2005, 229 figs., XXXI, 498 pages

105 Solid-State Random Lasers
By M. Noginov, 2005, 131 figs., XII, 238 pages

106 Coherent Sources of XUV Radiation
Soft X-Ray Lasers and High-Order Harmonic Generation
By P. Jaeglé, 2005, 150 figs., approx. 264 pages

107 Optical Frequency-Modulated Continuous-Wave (FMCW) Interferometry
By J. Zheng, 2005, 137 figs., XVIII, 254 pages

108 Laser Resonators and Beam Propagation
Fundamentals, Advanced Concepts and Applications
By N. Hodgson and H. Weber, 2005, 587 figs., XXV, 794 pages

109 Progress in Nano-Electro Optics IV
Characterization of Nano-Optical Materials and Optical Near-Field Interactions
By M. Ohtsu (Ed.), 2005, 123 figs., XIV, 206 pages

110 Kramers-Kronig Relations in Optical Materials Research
By V. Lucarini, J.J. Saarinen, K.-E. Peiponen, E.M. Vartiainen, 2005, 37 figs., X, 162 pages

111 Semiconductor Lasers
Stability, Instability and Chaos
By J. Ohtsubo, 2005, 169 figs., XII, 438 pages

112 Photovoltaic Solar Energy Generation
By A. Goetzberger and V.U. Hoffmann, 2005, 139 figs., XII, 234 pages

113 Photorefractive Materials and Their Applications 1
Basic Effects
By P. Günter and J.P. Huignard, 2005, 169 figs., XIV 426

114 Photorefractive Materials and Their Applications 2
Materials
By P. Günter and J.P. Huignard, 2006, 370 figs., approx. 664 pages

115 Photorefractive Materials and Their Applications 3
Applications
By P. Günter and J.P. Huignard, 2006, 316 figs., approx. 431 pages

116 Spatial Filtering Velocimetry
Fundamentals and Applications
By Y. Aizu and T. Asakura, 2006, 112 figs., approx. XII, 220 pages

117 Progress in Nano-Electro-Optics V
Nanophotonic Fabrications, Devices, Systems, and Their Theoretical Bases
By Motoichi Ohtsu, 2006, 122 figs., 3 tables, approx. 193 pages

118 Mid-infrared Semiconductor Optoelectronics
By A. Krier, 2006, 443 figs., approx. 776 pages

119 Optical Interconnects
The Silicon Approach
By L. Pavesi and G. Guillot, 2006, 265 figs., approx. 410 pages

120 Relativistic Nonlinear Electrodynamics
Interaction of Charged Particles with Strong and Super Strong Laser Fields
By H.K. Avetissian, 2006, 23 figs., XIII, approx. 333 pages

121 Thermal Processes Using Attosecond Laser Pulses
When Time Matters
By M. Kozlowski and J. Marciak-Kozlowska, 2006, 46 figs., XII, approx. 232 pages

122 Modeling and Analysis of Transient Processes in Open Resonant Structures
New Methods and Techniques
By Y.K. Sirenko, N.P. Yashina and S. Strom, 2006, 110 figs., XIV, approx. 346 pages

123 Wavelength Filters in Fibre Optics
By H. Venghaus, 2006, 210 figs., XXIV, approx. 451 pages

124 Light Scattering by Systems of Particles
Null-Field Method with Discrete Sources - Theory and Programs
By A. Doicu, T. Wriedt and Y.A. Eremin, 2006, 98 figs., XIII, approx. 318 pages

125 Electromagnetic and Optical Pulse Propagation 1
Spectral Representations in Temporally Dispersive Media
By K.E. Oughstun, 2006, 50 figs., IV, approx. 250 pages

126 Quantum Well Infrared Photodetectors
Physics and Applications
By H. Schneider, 2006, 153 figs, XVI, approx. 254 pages

127 Physics of Thin-Film Photovoltaics and Related Systems
By V. Karpov, 2006, 50 figs, approx. 304 pages

128 High-Power Diode-Lasers for Materials Processing
Principles and Applications
By R. Poprawe, P. Loosen, and F. Bachmann, 2006, 1,000 figs, approx. 700 pages

GPSR Compliance

The European Union's (EU) General Product Safety Regulation (GPSR) is a set of rules that requires consumer products to be safe and our obligations to ensure this.

If you have any concerns about our products, you can contact us on ProductSafety@springernature.com

In case Publisher is established outside the EU, the EU authorized representative is:

Springer Nature Customer Service Center GmbH
Europaplatz 3
69115 Heidelberg, Germany

Batch number: 10620078

Printed by Printforce, the Netherlands